Adenosine and Adenosine Receptors

The Receptors

Series Editor

David B. Bylund

University of Nebraska Medical Center, Omaha, NE

Board of Editors

S. J. Enna
Nova Pharmaceuticals
Baltimore, Maryland

Bruce S. McEwen
Rockefeller University
New York, New York

Morley D. Hollenberg
University of Calgary
Calgary, Alberta, Canada

Solomon H. Snyder
Johns Hopkins University
Baltimore, Maryland

Adenosine and Adenosine Receptors

Edited by

Michael Williams

Abbott Laboratories, Abbott Park, Illinois

 **The Humana Press**
Clifton, New Jersey

Dedication

To Geoff Burnstock and John Daly, whose vision, efforts and persistence allayed the sceptics and helped translate adenosine and ATP receptors from academic curiosities to viable drug targets.

Library of Congress Cataloging in Publication Data

©1990 The Humana Press Inc.
Crescent Manor
PO Box 2148
Clifton, NJ 07015

Printed in the United States of America

Adenosine and adenosine receptors/edited by Michael Williams.
        p. cm. — (The Receptors)
    ISBN 0-89603-163-2
    1. Adenosine—Receptors. 2. Adenosine—Physiological effect.
3. Cellular signal transduction. I. Williams, Michael, 1947 Jan.
3- II. Series.
QP625.A27A334   1989
612'.01579—dc20

90-4000
CIP

Preface

Historically, the major emphasis on the study of purinergic systems has been predominantly in the areas of physiology and gross pharmacology. The last decade has seen an exponential increase in the number of publications related to the role of both adenosine and ATP in mammalian tissue function, a level of interest that has evolved from a more molecular focus on the identity of adenosine and ATP receptor subtypes and the search for selective ligands and development of radioligand binding assays by Fred Bruns and colleagues (especially that for A_2 receptors) that played a highly significant role in advancing research in the area.

In the 60 years since adenosine was first shown to be a potent hypotensive agent, a considerable investment has been made by several pharmaceutical companies—including Abbott, Byk Gulden, Takeda, Warner-Lambert/Parke Davis, Boehringer Mannheim, Boehringer Ingelheim, Nelson/Whitby Research and CIBA-Geigy— as well as John Daly's laboratory at the National Institutes of Health, to design new adenosine receptor ligands, and both agonists and antagonists with the aim of developing new therapeutic entiities. Numerous research tools have derived from these efforts including: 2-chloroadenosine, R-PIA (N^6-phenylisopropyladenosine; NECA (5' N-ethylcarboxamidoadenosine); CV 1808; CI 936; PD 125,944; N^6-benzyladenosine; PACPX; CPX; CPT; XAC; CGS 15943 and CGS 21680. Yet in the realm of therapeutics it was only in 1989 that adenosine itself was approved for human use in the treatment of supraventricular arrythmias. The purine nucleoside is also undergoing clinical trials as a hypotensive agent for use in aneurysm surgery, while ATP is being evaluated for use in shock and cancer therapy. Though one or two companies are rumored to be actively pursuing agonists and selective xanthine antagonists in the clinic, other compounds for which substantiated data is available are 'indirect' adenosine agonists. The novel nucleoside transport inhibitor, mioflazine, is in clinical trials as a hypnotic, while the adeno-sine 'potentiator' AICA riboside is being evaluated for use in myocardial reperfusion injury.

This clinical situation appears in marked contrast to the wealth of potential uses for receptor selective agonists and antagonists that have been proposed over the past 10 years and may lead the casual observer to the area wondering why it has taken 60 years and considerable chemical effort for the parent entity, adenosine, to be approved for human use.

The major constraint in the therapeutic use of adenosine has been the ubiquitous nature of both its distribution and cellular actions. The latter encompasses both extracellular and intracellular-related processes that are fundamental to cell function. Thus while adenosine and its more stable analogs are effective hypotensive agents and the bradycardiac effects of the purine can be eliminated via the use of an A_2-receptor selective ligand, the effects of such agents on neutrophil, lymphocyte, renal, CNS and immune system function have not been systematically evaluated within the context of such ancillary actions. Conversely, the interesting actions of adenosine on striatal dopaminergic systems documented by the Parke Davis group have not been adequately addressed within the context of the constellation of adenosine receptor actions in the periphery, the latter not a reflection of inadequate experimental paradigms, but rather a global paucity of data, both acute and chronic. As a result, though adenosine, its stable analogs, and ATP each may elicit profound and potentially exciting actions in various systems, the inevitably reductionistic nature of such studies have little bearing on the clinical situation in which each and every adenosine receptors are potential targets for drug actions and do little to refute the inevitable expectations of lack of selectivity that appear unique to the purines, as opposed to peptides.

These limitations, though considerable, should not obscure the fact that hypotheses related to the discrete role of adenosine in tissue function and disease pathophysiology have yet to be adequately tested. For many of the selective ligands mentioned above, the data derived from their clinical evaluation was generated in the absence of knowledge of the adenosine receptor subtypes. For others, toxicity, lack of efficacy, or other reasons led to such compounds being 'killed' before they were evaluated in the clinic. As an example, it is important to remember that adenosine is not the only agent capable of reducing blood pressure. It is however, one of the oldest entities reported to have such activity and still has yet to reach the marketplace for this indication. In the face of the presently used diuretics, β-blockers and angiotensin-converting enzyme inhibitors, and the renin inhibitors and angiotensin II antagonists currently in development, it appears highly unlikely that adenosine will ever be marked as a classical antihypertensive. However, in other areas such as inflammation, epilepsy, and ischemia-related cell death, where therapies are either limited or nonexistent, present research findings suggest that drugs that modulate adenosine systems may have significant therapeutic potential. The only sure way of assessing such potential is to delineate the receptor subtypes/mechanisms involved, and then to design safe and selective agents and test these in humans.

In the present volume, these and other issues related to adenosine as a therapeutic target are reviewed in a consistent and comprehensive manner, emphasizing the progress to date and highlighting the challenges for the next decade. The book thus may be contrasted with the majority of those published since 1978, not only in that it highlights progress in a more pragmatic, therapeutic context, but also in that it is not the proceedings of a conference, but a more systematic collation and critical analysis of research in this area from the chemical, biochemical, and therapeutic vantage points.

The authors of the various chapters are distinguished authorities in the realm of adenosine research and have each focused on their areas of speciality from receptor concepts through receptor ligands and adenosine availability to biochemical and electrophysiological actions of the purine nucleoside. Considerable attention is also focused on the cardiovascular, CNS, renal, pulmonary, and immune/systemic actions of adenosine and the therapeutic potential in these areas.

The editor would like to extend sincere thanks to the authors for the excellence of their contributions as well as their patience during the editorial process. The reader may note that the editor has taken a pro-active role in several places in order to ensure that the book was as comprehensive as possible. His thanks go to Kevin Mullane, Ken Jacobson, and Mike Jarvis for their willingness to co-author their respective chapters and to Jonathan Geiger, David Bylund, and Tom Lanigan whose support (and persistence) helped in no small way to maintain momentum at times when the likelihood of this volume being completed appeared remote. Especial thanks also go to Holly Roth Williams for designing the book jacket and to the editor's former colleagues at CIBA-Geigy, Al Hutchison, Jen Chen, and John Francis, for their interest in the chemistry of adenosine and from whose efforts the nonxanthine antagonist, CGS 15943 and the A_2-selective agonist, CGS 21680 and related entities arose.

For the ever increasing body of 'adenophiles,' it is to be hoped that the 1990s will be the decade in which many of the hypotheses reviewed in the subsequent pages will be more fully tested and result in therapeutic agents that—acting directly or indirectly via adenosine (and ATP) receptors of whatever subtype—will represent significant improvements over existing drug therapies in a variety of disease states.

Michael Williams

Contents

Contributors

ANIL K. BIDANI • Division of Nephrology, Department of Medicine, Rush-Presbyterian St. Luke's Medical Center, Chicago, Illinois

ALEXANDER J. BRIDGES • Kisai Research Institute, Cambridge, Massachusetts

ROBERT F. BRUNS • Fermentation Products, Eli Lilly, Indianapolis, Indiana

KEVIN K. CALDWELL • Department of Pharmacology, University of Colorado Health Sciences Center, Denver, Colorado

PAUL C. CHURCHILL • Department of Physiology, Wayne State University School of Medicine, Detroit, Michigan

DERMOT M. F. COOPER • Department of Pharmacology, University of Colorado Health Sciences Center, Denver, Colorado

BRUCE N. CRONSTEIN • Department of Medicine, Divisions of Rheumatology and Medical Genetics, New York University Medical Center, New York, New York

THOMAS V. DUNWIDDIE • Department of Pharmacology, University of Colorado Health Sciences Center, Denver, Colorado; and, Veteran's Administration, Medical Research Service, Denver, Colorado

JONATHAN D. GEIGER • Department of Pharmacology, University of Manitoba, Faculty of Medicine, Winnipeg, Manitoba, Canada

TIMOTHY L. GRIFFITHS • Medicine I, Southampton General Hospital, Southampton, UK

ROCHELLE HIRSCHHORN • Department of Medicine, Divisions of Rheumatology and Medical Genetics, New York University Medical Center, New York, New York

STEPHEN T. HOLGATE • Medicine I, Southampton General Hospital, Southampton, UK

KENNETH JACOBSON • Laboratory of Chemistry, National Institutes of Diabetes, Digestive and Kidney Diseases, National Institutes of Health, Bethesda, Maryland

MICHAEL F. JARVIS • Department of Pharmacology, Rorer Central Research, King of Prussia, Pennsylvania

HILLARY G. E. LLOYD • Department of Pharmacology, Karolinska Institute, Stockholm, Sweden

KEVIN M. MULLANE • Gensia Pharmaceuticals, San Diego, California

JAMES I. NAGY • Department of Physiology, University of Manitoba, Faculty of Medicine, Winnipeg, Manitoba, Canada

A. C. NEWBY • Department of Cardiology, University of Wales College of Medicine, Cardiff, UK

TREVOR W. STONE • Department of Pharmacology, University of Glasgow, Glasgow, Scotland

BHARAT K. TRIVEDI • Department of Chemistry, Parke-Davis Pharmaceutical Research Division, Warner-Lambert Co., Ann Arbor, Michigan

MICHAEL WILLIAMS • Pharmaceutical Products Division, Abbott Laboratories, Abbott Park, Illinois

CHAPTER 1

Adenosine Receptors

An Historical Perspective

Michael Williams

1. Introduction

It is now 60 years since Drury and Szent-Györgyi (1929) found that adenosine produced profound hypotension and bradycardia as well as affecting kidney function in mammals. At a time when the thiazide diuretics, β-adrenergic receptor antagonists, and angiotensin converting enzyme (ACE) inhibitors were unknown, the potential of the nucleoside as an antihypertensive agent was attractive, and consequently, adenosine was rapidly evaluated in man (Honey et al., 1930; Jezer et al., 1933; Drury, 1936) with disappointing results because of the lability of the natural nucleoside. Accordingly, the purine was felt to have limited usefulness as an antihypertensive agent, and it was not until the 1950s that the effects of adenosine on mammalian cellular function were again evaluated (Green and Stoner, 1950; Feldberg and Sherwood, 1953; Winbury et al.,1953; Wolf and Berne, 1956). The hypotensive actions of adenosine were reconfirmed, as was the ability of the purine to modulate pulmonary and CNS function.

Berne and coworkers, in a seminal series studying the involvement of endogenous adenosine in dilating coronary resistance vessels (Berne, 1963;1985; Berne et al.,1974), proposed that the purine,

Adenosine and Adenosine Receptors Editor: Michael Williams ©1990 The Humana Press Inc.

via effects on local blood flow, regulated oxygen availability in accord with the metabolic demands of the heart. Evidence has accrued from other systems to support this "retaliatory" action of the purine (Newby, 1984); however, definitive evidence for this hypothesis is still lacking (Berne et al., 1987). Nontheless, the potential role of adenosine as a modulator of cardiovascular function is well established (Mullane and Williams, 1990).

Concommitant with the physiological studies of Berne and coworkers, Burnstock (1972), while studying nonadrenergic, noncholinergic (NANC) transmission processes in smooth muscle in peripheral tissues, identified Adenosine triphosphate (ATP) as the putative transmitter for a third, purinergic division of the autonomic nervous system. Like that related to the metabolic role of adenosine, this theory has been repeatedly challenged, with several peptides also being proposed as the mediator(s) of NANC transmission. Despite this controversy and the lack of selective antagonists for its actions, ATP is still the most likely contender for the role of mediator in NANC transmission.

The third major contribution to the definition of the role of purines in extracellular communication arose from the discovery of the second messenger, cyclic AMP, in the mid 1960s. In the flurry of papers studying the effects of a variety of putative neurotransmitter/ neuromodulator candidates on brain slice cyclic AMP levels, McIlwain and Rall and their coworkers (Kakiuchi et al.,1968; Sattin and Rall, 1970; Pull and McIlwain, 1972; Kuroda and McIlwain, 1973) found that adenosine was released when slices were electrically stimulated and that the purine could enhance the production of the cyclic nucleotide. This effect did not occur by an increase in ATP precursor pools for cyclic AMP, but rather by a receptor-mediated effect. The pivotal discovery resulting from these neurochemical studies was that the methylxanthines, caffeine and theophylline, were adenosine antagonists (Sattin and Rall, 1970). This finding thus provided an effective means to study adenosine-related effects in a variety of mammalian tissues. The exponential rise in publications devoted to the study of adenosine receptors (Fig. 1) resulted directly from the biochemical demonstration of the existence of a distinct adenosine receptor and the subsequent ability to radiolabel this entitity (Bruns et al., 1987a; Williams and Jacobson, 1990).

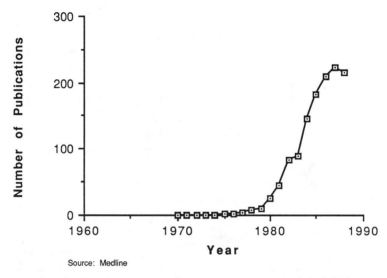

Source: Medline

Fig. 1. Publications in the area of adenosine receptors since 1970.

The metabolic lability of ATP in physiological solutions, the nucleotide giving rise to the nucleoside, adenosine, led many to equate ATP with adenosine, the effects of the former being mediated via formation of the latter. However, in the periphery, the actions of the two purinergic compounds were distinct (Burnstock, 1976), leading to the seminal proposal of two classes of purinergic receptor (Burnstock, 1978) termed P_1 and P_2 (Table 1), the P_1 receptor being sensitive to adenosine and the P_2 to ATP.

2. Purinergic Receptors

Further studies on the effects of adenosine and its more metabolically stable analogs on cyclic AMP formation in a variety of cell systems led Londos and Wolff (1977) and Van Calker et al. (1979) independently to describe two subtypes of the P_1 receptor. Initially described in terms of their effects on cyclic AMP formation, these were termed Ri and Ra, the "R" designation referring to the need for an intact ribose moiety to produce activity, and the "i" and "a" indicating that receptor activation led to inhibition or activation of adenylate cyclase activity. A third subclass of the P_1 receptor was also described and termed the P receptor (Haslam et al.,1978). In contrast to the R receptors, this receptor is located intracellularly on the cata-

Table 1
Purinergic Receptor Classification[a]

Receptor	Subtype	Ligand Potency
P1	A_1 (Ri)	CPA > CHA > R-PIA > 2-CADO > NECA > S-PIA > CV 1808 > CGS 21680
	A_2 (Ra)	NECA = CGS 21680 > MECA > 2-CADO > CV 1808 = R-PIA > CPA = CHA > S-PIA
	A_{2a}	CV 1674 > MADO[b]
	A_{2b}	MADO >> CV 1674
P2	P_{2x}	α-β-Methylene ATP = β-γ-Methylene ATP > ATP > 2 Methylthio ATP > adenosine
	P_{2y}	2-methylthio ATP >> ATP > α-β-methylene ATP = β-γ-methylene ATP >> adenosine

[a]Data from Bruns et al. (1987a); Burnstock and Kennedy, (1985) and Trivedi et al. (1990).
[b]MADO = N^6methyladenosine.

lytic subunit of adenylate cyclase and requires an intact purine ring to produce its effects. The physiological significance of this site is unknown, although it is pharmacologically distinct from the R receptors. When activated by high concentrations of adenosine, the P receptor inhibits cyclic AMP formation.

The Ri and Ra subtypes of the P_1 receptor are now called A_1 and A_2, as designated by Van Calker et al. (1979). Although adenosine receptor activation is traditionally associated with cyclic AMP formation, there have been many instances where a role for the cyclic nucleotide has not been proven. Accordingly, the A_1/A_2 receptor designation now relates to the pharmacology of the receptors rather than their second messenger system (Hamprecht and Van Calker, 1985; Fredholm and Dunwiddie, 1988), since cations, phosphatidylinositol turnover, and arachidonic acid metabolites may function as second messengers for adenosine.

Many selective agonist and anatagonist ligands exist for the A_1 receptor, including CHA (N^6-cyclohexyl adenosine), CPA (N^6-cyclopentyl adenosine), R-PIA (R-N^6-phenylisopropyl adenosine), CCPA (2-chloro-N^6-CPA), (S)-EBNA (N^6-endonorbornyladenosine), DPX (1,3-diethyl-8-phenylxanthine), 8-PT (8-phenyltheophylline), and

CPX (1,3-dipropyl-8-cyclopentylxanthine). These compounds have been extensively used either as "cold" ligands or as radioligands to characterize the distribution and function of this adenosine receptor subtype in a variety of tissues (Schwabe, 1988; Bruns et al., 1987a ; Lohse et al., 1988; Trivedi et al., 1989, 1990b; Williams and Jacobson, 1990). However, there have been few selective ligands for the A_2 receptor. NECA and CV 1808 (2-phenylamino adenosine) are active at the A_2 receptor, with the latter adenosine analog being some five-fold selective for the brain A_2 receptor (Bruns et al., 1986; Stone et al., 1988; Trivedi et al., 1989). Activity associated with the use of NECA where this adenosine agonist is more active than CPA or CHA can therefore be ascribed to A_2 receptor-related events (Table 1). At one time, an additional criterion for distinguishing between A_1 and A_2 receptor effects was the relative ratio of activities of R- and S-isomers of PIA. At the A_1 receptor, R-PIA usually has an activity greater than 10-fold that of the S-isomer, whereas at the A_2 receptor, this difference is less than 10-fold (Stone, 1985). In studies on the effects of R- and S-PIA on cyclic AMP production (Fredholm et al., 1982) and radio- ligand binding (Ferkany et al., 1986; Stone et al., 1988) in different species, it has become apparent that the R/S-PIA ratio cannot be used to delineate receptor subtypes. However, the discovery of the potent and selective A_2 agonist, CGS 21680 (Hutchison et al., 1989), has provided an important new tool to define A_2 receptor-mediated events.

Antagonist ligands selective for the A_2 receptor have also been identified including, DMPX (Ukena et al., 1986; Seale et al., 1988) and the nonxanthine CGS 15943A (Williams et al.,1987; Daly et al., 1988). Like the 8-phenylsulphonamide xanthine, PD 115,199, which is equiactive at both A_1 and A_2 receptors (Bruns et al., 1987b), these compounds are marginally useful as A_2 selective ligands.

Pharmacologically, in the rat, A_1 receptors have the rank order activity: CCPA = (S)-EBNA $\geq$ CPA > CHA $\geq$ R-PIA > NECA > S-PIA $\geq$ CV 1808 > CGS 21680C (Table 1). For A_2 receptors, the rank order is: NECA = CGS 21680C > CV 1808 > R-PIA > CHA = CPA > S-PIA.

High and low affinity subclasses of the A_2 receptor have been proposed (Daly et al., 1983; Bruns et al., 1986) based on adenylate cyclase and binding data. These have been termed A_{2a} and A_{2b} respectively (Bruns et al.,1986) and, despite the fact that antagonists are more active than agonists at the latter receptor, appear to represent distinct binding sites. Since many agonists have similar activities at A_1 and A_{2a} receptors, the use of N^6-methyl adenosine and CV 1674 (2 [p-methoxyphenyl] adenosine) to delineate interactions between the two has been proposed (Bruns et al., 1987a) and is shown in Table 1.

A third type of adenosine receptor, termed A_3, has also been proposed, based on the effects of the purine on calcium-related events in nervous tissue (Riberio and Sebastiao, 1986). In addition, there are many documented reports, usually in functional systems in which the rank order of potency of adenosine receptor ligands fits neither the A_1 nor either A_2 activity profile (Varney and Skidmore, 1985; Gilfillan et al.,1989). Although such findings may represent differences in the bioavailability of the respective agonists, they may also represent other potential adenosine receptor subtypes. One concern in ranking adenosine agonists that may contribute to potential receptor differences that has not been evaluated to a major extent is the measurement of efficacy in different functional systems (Bazil and Minneman, 1986).

Subclasses of the P_2 receptor have also been described (Burnstock and Kennedy,1985), termed P_{2x} and P_{2y} (Table 1). The P_{2x} receptor is more sensitive to α-β-methylene ATP than 2-methyl thioATP, whereas the converse is true for the P_{2y} receptor. Different functional responses to these receptors have been described, but again, selective antagonists are not available, and studies on ATP-related responses have been more physiological than pharmacological. Two attempts to label the P_2 receptor (Williams and Risley, 1980; Fedan et al.,1985) were somewhat inconclusive.

3. Adenosine as a Neuromodulator

Despite a large body of evidence showing receptor-mediated effects of adenosine, there is still considerable scepticism that the nucleoside can function as a physiologically important modulatory agent. The involvement of adenosine in nucleic acid synthesis, as a cofactor in many enzymic reactions, and as an integral part of cellular energy systems (Arch and Newsholme, 1978) has tended to argue against any discrete role for the nucleoside in cell function, as has its relatively high concentrations. In addition, the factors governing the availability of the purine in the extracellular mileau, although well studied (Stone et al., 1989; Geiger and Nagy, 1990), are poorly understood. Many contradictory studies exist relating neuron activity with alterations in adenosine availability (Williams, 1987). Electrical stimulation and K^+ depolarization have been reported to evoke adenosine release (Kuroda and McIlwain, 1973, 1974; Fredholm and Hedqvist, 1980; Stone, 1981); however, such release has also been described as a carrier-mediated "neuronal nonexocytotic" process (Jonzon and Fredholm, 1985). Despite many attempts to categorize adenosine availablity in terms of that of more conventional neurotransmitters, the issue, after nearly 20 years of research, remains unresolved. However, adenosine-containing neurons have been observed using antibodies against an adenosine conjugate (Braas et al., 1986), and a discrete adenosine deaminase containing neuronal network has been observed in brain tissue (Nagy et al., 1984). Furthermore, the distribution of both A_1 and A_2 receptors is heterogenous in a variety of tissues (Goodman and Snyder, 1982; Jarvis,1988). Nonetheless, the lack of a clear consensus on factors affecting adenosine availability plus evidence for the vesicular storage of the purine as distinct from that portion involved in intracellular energy metabolism has made the acceptance of adenosine as a *bona fide* neuromodulatory agent that much more difficult.

In studies on the effects of the xanthine adenosine antagonists on neurotransmitter release from brain tissue, Harms et al. (1978) coined the term "purinergic inhibitory tone." These authors found that xanthines could produce effects on their own that they ascribed to antagonism of the actions of endogenous adenosine. This led to the postulate that, under normal circumstances, the purine was available in sufficent quantities to affect receptor function and thus functioned as a homeostatic mediator (Williams, 1987), more akin to a hormone than a transmitter. Agents with such actions have been termed paracrine neuroactive substances (Schmitt, 1982) "exist(ing) in solution in the intercellular ambient mileu bathing...cells." The selectivity of action of such agents is dependent, not on transient changes in their availability but rather "their ability to bind with high affinity to specific re-

ceptors presumably arrayed on neuronal surfaces that interface with
the ambient intercellular mileu." Although Schmitt's comments related primarily to peptides, the concept proposed is not inconsistent
with that for adenosine, as deduced from available evidence.

Interestingly, although the concept of peptidic neurotransmission is readily accepted, many of the problems associated with the
acceptance of adenosine (or ATP) as a neuroeffector agent can be
extrapolated to the peptides. This has not diminished research interest in the latter, and it is thus somewhat lamentable that the concept
of purinergic neuromodulation has not received the same resources in
terms of chemical and biological research support as such peptides as
Substance P and cholecystokinin.

Like many other neurohumoral agents (Roth et al., 1986), adenosine is active in both vertebrates and unicellular organisms, and has
been proposed as the prototypic neuromodulator, preceeding the peptides in evolutionary terms (Williams, 1989). Both the cyclic nucleotide and adenosine are melanin-dispersing agents in fish (Novales and
Fuiji, 1970; Miyashita et al., 1984). Since the effects of the cyclic nucleotides can be blocked by theophylline, they have been proposed to
involve adenosine receptor activation (Oshima et al., 1986).

From a pathophysiological perspective, the paracrine neuroactive substance concept is attractive when considering both the retaliatory role of adenosine in response to tissue trauma (Newby, 1984)
and the concept of disease processes in general. Under conditions of
hypoxia and ischemia, tissue levels of adenosine are markedly increased because of ATP breakdown (Berne et al., 1974; Winn et al.,
1985; Phillis, 1989). The nucleoside can then act, via adenosine receptor activation, to attenuate host defensive responses, especially
those that are neutrophil-mediated, in order to facilitate tissue recovery (Engler,1987; Mullane and Williams, 1990). This concept of a
global homeostatic role for the purine is supported by the diverse
disease states in which adenosine has been implicated (Williams,
1987). In the CNS, considerable evidence exists to implicate the
nucleoside in mechanisms related to epilepsy, anxiety, and pain. Although purinergic inhibitory tone and the consequent effect of adenosine in inhibiting transmitter release is one explanation for the spectrum of CNS activity, the benzodiazepines share a similar profile of

activity, diazepam as an anxiolytic/anticonvulsant and tifluadom as an analgesic (Williams and Olsen, 1988). The benzodiazepines have been considered as universal pharmacophores or "privileged structures" for peptide receptors (Williams et al., 1987b; Evans et al., 1988) and may thus represent exogenous factors linking purine and peptide neuromodulation to disease amelioration.

A major objection to the use of adenosine receptor ligands as therapeutic agents is that the postulated involvement of the purine in a vertiable plethora of disease states is inconsistent with any conceivable role as a discrete agent. Yet, one of the oldest and most frequently used medications, aspirin, would theoretically suffer from similiar qualifications if cyclooxygenase inhibition were considered as a molecular target. Similarly, the HMG CoA reductase inhibitor, mevinolin (Stokker et al., 1986), which is used to lower plasma cholesterol levels, can be conceptually considered to be a poor drug target, since inhibition of this enzyme can lead to the impairment of membrane structure as well as disrupt steroid synthesis. Yet, the proscribed use of drug therapies implies atypical cellular function. Thus aspirin is an effective treatment for headache, since it presumably inhibits a cyclooxygenase that is hyperactive and thus acts in the manner of a paracrine mediator, so that the hyperactive enzyme rather than drug availability governs the response observed. In the case of mevinolin, under circumstances where plasma cholesterol is high, it may be assumed that the associated disease lesion is an excessive formation of the steroid and that the enzyme inhibitor at therapeutically effective doses can reduce cholesterol without affecting other steroid-dependent processes. These examples provide an important framework in further considerations of a therapeutic role for adenosine-related drugs.

Epilepsy and cardiac arrhythmias have been termed "dynamical diseases" (Pool, 1989) involving alterations in existing chaotic rhythmical processes that are homeostatic in nature. Both involve adenosine as an endogenous mediator of rhythm generation (Bellardinelli et al., 1983; Dragunow et al., 1985). The purine has been proposed as an endogenous anticonvulsant (Dragunow, 1988) and is used in the treatment of supraventricular tachycardia (DiMarco et al., 1983). According to the chaotic rhythm hypothesis, an entropic shift to a simple, more ordered state is indicative of an impaired biological system, with the attendent synchronization leading to an overt pathophysiology. The effectiveness of adenosine in treating the rhythmic disturbances associated with seizure and ventricular desynchronization and its underlying role in CNS and cardiac function may point to a pivotal role for the nucleoside in chaotic flexibility and thus provide an explanation for the ubiquitous role of adenosine in mammalian tissue function, purinergic homestasis providing the mileu for vital tissue function under both normal and traumatic conditions. In those instances where adenosine function (or availability) become compromised, as may occur in chronic disease states with a genetic or progressive etiology, both exogenous adenosine and certain drug classes may prove beneficial in the rectification of cell function. This global hypothesis, built on the as yet unproven proposal of chaotic health (Pool, 1989), may provide a focal point in continuing studies related to the role of adenosine in cell function.

References

Arch, J. R. S. and Newsholme, E. A. (1978) The control of the metabolism and the hormonal role of adenosine. *Essays Biochem.* **14**, 82–123.

Bazil, C. W. and Minneman, K. P. (1986) An investigation of the low intrinsic activity of adenosine and its analogs at low affinity (A2) adenosine receptors in rat cerebral cortex. *J. Neurochem.* **47**, 547–553.

Bellardinelli, L., West, A., Crampton, R., and Berne, R. M. (1983) Chronotropic and dromotropic effects of adenosine, in *Regulatory Function of Adenosine* (Berne, R. M., Rall, T. W., and Rubio, R., eds.), Nijhoff, Boston, pp 377–398.

Berne, R. M. (1963) Cardiac nucleotides in hypoxia—possible role in regulation of coronary flow. *J. Physiol. (Lond.)* **204**, 317–322.

Berne, R. M. (1985) Criteria for the involvement of adenosine in the regulation of blood flow, in *Methods in Pharmacology* vol. 6 *Methods Used in Adenosine Research* (Paton, D. M., ed.) Plenum, New York, pp. 331–336.

Berne, R. M., Rubio, R., and Curnish, R. R. (1974) Release of adenosine from ischemic brain: Effect on cerebral vascular resistance and incorporation into cerebral adenine nucleotides. *Circ. Res.* **35**, 262–271.

Berne, R. M., Gidday, J. M., Hill, H. E., Curnish, R. R., and Rubio, R. (1987) Adenosine in the local regulation of blood flow: Some controversies, in *Topics and Perspectives in Adenosine Research* (Gerlach, E. and Becker, B. F., eds.), Springer-Verlag, Berlin, pp. 395–405.

Braas, K. M., Newby, A. C., Wilson, V. S., and Snyder, S. H. (1986) Adenosine containing neurons in the brain localized by immunocytochemistry. *J. Neurosci.* **6**, 1952–1961.

Bruns, R. F., Lu, G., and Pugsley, T. A. (1986) Characterization of the A$_2$ adenosine receptor labeled by [³H]NECA in rat striatal membranes. *Mol. Pharmacol.* **29,** 331–346.

Bruns, R. F., Lu, G. H., and Pugsley, T. A. (1987a) Adenosine receptor subtypes, in *Topics and Perspectives in Adenosine Research* (Gerlach, E. and Becker, B. F., eds.), Springer-Verlag, Berlin, pp. 59–73.

Bruns, R. F., Fergus, J. H., Badger, E. W., Bristol, J. A., Santay, L. A., and Hays, S. J. (1987b) PD115,199: an antagonist ligand for A_2 receptors. *Naunyn-Schmiedebergs Arch. Pharmacol.* **335,** 64–69.

Burnstock, G. (1972) Purinergic nerves. *Pharmacol. Rev.* **24,** 509–581.

Burnstock, G. (1976) Purinergic receptors. *J. Theor. Biol.* **62,** 491–503.

Burnstock, G. (1978) A basis for distinguishing two types of purinergic receptor, in *Cell Membrane Receptors for Drugs and Hormones* (Bolis, L. and Straub, R. W., eds.) Raven, New York, pp. 107–118.

Burnstock, G. and Kennedy, C. (1985) Is there a basis for distinguishing two types of P_2-purinoceptor? *Gen. Pharmacol.* **16,** 433–440.

Daly, J. W., Butts-Lamb, P., and Padgett, W. (1983) Subclasses of adenosine receptors in the central nervous system: Interaction with caffeine and related methylxanthines. *Cell. Mol. Neurobiol.* **3,** 69–80.

Daly, J. W., Horng, O., Padgett, W. L., Shamim, W. T., Jacobson, K. A., and Ukena, D. (1988) Non-xanthine heterocycles: Activity as antagonists of A_1 and A_2 receptors. *Biochem. Pharmacol.* **37,** 655–664.

DiMarco, J. P., Sellers, T. D., Berne, R. M., West, G. A., and Bellardinelli, L. (1983) Adenosine: Electrophysiologic effects and therapeutic use for terminating paroxysmal supraventricular tachycardia. *Circulation* **68,** 1254–1263.

Dragunow, M. (1988) Purinergic mechanisms in eplilepsy. *Prog. Neurobiol.* **31,** 85–107.

Dragunow, M., Goodard, G. V., and Laverty, R. (1985) Is adenosine an endogenous anticonvulsant? *Epilepsia* **26,** 480–487.

Drury, A. N. (1936) Physiological activity of nucleic acid and its derivatives. *Physiol. Rev.* **16,** 292–325.

Drury, A. N. and Szent-Györgyi, A. (1929) The physiological action of adenine compounds with especial reference to their action on the mammalian heart. *J. Physiol. (Lond.)* **68,** 214–237.

Engler, R. (1987) Consequences of activation and adenosine-mediated inhibition of granulocytes during myocardial ischemia. *Fed. Proc.* **46,** 2407–2412.

Evans, B. E., Rittle, K. E., Bock, M. G., DiPardo, R. M., Freidinger, R. M., Whitter, W. L., Lundell, G. F., Veber, D.F., Anderson, P. S., Chang, R. S. L., Lotti, V. L., Cerino, D. J., Chen, T. B., Kling, P. J., Kunkel, K. A., Springer, J. P., and Hirshfield, J. (1988) Methods for drug discovery: Development of potent, selective, orally effective cholecystokinin antagonists. *J. Med. Chem.* **31,** 2235–2246.

Fedan, J. S., Hogaboom, G. K., O'Donnell, J. P., Jeng, C. J., and Guillory, G. (1985) Interaction of [³H]ANAPP₃ with P_2-purinergic receptors in the smooth muscle of of the isolated guinea pig vas deferens. *Eur. J. Pharmacol.* **108,** 49–61.

Feldberg, W. and Sherwood, S. L. (1954) Injections of drugs into the lateral ventricles of the cat. *J. Physiol. (Lond.)* **123,** 148–167.

Ferkany, J. W., Valentine, H., Stone, G., and Williams, M. (1986) Adenosine A-₁ receptors in mammalian brain: Species differences in their interactions with agonists and antagonists. *Drug Dev. Res.* **9,** 85–93.

Fredholm, B. and Dunwiddie, T. V. (1988) How does adenosine inhibit neurotransmitter release? *Trends Pharmacol. Sci.* **9,** 130–134.

Fredholm, B. B. and Hedqvist, P. (1980) Modulation of neurotransmission by purine nucleotides and nucleosides. *Biochem. Pharmacol.* **29,** 1633–1643.

Fredholm, B. B., Jonzon, B., Lindgren, E., and Lindstrom, K. (1982) Adenosine receptors mediating cyclic AMP production in the rat hippocampus. *J. Neurochem.* **39,** 165–175.

Geiger, J. D. and Nagy, J. I. (1989) Adenosine deaminase and [³H]nitrobenzylthioinosine as markers of adenosine metabolism and transport in central purinergic systems, in *The Adenosine Receptors* (Williams, M., ed.), Humana, Clifton, New Jersey, in press.

Gilfillan, A. M., Wiggan, G. A., and Welton, A. F. (1989) The release of leukotriene C₄ (LTC₄) and histamine (H) from RBL 2H3 cells in response to adenosine (A) analogs. *FASEB J.* **3,** A907.

Goodman, R. R. and Snyder, S. H. (1982) Autoradiographic localization of adenosine receptors in rat brain using [³H]cyclohexyladenosine. *J. Neurosci.* **2,** 1230–1241.

Green, H. N. and Stoner, H. B. (1950) *Biological Actions of Adenine Nucleotides,* (Lewis, London).

Hamprecht, B. and Van Calker, D. (1985) Nomenclature of adenosine receptors. *Trends Pharmacol. Sci.* **6,** 153, 154.

Harms, H. H., Wardeh, G. and Mulder, A. H. (1978) Adenosine modulates depolarization-induced release of ³H-noradrenaline from slices of rat brain neocortex. *Eur. J. Pharmacol.* **49,** 305–309.

Haslam, R. J., Davidson, M. L. and Desjardins, T. (1978) Inhibition of adenylate cyclase by adenosine analogues in preparations of broken and intact human platelets. Evidence for the unidirectional control of platelet function by cyclic AMP. *Biochem. J.* **176,** 83–95.

Honey, R. M., Ritchie, W. T., and Thomson, W. A. R. (1930) The action of adenosine upon the human heart. *J. Med.* **23,** 485–490.

Hutchison, A. J., Webb, R. L., Oei, H. H., Ghai, G. R., Zimmerman, M., and Williams, M. (1989) CGS21680C, an A₂ selective adenosine receptor agonist with preferential hypotensive activity. *J. Pharmacol. Exp. Ther.* **251,** 47–55.

Jarvis, M. F. (1988) Autoradiographic localization and characterization of brain adenosine receptor subtypes, in *Receptor Localization: Ligand Autoradiography* (Leslie, F. M. and Altar, C. A., eds.), Liss, New York, pp. 95–111.

Jezer, A., Oppenheimer, B. S., and Schwartz, S. P. (1933) The effect of adenosine on cardiac irregularities in man. *Am. Heart J.* **9,** 252–258.

Jonzon, B. and Fredholm, B. B. (1985) Release of purines, noradrenaline and GABA from rat hippocampal slices by field stimulation. *J. Neurochem.* **44,** 217–224.

Kakiuchi, S., Rall, T. W., and McIlwain, H. (1968) The effect of electrical stimulation upon the accumulation of adenosine 3',5'-phosphate in isolated cerebral tissue. *J. Neurochem.* **16**, 485–491.

Kuroda, Y. and McIlwain, H. (1973) Subcellular localization of [^{14}C]adenine derivatives newly formed in cerebral tissues and the effects of electrical stimulation. *J. Neurochem.* **21**, 889–900.

Kuroda, Y. and McIlwain, H. (1974) Uptake and release of [^{14}C]adenine derivatives in beds of mammalian cortical synaptosomes in a superfusion system. *J. Neurochem.* **22**, 691–699.

Lohse, M. J., Klotz, K.-N., Schwabe, U., Cristalli, G., Vittori, S., and Grifantini, M. (1988) 2-Chloro-N^6-cyclopentyladenosine: A highly selective agonist at A$_1$ adenosine receptors. *Naunyn-Schmiedebergs Arch. Pharmacol.* **337**, 687–689.

Londos, C. and Wolff, J. (1977) Two distinct adenosine-sensitive sites on adenylate cyclase. *Proc. Natl. Acad. Sci. USA* **74**, 5482–5486.

Miyashita, Y., Kumazawa, T., and Fuiji, R. (1984) Receptor mechanisms in fish chromatophores—IV. Adenosine receptors mediate pigment dispersion in guppy and catfish melanophores. *Comp. Biochem. Physiol.* **77C**, 471–476.

Mullane, K. M. and Williams, M. (1990) Adenosine and cardiovascular function, in *Adenosine and Adenosine Receptors* (Williams, M., ed.), Humana, Clifton, New Jersey, in press.

Nagy, J. I., La Bella, F., Buss, M., and Dadonna, P. E. (1984) Immunohistochemistry of adenosine deaminase: Implications for adenosine neurotransmission. *Science* **224**, 166–168.

Newby, A. C. (1984) Adenosine and the concept of "retaliatory metabolites." *Trends Biochem.* **9**, 42–44.

Novales, R. R. and Fuiji, R. (1970) A melanin-dispersing effect of cyclic adenosine monophosphate on *Fundulus* melanophores. *J. Cell. Physiol.* **75**, 133–136.

Oshima, N., Furuuchi, T., and Fujii, R. (1986) Cyclic nucleotide action is mediated through adenosine receptors in damselfish motile iridophores. *Comp. Biochem. Physiol.* **85C**, 89–93.

Pool, R. (1989) Is it healthy to be chaotic? *Science* **243**, 604–607.

Phillis, J. W. (1989) Adenosine in the control of the cerebral circulation. *Cerebrovasc. Brain Metab. Rev.* **1**, 26–54.

Pull, I. and McIlwain, H. (1972) Adenine derivatives as neurohumoral agents in the brain. The quantities liberated on excitation of superfused cerebral tissues. *Biochem. J.* **130**, 975–981.

Riberio, J. A. and Sebastiao, A. M. (1986) Adenosine receptors and calcium: Basis for proposing a third (A$_3$) adenosine receptor. *Prog. Neurobiol.* **26**, 179–209.

Roth, J., LeRoith, D., Collier, E. S., Watkinson, A., and Lesniak, M. A. (1986) The evolutionary origins of intercellular communication and the Maginot Lines of the mind. *Ann. NY Acad. Sci.* **463**, 1–11.

Sattin, A. and Rall, T. W. (1970) The effect of adenosine and adenine nucleotides on the cyclic adenosine 3'-5-phosphate content of guinea pig cerebral cortex slices. *Mol. Pharmacol.* **6**, 13–23.

Schmitt, F. O. (1982) A protocol for molecular neuroscience, in *Molecular Genetic*

Neuroscience (Schmitt, F. O., Bird, S. J., and Bloom, F. E., eds.) Raven, New York, pp. 1–9.

Schwabe, U. (1988) Use of radioligands in the study of adenosine receptors, in *Adenosine and Adenine Nucleotides. Physiology and Pharmacology.* (Paton, D. M., ed.) Taylor and Francis, London, pp. 3–16.

Seale, T. W., Abla, K. A., Shamim, M. T., Carney, J. M., and Daly, J. W. (1988) 3,7-Dimethyl-1-propargylxanthine: A potent and selective in vivo agonist of adenosine analogs. *Life Sci.* **43,** 1671–1684.

Stokker, G. E., Alberts, A. W., Gilfillan, J. L., Huff, J. W., and Smith, R. L. (1986) 3-Hydroxy-3–methylglutaryl-coenzyme A reductase inhibitors. 5, 6-(fluren-9-yl) and 6-(fluoren-9-ylidenyl)-3,5-dihydroxyhexanoic caids and their lactone derivatives. *J. Med. Chem.* **29,** 852–855.

Stone, G. A., Jarvis, M. F., Sills, M. A.,Weeks, B., Snowhill, E. A., and Williams, M. (1988) Species differences in high affinity adenosine A_2 binding sites in strital membranes from mammalian brain. *Drug Dev. Res.* **15,** 31–46.

Stone, T. W. (1981) Physiological roles for adenosine and adenosine 5'-triphosphate in the nervous system. *Neurosci.* **6,** 523–545.

Stone, T. W. (1985) Summary of a discussion on purine receptor nomenclature. In *Purines: Pharmacology and Physiological Roles* (Stone, T. W., ed.), VCH, Deerfield Beach, Florida, pp. 1–4.

Stone, T. W., Newby, A. C., and Lloyd, H. G. E. (1990) Adenosine release, in *Adenosine and Adenosine Receptors* (Williams, M., ed.), Humana, Clifton, New Jersey, in press.

Trivedi, B. K., Bridges, A. J., Patt, W. C., Priebe, S. R., and Bruns, R. F. (1989) N^6-bicycloalkyl adenosines with unusually high potency and selectivity for the adenosine A^1 receptor. *J. Med. Chem.* **32,** 8–11.

Trivedi, B. K., Bridges, A. J., and Bruns, R. F. (1990) Structure activity relationships of adenosine A_1 and A_2 receptors, in *Adenosine and Adenosine Receptors* (Williams, M., ed.), Humana, Clifton, New Jersey, in press.

Ukena, D., Shamim, M., Padgett, W., and Daly, J. W. (1986) Analogs of caffeine with selectivity for A_2 adenosine receptors. *Life Sci.* **39,** 743–750.

Van Calker, D., Muller, M., and Hamprecht, B. (1979) Adenosine regulates via two different types of receptors the accumulation of cyclic AMP in cultured brain cells. *J. Neurochem.* **33,** 999–1005.

Varney, C. J. and Skidmore, I. F. (1985) Characterization of the adenosine receptor responsible for the enhancement of mediator release from rat mast cells, in *Purines: Pharmacology and Physiological Roles* (Stone, T.W., ed.), VCH, Deerfield Beach, Florida, p. 275.

Williams, M. (1987) Purine receptors in mammalian tissues: Pharmacology and functional significance. *Annu. Rev. Pharmacol. Toxicol.* **27,** 315–345.

Williams, M. (1989) Adenosine: The prototypic neuromodulator. *Neurochem. Inter.* **14,** 249–264.

Williams, M. and Jacobson, K. A. (1990) Radioligand binding assays for adenosine receptors, in *Adenosine and Adenosine Receptors* (Williams, M., ed.), Humana, Clifton, New Jersey, this volume.

Williams, M. and Olsen, R. A. (1988) Benzodiazepine receptors and tissue function, in *Receptor Pharmacology and Function* (Williams, M., Glennon, R. A., and Timmermans, P. B. M. W. M., eds.), Dekker, New York, pp. 385–413.

Williams, M. and Risley, E. A. (1980) Binding of ^{3}H-adenyl-5-imidodiphosphate (AppNHp) to rat brain synaptic membranes. *Fed. Proc.* **39,** 1009 Abstr.

Williams, M., Wasley, J. F., and Petrack, B. A. (1987b) Antianxiety agents: The benzodiazepine receptor and beyond. Chimica Oggi, September 1987, 11–16.

Williams, M., Francis, J., Ghai, G., Braunwalder, A., Psychoyos, S., Stone, G. A., and Cash, W. D. (1987a) Biochemical characterization of the triazoloquinazoline, CGS 15943A, a novel, non-xanthine adenosine antagonist. *J. Pharmacol. Exp. Ther.* **241,** 415–420.

Winbury, M. M., Papierski, D. H., Hemmer, M. L., and Hambourger, W. E. (1953) Coronary dilator action of the adenine-ATP series. *J. Pharmacol. Exp. Ther.* **109,** 255–260.

Winn, H. R., Morii, S., and Berne, R. M. (1985) The role of adenosine in autoregulation of cerebral blood flow. *Ann. Biomed. Eng.* **13,** 321–328.

Wolf, M. M. and Berne, R. M. (1965) Coronary vasodilator properties of purine and pyrimidine derivatives. *Circ. Res.* **4,** 343–348.

CHAPTER 2

Radioligand Binding Assays for Adenosine Receptors

Michael Williams and Kenneth A. Jacobson

1. Introduction

The technique of radioligand binding is one that has been of considerable importance in defining receptor function at the molecular level. In addition to providing evidence for the potential existence of receptor subtypes, subclasses, and states (Williams and Enna, 1986; Williams and Sills, 1989), the ease and resource efficiency of the technique makes it an indispensable part of the drug discovery process, facilitating the targeted screening of many thousands of new chemical entities (NCEs) to allow for the discovery of new structural classes that interact with the receptor target (Williams and Jarvis, 1989). In the adenosine area, the technique has permitted the discovery of several classes of heterocyclic compounds that function as adenosine antagonists (Psychoyos et al., 1982; Williams et al., 1987a; Daly et al., 1988; Trivedi and Bruns, 1988; Peet et al., 1988; Williams, 1989) but to date, the only known agonists at adenosine receptors retain properties of the purine nucleoside structure.

Adenosine and Adenosine Receptors Editor: Michael Williams ©1990 The Humana Press Inc.

2. General Principles of Radioligand Binding

There are several excellent monographs and reviews in which the subject of radioligand binding is covered in great depth (Yamamura et al., 1985; Limbird, 1986; Enna, 1987; Williams et al., 1988). For the purposes of the present chapter, it is sufficient to note that a radioligand of specific activity 20 Ci/mmol or greater is opti-mal for the labeling of a receptor by binding to it in a competitive, reversible and saturable manner, according to the law of mass action.

A radioactively labeled ligand can thus be used to recognize a neurotransmitter receptor. The basic properties of this binding site can be studied by kinetic and saturation analysis to derive a measure of the affinity of the ligand for its recognition site as well as an estimate of the number of binding sites in a given tissue. The affinity is usually measured as the dissociation constant, or K_d, and is in molar units, usually in the nanomolar range. The density of binding sites is measured as the apparent B_{max}, which has mol U/tissue weight, the latter usually expressed in terms of mol U/mg protein.

Saturation data are usually transformed for analysis in terms of what is generically known as a "Scatchard Plot," in which the amount of specifically bound radioligand is linearly transformed and plotted against the ratio of bound radioactivity divided by the added amount. In a simple system, this plot is linear, and the slope and abscissa intercept can be used to derive the K_d and B_{max} values for an experiment in which, typically, 6–12 concentrations of ligand are incubated with fixed amounts of tissue. More often, the plot obtained from the Scatchard derivation is curvilinear, suggesting the presence of multiple binding sites. The routine use of computerized nonlinear regression programs that use untransformed data can more objectively assess the true situation and provide a statistical fit of the data (Williams and Sills, 1989). This of course does not ascribe any functionality to binding sites with different affinities or densities, a major point of controversy that has led to the epithet "grind and bind" when binding data are generated in a vacuum devoid of any functional or physiological relevance.

There are technical constraints dependent on the parameters derived from saturation analysis. The first of these is that, unless a ligand interacts with a given binding site with high affinity, the means by which the receptor ligand complex is isolated can affect the data derived. For a high affinity ligand (K_d ~ 20 nM), filtration is an effective and reliable means by which to separate bound radioactivity from free. As the affinity decreases (an increase in the K_d value), filtration tends to remove bound as well as free radioactivity, leading to an underestimate of both parameters and providing extremely variable data. This problem can be circumvented by the use of centrifugation to pellet the receptor–ligand complex. This generally allows a more reliable estimate of radioactivity bound with low affinity, but is also more time consuming and results in higher background radioactivity. In those instances where the "ideal" high-affinity ligand is unavailable, the centrifugation assay can be used as a screen to provide such ligands (Williams and Jarvis, 1989).

Similarly, a low density of binding sites in a given tissue can limit the amount of useful data obtained. In many instances, brain tissue can be used as a model screening system for potential NCEs that will be used in peripheral tissues. This tissue has adenosine receptor densities in the pmol/mg protein range. In peripheral tissues, such values may be in the fmol range, reflecting the discrete innervation of such tissues as compared to the central nervous system. Binding sites in such tissues can, however, be characterized by the use of iodinated radioligands that have specific radioactivities of up to 2200 Ci/mmol. Thus the signal to noise ratio for such ligands is such that the binding of a few molecules can be detected. Alternatively, the technique of receptor autoradiography can be used (Jarvis, 1988).

Once the binding parameters (K_d and B_{max}) for a radioligand have been determined, its pharmacology can then be determined using a variety of other drugs and NCEs and compared with similar profiles derived in functional assays (adenylate cyclase, phosphatidy-linositol turnover, calcium rise) and tissue models to determine whether the binding site is indeed a receptor. It is customary to evaluate 20–30 compounds to determine the structure–activity profile for a binding site. Inactive compounds can be as important as those that are active in assessing the validity of a binding assay.

3. Adenosine Receptor Binding Assays

Early studies related to defining adenosine receptors using radioligand binding technology relied on [³H]adenosine and [³H]2-chloroadenosine (Fig.1) as ligands (Malbon et al., 1978; Schwabe et al., 1979; Williams and Risley, 1980; Newman et al., 1981; Wu et al., 1980; Schütz and Brugger, 1982). The use of [³H] adenosine was limited by its susceptibility to degradation by tissue enzymes such as adenosine deaminase (ADA) and was further confounded by apparently low binding affinity (K_d values = 0.6–9.5 μM), high-capacity binding sites (B_{max} values = 2–30 pmol/mg protein), and inconsistent structure activity data. The ability of tissues to produce large quantities of endogenous adenosine also proved to be a major factor in confusing the data obtained. The more metabolically stable 2-chloroadenosine (2-CADO), although an unstable radioligand, gave better data, with K_d values in the order of 100 nM (Williams and Risley, unpublished data). It was, however, not until the endogenous adenosine was removed by treatment of the tissue (in this instance, rat brain) with the enzyme (ADA) that high affinity, nanomolar binding of this ligand was observed (Williams and Risley, 1980). However, at this juncture, four independent groups reported high-affinity binding of agonist radioligands in brain tissue (Bruns et al., 1980; Schwabe and Trost, 1980; Williams and Risley, 1980; Wu et al., 1980).

For 2-CADO, two binding sites (K_d values = 1.3 and 16 nM) were reported in the presence of ADA (Williams and Risley, 1980) and one (K_d = 23 nM) in the absence of enzyme treatment (Wu et al., 1980), the identity of the latter site confounding the apparent need to remove endogenous adenosine. Although this ligand has been reported to bind to adenosine receptors in the cerebral microvasculature (Beck et al., 1984), human placenta (Fox and Kurpis, 1983), human brain (John and Fox, 1985), rat kidney membranes (Wu and Churchill, 1985), rat hippocampal membranes (Chin et al., 1985), and bovine atrial sarcolemmal membranes (Michaelis et al., 1985), the data obtained from these various studies are open to critique based on the low specific activity (~ 12Ci/mmol or less), low affinity of binding and the pharmacological profile as compared to other A_1-selective radioligands.

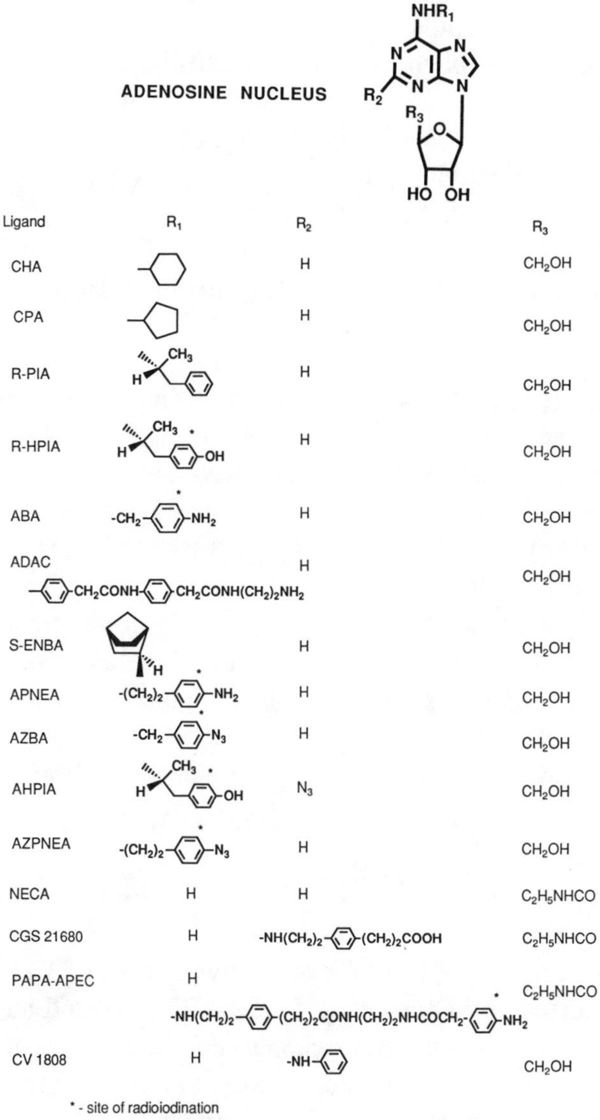

ADENOSINE NUCLEUS

Ligand	R₁	R₂	R₃
CHA	cyclohexyl	H	CH_2OH
CPA	cyclopentyl	H	CH_2OH
R-PIA		H	CH_2OH
R-HPIA		H	CH_2OH
ABA	$-CH_2$—⬡—NH_2	H	CH_2OH
ADAC	—⬡—CH_2CONH—⬡—$CH_2CONH(CH_2)_2NH_2$	H	CH_2OH
S-ENBA		H	CH_2OH
APNEA	$-(CH_2)_2$—⬡—NH_2	H	CH_2OH
AZBA	$-CH_2$—⬡—N_3	H	CH_2OH
AHPIA		N_3	CH_2OH
AZPNEA	$-(CH_2)_2$—⬡—N_3	H	CH_2OH
NECA	H	H	C_2H_5NHCO
CGS 21680	H	$-NH(CH_2)_2$—⬡—$(CH_2)_2COOH$	C_2H_5NHCO
PAPA-APEC	H	$-NH(CH_2)_2$—⬡—$(CH_2)_2CONH(CH_2)_2NHCOCH_2$-⬡-$NH_2$	C_2H_5NHCO
CV 1808	H	$-NH$—⬡	CH_2OH

* - site of radioiodination

Fig. 1. Structures of adenosine agonist radioligands.

Whereas the possibility cannot be excluded that adenosine receptors are being labeled, this ligand has been replaced in routine use with the N^6-substituted ligands, N^6-cyclohexyladenosine (CHA; Bruns et al., 1980) and R-N^6 phenylisopropyladenosine (R-PIA; Schwabe and Trost, 1980).

Both ligands are selective for the A_1 receptor subtype, have high affinity and selectivity (Table 1) and their binding is usually measured in the presence of ADA. Concomitantly, the Bruns-Daly-Snyder group (Bruns et al., 1980), reported [^{3}H]DPX (1,3-diethyl-8-phenylxanthine; Fig. 2) as the first antagonist ligand for adenosine receptors.

In general, therefore, the key to detecting specific, selective and high-affinity binding of ligands to adenosine receptors was the use of either adenosine analogs resistant to degradation by ADA or xanthine analogs that were potent and selective adenosine antagonists in tissues that had been depleted of endogenous adenosine.

The next major breakthrough, again from Bruns and co-workers (Bruns et al., 1986) was the use of the nonselective ligand, [^{3}H]NECA (5'-N-ethylcarboxamidoadenosine; Fig. 1) in the presence of an A_1 receptor blocking agent to specifically label A_2 receptors. Subsequently, [^{3}H]XAC (xanthine amine congener) was developed as the first high-affinity antagonist radioligand (Jacobson et al., 1986).

4. Agonist Radioligands

4.1. A_1 Selective Ligands

4.1.1. CHA

The binding of [^{3}H]N^6-cyclohexyladenosine (CHA; Fig.1) was first described by Bruns et al. (1980). This ligand has become the standard for defining the A_1 receptor and is approximately 300–600-fold selective for this receptor (Bruns et al., 1986; Stone et al., 1988; Jarvis et al., 1989a). ADA-pretreatment of tissues is essential to define high-affinity specific binding. K_d values varying between 0.4 and 6 nM have been reported in brain tissue (Bruns et al., 1980; Marangos et al., 1983, Williams et al., 1986). Binding equilibrium is reached in approximately 90 min at 23°C. Saturation analysis of CHA binding, like that seen for 2-CADO, can reveal two binding sites with

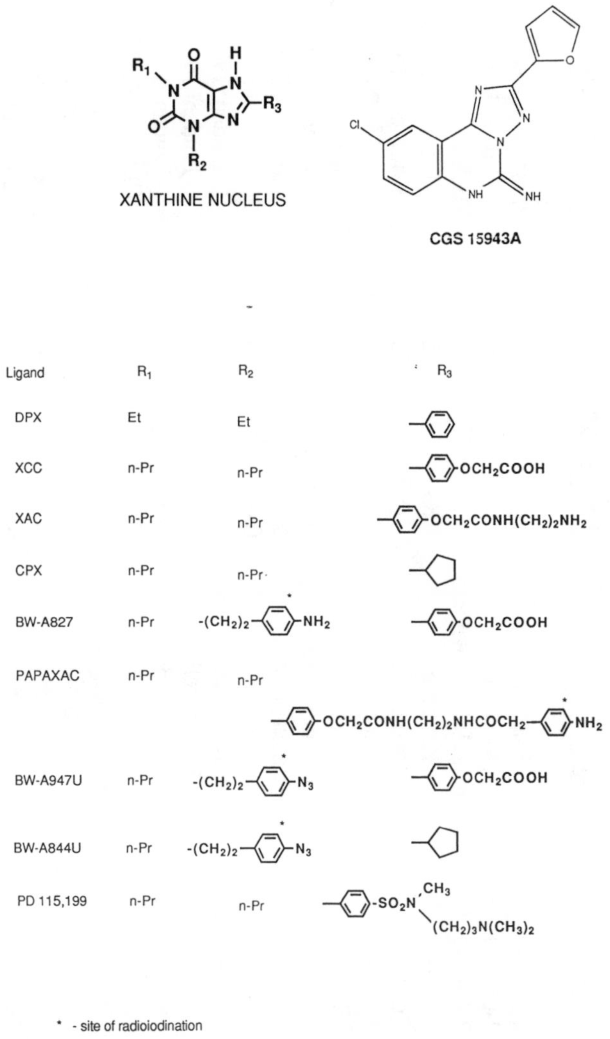
XANTHINE NUCLEUS
CGS 15943A
Ligand
R_1
R_2
R_3
DPX
Et
Et
XCC
n-Pr
n-Pr
OCH_2COOH
XAC
n-Pr
n-Pr
$OCH_2CONH(CH_2)_2NH_2$
CPX
n-Pr
n-Pr
BW-A827
n-Pr
$-(CH_2)_2-$
NH_2
OCH_2COOH
PAPAXAC
n-Pr
n-Pr
$OCH_2CONH(CH_2)_2NHCOCH_2-$
NH_2
BW-A947U
n-Pr
$-(CH_2)_2-$
N_3
OCH_2COOH
BW-A844U
n-Pr
$-(CH_2)_2-$
N_3
PD 115,199
n-Pr
n-Pr
$-SO_2N$
CH_3
$(CH_2)_3N(CH_3)_2$
* - site of radioiodination

Fig. 2. Structures of adenosine antagonist radioligands.

K_d values of 0.4 and 4 n*M* (Marangos et al., 1983; Bruns, 1988). CHA binding, like that for most adenosine agonist radioligands, is decreased by GTP and sodium, and enhanced by divalent cations (Goodman et al., 1982). The effects of GTP and NaCl appear to involve distinct mechanisms (Stiles, 1988).

Table 1
Adenosine Ligands—Binding Parameters

Compound	Ligand	Tissue	K_d nM	B_{max} fmol/mg protein	Reference
A_1 Receptor Ligands					
Agonists					
	2-CADO	Rat brain	1.3	207	Williams and Risley, 1980
	2-CADO	Rat brain	24	476	Wu et al., 1980
	2-CADO	Bovine atrial sarcolemma	18	100	Michaelis et al., 1985
	CHA	Rat brain	0.89	662	Williams et al., 1986
	CHA	Bovine brain	0.3/1.8	540	Bruns et al. 1980
	CHA	Rat testes	2	200	Murphy and Snyder, 1982
	R-PIA	Rat brain	5.2	810	Schwabe and Trost, 1980
	R-PIA	Canine myocardium	52–85	610	Lee et al., 1986
	R-PIA	Guinea pig lung	250	12,000	Ukena et al., 1985
	CPA	Rat brain	0.48	416	Williams et al., 1986
	HPIA	Rat brain	0.94	871	Linden, 1984
	HPIA	Rat fat cells	2.95	379	Linden, 1984
	HPIA	Rat heart	3.1	22	Linden, 1984
	HPIA	Rat fat cells	0.7, 7.6	940,950	Ukena et al., 1984
	HPIA	Bovine myocardium	1.1	305	Lohse et al., 1985
	ABA	Rat brain	1.9	585	Linden et al., 1985
	ADAC	Rat brain	1.4	570	Jacobson et al., 1987
	S-EBNA	Rat brain	0.3	160	Trivedi et al., 1989
Antagonists					
	DPX	Rat brain	68	1220	Bruns et al., 1980
	XAC	Rat brain	1.2	580	Jacobson et al., 1986
	XCC	Rat brain	2–3	600	Jarvis et al., 1987
	CPX	Rat brain	0.42	460	Bruns et al., 1987a

Ligand	Preparation			Reference
CPX	Rat brain solubilized + caffeine*	0.55	911	Helmke and Cooper, 1989
CPX	Rat brain solubilized + adenosine*	0.91	1168	Helmke and Cooper, 1989
PAPAXAC	Bovine brain	0.13	986	Stiles and Jacobson, 1987
BW A-827	Rat brain	1.25	516	Linden et al., 1988
A_2 Receptor Ligands				
Agonists				
NECA	Rat striatum	1.8	118	Bruns et al., 1986
NECA	Rat striatum	4.5	813	Stone et al., 1988
NECA	Human platelet—solubilized	46	510	Lohse et al., 1988
CGS 21680	Rat striatum	15	375	Jarvis et al., 1989
PAPA-APEC	Bovine striatum	1.8	222	Barrington et al., 1989b
Antagonists				
PD 115,199	Rat striatum	2.6	560	Bruns et al., 1987c
CGS 15943	Rat striatum	4	1500	Jarvis et al., 1988
Uptake sites				
NBI	Guinea pig brain	0.25	225	Marangos et al., 1987
Dipyridamole	Guinea pig brain	3.5	900	Marangos et al., 1987

*Receptor solubilized in the presence of the ligand indicated.

4.1.2. R-PIA

As noted by Bruns (1988), [^{3}H]N^6-R-phenylisopropyladenosine (R-PIA) has nearly identical binding characteristics to CHA, being A$_1$-receptor selective. In rat brain membranes, this ligand binds to two sites with K_d values of 1.4 and 139 nM (Lohse et al., 1984). Under some circumstances, the data obtained with this ligand can differ from that seen with CHA, a fact that may be attributable to the presence of an amphetamine moiety in the N^6 position. The S-diastereomer of PIA is several times less active than the R-isomer, the magnitude of delineation being dependent on the test paradigm measured. High and low affinity states of the A$_1$ receptor have been reported using [^{3}H]R-PIA (Green and Stiles, 1986).

4.1.3. CPA

N^6-Cyclopentyladenosine (CPA; Fig. 1) is some 700-fold selective for the A$_1$ receptor (Bruns et al.,1986). [^{3}H]CPA binds to A$_1$ receptors in rat brain membranes with a K_d value of 0.5 nM (Williams et al., 1986). Although minor differences in binding compared to CHA were observed, notably the labeling of only half the number of binding sites—a finding potentially indicative of preferential labeling of a higher-affinity subtype of the A$_1$ receptor (Bruns, 1988)—this ligand has not been extensively used.

4.1.4. HPIA

[^{125}I]N^6-p-hydroxyphenylisopropyladenosine (HPIA; Fig. 1) was the first iodinated radioligand for adenosine receptors (Munshi and Baer, 1982; Schwabe et al., 1982). While the K_d seen with [^{125}I]HPIA is about twice that seen with [^{3}H]PIA (Schwabe et al., 1982; Lohse et al., 1984), the B_{max} is 60% less. The higher specific radioactivity of the iodinated ligand has permitted the labeling of A$_1$ receptors in heart, brain, and fat cells (Linden, 1984; Ukena et al., 1984; Lohse et al., 1985; Lied et al., 1988).

4.1.5. ABA

[^{125}I]N^6-(4-amino-3-iodobenzyl)adenosine (ABA; Fig. 1) was introduced as an alternate ligand to HPIA because of problems with nonspecific binding of the former ligand (Linden et al., 1985). ABA had a K_d of 2 nM in brain tissue and 11 nM in rat heart.

4.1.6. S-ENBA

[^{3}H]S-ENBA (1R,2S,4S-N^6-2-*endo*-norbornyladenosine; Fig. 1) is 4700-fold selective for the A$_1$ receptor and binds to rat brain membranes with a K_d value of 0.3 nM (Trivedi et al., 1989). It has been suggested as a substitute for CHA in peripheral tissues.

4.1.7. APNEA, AZBA, AHPIA, and AZPNEA

N^6-(2-(4-Aminophenyl)ethyl)adenosine (APNEA; Fig. 1), N^6-(4-azidobenzyl)adenosine (AZBA; Fig.1), 2-azido-PIA (AHPIA; Fig. 1), and N^6-(2-(4-azidophenyl)ethyl)adenosine (AZPNEA; Fig. 1) are iodinated ligands developed to photoaffinity label A$_1$ receptors (Choca et al., 1985; Klotz et al., 1985; Stiles et al., 1985, 1986). All four ligands are A$_1$ receptor selective with K_d values of 1–2 nM and can be used to covalently label the target receptor.

4.1.8. ADAC

N^6-[4-[[[4-[[[(2-Aminoethyl)amino]carbonyl]methyl]anilino]-carbonyl]-methyl]phenyl]adenosine (ADAC, Fig. 1) is a functionalized congener with high affinity for A$_1$ receptors and has been derivatized to obtain spectroscopic probes. [^{3}H]ADAC binds to rat and bovine brain adenosine receptors with K_d-values of 1.4 and 0.34 nM, respectively (Jacobson et al., 1987).

4.2. A$_2$ Selective Ligands

4.2.1. NECA

For many years, NECA (5'-N(-ethylcarboxaamidoadenossine) (Fig. 1) was used in functional assays as an "A$_2$-selective" ligand. This adenosine analog is, in fact, nonselective (Bruns et al., 1986) and has approximately equal affinity (~15 nM) for both A$_1$ and A$_2$ receptors. Thus, effects seen with NECA, but not with CHA or R-PIA, can be ascribed to A$_2$ receptor interactions. Effects examined solely with NECA cannot be ascribed to A$_2$ receptor activation (Williams, 1987).

[^{3}H]NECA binding was found in liver (Schütz et al., 1982), uterus (Ronca-Testoni et al., 1984), platelets (Hütteman et al., 1984), and brain (Yeung and Green 1984). The demonstration of biphasic binding curves for [^{3}H]NECA in rat striatum suggested the presence of A$_1$ and A$_2$ receptor binding components (Yeung and Green, 1984).

N-ethylmaleimide (NEM), a sulfhydryl alkylating agent, was then used to eliminate the A_1 binding component, leaving a [^{3}H]NECA binding site with characteristics of an A_2 receptor. Subsequent examination of the effects of NEM suggested that it might in fact be modulating the pharmacological properties of the remaining binding sites, causing an increase in their number (Bruns et al., 1986). As a result, Bruns et al. (1984) removed the A_1 component of [^{3}H]NECA binding with a 50 nM concentration of the A_1-selective agonist, CPA. Thus, in rat striatum in the presence of CPA, [^{3}H]NECA can selectively label A_2 receptors. This concentration of CPA appears applicable to [^{3}H]NECA binding in a variety of mammalian species (Stone et al., 1988). The K_d for NECA in the presence of CPA was 11–12 nM in rat striatum (Bruns et al., 1986). Four binding sites for [^{3}H] NECA could be identified by computer analysis in the absence of CPA: a high-affinity A_1 site; a high-affinity A_2 site; a low-affinity A_2 site, and a high-capacity "nonreceptor" site that had been previously noted by Yeung and Green (1984). In the presence of CPA, 67% of the specific binding of 4 nM [^{3}H]NECA was estimated to bind to the high-affinity site, termed A_{2a} and the remaining 33% to the lower-affinity A_{2b} site. Examination of a series of adenosine agonists and antagonists using [^{3}H]NECA binding to determine A_{2a} affinity and human fibroblast cyclic AMP accumulation as an estimate of A_{2b} affinity showed that most agonists were selective for the "a" site, whereas antagonists were more effective at the "b" site.

The "nonreceptor" site to which [^{3}H]NECA binds is present in several tissues (Hütteman et al., 1984; Reddington et al., 1986; 1987). This site is sensitive to $MgCl_2$ (Bruns et al., 1986; Stone et al., 1988) and can be physically distinguished from the A_2 receptor present in human platelets (Lohse et al., 1988) and bovine striatum (Barrington et al., 1989b). The identity and physiological significance of this site remains unknown.

4.2.2. CGS 21680

CGS 21680 (2-(p-(carboxyethyl)phenethylamino)-5'-N-ethyl-carboxamidoadenosine; Fig.1) is a 2-substituted NECA derivative that is 140-fold selective for the striatal A_2 receptor with an IC_{50} value of 22 nM (Hutchison et al., 1989). In rat brain membranes pretreated with ADA, [³H]CGS 21680 had a K_d value of 15 nM (Jarvis et al., 1989a) and an apparent B_{max} of 375 fmol/mg protein. CPA had no effect on the binding of [³H]CGS 21680. A correlation coefficient of 0.98 was obtained when the activity of a series of A_2-selective agonists and antagonists in the [³H]CGS 21680 assay was compared to data obtained using [³H]NECA in the presence of CPA. CGS 21680 is thus the first selective adenosine A_2 receptor ligand that can be used without the need to chemically or pharmacologically block A_1 receptor binding.

4.2.3. PAPA-APEC

[¹²⁵I]-2-[4-[2-[2-[(4-aminophenyl)methylcarbonylamino]-ethylaminocarbonyl]ethyl]phenyl]ethylamino ethylamino-5'-N-ethylcarboxamidoadenosine (Fig. 1) is a derivative of CGS 21680 that is greater than 500-fold selective for the A_2 receptor and that binds to this receptor in bovine striatum with a K_d value of 1.5 nM (Barrington et al., 1989b). PAPA-APEC structurally contains a p-aminophenylacetyl (PAPA) moiety that serves as a prosthetic group for both radioiodination and crosslinking to the receptor. As with CGS 21680, CPA has no effect on either the binding characteristics or pharmacology of PAPA-APEC. The ligand can be used to photoaffinity label the A_2 receptor in the presence of the crosslinking agent, SANPAH (Barrington et al., 1989b), as described in section 9.

4.2.4. CV 1808

CV 1808 (2-(Phenylamino)adenosine) is a relatively weak, fivefold A_2 selective adenosine receptor agonist (Bruns et al., 1986; Hutchison et al., 1989). Tritiated CV 1808 binds to two sites in rat brain membranes with K_d values of 20 and 400 nM (Sills et al., 1990), the latter of which is insensitive to adenosine agonists.

5. Antagonist Radioligands

Antagonist radioligands bind with high affinity to both agonist-defined coupling states (which would have high and low affinity for agonist) of a given receptor and are thus preferable (when available) to agonist ligands (Bruns, 1988; Williams and Sills, 1989). Several xanthine and one nonxanthine adenosine antagonist radioligands have been developed.

5.1. A_1 Selective Ligands

5.1.1. DPX

[³H] 1,3-Diethyl-8-phenylxanthine (DPX; Fig. 2) was the first adenosine antagonist radioligand developed (Bruns et al.,1980). Although initially described as binding to both A_1 and A_2 sites, examination of [³H]DPX binding in different species indicated that this ligand was A_1 selective (Murphy and Snyder, 1982). The xanthine had a K_d of 68 nM at rat brain adenosine receptors.

5.1.2. PACPX

PACPX (1,3-dipropyl-8-(2-amino-4-chloro)xanthine; Fig. 2) was originally described as the most potent and selective A_1 receptor antagonist known (Bruns et al., 1983a), but unfortunately such activity was species dependent (Ferkany et al., 1986; Stone et al., 1988). The use of this xanthine analog as a radioligand was limited by its high lipophilicity. PACPX has been characterized as a noncompetitive antagonist (Williams et al., 1987b).

5.1.3. Functionalized Xanthine Congeners (XAC,XCC)

By the functionalized congener approach (Jacobson et al., 1986) insensitive sites on a receptor ligand are utilized for chemical derivatization through chain attachment. The incorporation of distal functional groups that are ionized at pH 7.4 led to 8-phenyl substituted xanthines that lacked the inherent insolubility seen with other compounds in this series. XAC (Xanthine amine congener or 1,3-dipropyl-8-(4-(2-aminoethyl)aminocarbonylmethyloxyphenyl)xanthine; Fig. 2) can be labeled to a specific activity of 120-140 Ci/mmol (Jacobson et al., 1986). The compound had a K_d of 1.2 nM in rat brain and labeled approximately half as many sites as DPX, a more accurate reflection of A_1 receptor labeling. High levels of filter binding because of hydrophobicity and the charged amine moiety could be overcome by the use of PEI pretreated filters (Bruns et al., 1983b). The pharmacology of binding was similar to that seen with CHA. Whereas XAC has been reported as moderately A_1 selective, the xanthine can also label A_2 receptors in platelets (Ukena et al., 1986) with a K_d of 12 nM. XCC (Xanthine carboxylic acid congener or 1,3-dipropyl-8-(4-carboxymethyloxy)phenylxanthine; Fig. 2) has also been radiolabeled and appears to be an A_1-selective ligand. (Jarvis et al., 1987).

5.1.4. PD 116,948 (CPX)

PD 116, 948 or 1,3-dipropyl-8-cyclopentylxanthine (CPX or DPCPX; Fig. 2) is approximately 700-fold selective for the A_1 receptor (Bruns et al., 1987a), the radiolabeled entity having a K_d of 0.5 nM (Lee and Reddington, 1986; Bruns et al., 1987a). The pharmacology of binding was similar to that observed for CHA. [^{3}H]CPX has also been used to label A_1 receptors in cardiac myocyte cultures (Liang, 1989).

5.1.5. BW-A827

BW-A827[^{125}I]-(1-propyl-3-[(4-amino-3-iodophenyl)ethyl]-8-[p-(carboxymethyloxy)phenyl]xanthine) is an A_1-selective ligand. It binds to rat brain membranes with a K_d of 1.25 nM (Linden et al., 1988). The corresponding aryl azido derivative, BW-A947U (K_d = 1.2 nM), has been used to photoaffinity label this receptor (Earl et al., 1988).

5.1.6. PAPAXAC

PAPAXAC (p-aminophenylacetyl-XAC, Fig. 2) is a covalent conjugate of the functionalized congener XAC and a prosthetic group for radioiodination (conceptually similar to PAPA-APEC, above) linkked through an amide bond. The ^{125}I form of PAPAXAC is a high affinity (K_d = 0.1–1 nM) radioligand with A_1 selectivity, which may also be bound in an irreversible fashion upon addition of the photoaffinity crosslinker SANPAH followed by irradiation (Stiles and Jacobson, 1987) or via conversion of the amine to a photoreactive azide (Barrington et al., 1989a).

5.1.7. m-DITC-XAC

The 1,3-phenylenediisothiocyanate derivative of XAC, m-DITC-XAC, and its isomer p-DITC-XAC, are high affinity A_1-selective antagonists that spontaneously form a covalent bond with a nucleophilic group on the receptor protein. m-DITC-XAC has an IC_{50} value of 52 nM for irreversible inhibition and has been successfully used in its tritiated form to radioactively label the A_1 receptor for direct detection on an electrophoretic gel (Stiles and Jacobson, 1988). These xanthines are unique site-specific irreversible inhibitors of adenosine receptors that do not require an irradiation step.

5.1.8. BW-A844U

BW-A844U(3-[(4-amino)phenethyl]-8-cyclopentylxanthine, Fig. 2) differs from BW-A827 in the presence of a cycloalkyl group rather than an aryl group at the 8-position, resulting in a higher affinity (K_d = 0.23 nM in bovine brain membranes) at A_1 receptors. The radioiodinated form has been converted to the corresponding azido derivative (^{125}I-azido-BW-A844U), which was used to photoaffinity label the A_1 receptor (Patel et al., 1988b).

5.2. A_2 Selective Ligands

5.2.1. PD 115,199

PD 115, 199 is a sulfonamide congener of 1,3-dipropyl-8-phenylxanthine that like XAC, bears a charged amino group resulting in water-soluble xanthine adenosine antagonists (Hamilton et al., 1985). PD 115,199 (Fig. 2) was found to be almost equiactive at A_1 and A_2 receptors with IC_{50} values in the range of 15 nM (Bruns et al., 1987b). In the presence of 20 nM CPX, the A_1 selective xanthine [3H]PD 115,199 bound to striatal membranes with a K_d value of 2.6 nM and a pharmacology consistent with the labeling of an A_2 receptor. Thus, [3H] PD 115,199 is the xanthine equivalent of [3H] NECA, selectively labeling A_2 receptors when the A_1 component is selectively blocked. As would be expected for an antagonist radioligand, agonists were about five times less potent in displacing PD115,199, whereas antagonists were three times more potent than comparable data obtained using [3H]NECA in the presence of CPA. The stability of this xanthine radioligand, however, is poor.

5.2.2. CGS 15943

One of the first nonxanthine adenosine receptor antagonists identified was CGS 8216 (Czernik et al., 1982), a triazolquinazoline whose predominant activity was at the benzodiazepine receptor. Subsequent SAR work on this novel heterocycle led to the identification of CGS 15943 (Fig. 2), a novel and potent adenosine receptor antagonist with sevenfold selectivity for, and an IC_{50} of 3 nM at the A_2 receptor (Williams et al., 1987a). Labeling of this adenosine receptor antagonist has been confounded by ligand instability. However, a K_d value of 4 nM for [3H] CGSS15943 in rat cortical membranes has been reported (Jarvis et al., 1988). Whereas binding curves for antagonists were monophasic, those for agonists were biphasic. Like XAC, however, it appears that this ligand can also interact with A_1 receptors.

6. Species Differences

Like many receptor binding assays, those for adenosine have been carried out in a variety of species with often conflicting data. Although this may be ascribed to interexperimental variations, side by side comparisons in different species have shown interspecies differences in the potency of various adenosine ligands for the respective binding sites (Murphy and Snyder, 1982; Bruns et al., 1983a; Schwabe et al., 1985). These differences were especially marked for the antagonists, PACPX being reported to have a K_i value of 0.2 nM in calf cortical membranes, but activity of 50 nM in guinea pig membranes (Bruns et al., 1983). Thus the species chosen for binding can markedly affect the data obtained.

In-depth evaluations of the binding of a series of adenosine agonists and antagonists to A_1 (Ferkany et al., 1986) and A_2 (Stone et al., 1988) receptors in six species allowed for a direct comparison of these differences. At the A_1 receptor, using [^{3}H]CHA, the number of binding sites in cortical membranes from cow, rabbit, rat, mouse, guinea pig, and human brain showed little variability (Ferkany et al., 1986). Differences in receptor affinity were noted, however, with bovine brain having a K_d value for [^{3}H] CHA binding (0.6 nM) 3–4× greater than that in rat, rabbit, or mouse brain (K_d values = 1–2 nM). In human and guinea pig brain, K_d values were 3.7 and 6.6 nM, respectively. Evaluation of the pharmacology of [^{3}H]CHA binding in the six species showed that, in general, all agonists had approximately the same rank order of activity within each species; R-PIA > CHA > NECA > 2-CADO > S-PIA. In bovine brain, S-PIA was more active than NECA. Across species, the N^6-substituted analogs were most active in cow and least active in guinea pig. NECA was generally equiactive across all species, whereas 2-CADO was more active in guinea pig and human brain, and least active in cow brain. More marked interspecies differences were seen for the xanthine antagonists. PACPX, the most active of these, had a K_i value of 0.18 nM in bovine brain and was some 390-fold less active in guinea pig ($K_i = 71$ nM). Within species, the differences in activities for theophylline and four 8-substituted xanthines were variable. In all six species, PACPX was the most active xanthine studied, with theophylline the least active. Theophylline was 2000– to 5450-fold less active than PACPX in cow, rabbit, rat, and mouse. In human brain, the ratio of K_i values for theophylline and PACPX was 243, and in this difference in activity was 243 and in guinea pig, 82. Based on these data, it appeared that the structure activity requirements of xanthines at the A_1 receptor was highly species dependent.

A subsequent study of A_2 receptor differences between species using [^{3}H]NECA in the presence of 50 nM CPA (Stone et al., 1988) indicated the existence of separate high- and low-affinity binding sites. The high-affinity site had K_d values between 3.2 nM (rabbit) and 9.6 nM (human) and the lower affinity, 54 nM (rabbit) and 232 nM (human). Comparison of the pharmacological profiles of the higher-affinity site showed a rank order of activity for agonists of: NECA > 2-CADO > R-PIA ≥ CHA ≥ CPA >> S-PIA. NECA was equiactive in all species (IC_{50} = 11–19 nM). The N^6-substituted analogs (R- and S- PIA, CPA and CHA) showed a species variation of 1.6–4.6 fold. As in the A_1 studies, there were marked species differences in the ability of xanthine adenosine receptor antagonists to displace [^{3}H] NECA. In rabbit and human brain, PACPX had an IC_{50} value of 14–15 nM. The activity was threefold less in calf brain and 20–24-fold less in guinea pig, mouse, and rat. Similar species differences were observed for the other xanthines studied. These side by side comparisons of A_1 and A_2 receptor pharmacology in six different mammalian species emphasize the need for caution in comparing data obtained from different species. In addition, they raise concerns as to the appropriate preclinical model for profiling new adenosine ligands for use in humans.

7. Receptor Autoradiography

Using computer assisted densitometry, the technique of quantitative autoradiography developed by Kuhar et al. (1985), has provided the means to localize both A_1 and A_2 receptors in intact tissue sections. Initial studies using [^{3}H] CHA to label the A_1 receptor (Lewis et al., 1981; Goodman and Snyder, 1982; Lee and Reddington,

1986) were essentially qualitative in nature. A_1 receptors were found to be highly concentrated in the molecular level of the cerebellum and the hippocampal CA-1 and CA-3 regions located on intrinsic pyramidal cells (Onodera and Kogure, 1988). Moderate receptor densities were observed in cerebral cortex, striatum, and thalamus with minimal binding in the brainstem. Later studies using a more quantitative approach (Snowhill and Williams, 1986) defined binding and pharmacological parameters for CHA binding in seven rat brain regions. Slight differences in binding affinity (K_d) were noted with significant differences in receptor density. Minimal differences in regional pharmacology were noted for five agonist and two antagonist radioligands. The rank order of activity for all seven brain regions was CPA > R-PIA > NECA >> PACPX $\geq$ 2-CADO $\geq$ S-PIA >> DPX. Comparison of [³H]CHA binding in the cerebellum of Weaver and Reeler mice and in control vs unilaterally enucleated rats indicated that A_1 receptors were located on axon terminals of excitatory cells (Goodman et al., 1983). The association of A_1 receptors with retinal ganglion cells, apparently of the W type (Kocis et al., 1984; Braas et al., 1987), is consistent with evidence for a retinocollicular pathway in which adenosine is an important effector substance (Heller and McIlwain, 1973; Kostopoulos and Phillis, 1977).

The regional distribution of A_2 receptors in brain has received limited attention again because of problems with available ligands. Early studies with [³H]NECA (Lee and Reddington, 1986) resulted in the labeling of "nonreceptor" sites localized to the striatum, thalamus, and cerebral cortex. More recently however, using the NECA binding conditions developed by Bruns (Bruns et al., 1986), specific, high-affinity $(K_d = 9 \text{ n}M)$ NECA binding sites were identified in rat striatum and olfactory tubercule (Jarvis et al., 1989b). Evaluation of these sites with five agonist and four antagonist ligands showed them to be A_2 in nature. The resolution of the identity of the "nonreceptor" site labeled by NECA has not been determined.

The xanthine carboxylic congener adenosine antagonist, [³H]XCC, and the amine congener, [³H]XAC, have also been used to label A_1 receptors in autoradiographic studies (Jarvis, 1988; Jarvis et al., 1987; Deckert et al., 1988).

8. Labeling
of Other Adenosine Recognition Sites

Nucleoside transport systems in endothelial and erythrocyte membranes mediate removal of adenosine from the extracellular space, thus terminating the actions of the nucleoside (Pearson and Gordon, 1985; Geiger and Nagy, 1990; Stone et al., 1990).

Inhibition of adenosine transport systems can potentiate the effects of adenosine and in some situations, a transport inhibitor and a receptor agonist can appear similar in their physiological actions. Dipyridamole, the coronary vasodilator, and CV 1808 have activity at both the receptor and transporter site (Williams and Risley, 1980; Taylor and Williams, 1982; Balwierczak et al., 1989).

Adenosine transport occurs by a number of facilitated diffusion processes, some of which can be blocked by 6-thiopurine nucleosides in the nanomolar range (Young and Jarvis, 1983; Jarvis, 1988). Both [³H] nitrobenzylthioinosine (NBTI;NBMPR; Marangos et al., 1982) and [³H]dipyridamole (Marangos et al., 1985) bind to adenosine transporter sites. However, the binding properties of these two ligands are different. [³H]NBI has a K_d value of 0.25 nM in guinea pig cortical membranes with an apparent B_{max} of 225 fmol/mg protein (Marangos et al., 1987). Most adenosine receptor ligands are weak inhibitors of transporter ligand binding (Table 2). Additionally, the activity of R- and S-PIA are reversed (Clanachan et al., 1987). For [³H]dipyridamole, the K_d and B_{max} values were 3.5 nM and 800 fmol/mg protein, respectively. Thus dipyridamole labels four times as many sites as NBI with a lower affinity. Marked species differences in the binding of the two transporter ligands also exist (Verma and Marangos, 1985; Hammond and Clanachan, 1985). On a regional basis, [³H]dipyridamole binding in guinea pig brain correlates well with that of [³H] CHA, showing some degree of overlap with [³H]NBI binding. These data and the finding that NBI effects on [³H]dipyridamole binding are biphasic in nature have led to the conclusion (Davies and Hambley, 1986; Marangos et al., 1987) that the two ligands differentially label a heterogeneous population of adenosine transporter sites.

Table 2
Ligand Pharmacology

Compound	Ligand	CHA	2-CADO	NECA	CV 1808	R-PIA	S-PIA	DPX	PACPX	8-PT	CPT	Theophylline	Reference
Agonists													
A1	2-CADO	–	2.6	42.5	–	1.1	29.3	–	–	116	–	8775	Williams & Risley, 1980
A1	2-CADO	38	13	–	–	60	3180	–	–	–	–	2100	Michaelis et al., 1985
A1	CHA	1.3	9.3	6.3	561	1.2	49	45	25	86	10.9	8470	Bruns et al, 1986
A1	CPA	4.5	–	17.3	–	4.1	66.3	–	21.2	–	–	–	Williams et al., 1986
A1	R-PIA	7.2	8.7	–	–	4.9	54	–	–	76	–	12,800	Ukena et al., 1984
A1	R-PIA	14	16	86	–	39	353	1500	–	14,100	–	26,700	Lee et al., 1986
A1	HPIA	0.5	51	14.3	–	0.53	4.6	–	–	99	–	12,300	Lohse et al., 1985
A2	NECA	514	63	10.3	119	124	1820	863	92	848	1440	25,300	Bruns et al, 1986
A2	NECA	885	85	12	115	238	3447	4822	579	2034	–	26,000	Stone et al, 1988
A2	NECA	380,000	1400	190	–	240,000	580,000	>100,000	–	>100,000	–	>270,000	Ukena et al., 1985
A2	CGS 21,680	790	160	15	130	530	3930	640	70	720	2500	27,100	Jarvis et al, 1989
A2	PAPA-APEC	–	–	55	–	870	10,300	–	–	–	–	20,300	Barrington et al., 1989b
Antagonists													
A1	DPX	55	74	55	–	22	1000	–	–	–	–	–	Bruns et al.
A1	XCC	–	10	4.3	–	0.7	3.2	63	24	–	–	–	Jaevis, 1988
A1 + A2	XAC	–	–	–	–	1.8	–	–	–	74	–	7600	Jacobson et al, 1986
A1	CPX	3.1	19	16	170	3.8	120	37	1.4	84	–	–	Bruns et al, 1987a
A1	BW A827	–	–	2.4	–	0.9	–	32	–	8.8	–	–	Linden et al, 1988
A2	PD 115,199	4100	–	49	470	980	15,000	370	33	–	–	–	Bruns et al, 1987b
A1 + A2	CGS 15,943	–	41	40	>10,000	20	91	393	20	128	–	>10,000	Jarvis et al, 1988
Uptake site													
NBI		2800	31,600	>100,000	15,500	9500	–	–	–	–	–	>300,000	Clanachan et al., 1987

9. Characterization of Receptor Structure

Although the techniques of molecular biology have contributed significantly to understanding receptor structure and function, permitting the isolation, cloning, and expression of a variety of receptor superfamilies, studies related to adenosine receptors have been confined to molecular sizing of the A_1 receptor following detergent solubilization either through photoaffinity labeling (Stiles, 1985; Stiles et al., 1985; Green et al., 1986) or through target size analysis (Reddington et al., 1987).

The A_1 receptor was reported to be unstable in its solubilized form and rapidly lost binding activity once removed from the membranes (Cooper, 1988; Helmke and Cooper, 1989). Although Mg^{2+} can stabilize the solubilized receptor (Klotz et al., 1985), the purification of the receptor, together with a ligand, can facilitate stability. If an agonist is used, the high-affinity form of the A_1 receptor that is associated with the G protein is favored. With the slowly dissociating agonist, R-PIA, 89–90% of A_1 receptors can be solubilized with detergent. When the more rapidly dissociating endogenous ligand, adenosine, is used, only 4–10% of the receptors are conserved (Yeung et al., 1987). Cosolubilization with caffeine can also be used to stabilize free A_1 receptors (those unassociated with a G protein [Helmke and Cooper, 1989]). The potentially heterogeneous nature of the receptors that are solubilized when adenosine concentration is used can obviously influence the data obtained during subsequent molecular sizing studies, as can the presence of receptors not associated with a cyclase-linked G protein.

Sedimentation analysis of detergent-solubilized A_1 receptors gave a M_r of 250,000 – >500,000 (Stiles, 1985; Martini et al., 1985; Yeung et al., 1987), this value being dependent on the detergent concentration and micelle formation (Cooper, 1988).

Agonist and antagonist ligands that bind to adenosine receptors irreversibly through a covalent complex have been used to characterize the receptor protein. Ligands containing a photolabile arylazide group (e.g., AZPNEA, AZBA, AHPIA, BW-A947U, azido-PAPAXAC, or azido-BW-A844U) or a ligand (e.g., APNEA, PAPAXAC, or PAPA-APEC) containing an amino group that can be derivatized by a heterobifunctional agent, such as SANPAH (*N*- succinimidyl-6(4'-azido-2-nitrophenylamino)-hexanoate), provide a means to covalently label the receptor. In the latter case of photoaffinity crosslinking experiments, SANPAH has been used to acylate an aryl amine present on an adenosine receptor-bound radioligand, by virtue of its active ester (*N*-hydroxysuccinimide) moiety. Then a second reaction, initiated by UV irradiation acting on the azido (N_3) group at the distal end of the SANPAH molecule, leads to the formation of a covalent bond with the receptor protein. Similarly, the isothiocyanate derivatives *m*- and *p*-DITC-XAC form a stable covalent bond with the receptor protein, yet require neither the addition of a crosslinking reagent nor irradiation. Thus, they are potentially of use in organ and in vivo studies. The irreversible step in binding consists of the reaction of a yet unidentified nucleophilic group at or near the receptor binding site with the isothiocyanate ($-N=C=S$) group, which acts as an acylating agent. When the photoaffinity or chemical affinity ligand is radioactive, as in the case of APNEA, AZBA, AHPIA, AZPNEA, PAPAXAC, BW-A947U, BW-A844U, *m*-DITC-XAC, or *p*-DITC-XAC, receptor-ligand complexes solubilized in detergents, such as CHAPS, can be analyzed by SDS-gel electrophoresis, thus permitting a M_r determination of the receptor covalently linked to the radioligand.

The A_1 receptor in cerebral cortex and adipocyte membranes has an M_r of between 34,000 and 38,000 (Choca et al., 1985; Green et al., 1986; Klotz et al., 1985; Linden et al., 1985; Stiles et al., 1985,1986; Lohse et al., 1986; Stiles and Jacobson, 1987,1988; Earl et al., 1988) as determined by photoaffinity labeling. This M_r was obtained irrespective of whether an agonist or antagonist photoaffinity ligand was used. Target size analysis of the A_1 receptor using high-energy irradiation inactivation gave an M_r between 65,000 and 80,000 (Frame et al., 1986; Reddington et al.,1987), approximately twice that seen using photoaffinity labeling. Whether this form of the receptor represents a dimer remains to be determined (Reddington et al., 1987). Evaluation of the binding of the photoaffinity probe [^{125}I]AZBA in several species gave essentially identical data (Patel and Linden, 1988), suggesting that although these receptors show pharmacological differences (Murphy and Snyder, 1982; Schwabe et al., 1985; Ferkany et al., 1986; Stone et al., 1988), these are not manifest at the level of gel analysis.

Examination of the carbohydrate content of A_1 receptors solubilized from rat and hamster brain and fat cells showed a glycoprotein component of M_r of approximately 3000–4000 (Klotz and Lohse, 1986; Ramkumar et al., 1988) with terminal neuraminic acid residues indicating the presence of a complex carbohydrate chain. Histidine residues also appear to be important in defining the conformation of the A_1 receptor (Klotz et al., 1988).

Electrophoresis gels of the radioactive products of photoaffinity labeling using ^{125}I-PAPA-APEC of detergent-solubilized A_2 receptors from bovine striatum (Barrington et al., 1989b) showed that this receptor was distinct from the A_1 receptor and had an M_r of 45,000. ^{125}I-PAPA-APEC is an analog of NECA, which is known to label "nonreceptor" sites. It appears that a minor "nonreceptor" site labeled by ^{125}I-PAPA-APEC has a M_r of 55,000.

Monoclonal antibodies to the A_1 receptor have been described (Perez-Reyes et al., 1987), although these were found to also inhibit binding of a dopamine D_2 receptor ligand.

Photoaffinity labeling of A_1 receptor in isolated fat cells using $[^{125}I]$AHPIA has also provided evidence to suggest the existence of spare receptors (Lohse et al., 1986) as well as a persistent activation of photolabeled receptors.

Barrington et al. (1989a) conducted photoaffinity labeling using azido-derivatized agonists (AZPNEA) and antagonists (preformed PAPAXAC-SANPAH and azido-PAPAXAC) in parallel followed by peptide mapping. It was shown that both agonists and antagonists incorporate into the same region of the receptor protein, but that the agonist- and antagonist-occupied receptor conformations are different.

Based on structure activity profiles for various agonist and antagonist ligands in A_1 and A_2 receptor binding assays and functional assays, attempts have been made to describe the structural regions thought to be responsible for binding. Such models provide a basis for further synthetic work and provide the framework for assessing the binding conformation of the various nonxanthine adenosine antagonists that are being identified using receptor-based screening (Francis et al., 1988; Daly et al., 1988; Peet et al., 1988; Williams, 1990b).

Daly (1985), summarizing data available in 1985, described two subdomains for agonist binding to the adenosine receptor, one binding the purine moiety and the other the ribose group. These domains were proposed, based on earlier work by Bruns (Bruns, 1980, 1981), to be oriented for the anti configuration of the purine nucleoside. 8-bromoadenosine was cited as an instance of an adenosine analog that is unstable in the anti configuration and lacks activity at adenosine receptors. The ribose group is an important determinant of activity. 5'-substitutions can enhance activity, especially at the A_2 receptor, as in the case of NECA. Activity of 5' carboxamides has been taken as evidence of an "ethyl sized hydrophobic pocket" in the A_2 receptor. When a nonhydrogen donor group is present at the 5' position, as in the case of 5'-methylthioadenosine, efficacy is lost, resulting in an antagonist ligand at the A_2 receptor. Groups larger than propyl decrease activity. Carbocyclic analogs such as aristeromycin, are also active (Dunham and Vince, 1986; Chen et al., 1988). Alterations in the 2'- and 3'-hydroxyl groups are, in general, unfavorable (Taylor et al., 1986). Hydrogen bonding at the 2',3', and 5' positions appears important for receptor activation. Adenosine analogs modified in the 7, 8, and 9 positions have reduced or no activity at A_2 receptors whereas N^6-substituted analogs are potent A_1 receptor agonists. Of a series of pyrazolopyrimidine ribosides, only 2-aza-adenosine retains full agonist activity (Hamilton and Bristol, 1983). Certain 1-deazaadenosine analogs, such as the N^6-cyclopentyl-2-chloro derivative, are highly A_1 selective (Cristalli et al., 1988). Numerous adenosine derivatives modified only in the N^6-position, such as CPA (N^6-cyclopentyl), CHA (N^6-cyclohexyl), and various bicycloalkyl derivatives (Trivedi et al., 1989), are potent A_1 agonists. A model involving four hydrophobic aliphatic domains for the N^6 group, termed S-1–S-4, based on studies with 145 N^6-substituted adenosine analogs, has been developed by Olsson and coworkers (Kusachi et al., 1985; Paton et al., 1986). The S-1 region can accommodate three carbon atoms, the S-2 region favors the R-configuration, and the S-3 region is of importance in defining the degree of A_2 activity seen with N^6-substituted adenosine analogs. Bridges et al. (1988) have identified an N^6-substituent, the 2-(3,5-dimethoxyphenyl)-2-(2-methylphenyl)ethyl group, that confers A_2 selectivity upon adenosine and its 5'-ethyl-uronamide derivatives. The S-4 region is a sterically constrained hydrogen bonding region. This model provided the basis for the design and synthesis of S-ENBA, a subnanomolar A_1 agonist with 4700-fold selectivity (Trivedi et al., 1989). The N^6-stereoselectivity of the brain A_1 receptor is, however, somewhat different from that of the coronary receptor (Paton et al., 1986).

Using computer graphic modeling of 26 N^6-substituted adenosine analogs, Van Galen et al. (1989), have extended the Kusachi model to identify five new subregions of the N^6 "pocket." These are termed C (cycloalkyl), B (bulk), and F1–F3, the "forbidden areas." Ligands showing favorable interactions with the latter "forbidden areas" have reduced affinity for the A_1 receptor.

Studies in fibroblast A_2 receptors (Bruns et al., 1980) suggested that there was a low degree of bulk tolerance at the 2-position of the purine nucleus. Based on the activity of CV-1808 and 2-CADO in brain binding assays, it appears that this requirement is not shared by all A_2 receptors, the 2-phenethylamino substituted NECA analog (Hutchison et al., 1989) and related structures (Hutchison et al., 1990) being among the most potent A_2 selective agonists reported.

For antagonists, most of which are xanthine in structure, 8-phenyl substitution decreases phosphodiesterase inhibitory activity and increases adenosine receptor antagonism (Daly, 1985). Whereas substitution at the 9 position completely eliminates receptor affinity, 1, 3, and 7 substitutions can further improve or modify activity with respect to adenosine receptor interactions (Jacobson, 1988). The majority of 1,3-dialkyl-8-phenylxanthines synthesized to date are A_1 selective (Daly et al., 1985,1986b; Schwabe et al., 1985; Ukena et al., 1986); however, several 7-substituted xanthines have weak albeit extant A_2-receptor selectivity (Daly et al., 1986a). By combining 7-alkyl and 8-cycloalkyl modifications in xanthines, a higher degree of A_2 selectivity may be achieved, as was shown recently for the 8-cyclohexyl analog of caffeine (Shamim et al., 1989). In general, the 8-phenylxanthines have poor solubility, restricting their use as research tools. The functionalized congeners, XCC and XAC (Jacobson, 1988), and the 8-phenylsulfonamide xanthine analogs, PD

113,297 and 115,199 (Hamilton et al., 1985; Bruns et al., 1987a,c), have improved solubility.

Members of diverse classes of nonxanthine heterocycles have been shown to be adenosine receptor antagonists (Daly et al., 1988). A series of triazolol[4,3-a]quinoxalinamines, triazoloquinazolines, pyrazolol[4,3-d]pyrimidin-7-ones, and benzo[1.2-c:5,4-c']dipyrazoles have been identified as adenosine receptor antagonists (Williams, 1990; Peet et al., 1988).

Comparisons between theophylline and adenosine suggest (Bruns et al., 1986; Jacobson, 1988) that the xanthines and purines bind in a diametrically opposed manner, since the purine C-4/C-5 dipole is rotated 180°. The N-7 position of theophylline may then correspond to the N-9 binding region of adenosine. Inclusion of caffeine in this hypothesis would then indicate that the 7-methyl group would occupy a similar position to the ribose C-1. Furthermore, it can be proposed that the N^6 region of the purine nucleus and the 8-phenyl group on the xanthine nucleus occupy similar hydrophobic pockets. As previously noted (Williams, 1989), this model does not account for the importance of the 7 position in conferring A_2 selectivity nor for the steric interference seen between substituents at the 7 and 8 positions.

Modeling of the triazoloquinazoline adenosine receptor antagonists has indicated that the furyl moiety of CGS 15943 can overlap the phenyl group at position 8 of substituted xanthines (Francis et al., 1988). The triazolopyrimidine portion of the triazoloquinazoline nucleus may overlap the imidazopyrimidine portion of the xanthine nucleus or, alternately, the benzene ring of CGS 15943, when overlapped with the 8-phenyl group of the 8-phenylxanthines, allows the furyl group to assume a position adjacent to that of the xanthine benzene ring. The 4 position of the triazoloquinazoline nucleus can then be seen as proximal to the N-7 group on the xanthine. This model might then account for the A_2 selectivity of CGS 15943 (Williams et al., 1987a) and DMPX (Ukena et al., 1986) in terms of electron-donating substituents. A further discussion of the structure activity requirements of A_1 and A_2 receptors can be found in Trivedi et al. (1990).

10. Conclusions

Since the development of adenosine receptor binding assays in 1980, many different agonist and antagonist radioligands have been developed, most of which are A_1-receptor selective. Only two selective A_2 receptor agonists have been reported, CGS 21680 and its analog for radioiodination and photoaffinity crosslinking, PAPA-APEC.

Using these radioligands, the majority of studies have been carried out using various mammalian brain membranes because of the high density of receptors in this tissue. This is somewhat paradoxical, inasmuch as the role of adenosine in brain function is less well-defined than the role of the purine nucleoside in heart, fat cell, platelet, and kidney function. Studies in peripheral tissues have been limited by the paucity of binding sites in these tissues. High specific radioactivity-iodinated ligands have permitted characterization of receptors in myocardium, lung, fat cells, platelets and kidney, but high non-specific binding has tended to limit the signal-to-noise ratio and, consequently, the reproducibility of the data obtained. This has been compounded by the lipophilic nature of some of the ligands used.

Comparative data on the binding properties and pharmacological profiles of the various ligands described are shown in Tables 1 and 2. Variations in the K_d values, although present, are minor, irrespective of whether the ligand used is A_1 or A_2 selective or agonist/antagonist. [³H] R-PIA binding in canine myocardium (Lee et al., 1986) is an exception (Table 1) and contrasts markedly with data obtained for [¹²⁵I] HPIA in bovine myocardium (Lohse et al., 1985). The density of binding sites is more variable. For the A_1 receptor, [³H]S-ENBA gives a density of approximately 160 fmol/mg protein (Trivedi et al., 1989) whereas [³H]CPX binding to solubilized adenosine receptors has a B_{max} of 1168 fmol/mg protein (Helmke and Cooper, 1989). Similarly, using A_2 selective ligands, [¹²⁵I] PAPA-APEC labels sites with a density of 222 fmol/mg protein (Barrington et al., 1989b), whereas the antagonist ligand, [³H]CGS 15943A labels 1500 fmol/mg protein. Although these data may reflect the ability of the various ligands to label the free receptor (Perez-Reyes et al., 1987; Cooper, 1988), it is equally possible that the derived data reflect potential interactions with receptor subtypes. The pharmacology of the binding of those radioligands on which data are available is shown in Table 2. Whether the differences observed reflect interexperimental variation or receptor subtype selectivity cannot be determined at this time. Similarly, the differences in receptor density noted with different ligands (Table 1) might be suggestive of the existence of multiple receptors, but they more probably reflect both species and interexperimental variation.

The seminal work of Bruns (Bruns et al., 1986,1987b) using [^{3}H]NECA provided the framework for the description of A_{2a} and A_{2b} receptors. These have been delineated on the basis of the effects of N^6-methyladenosin andCV1674 (2-(4-methyoxyphenyl) amino) adenosine (Bruns et al., 1987b), CV 1674 being four-times more active that N^6-methyladenosine at the A_{2a} receptor and 50 times less active at the A_{2b} receptor.

As more effort is expended in developing analogs of the various stable adenosine agonists and in screening novel structures for potential adenosine receptor activity, the ability to delineate the structure activity requirements for the receptor subtypes will permit the use of computer-assisted molecular modeling to design new ligands. This approach should be complemented with the cloning of both A_1 and A_2 receptors in the near future.

The two major challenges will be to discover agonists that are neither purine nor riboside in nature and to seek selective ligands that will permit the delineation of receptor subtypes.

References

Balwierczak, J. L., Krulan, C. M., Wang, Z. C., Chen, J., and Jeng, A. Y. (1989) The effects of adenosine A_2 receptor agonists on nucleoside transport. *J. Pharmacol. Exp. Ther.* **251,** 279-287

Barrington, W. W., Jacobson, K. A., and Stiles, G. L. (1989a) Demonstration of distinct agonist and antagonist conformations of the A_1 adenosine receptor. *J. Biol. Chem.* **264,** 13157-13164.

Barrington, W. W., Jacobson, K. A., Williams, M., Hutchison, A. J., and Stiles, G. L. (1989b) Identification of the A_2 adenosine receptor binding subunit by photoaffinity crosslinking. *Proc. Natl. Acad. Sci. USA* **86,** 6572-6576.

Beck, D. W., Vinters, H. V., Moore, S. A., Hart, M. N., Henn, F. A., and Cancilla, P. A. (1984) Demonstration of adenosine receptors on mouse cerebral smooth muscle membranes. *Stroke* **15,** 725-727.

Braas, K. M., Zarbin, M. A., and Snyder, S. H. (1987) Endogenous adenosine and adenosine receptors localized to ganglion cells of the retina. *Proc. Natl. Acad. Sci. USA* **84,** 3906–3920.

Bridges, A. J., Bruns, R. F., Ortwine, D. G., Priebe, S. R., Szotek, D. L. and Trivedi, B. K. (1988) N^6-[2(3,5-dimethoxyphenyl)-2-(2-methylphenyl)ethyl] adenosine and its uronamide derivatives. Novel adenosine agonists with both high affinity and high selectivity for the adenosine A_2 receptor. *J. Med. Chem.* **31,** 1282–1285.

Bruns, R. F. (1980) Adenosine receptor activation in human fibroblasts: nucleoside agonists and antagonists. *Can. J. Physiol. Pharmacol.* **58,** 673–691.

Bruns, R. F. (1981) Adenosine antagonism by purines, pteridines and benzopteridines in human fibroblasts. *Biochem. Pharmacol.* **30,** 325–333.

Bruns, R. F. (1988) Adenosine receptor binding assays, in *Adenosine Receptors* (Cooper, D. M. F. and Londos, C. eds.), Liss, New York, pp. 43–62.

Bruns, R. F., Daly, J. W., and Snyder, S. H. (1980) Adenosine receptors in brain membranes: Binding of N^6-cyclohexyl[^{3}H]adenosine and 1,3-diethyl-8-[^{3}H] phenylxanthine. *Proc. Natl. Acad. Sci. USA* **77,** 5547–5551.

Bruns, R. F., Daly, J. W., and Snyder, S. H. (1983a) Adenosine receptor binding: Structure-activity analysis generates extremely potent xanthine antagonists. *Proc. Natl. Acad. Sci. USA* **80,** 2077–2080.

Bruns, R. F., Lawson-Wendling, K., and Pugsley, T. A. (1983b) A rapid filtration assay for soluble receptors using polyethylenimine-treated filters. *Anal. Biochem.* **132,** 74–81.

Bruns, R. F., Lu, G. H., and Pugsley, T. A. (1984) Binding of ^{3}H-NECA to a subtype of A_2 adenosine receptor in rat striatal membranes. *Soc. Neurosci. Abstr.* **10,** 957.

Bruns, R. F., Lu, G. H., and Pugsley, T. A. (1986) Characterization of the A_2 adenosine receptor labeled by [^{3}H]NECA in rat striatal membranes. *Mol. Pharmacol.* **29,** 331–346.

Bruns, R. F., Fergus, J. H., Badger, E. W., Bristol, J. A., Santay, L. A., Hartman, J. D., Hays, S. J., and Huang, C. C. (1987a) Binding of the A_1-selective adenosine antagonist 8-cyclopentyl-1,3-dipropylxanthine to rat brain membranes. *Naunyn-Schmiedebergs Arch. Pharmacol.* **335,** 59–63.

Bruns, R, F., Lu, G. H., and Pugsley, T. A. (1987b) Adenosine receptor subtypes, in *Topics and Perspectives in Adenosine Research* (Gerlach, E. and Becker, B. F., eds.), Springer-Verlag, Berlin, pp. 59–73.

Bruns, R. F., Fergus, J. H., Badger, E. W., Bristol, J. A., Santay, L. A., and Hays, S. J. (1987c) PD115,199: an antagonist ligand for adenosine A_2 receptors. *Naunyn-Schmiedebergs Arch. Pharmacol.* **335,** 64–69.

Bruns, R. F., Davis, R. E., Ninteman, F. W., Poschel, B. P., Wiley, J. N., and Heffner, T. G. (1988) Adenosine antagonists as pharmacological tools, in *Adenosine and Adenine Nucleotides: Physiology and Pharmacology.* (Paton, D. M., ed.), Taylor and Francis, London, pp. 39–49.

Chen, J., Rock, C., Clarke, F., Webb, R., Gunderson, K., and Hutchison, A. J. (1988) A convenient route to carbocyclic N-ethyladenosine-5'-carboxamide (c-NECA) 196th ACS Annual Meeting, Los Angeles, California.

Chin, J. H., Mashman, W. E., and DeLorenzo, R. J. (1985) Novel adenosine receptors in rat hippocampus. Identification and characterization. *Life Sci.* **36,** 1751–1760.

Choca, J. I., Kwatra, M. M., Hosey, M. M., and Green, R. D. (1985) Specific photoaffinity labeling of inhibitory adenosine receptors. *Biochem. Biophys. Res. Commun.* **131,** 115–121.

Clanachan, A. S., Heaton, T. P., and Parkinson, F. E. (1987) Drug interactions with nucleoside transport systems, in *Topics and Perspectives in Adenosine Research* (Gerlach, E. and Becker, B. F., eds.), Springer-Verlag, Berlin, pp. 118–130.

Cooper, D. M. F. (1988) Structural studies of adenosine A_1 receptors, in *Adenosine Receptors* (Cooper, D. M. F. and Londos, C., eds.), Liss, New York, pp. 63–74.

Cristalli, G., Franchetti, P., Grifantini, M., Vittori, S., Klotz, K.-N., and Lohse, M. J. (1988) Adenosine receptor agonists: Synthesis and biological evaluation of 1-deaza analogues of adenosine derivatives. *J. Med. Chem.* **31,** 1179–1183.

Czernik, A., Petrack, B., Kalinsky, H. J., Psychos, S., Cash, W. D., Tsai, C., Rinehart, R. K., Granat, F. R., Lovell, R. A., Brundish, D. E., and Wade, R. (1982) CGS 8216: Receptor binding characteristics of a potent benzodiazepine receptor antagonist. *Life Sci.* **30,** 363–372.

Daly, J. W. (1985) Adenosine receptors in the central nervous system: structure-activity relationships for agonists and antagonists, in *Purines: Pharmacology and Physiological Roles.* (Stone, T.W., ed.), Macmillan, London, pp. 5–15.

Daly, J. W., Padgett, W. M., Shamim, M. T., Butts-Lamb, P., and Waters, J. (1985) 1,3-Dialkyll-8(p-sulfophenyl)xanthines: Potent water soluble antagonists for A_1 and A_2 receptors. *J. Med. Chem* **28,** 487–492.

Daly, J. W., Padgett, W. L., and Shamim, M. T. (1986a) Analogues of caffeine and theophylline: Effect of structural alterations in affinity at adenosine receptors. *J. Med. Chem.* **29,** 1305–1308.

Daly, J. W., Padgett, W. L., and Shamim, M. T. (1986b) Analogs of 1,3-dipropyl-8-phenylxanthine: Enhancement of selectivity at A_1 adenosine receptors by aryl substituents. *J. Med. Chem.* **29,** 1520–1524.

Daly, J. W., Hong, O., Padgett, W. L., Shamim, W. T., Jacobson, K. A., and Ukena, D. (1988) Nonxanthine heterocycles: Activity as antagonists of A_1 and A_2 receptors. *Biochem. Pharmacol.* **37,** 655–664.

Davies, L. P. and Hambley, J. W. (1986) Regional distribution of adenosine uptake in guinea-pig brain slices and the effects of some inhibitors: Evidence for nitrobenzylthioinosine-sensitive and insensitive sites? *Neurochem. Inter.* **8,** 103–108.

Deckert, J., Morgan, P. F., Bisserbe, J.-C., Jacobson, K. A., Kirk, K. L., Daly, J. W., and Marangos, P. J. (1988) Autoradiographic localization of mouse brain adenosine receptors with an antagonist ([³H] xanthine amine congener) ligand probe. *Neurosci. Lett.* **86,** 121–126.

Dunham, E. W. and Vince, R. (1986) Hypotensive and renal vasodilator effects of carbocyclic adenosine (aristeromycin) in anesthetized spontaneously hypertensive rats. *J. Pharmacol. Exp. Ther.* **238,** 954–959.

Earl, C. Q., Patel, A., Craig, R. H., Daluge, S. M., and Linden, J. L. (1988) Photoaffinity labeling adenosine A_1 receptors with an antagonist ^{125}I-labeled aryl azide derivative of 8-phenylxanthine. *J. Med. Chem.* **31,** 752–756.

Enna, S. J. (1987) Biochemical approaches for evaluating drug-receptor interactions, in *Drug Discovery and Development.* (Williams, M. and Malick, J. B., eds.), Humana, Clifton, New Jersey, pp. 151–176.

Ferkany, J. W., Valentine, H. L., Stone, G. A., and Williams, M. (1986) Adenosine A_1 receptors in mammalian brain: Species differences in their interactions with agonists and antagonists. *Drug Dev. Res.* **9,** 85–93.

Fox, I. H. and Kurpis, L. (1983) Binding characteristics of an adenosine receptor in human placenta. *J. Biol. Chem.* **258,** 6952–6955.

Frame, L. T., Yeung, S.-M. H., Venter, J. C., and Cooper, D. M. F. (1986) Target size of the adenosine Ri receptor. *Biochem. J.* **235,** 621–624.

Francis, J. E., Cash, W. D., Psychoyos, S., Ghai, G., Wenk, P., Friedmann, R. C., Atkins, C., Warren, V., Furness, P., Hyun, J. L., Stone, G. A., Desai, M., and Williams, M. (1988) Structure activity profile of a series of novel triazoloquinazoline adenosine agonists. *J. Med. Chem.* **31,** 1014–1020.

Geiger, J. D. and Nagy, J. I. (1990) Adenosine deaminase and [³H]nitrobenzylthioinosine as markers of adenosine metabolism and transport in central purinergic systems, in *Adenosine and Adenosine Receptors* (Williams, M., ed.), Humana, Clifton, New Jersey,

Goodman, R. R. and Snyder, S. H. (1982) Autoradiographic localization of adenosine receptors in rat brain using [³H]cyclohexyl adenosine. *J. Neurosci.* **2,** 1230–1241.

Goodman, R. R., Cooper, M. J., Gavish, M., and Snyder, S. H. (1982) Guanine nucleotide and cation regulation of the binding of [³H]cyclohexyl adenosine and [³H]diethylphenylxanthine to adenosine A_1 receptors in brain membranes. *Mol. Pharmacol.* **21,** 329–335.

Goodman, R. R., Kuhar, M. J., Hester, L., and Snyder, S. H. (1983) Adenosine receptors: Autoradiographic evidence for their location on axon terminals of excitatory neurones. *Science* **220,** 967–969.

Green, A., Stuart, C. A., Pietrzyk, R. A. and Partin, M. (1986) Photochemical cross-linking of ^{125}I-hydroxyphenylisopropyl adenosine to the A_1 adenosine receptor of rat adipocytes. *FEBS Lett.* **206,** 130–134.

Green, R. and Stiles, G. L. (1986) Chronic caffeine ingestion sensitizes the A_1 adenosine receptor-adenylate cyclase system in rat cerebral cortex. *J. Clin. Invest.* **77,** 222–227.

Hamilton, H. W. and Bristol, J. A. (1983) C4-substituted 1-b-D-ribofuranosyl-pyrazolol[3,4-]pyrimidines as adenosine agonist analogs. *J. Med. Chem.* **26,** 1601–1606.

Hamilton, H. E., Ortwine, D. F., Worth, D. F., Badger, E. W., Bristol, J. A., Bruns, R. F., Haleen, S. J., and Steffen, R. P. (1985) Synthesis of xanthines as adenosine antagonists, a practical quantitative structure-activity relationship. *J. Med. Chem.* **28,** 1071–1079.

Hammond, J. R. and Clanachan, A. S. (1985) Species differences in the binding of [³H]nitrobenzylthioinosine to the nucleoside transport system in mammalian CNS membranes: Evidence for interconvertible conformations of the binding site/transporter complex. *J. Neurochem.* **45,** 527–535.

Heller, I. H. and McIlwain, H. (1973) Release of [¹⁴C]adenine derivatives from isolated subsystems of the guinea pig brain: actions of electrical stimulation and of papaverine. *Brain Res.* **53,** 105–116.

Helmke, S. M. and Cooper, D. M. F. (1989) Solubilization of stable adenosine A1 receptors from rat brain. *Biochem. J.* **257,** 413–418.

Hutchison, A. J.,Webb, R. L., Oei, H. H., Ghai, G. R., Zimmerman, M. B., and Williams, M. (1989) CGS 21680C, an A_2 selective adenosine receptor agonist with preferential hypotensive activity. *J. Pharmacol. Exp. Ther.* **251,** 47–55

Hutchison, A. J., Williams, M., deJesus, R., Oei, H. H., Ghai, G. R., Webb, R. L., Zoganas, H. C., Stone, G. A., and Jarvis, M. F. (1990b) 2-Arylalkylaminoadenosine 5'-uronamides: A new class of highly selective adenosine A_2 receptor agonists. *J. Med. Chem.*, in press.

Hütteman, E., Ukena, D., Lenschow, V., and Schwabe, U. (1984) R_a adenosine receptors in human platelets. Characterization by 5-*N*-ethylcarboxamido-[³H]adenosine binding in relation to adenylate cyclase activity. *Naunyn-Schmiedebergs Arch. Pharmacol.* **325,** 226–233.

Jacobson, K. A. (1988) Chemical approaches to the definition of adenosine receptors, in *Adenosine Receptors* (Cooper, D. M. F. and Londos, C., eds.), Liss, New York, pp. 1–26.

Jacobson, K. A., Ukena, D., Kirk, K. L., and Daly, J. W. (1986) [³H]Xanthine amine congener of 1,3-dipropyl-8-phenylxanthine: An antagonist radioligand for adenosine receptors. *Proc. Natl. Acad. Sci. USA* **83,** 4089–4093.

Jacobson, K. A., Ukena, D., Padgett, W., Kirk, K. L., and Daly, J. W. (1987) Molecular probes for extracellular adenosine receptors. *Biochem. Pharmacol.* **36,** 1697–1707.

Jarvis, M. F. (1988) Autoradiographic localization and characterization of brain adenosine receptors subtypes, in *Receptor Localization: Ligand Autoradiography.* (Leslie, F. M. and Altar, C. A., eds.), Liss, New York, pp. 95–111.

Jarvis, M. F., Jackson, R. H., and Williams, M. (1989b) Autoradiographic characterization of high-affinity adenosine A_2 receptors in the rat brain. *Brain Res.* **484,** 111–118.

Jarvis, M. F., Jacobson, K. A., and Williams, M. (1987) Autoradiographic localization of adenosine A_1 receptors in rat brain using [³H]XCC, a functionalized congener of 1,3-dipropylxanthine. *Neurosci. Lett.* **81,** 69–74.

Jarvis, M. F., Schutz, R., Hutchison, A. J., Do, E., Sills, M. A., and Williams, M.

(1989a) [^{3}H]CGS 21680, an A$_2$ selective adenosine receptor agonist directly labels A$_2$ receptors in rat brain tissue. *J. Pharmacol. Exp. Ther.* **251,** in
press.

Jarvis, M. F., Williams, M., Stone, G., and Sills, M. (1988) Characterization of [^{3}H]
CGS 15943A binding to adenosine A$_1$ receptors in rat cortex. *Pharmacologist,* **30,** A210.

Jarvis, S. M. (1988) Adenosine Transporters, in *Adenosine Receptors* (Cooper,
D. M. F. and Londos, C., eds.), Liss, New York, pp. 113–123.

John, D. and Fox, I. H. (1985) Characteristics of high-affinity and low-affinity
adenosine binding sites in human cerebral cortex. *J. Lab. Clin. Med.* **106,**
401–407.

Klotz, K.-N., Cristalli, G., Grifantini, M., Vittori, S., and Lohse, M. J. (1985) Photoaffinity labeling of A$_1$ adenosine receptors. *J. Biol. Chem.* **260,** 14659–
14664.

Klotz, K.-N. and Lohse, M. J. (1986) The glycoprotein nature of A$_1$ adenosine receptor. *Biochem. Biophys. Res. Commun.* **140,** 406–413.

Klotz, K.-N., Lohse, M. J., and Schwabe, U. (1988) Chemical modification of A$_1$
adenosine receptors in rat brain membranes. Evidence for histidine in different domains of the ligand binding site. *J. Biol. Chem.* **263,** 17522–17526.

Kocis, J. D., Eng, D. L., and Bhistkul, R. B. (1984) Adenosine selectively blocks
parallel-fiber-mediated synaptic potentials in rat cerebellar cortex. *Proc.
Natl. Acad. Sci. USA* **81,** 6531–6534.

Kostopoulos, G. K. and Phillis, J. W. (1977) Purinergic depression of neurons in
different areas of rat brain. *Exp. Neurol.* **55,** 719–724.

Kuhar, M. J., De Souza, E. B., and Unnerstall, J. M. (1985) Neurotransmitter receptor mapping by autoradiography and other methods. *Annu. Rev. Neurosci.* **9,**
27–58.

Kusachi, S., Thompson, R. D., Bugani, W. J., Yamada, N., and Olsson, R. A. (1985)
Dog coronary artery adenosine receptor: Structure of the N^6-alkyl subregion.
J. Med. Chem. **28,** 1636–1643.

Lee, J. T., Jr., Newman, W. H., and Webb, J. G. (1986) Adenosine receptors in an
enriched fraction of plasma membranes from canine ventricular myocardium. *J. Cardiovasc. Pharmacol.* **8,** 621–628.

Lee, K. and Reddington, M. (1986) 1,3-Dipropyl-8-cyclopentylxanthine (DPCPX)
inhibition of [^{3}H]*N*-ethylcarboxamido adenosine (NECA) binding allows the
visualization of putative non-A$_1$ adenosine receptors. *Brain Res.* **368,** 394–
398.

Lewis, M. E., Patel, J., Edley, S. M., and Marangos, P. J. (1981) Autoradiographic
visualization of rat brain adenosine receptors using N^6-cyclohexyl[^{3}H]
adenosine. *Eur. J. Pharmacol.* **73,** 109, 110.

Liang, B. T. (1989) Characterization of the adenosine receptor in cultured embryonic chick atnal mycoytes: Coupling to modulation of contractility and
adenylate cyclase actuation and identification by direct radioligand binding.
J. Pharmacol. Exp. Ther. **249,** 775–784.

Lied, M., Schimerlik, M. I., and Murray, T. F. (1988) Characterization of agonist radioligand binding interactions with porcine atrial A_1 adenosine receptors. *Mol. Pharmacol.* **34,** 334–339.

Limbird, L. E. (1986) *Cell Surface Receptors: A Short Course on Theory and Methods* (Nijhoff, Boston).

Linden, J. (1984) Purification and characterization of (-)[^{125}I]hydroxyphenylisopropyladenosine, an adenosine R-site agonist radioligand and theoretical analysis of mixed stereoisomer radioligand binding. *Mol. Pharmacol.* **26,** 414–423.

Linden, J. Patel, A., and Sadek, S. (1985) [^{125}I]Aminobenzyl adenosine, a new radioligand with improved specific binding to adenosine receptors in heart. *Circ. Res.* **56,** 279–284.

Linden, J., Patel, A., Earl, C. Q., Craig, R. H., and Daluge, S. M. (1988) ^{125}I-Labeled 8-phenylxanthine derivatives: Antagonist radioligands for adenosine A_1 receptors. *J. Med. Chem.* **31,** 745–751.

Lohse, M. J., Klotz, K.-N., and Schwabe, U. (1986) Agonist photoaffinity labeling of A_1 adenosine receptors: Persistent activation reveals spare receptors. *Mol. Pharmacol.* **30,** 403–409.

Lohse, M. J., Lenschow, V., and Schwabe, U. (1984) Two affinity states of Ri receptors in brain membranes. Analysis of guanine nucleotide and temperature effects on radioligand binding. *Mol. Pharmacol.* **26,** 1–9.

Lohse, M. J., Ukena, D., and Schwabe, U. (1985) Demonstration of Ri-type adenosine receptors in bovine myocardium by radioligand binding. *Naunyn-Schmiedebergs Arch. Pharmacol.* **328,** 310–316.

Lohse, M. J., Elger, B., Lindenborn-Fotinos, J., Klotz, K-L., and Schwabe, U. (1988) Separation of solubilized A_2 adenosine receptors of human platelets from non-receptor [^{3}H]NECA binding sites by gel filtration. *Naunyn-Schmiedebergs Arch. Pharmacol.* **337,** 64–68.

Malbon, C. C., Hert, R. C., and Fain, J. N. (1978) Characterization of [^{3}H] adenosine binding to fat cell membrane. *J. Biol. Chem.* **253,** 3114–3122.

Marangos, P. J., Deckert, J., and Bisserbe, J.-C. (1987) Central sites of adenosine action and their interaction with various drugs, in *Topics and Perspectives in Adenosine Research* (Gerlach, E. and Becker, B. F., eds.), Springer-Verlag, Berlin, pp. 74–88.

Marangos, P. J., Montgomery, P., and Houston, M. (1985) [^{3}H]Dipyridamole: a new ligand probe for brain adenosine uptake sites. *Eur. J. Pharmacol.* **117,** 393–395.

Marangos, P. J., Patel, J., Clark-Rosenberg, R., and Martino, A. M. (1982) [^{3}H]Nitrobenzylthioinosine binding as a probe for the study of adenosine uptake sites in the brain. *J. Neurochem.* **39,** 184–191.

Marangos, P. J., Patel, J., Martino, A. M., Dilli, M., and Boulanger, J. P. (1983) Differential binding properties of adenosine receptor agonists and antagonists in brain. *J. Neurochem.* **41,** 367–374.

Martini, C., Pennacchi, E., Poli, M. G., and Lucacchini, A. (1985) Solubilization of adenosine A_1 binding sites from sheep cortex. *Neurochem. Int.* **6,** 1017–1020.

Michaelis, M. L., Kitos, T. E., and Mooney, T. (1985) Characteristics of adenosine binding sites in atrial sarcolemmal membranes. *Biochim. Biophys. Acta.* **816,** 241–250.

Munshi, R. and Baer, H. P. (1982) Radioiodination of *p*-hydroxylphenyliosopropyl adenosine: Development of a new ligand for adenosine receptors. *Can. J. Physiol. Pharmacol.* **60,** 1320–1322.

Murphy, K. M. M. and Snyder, S. H. (1982) Heterogeneity of adenosine A_1 receptor binding in brain tissue. *Mol. Pharmacol.* **22,** 250–257.

Newman, M. E., Patel, J., and McIlwain, H. (1981) The binding of [^{3}H]adenosine to synaptosomal and other preparations from the mammalian brain. *Biochem. J.* **194,** 611–620.

Onodera, H. and Kogure, K. (1988) Differential localization of adenosine A_1 receptors in the rat hippocampus: Quantitative autoradiographic study. *Brain Res.* **458,** 212–217.

Patel, A. and Linden, J. (1988) Photoaffinity labeling of adenosine receptors, in *Adenosine Receptors,* (Cooper, D. M. F. and Londos, C., eds.), Liss, New York, pp. 27–42.

Patel, A., Craig, R. C., Daluge, S. M., and Linden, J. (1988b) [^{125}I]BW-A844U, an antagonist radioligand with high afffinity and selectivity for adenosine A_1 receptors, and ^{125}I-azido-BW-A844U, a photoaffinity label. *Mol. Pharmacol.* **33,** 585–591.

Paton, D. M., Olsson, R. A., and Thompson, R. T. (1986) Nature of the N^6 region of the adenosine receptor in guinea-pig ileum and rat vas deferens. *Naunyn-Schmiedebergs Arch. Pharmacol.* **333,** 313–322.

Pearson, J. and Gordon, J. L. (1985) Nucleotide metabolism by endothelium. *Annu. Rev. Physiol.* **47,** 617–627.

Peet, N., Dickerson, G. A., Abdallah, Daly, J. W., and Ukena, D. (1988) Benzo [1,2-c:5,4]dipyrazoles: Non-xanthine adenosine antagonists. *J. Med. Chem.* **31,** 2034–2039.

Perez-Reyes, E., Yeung, S.-M. H., Lasher, R. L., and Cooper, D. M. F. (1987) Structural studies on adenosine Ri receptors from rat cerebral cortex, in *Topics and Perspectives in Adenosine Research* (Gerlach, E. and Becker, B. F., eds.), Springer-Verlag, Berlin, pp. 37–48.

Psychoyos, S., Ford, C. J., and Phillips, M. A. (1982) Inhibition by etazolate (SQ 20009) and cartazolate (SQ 65396) of adenosine-stimulated [^{3}H]cAMP formation in 2-[^{3}H]-adenosine prelabeled vesicles prepared from guinea pig cerebral cortex. *Biochem. Pharmacol.* **31,** 1441, 1442.

Ramkumar, V., Pierson, G., and Stiles, G. L. (1988) Adenosine receptors: Clinical implications and biochemical mechanisms. *Prog. Drug. Res.* **32,** 195–247.

Reddington, M., Erfurth, A., and Less, K. S. (1986) Heterogeneity of binding sites for *N*-ethylcarboxamido[^{3}H]adenosine in rat brain: Effects of *N*-ethylmaleimide. *Brain Res.* **399,** 232–239.

Reddington, M., Alexander, S. P., Erfuth, A., Lee, K. S., and Kreutzberg, G. W. (1987) Biochemical and autoradiographic approaches to the characterization of adenosine receptors in the brain, in *Topics and Perspectives in Adenosine Research* (Gerlach, E. and Becker, B. F., eds.), Springer-Verlag, Berlin, pp. 49–58.

Ronca-Testoni, S., Galbani, P., and Gambacciani, M. (1984) Some properties of a purinergic receptor solubilized from human uterus membranes. *FEBS Lett.* **172,** 335–338.

Schütz,W. and Brugger, G. (1982) Characterization of [^{3}H]adenosine to media membranes of hog carotid arteries. *Pharmacology* **24,** 26–34.

Schütz, W., Tuisl, E., and Kraupp, O. (1982) Adenosine receptor agonists: Binding and adenylate cyclase stimulation in rat liver plasma membranes. *Naunyn-Schmiedebergs Arch. Pharmacol.* **319,** 34–39.

Schwabe, U. and Trost, T. (1980) Characterization of adenosine receptors in rat brain by (-)[^{3}H]N^6-phenylisopropyl adenosine. *Naunyn-Schmiedebergs Arch. Pharmacol.* **313,** 179–187.

Schwabe, U., Ukena, D., and Lohse, M. J. (1985) Xanthine derivatives as antagonists at A$_1$ and A$_2$ adenosine receptors. *Naunyn-Schmiedebergs Arch. Pharmacol.* **330,** 212–221.

Schwabe, U., Kiffe, H., Puchstein, C., and Trost, T. (1979) Specific binding of [^{3}H] adenosine to rat brain membranes. *Naunyn-Schmiedebergs Arch. Pharmacol.* **310,** 59–67.

Schwabe, U., Lenschow, V., Ukena, D., Ferry, D. R., and Glossman, H. (1982) [^{125}I]N^6-*p*-hydroxyphenylisopropyl adenosine, a new ligand for Ri receptors. *Naunyn-Schmiedebergs Arch. Pharmacol.* **321,** 84–87.

Shamim, M. T., Ukena, D., Padgett, W. L., and Daly, J. W. (1989) Effects of 8-phenyl and 8-cycloalkyl substituents on the activity of 1,3,7-mono-, di-, and trisubstituted alkylxanthines: Selective antagonists for A$_2$ adenosine receptors. *J. Med. Chem.* **32,** 1231–1237.

Sills, M. A., Williams, M., Hurt, S. D. and Loo, P. A. (1990) Characterization of the binding of [^{3}H] CV 1808 in rat stimulation. In *Punne Nucleotides and Nucleotides in Cell Signalling: Targets for New Drugs.* Jacobson, K.A. , Daly, J. W. and Manganiello, V. C. eds.). Springer-Verlag, New York, in press.

Snowhill, E. W. and Williams, M. (1986) [^{3}H]Cyclohexyl adenosine binding in rat brain: A pharmacological analysis using quantitative autoradiography. *Neurosci. Letts.* **68,** 41–46.

Stiles, G. A. (1985) The A$_1$ adenosine receptor. Solubilization and characterization of a guanine-nucleotide-sensitive form of the receptor. *J. Biol. Chem.* **260,** 6728–6732.

Stiles, G. A. (1988) A$_1$ adenosine receptor-G protein coupling in bovine brain membranes: Effects of guanine nucleotides, salt, and solubilization. *J. Neurochem.* **51,** 1592–1598.

Stiles, G. A. and Jacobson, K. A. (1987) A new high affinity iodinated adenosine receptor antagonist as a radioligand/photoaffinity crosslinking probe. *Mol. Pharmacol.* **32,** 184–188.

Stiles, G. A. and Jacobson, K. A. (1988) High affinity acylating antagonists for the A$_1$ adenosine receptor: Identification of binding subunit. *Mol. Pharmacol.* **34,** 724–728.

Stiles, G. A., Daly, D. T., and Olsson, R. A. (1985) The A$_1$ adenosine receptor. Identification of the binding subunit by photoaffinity crosslinking. *J. Biol. Chem.* **260,** 10806–10811.

Stiles, G. L., Daly, D. T., and Olsson, R. A. (1986) Characterization of the A_1 adenosine receptor-adenylate cyclase system of cerebral cortex using an agonist photoaffinity probe. *J. Neurochem.* **47**, 1020–1025.

Stone, G. A., Jarvis, M. F., Sills, M. A., Weeks, B., Snowhill, E. W., and Williams, M. (1988) Species differences in high-affinity adenosine A_2 binding sites in striatal membranes from mammalian brain. *Drug Dev. Res.* **15**, 31–46.

Stone, T. W., Newby, A. C., and Lloyd, H. G. E. (1990) Adenosine release, in *Adenosine and Adenosine Receptors* (Williams, M., ed.), Humana, Clifton, New Jersey,

Taylor, D. A. and Williams, M. (1982) Interaction of 2-phenylamino adenosine (CV 1808) with adenosine systems in rat tissues. *Eur. J. Pharmacol.* **85**, 335–338.

Taylor, M. D., Moos, W. H., Hamilton, H. W., Szotek, D. S., Patt, W. C., Badger, E. W., Bristol, J. A., Bruns, R. F., Heffner, T. G., and Mertz, T. E. (1986) Ribose-modified adenosine analogues as adenosine receptor agonists. *J. Med. Chem.* **29**, 346–353.

Trivedi, B. K. and Bruns, R. F. (1988) [1,2,4]Triazolo[4,3-a]quinoxalin-4 amines:- A new class of A_1 receptor selective adenosine antagonists. *J. Med. Chem.* **31**, 1011–1014.

Trivedi, B. K., Bridges, A. J., and Bruns, R. F. (1990) Structure activity relationships of adenosine A_1 and A_2 receptors, in *Adenosine and Adenosine Receptors* (Williams, M., ed.), Humana, Clifton, New Jersey,

Trivedi, B. K., Bridges, A. J., Patt, W. C., Priebe, S. R., and Bruns, R. F. (1989) N^6-bicycloalkyladenosines with unusually high potency and selectivity for the adenosine A_1 receptor. *J. Med. Chem.* **32**, 8–11.

Ukena, D., Jacobson, K. A., Kirk, K. L., and Daly, J. W. (1986) A [^3H]amine congener of 1,3-dipropyl-8-phenyl xanthine—a new radioligand for A_2 adenosine receptors of human platelets. *FEBS Lett.* **199**, 269–274.

Ukena, D., Furler, R., Lohse, M. J., Engel, G., and Schwabe, U. (1984) Labelling of Ri adenosine receptors in rat fat cell membranes with (-)[^{125}iodo]N^6-hydroxyphenyl isopropyl adenosine. *Naunyn-Schmiedebergs Arch. Pharmacol.* **326**, 233–240.

Van Galen, P. J. M., Leusen, F. J. J., Ijzerman, A. P., and Soudijn, W. (1989) Mapping the N^6-region of the adenosine A_1 receptor with computer graphics. *Eur. J. Pharmacol.* **172**, 19–27.

Verma, A. and Marangos, P. J. (1985) Nitrobenzylthioinosine binding sites in brain: An interspecies study. *Life Sci.* **36**, 283–290.

Williams, M. (1987) Purine receptors in mammalian tissue: Pharmacology and functional significance. *Annu. Rev. Pharmacol. Toxicol.* **27**, 315–345.

Williams, M. (1989) Adenosine antagonists. *Med. Res. Rev.* **9**, 219–243.

Williams, M. (1990) Adenosine receptors: An historical perspective, in *Adenosine and Adenosine Receptors* (Williams, M., ed.), Humana, Clifton, New Jersey,

Williams, M. and Enna, S. J. (1986) The receptor: From concept to function. *Ann. Rep. Med. Chem.* **19**, 283–292.

Williams, M. and Jarvis, M. F. (1989) Biochemical approaches to drug discovery and characterization, in *Modern Drug Discovery Technologies* (Clark, C. R. and Moos, W. H., eds.), VCH/Ellis Horwood, 129–166.

Williams, M. and Risley, E. A. (1980) Biochemical characterization of putative central purinergic receptors by using 2-chloro[³H]adenosine, a stable analog of adenosine. *Proc. Natl. Acad. Sci. USA* 77, 6892–6896.

Williams, M. and Sills, M. A. (1989) Quantitative analysis of ligand receptor interactions, in *Comprehensive Medicinal Chemistry,* vol. 3 (Emmett, J. C., ed.), Pergamon, Oxford, in press.

Williams, M., Braunwalder, A., and Erickson, T. E. (1986) Evaluation of the binding of the A-1 selective adenosine radioligand, cyclopentyladenosine to rat brain tissue. *Naunyn-Schmiedebergs Arch. Pharmacol.* 332, 179–183.

Williams, M., Glennon, R. A., and Timmermans, P. B. M. W. M. (1988) *Receptor Pharmacology and Function,* Marcel Dekker, New York.

Williams, M., Francis, J. E., Ghai, G. R., Braunwalder, A., Psychyos, S., Stone, G. A., and Cash, W. D. (1987a) Biochemical characterization of the triazoloquinazoline, CGS 15943, a novel non-xanthine adenosine antagonist. *J. Pharmacol. Exp. Ther.* 241, 415–420.

Williams, M., Jarvis, M. F., Sills, M. A., Ferkany, J. W., and Braunwalder, A. (1987b) Biochemical characterization of the antagonist actions of the xanthines, PACPX (1,3-dipropyl-8-(2-amino-4-chloro)phenylxanthine) and 8-PT (8-phenyltheophylline) at adenosine A_1 and A_2 receptors in rat brain tissue. *Biochem. Pharmacol.* 36, 4024–4027.

Wu, P. H. and Churchill, P. C. (1985) 2-Chloro[³H] adenosine binding in isolated rat kidney membranes. *Arch. Int. Pharmacodyn. Ther.* 273, 83–87.

Wu, P. H., Phillis, J. W., Balls, K., and Rinaldi, B. (1980) Specific binding of 2[³H]chloroadenosine to rat brain cortical membranes. *Can. J. Physiol. Pharmacol.* 58, 576–579.

Yamamura, H. I., Enna, S. J., and Kuhar, M. J. (1985) *Neurotransmitter Receptor Binding,* 2nd Ed. (Raven, New York).

Yeung, S.-M. H. and Green, R. (1984) [³H]5'-N-Ethylcarboxamide adenosine binds to both Ra and Ri adenosine receptors in rat striatum. *Naunyn-Schmiedebergs Arch. Pharmacol.* 325, 218–225.

Yeung, S.-M. H., Perez-Reyes, E., and Cooper, D. M. F. (1987) Hydrodynamic properties of adenosine Ri receptors from rat cerebral-cortex membranes. *Biochem. J.* 248, 635–642.

Young, J. D. and Jarvis, S. M. (1983) Nucleoside transport in animal cells. *Rev. Biosci. Rep.* 3, 309–322.

CHAPTER 3

Structure–Activity Relationships of Adenosine A_1 and A_2 Receptors

Bharat K. Trivedi,
Alexander J. Bridges,
and Robert F. Bruns

1. Introduction

Philidor once said that pawns are the soul of chess. In a similar way, it might be said that structure–activity relationships are the soul of receptorology. Structure–activity relationships are used to define receptors, to identify the selective agonists and antagonists that are used as pharmacological tools to study the roles of receptors, and ultimately to develop drugs that act via modulation of receptors.

A major advance in the adenosine receptor field was the observation that extracellular adenosine receptors could be divided into two major subclasses, A_1 and A_2 (van Calker et al., 1979; Londos et al., 1980). A_1 and A_2 receptors were originally defined in terms of their opposite effects on adenylate cyclase, with A_1 receptor activa-

Adenosine and Adenosine Receptors Editor: Michael Williams ©1990 The Humana Press Inc.

tion causing inhibition and A_2 activation causing stimulation of the enzyme. However, in light of later studies showing that some of the responses to adenosine were independent of cyclic AMP (Dunwiddie and Proctor, 1987; Scott and Dolphin, 1987), the subtypes have been redefined on the basis of structure–activity relationships in order to achieve greater universality (Hamprecht and van Calker, 1985). Further refinements in these definitions have been proposed as new compounds with greater selectivity have been discovered (Bruns et al., 1987a; Ukena et al., 1987a). This chapter focuses on the efforts to identify adenosine agonists and antagonists with greater selectivity for the different subtypes of adenosine receptors.

Another development that has facilitated the study of adenosine structure–activity relationships is the advent of binding assays for the adenosine A_1 and A_2 receptors (Bruns et al., 1980; Schwabe and Trost, 1980; Williams and Risley, 1980; Yeung and Green, 1984; Bruns et al., 1986). Compared to other in vitro or in vivo measures of receptor affinity, binding assays have the advantage of being insensitive to many confounding influences, including differences in intrinsic activity, differences in pharmacokinetics, and the existence of extraneous postreceptor effects, such as inhibition of adenylate cyclase via the intracellular "P site." The A_2 binding assay has been especially helpful in the study of A_1/A_2 selectivity, because it allows A_1/A_2 affinity ratios to be determined under almost identical experimental paradigms in the same species and tissue (usually rat brain) (Moos et al., 1985; Taylor et al., 1986; Kwatra et al., 1987). In order to achieve the greatest degree of consistency in the comparison of different compounds, the present chapter will attempt (whenever possible) to use receptor binding affinities as the basis for structure–activity comparisons.*

---

*For some compounds without published A_2 binding affinities, we have attempted to infer approximate binding affinities from dog coronary artery potencies. After adjustment of the dog data for albumin binding (Kusachi et al., 1985, 1986), a high correlation ($r^2 = 0.74$) is seen between A_2 binding affinity and dog flow activity (Bruns, R. F., unpublished results) indicating that potency in the dog system is a reasonably valid predictor of A_2 binding affinity.

2. Modified Adenosine Analogs

2.1. N⁶-Alkyladenosines

The ability of N^6-alkyl substituents to boost A_1 affinity and selectivity in functional assays has been known for some time (Trost and Stock, 1977). More recently, many N^6-alkyladenosines have been synthesized and evaluated in the A_1 receptor binding assay (Daly et al., 1986a).[†] When the A_2 binding assay became available, it was found that these compounds also strongly favored the A_1 receptor in their binding profiles (Bruns et al., 1986).

The chain length in this series of compounds is critical for A_1 receptor affinity and selectivity. A three- or four-carbon chain extension is optimum for A_1 receptor interaction, giving a moderately A_1-selective compound (compare compounds *1–6*, Table 1). Further extension of chain length diminishes the A_1 affinity and selectivity. The N^6-isopropyl derivative *7* has considerably improved A_1 selectivity compared to the ethyl derivative *2* and slightly better selectivity than the *n*-propyl *3*, implying that a branched α-carbon may enhance A_1 selectivity. Interestingly, although incorporation of a heteroatom at the terminus of long-chain alkyl groups lowers binding affinity at both receptors (*14–17*), 2-hydroxypropyl or 3-hydroxypropyl substitution maintains A_1 affinity while reducing A_2 affinity, resulting in highly A_1-selective agents (*9–11*) (Bruns et al., 1986; Hamilton et al., 1987a). This is also true for the 2,3-dihydroxypropyl derivatives (*12* and *13*). Furthermore, the well-known stereochemical differences shown by *R*-PIA and *S*-PIA (*see* compounds *46* and *47* below) are also demonstrated by these short-chain hydroxyalkyl derivatives. The 2,3-dihydroxypropyl derivative having *R* stereochemistry at the N^6-side chain

---

[†]For maximum consistency, A_1 binding affinities from our own laboratories are given in the tables even when values from other groups have been previously reported in the literature (e.g., Daly et al., 1986a). Our A_1 affinities generally parallel those from the Daly group quite closely ($r^2 = 0.94$).

Table 1
A_1 and A_2 Binding Affinities of N^6-Alkyladenosines*

Compound	N^6-substituent	K_i, nM A_1	A_2	Ratio A_2/A_1	Data source* A_1	A_2
1	Methyl	360	4500	12.5	e	e
2	Ethyl	15.9	1180	74	h	h
3	n-propyl	2.4	1050	430	h	h
4	n-butyl	1.9	–	–	j	–
5	n-pentyl	15.1	480	32	h	h
6	n-hexyl	66	770	11.6	h	h
7	1-methylethyl	2.4	1430	600	h	h
8	2-aminoethyl	144	15,600	108	h	h
9	(R)-2-hydroxypropyl	13.3	10,200	760	l	l
10	(S)-2-hydroxypropyl	5.0	9300	1870	l	l
11	3-hydroxypropyl	7.0	4900	700	e	e
12	(R)-2,3-dihydroxy-propyl	4.4	7300	1640	h	h
13	(S)-2,3-dihydroxy-propyl	26	13,200	500	h	h
14	4-aminobutyl	155	2300	14.7	h	h

(continued)

Compound	N^6-substituent	A_1	A_2	A_2/A_1	A_1	A_2
15	5-aminopentyl	38	1060	28	h	h
16	5-hydroxypentyl	76	2700	35	h	h
17	6-hydroxyhexyl	72	2400	34	h	h

*Binding affinities were determined using [^{3}H]CHA in rat whole brain membranes (A_1) (Bruns et al., 1980, 1986; Daly et al., 1986a) and [^{3}H]NECA in rat striatal membranes in the presence of 50 n*M* N^6-cyclopentyladenosine (A_2) (Bruns et al., 1986). Sources of affinity data are listed below: [a]Bridges et al., 1987; [b]Bristol et al., 1988; [c]Bruns and Coughenour, 1987; [d]Bruns and Hamilton, 1987; [e]Bruns et al., 1986; [f]Bruns et al., 1987b; [g]Bruns et al., 1987c; [h]unpublished data using the method of Bruns et al. (1986); [i]Cristalli et al., 1985; [j]Daly et al., 1986a; [k]Daly et al., 1986b; [l]Hamilton et al., 1987a; [m]Hamilton et al., 1987b; [n]Jacobson et al., 1985a; [o]Jacobson et al., 1985b; [p]Jacobson et al., 1985c.

[q]An approximate A_2 binding affinity was calculated by regression analysis from dog coronary artery data (Kusachi et al., 1985, 1986) after adjustment for albumin binding (Bruns, R. F., manuscript in preparation); [r]Kwatra et al., 1987; [s]Moos et al., 1985; [t]Taylor et al., 1986; [u]Trivedi et al., 1988a; [v]Trivedi and Bruns, 1988; [w]Ukena et al., 1986a; [x]Ukena et al., 1986b; [y]Ukena et al., 1987b; [z]Williams et al., 1987.

is about six times as potent and three times as selective for the A_1 receptor as its *S* counterpart. The rationale for such significant differences in binding affinities has been incorporated by Kusachi et al. (1985) into a model in which affinity is controlled by stereospecific interactions between the alkyl group and a hydrophobic site on the receptor. The stereoselectivity of the dihydroxypropyl derivatives *12* and *13* suggests that asymmetric hydrophilic binding sites may also exist.

2.2. N^6-*Cycloalkyladenosines*

Unlike the N^6-alkyl derivatives of adenosines, the N^6-cycloalkyladenosines are conformationally rather rigid molecules. However, like the straight-chain alkyl derivatives, they are generally highly selective for the A_1 receptor (Trost and Stock, 1977; Moos et al., 1985). Among the simple N^6-cycloalkyladenosines, N^6-cyclopentyladenosine (CPA, *20*, Table 2) has been reported to be the most potent and A_1-selective agonist, with K_i values of 0.59 and 460 n*M* for the A_1 and A_2 receptors, respectively (Moos et al., 1985).

Table 2
A$_1$ and A$_2$ Binding Affinities of N^6-Cycloalkyladenosines

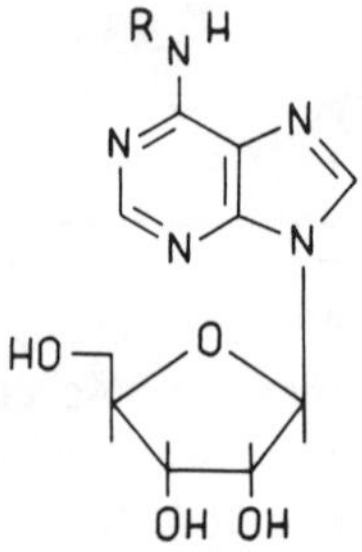

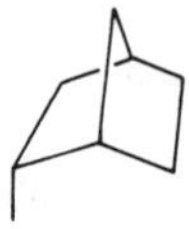

24 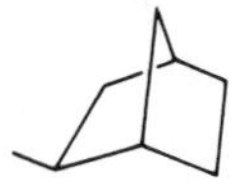   25

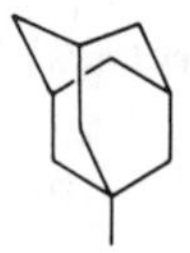

26 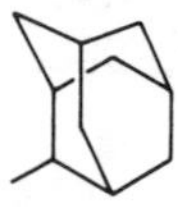   27

Compound	N^6-substituent	K_i, nM		Ratio	Data source*	
		A$_1$	A$_2$	A$_2$/A$_1$	A$_1$	A$_2$
18	Cyclopropyl	3.2	1240	390	s	s
19	Cyclobutyl	0.79	260	330	s	s
20	Cyclopentyl	0.59	460	780	s	s
21	Cyclohexyl	1.31	360	277	s	s
22	Cycloheptyl	2.80	1700	610	s	s
23	Cyclooctyl	4.0	2400	610	s	s
24	2-endo-norbornyl	0.42	750	1790	h	h
25	2-exo-norbornyl	0.91	970	1070	h	h
26	1-adamantyl	290	87,000	300	h	h
27	2-adamantyl	142	40,000	280	h	h

*Data sources are listed in Table 1.

Recently, several N^6-bicycloalkyladenosines have been identified as potent A_1 receptor agonists (Daly et al., 1986a, Paton et al., 1986). These compounds (which are essentially bridged derivatives of CPA or CHA) are extremely selective for the A_1 receptor. The N^6-(2-*endo*-norbornyl) derivative (*24*) is twice as potent as the *exo* isomer (*25*) and is 1800-fold selective for the A_1 receptor. When the size of the bicyclic ring structure is increased, as in the N^6-adamantyladenosines *26* and *27*, the affinity at both receptors diminishes significantly.

2.3. N⁶-Aryladenosines

The aryl-containing side chains are the most studied class of N^6-substituents. Until recently, all of the examples reported were highly A_1-selective agonists, supporting the premise that the N^6-binding region could only be used to enhance binding to the A_1 receptor. However, as we shall discuss shortly, suitably placed N^6-aryl groups can greatly augment binding to the A_2 receptor, and even lead to agonists with greater A_2 selectivity and receptor affinity than previously available agents.

2.3.1. A₁ Receptor Selective Agents

N^6-Phenyladenosine (*28*) is a potent, highly A_1-selective agonist (A_1 K_i 4.6 nM, 144-fold selective; Table 3), presumably through the binding of the aryl ring to the same hydrophobic region as the alkyl substituents. Simple substituents do not change A_1 affinity very much, generally leading to slightly lower affinity (*29–31*, Kwatra et al., 1987). However, substitution with *m*-hydroxy or *m*-iodo also reduces A_2 affinity, resulting in compounds with over 600-fold A_1 selectivity, equaling CPA.

The "functionalized congener" approach (Jacobson et al., 1985a,b), whereby the *para* position of phenyl ring of *28* is linked through large spacers to biomolecules, such as amino acids or biotin (*32–34*), demonstrates the steric tolerance of the A_1 hydrophobic binding region. In addition, as some of these compounds have stronger A_1 affinity than *28*, they appear to imply the existence of another binding region further out from the nitrogen. The most potent congener (*33*) has a positively charged amino function at physiological pH and is about four times as potent as *28*, suggesting the possible

Table 3
A_1 and A_2 Binding Affinities of N^6-Aryladenosines

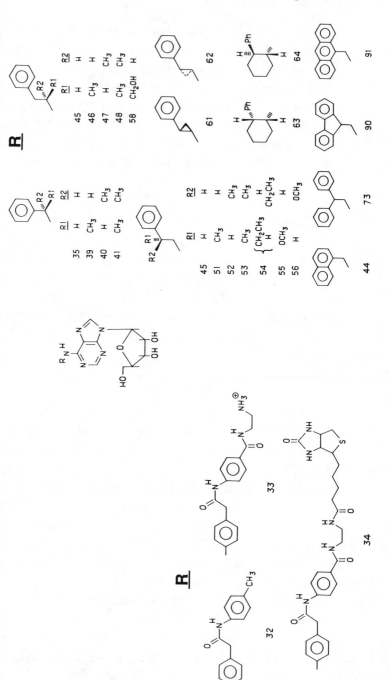

Compound	N⁶-substituent	K_i, nM		Ratio	Data source*	
		A_1	A_2	A_2/A_1	A_1	A_2
28	Phenyl	4.6	660	144	e	e
29	m-hydroxyphenyl	5.6	3800	680	r	r
30	m-iodophenyl	9.5	6000	630	r	r
31	p-iodophenyl	4.2	640	152	r	r
32	p-{[[(p-methylphenyl) amino]carbonyl] methyl}phenyl	1.70	—	—	n	—
33	ADAC	0.85	—	—	n	—
34	Biotinyl-ADAC	11.4	—	—	o	—
35	Benzyl	120	280	2.4	e	e
36	2-chlorobenzyl	15.5	290	18.7	j	q
37	3-methylbenzyl	43	440	10.2	j	q
38	4-phenylbenzyl	8.0	1570	196	h	h
39	(R)-1-phenylethyl	5.5	460	83	h	h
40	(S)-1-phenylethyl	189	6300	34	h	h
41	1-methyl-1-phenylethyl	600	—	—	j	—
42	2-thienylmethyl	36	3500	97	j	q
43	2-pyridylmethyl	220	1800	8.2	j	q
44	1-naphthylmethyl	24	9.3	0.38	u	u
45	2-phenylethyl	12.7	161	12.6	a	a
46	(R)-1-methyl-2-phenylethyl	1.17	124	106	e	e
47	(S)-1-methyl-2-phenylethyl	49	1820	37	e	e
48	1,1-dimethyl-2-phenylethyl	28	5300	189	j	q
49	(R)-1-ethyl-2-phenylethyl	0.26	13.4	52	j	q

(continued)

Table 3 (continued)

Compound	N^6-substituent	K_i, nM		Ratio	Data source*	
		A_1	A_2	A_2/A_1	A_1	A_2
50	(S)-1-ethyl-2-phenylethyl	13.0	350	27	j	q
51	(R)-2-phenylpropyl	1.40	157	112	j	q
52	(S)-2-phenylpropyl	3.5	120	34	j	q
53	2-methyl-2-phenylpropyl	16	320	19.9	h	h
54	(RS)-2-phenylbutyl	3.5	33	9.4	j	q
55	(R)-2-methoxy-2-phenylethyl	43	730	16.7	h	h
56	(S)-2-methoxy-2-phenylethyl	95	350	3.7	h	h
57	(RS)-2-cyclohexyl-2-phenylethyl	16.0	158	9.9	h	h
58	(S)-1-(hydroxymethyl)-2-phenylethyl	2.7	390	144	j	q
59	(RS)-1-methyl-2-hydroxy-2-phenylethyl	4.7	390	83	j	q
60	1-(hydroxymethyl)-2hydroxy-2-phenylethyl	1180	42,000	35	e	e
61	(1R,2S)-2-phenylcyclopropyl	5.0	670	134	j	q
62	(1S,2R)-2-phenylcyclopropyl	7.0	1470	210	j	q
63	(1R,2R)-2-phenylcyclohexyl	15.0	2900	193	j	q
64	(1R,2S)-2-phenylcyclohexyl	6.0	230	38	j	q
65	2-(1-naphthyl)ethyl	12.3	230	18.8	u	u
66	2-(2-naphthyl)ethyl	1.13	510	440	h	h
67	3-phenylpropyl	23	420	18.2	a	a
68	(R)-1-methyl-3-phenylpropyl	8.0	10,600	1325	j	q
69	(S)-1-methyl-3-phenylpropyl	43	35,000	810	j	q
70	4-phenylbutyl	15.9	1230	77	a	a
71	Diphenylmethyl	480	3800	7.9	a	a

72	(RS)-1,2-diphenylethyl	156	920	5.9	h h
73	2,2-diphenylethyl	6.8	25	3.6	a a
74	(RS)-2,3-diphenylpropyl	8.5	20	2.4	h h
75	2-phenyl-1-(phenylmethyl)ethyl	14.3	195	13.6	h h
76	3,3-diphenylpropyl	240	290	1.19	a a
77	4,4-diphenylbutyl	290	4700	16.1	a a
78	5,5-diphenylpentyl	83	2200	27	a a
79	2,2,2-triphenylethyl	14,400	26,000	1.82	a a
80	(RS)-2-(3-chlorophenyl)-2-phenylethyl	1.72	5.6	3.2	b b
81	(RS)-2-(3-bromophenyl)-2-phenylethyl	1.46	4.9	3.2	b b
82	(RS)-2-(2-furanyl)-2-phenylethyl	4.1	13	3.1	h h
83	(RS)-2-(3-chlorophenyl)-2-(2-furanyl)ethyl	1.16	6.1	5.2	h h
84	2,2-bis-(3-chlorophenyl)ethyl	2.8	6.0	2.1	b b
85	(RS)-2-(3,4-dimethoxyphenyl)-2-phenylethyl	26	11.5	0.44	b b
86	(RS)-2-(2,6-dichlorophenyl)-2-phenylethyl	58	19.0	0.33	b b
87	(RS)-2-(2,5-dimethoxyphenyl)-2-phenylethyl	162	32	0.196	b b
88	(RS)-2-(3,5-dimethoxyphenyl)-2-phenylethyl	30	6.1	0.20	b b
89	(RS)-2-[3,5-bis-(trifluoromethyl)-phenyl]-2-phenylethyl	147	10.9	0.074	h h
90	9H-fluoren-9-ylmethyl	5.2	4.9	0.95	u u
91	9-anthracenylmethyl	9000	29,000	3.2	u u

*Data sources are listed in Table 1

existence of a negatively charged site distal to N^6 on the receptor. A distal hydrophobic site may also contribute to the affinities of *32–34*.

Although N^6-benzyladenosine *35* itself is a weak, balanced agonist (A_1 K_i 120 nM, A_2 K_i 280 nM), it has more varied structure–activity relationships than *28*, and exhibits several of the trends that are also seen with the 2-phenylethyl substituent. Simple aryl substitutions at the *ortho* and *meta* positions usually improve A_1 affinity moderately (*36* and *37*), whereas substitution of the *para* position usually has less effect. An exception is the phenyl substituent, which produces *38* with 8 nM A_1 affinity, again suggesting the existence of a distant binding site. A more dramatic effect can be seen with alkyl substitution of the methylene spacer. Interestingly, there is a strong stereoisomer effect with N^6-[(R)-1-phenethyl]adenosine *39* (K_i 5.5 nM) being considerably more potent than *35*, whereas the S-isomer *40* (K_i 189 nM) has attenuated binding compared to *35*. Dimethyl substitution of the methylene spacer decreases binding drastically (*41*). Substitution of 2-pyridyl for phenyl has little effect, whereas 2-thienyl or 1-naphthyl substitution improves A_1 affinity (*42–44*) (however, the 1-naphthyl substitution increases A_2 affinity to an even greater extent, as discussed later).

Although N^6-(2-phenethyl)adenosine *45* has only moderately strong A_1 affinity and selectivity (K_i 12.7 nM, 13-fold selectivity), it is probably the most intensely studied of all adenosine modifications because of the fact that suitable substitution can result in A_1 affinities ranging from 0.26 nM to over 1 μM. This entire 4000-fold range of A_1 affinities can be achieved by substitution of the ethyl chain, suggesting that groups at this position interact intimately with the N^6 domain of the receptor. Alkyl substitution of the ethyl chain has been examined in great detail (Daly et al., 1986a), and the results have been used to construct a detailed model of the A_1 receptor N^6 binding region.

Alkyl substitution at the a-position (i.e., next to the N^6 nitrogen) has effects that are similar to, but greater than, those seen in the benzyl series. Addition of a methyl group in the R configuration results in the potent A_1 reference agonist R-PIA (*46*), with a K_i of 1.17 nM, whereas the S-isomer is considerably weaker (S-PIA, *47*), with a K_i of 49 nM. It should be noted that the phenyl moiety of R-PIA contributes very little to A$_1$ affinity, as the simple N^6-isopropyl derivative 7 retains half the affinity of 46. 1,1-Dimethyl substitution gives 48, which is intermediate between the PIAs in affinity (28 nM, contrast to the 600 nM affinity of the 1,1-dimethylbenzyl derivative 41). R-Ethyl substitution at the a-position gives 49, which with A$_1$ affinity of 0.26 nM is the most potent A$_1$ agonist reported, whereas the S-isomer 50 (K_i 13 nM) is quite similar to 45 in affinity. Similar trends occur at the b-position. Methyl substitution improves the A$_1$ affinity with the R-isomer 51 being as potent as R-PIA, and the S-isomer 52 having 3.5 nM affinity, whereas the β, β-dimethyl compound 53 is similar in potency to the parent compound 45. RS-Ethyl modification at the B-position (54) also gives a potent agonist (3.5 nM), but methoxy substitution reduces binding moderately in both stereoisomers (55 and 56), more than a cyclohexyl substituent (57). Hydroxy substitution of R-PIA at either the β-position or the R-methyl group diminishes binding two- to fivefold (58 and 59), but hydroxy modification of both positions nearly destroys receptor affinity (60). Incorporation of the ethyl link into a three-membered ring has little effect on affinity for either *trans* diastereomer (61 and 62), suggesting an extended conformation for the 2-phenylethyl substituent on the receptor. However, an extended conformation is not essential for affinity, since the *cis* and *trans* 1,2-cyclohexyl isomers 63 and 64 also have similar affinities; this may reflect a binding mode similar to N^6-cyclohexyladenosine. Substitution on the phenyl ring has very little effect generally, but can lead to a loss in A$_1$ affinity of up to tenfold (Daly et al., 1986a). Similarly, replacement of phenyl with other aromatic rings generally leads to small losses of affinity (Daly et al., 1986a), although the 2-naphthyl substituent (66) leads to an agonist with 1.13 nM A$_1$ affinity.

Longer side chains have been briefly examined. The 3-phenylpropyl compound 67 has 23 nM affinity, and methyl substitution next to the nitrogen leads to the same stereochemical effects as earlier in the series, but in a more attenuated form (68 and 69). The A$_1$ selectivity of 68 is high (estimated at 1000-fold from dog coronary data), perhaps because of binding of the phenyl moiety to a distal hydrophobic domain specific to the A$_1$ receptor (compounds 38 and 66 may also interact with this binding site). The 4-phenylbutyl compound 70 has 15.9 nM affinity.

Diaryl side chains have also been examined (Daly et al., 1986a; Bridges et al., 1987). N^6-(Diphenylmethyl) adenosine (71) is weak, with a K_i value of 480 nM, and the RS-1,2-diphenylethyl compound 72 is also weak, with 156 nM A_1 affinity. This strong loss of affinity seems to be associated with phenyl substitution in the α-position, since N^6-(2,2-diphenylethyl) adenosine (73, CI-936) has 6.8 nM affinity, tl 2,3-diphenylpropyl analog 74 has 8.5 nM affinity, and the 2-phenyl-1-(phenylmethyl)ethyl compound 75 has 14.3 nM affinity, all comparable to 2-phenylethyl (45). However, the 3,3- and 4,4-di-phenylalkyl compounds 76 and 77 are rather weaker than their mono-phenyl analogs, although binding does return to some extent with the 5,5-diphenylpentyl side chain 78 (83 nM A_1 affinity). The 2,2,2-tri-phenylethyl analog 79 has essentially no receptor affinity.

The 2,2-diarylethyl side chain has been studied extensively (Bristol et al., 1988; Bridges et al., 1988). Extension of the chain is moderately deleterious to A_1 affinity, as are most *ortho* and *para* sub-stituents. *Meta* halogenation improves the A_1 affinity to below 2 nM (80 and 81), and replacement of a phenyl with 2-furanyl also improves binding (82). The combination 83, with K_i value of 1.16 nM, is as potent as R-PIA.

2.3.2. A_2 Receptor Selective Agents

N^6-Phenyladenosine (28) has rather poor A_2 affinity (K_i 660 nM), which is not improved significantly by substitution (Kwatra et al., 1987). The benzyl derivative 35 has slightly better affinity (K_i 280 nM), and again ring substituents can have minor effects. Methyl sub-stitution at the benzylic methylene has a deleterious effect (39 and 40), with the S isomer considerably weaker than the R isomer and the dimethyl compound 41 too weak to measure. Changing the aromatic ring has a major effect. The heteroaromatics 42 and 43 are much weaker, but the 1-naphthylmethyl analog 44 (Jahn, 1969; Kusachi et al., 1986; Trivedi et al., 1988a) is a very potent agonist (K_i 9.3 nM), with 2.6-fold selectivity for the A_2 receptor, making it the most potent of the "first-generation" A_2-selective agonists.

N^6-(2-phenylethyl)adenosine (45) has moderate A_2 affinity (K_i 161 nM). R-PIA (46) has slightly improved affinity compared to 45, whereas S-PIA 47 is markedly weaker, and the α,α-dimethyl analog

48 is weaker yet. Surprisingly, the R-ethyl isomer 49 appears to be rather potent. Although an actual A_2 binding affinity is not yet available, the molar potency ratio of 49 in the dog coronary artery (Kusachi et al., 1985) appears to imply an A_2 binding affinity of about 13 nM. However, 49 has high affinity at the A_1 receptor (K_i 0.26 nM; Daly et al., 1986a) and, therefore, is not A_2-selective. The S isomer 50 is five times as active as S-PIA. Alkyl substitution at the β-position of the ethyl chain has little effect, except for the β-ethyl compound 54, which again appears to be rather more potent. It seems likely that the β-ethyl group of 54 may bind to the same receptor domain as the second phenyl of N^6-(2,2-diphenylethyl)adenosine (73), since both compounds have enhanced A_2 affinity compared to the parent 45. Compound 54 has about 10-fold A_1-selectivity.

The β-cyclohexyl compound 57 (A_2 K_i 158 nM) is of similar affinity to 45. Incorporation of the ethyl chain into a three-membered ring diminishes the affinity considerably (61 and 62). Aromatic substituents and heteroaromatic rings have modest effects on binding affinity, ranging from twofold enhancement to about fivefold diminution, (Daly et al., 1986a). Extension of the alkyl chain of 45 leads to a rapid falloff in A_2 affinity, as illustrated by compounds 67 and 70.

In the diarylalkyl series two compounds, the 2,2-diphenylethyl (Kusachi et al., 1986; Bridges et al., 1987) and 2,3-diphenylpropyl derivatives (73 and 74), have good A_2 affinity, with 25 and 20 nM affinity, respectively. Meta substitution of the aromatic ring(s) of 73 improves the A_2 affinity of agonists considerably with several agonists possessing about 6 nM A_2 affinity, while retaining around threefold A_1 selectivity. The interesting agonist 84 has m-chloro substituents in both rings, and is a very potent balanced agonist, with A_1 and A_2 affinities of 2.8 and 6.0 nM, respectively. Good A_2 affinity is also obtained with a series of disubstituted phenyl compounds 85–89, but the major interest of these compounds is that the disubstituted ring is poorly tolerated by the A_1 receptor, making them all A_2-selective, with 88 being as A_2-selective as CV-1808 (106), but 19 times as potent (Bristol et al., 1988). Compound 89 is the most A_2-selective (13.5-fold) of the disubstituted derivatives.

Another very potent A_2 agonist is N^6-(9H-fluoren-9-ylmethyl)adenosine (90) (Trivedi et al., 1988a). This balanced agonist (A_1 5.2

nM, A$_2$ 4.9 nM) can be considered a cyclized derivative of CI-936 (*73*) (*see* structures in Table 3). However, whereas in *73* the two phenyl rings are highly skewed, in *90* the tricyclic structure is flat and the two phenyl rings are coplanar. In this sense, *90* is similar to the N^6-(1-naphthylmethyl) derivative *44*, although the two phenyl rings of *90* are somewhat more distant from N^6. Like *90*, *44* is a very potent A$_2$ agonist. A very informative compound that is closely related to both *44* and *90* is N^6-(9-anthracenylmethyl) adenosine (*91*), which is virtually inactive at both receptors (A$_1$ 9000 nM, A$_2$ 29,000 nM). Compound *91* differs from *44* in possessing an additional fused phenyl ring, which presumably projects into a portion of space that is already occupied by the receptor. Compounds *90* and *91* differ mainly in the angle between the planar polycyclic aromatic hydrocarbon and the methylene bridge, implying that this angle is important for A$_2$ affinity.

2.4. N^6-Benzocycloalkyladenosines

Although many N^6-alkyl, cycloalkyl, aryl, and arylalkyl adenosines have been studied for their ability to interact at the adenosine A$_1$ receptor, there are only a few examples of N^6-benzocycloalkyl adenosines reported in the literature (Kusachi et al., 1986). We have synthesized and evaluated several benzocycloalkyl adenosines for their affinities at both A$_1$ and A$_2$ receptors (Trivedi et al., 1988b) (Table 4). Several key features of these molecules appear to be important for A$_1$ receptor affinity and selectivity. Compounds such as N^6-[(R)-1-indanyl]adenosine (*92*) and N^6-[(R)-1-tetralinyl]adenosine (*96*) show moderate A$_1$ affinity and selectivity, with K_i values of 22 nM and 23 nM, respectively. The regioisomeric compounds N^6-(2-indanyl)-adenosine (*94*) and N^6-(2-tetralinyl)adenosine (*98*) show similar A$_1$ potencies (K_i values 24 and 8.4 nM, respectively) but greater selectivity for the A$_1$ receptor. Furthermore, the higher homologs of these derivatives are not only very potent at the A$_1$ receptor, but also at the A$_2$ receptor (*95, 99*). One can postulate from this series and other aralkyl series that the phenethyl substituted derivatives that are selective for the A$_1$ receptor (e.g., *46, 94, 98*) achieve this selectivity because of favorable interactions from the alkyl side chain at the A$_1$ receptor combined with deleterious effects at the A$_2$ receptor resulting

Table 4
A_1 and A_2 Binding Affinities of N^6-Benzocycloalkyladenosines

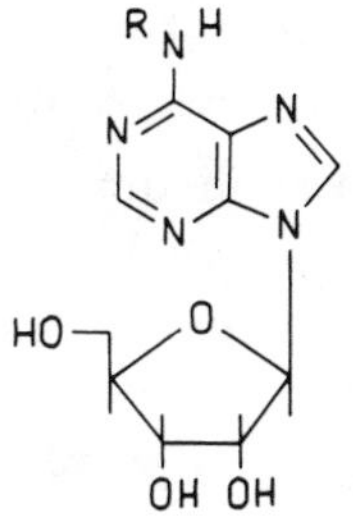

R

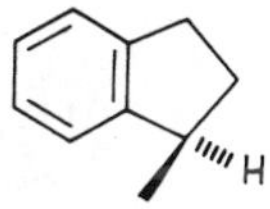

92 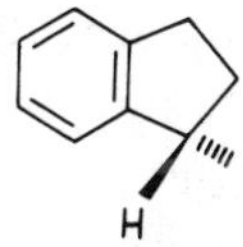   93 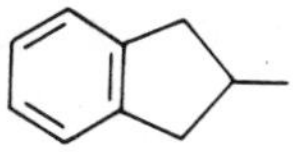   94

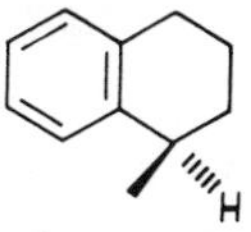

96 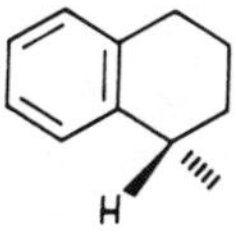   97 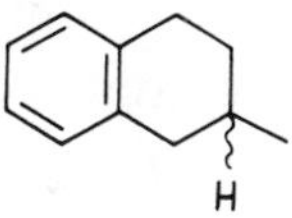   98

Compound	N^6-substituent	K_i, nM		Ratio	Data source*	
		A_1	A_2	A_2/A_1	A_1	A_2
92	(R)-1-indanyl	22	410	18.6	h	h
93	(S)-1-indanyl	310	23,000	74	h	h
94	2-indanyl	24	1640	69	h	h
95	(RS)-1-indanylmethyl	5.2	67	12.9	h	h
96	(R)-1-tetralinyl	23	163	7.1	h	h
97	(S)-1-tetralinyl	300	4000	13.3	h	h
98	2-tetralinyl	8.4	340	40	h	h
99	(RS)-1-tetralinyl- methyl	17.2	92	5.3	h	h

*Data sources are listed in Table 1.

from the branching at the α-carbon. Adenosine derivatives in which the phenyl is two carbon atoms away from the N^6 nitrogen without branching (the simplest example being 45) exhibit significant potency at both A_1 and A_2 receptors. Additionally, as previously illustrated by R-PIA and S-PIA, these N^6-benzocycloalkyl adenosines also indicate that both receptors interact with the N^6 region in a stereospecific manner. Thus, N^6-[(R)-1-indanyl]adenosine (92) is more potent than the corresponding S isomer (93) at both receptors.

2.5. Model of the N⁶-Domain

The structure–activity relationships described above suggest that there are two separate binding regions at N^6 controlling adenosine receptor potency: a close-in site at which such compounds as CPA bind to the receptor via hydrophobic interactions from the methylene groups, and a more distal site (or sites) at which such compounds as R-PIA bind to the receptor via hydrophobic interactions from the aryl function. Occupation of the aryl-binding region seems to be more necessary for A_2 affinity than for A_1 affinity (see Bridges et al., 1987; also compare compounds 7 and 46), whereas the reverse is the case for the alkyl site, explaining in part why the N^6-alkyl- and N^6-cycloalkyl-adenosines tend to be highly A_1-selective, whereas N^6-aryl derivatives tend to have better affinity and selectivity for the A_2 receptor. These stereo- and regiochemical differences are the basis of the N^6-domain model of Olsson and colleagues (Kusachi et al., 1985, 1986; Daly et al., 1986a).

3. 2-Substituted Adenosine Analogs

Although large numbers of C2-substituted adenosines have been synthesized, very few have been evaluated for their binding affinities at A_1 and A_2 receptors. This series definitely bears further examination in the future, as demonstrated by the fact that one member, CV-1808, until recently was the most A_2-selective agonist known (Bruns et al., 1986). CV-1808 [2-(phenylamino)adenosine, 106] is the prototype of an extensive group of 2-(arylamino)adenosines synthesized as selective coronary vasodilators (Marumoto et al., 1975, 1985). CV-

1808 itself has moderate affinity in A_2 binding (K_i 116 nM) (Table 5), but because of its poor A_1 affinity (K_i 600 nM) possesses fivefold selectivity for the A_2 receptor (Bruns et al., 1986). Other derivatives with bulky 2-substituents (for example, *105*, CV-1674) also show A_2 selectivity, suggesting that the 2-position domain of the high affinity A_2 receptor has greater bulk tolerance than that of the A_1 receptor. Einstein et al. (1972) also noted that agonists with bulky 2-position substituents showed selectivity for stimulation of coronary flow (later shown to be an A_2 response) over inhibition of cardiac contractility (later shown to be an A_1 response). Interestingly, a low affinity subtype of A_2 receptor also has been proposed to exist (Daly et al., 1983), and this receptor apparently has almost no measurable affinity for bulky 2-position derivatives, such as *105* (for a discussion, *see* Bruns et al., 1986, 1987a).

In an attempt to evaluate the significance of the anilino function, corresponding C2-phenylthio (Trivedi, 1988) and C2-benzyl adenosines (*102–104*) were synthesized. The latter compounds bound rather weakly at both receptors. These results suggest that, for high affinity at the A_2 receptor, there must be a binding site near the C2 position that requires a moderately acidic proton (i.e., NH-). The exact orientation of the phenyl ring may also be important.

Compounds with small groups at the 2-position (*100* and *101*) tend to be balanced agonists. 2-Chloroadenosine is probably a good substitute for adenosine when a metabolically stable agonist with the same A_1/A_2 balance is needed (Bruns et al., 1986). More recently, the 2-substituted NECA analog, CGS 21680 has been reported as 140-fold selective for the A_2 receptor (Hutchison et al., 1989).

4. C2, N^6-Disubstituted Adenosine Analogs

To explore the interactions between C2 and N^6-substitution, various C2, N^6-disubstituted adenosines were synthesized (Trivedi, 1988). Most of these compounds had lower binding affinities than the parent N^6 derivatives at both receptors (Table 5). These results suggest that there is a less than additive interaction between C2 and N^6 substitution.

Table 5

A_1 and A_2 Binding Affinities of C2-Substituted and C2,N^6-Disubstituted Adenosines

Compound	N^6-substituent	2-substituent	K_i, nM		Ratio A_2/A_1	Data source*	
			A_1	A_2		A_1	A_2
100	H	Chloro	9.3	63	6.8	e	e
101	H	Hydroxy	94	330	3.5	e	e
102	H	Phenylthio	2200	2000	0.91	h	h
103	H	Phenylsulfonyl	2600	1230	0.47	h	h
104	H	Benzyl	4300	7000	1.64	h	h
105	H	4-methoxyphenyl	1320	600	0.46	e	e
106	H	Phenylamino	600	116	0.193	e	e
107	Cyclopentyl	Amino	2.90	6100	2100	h	h
108	Cyclopentyl	Phenylthio	37	4000	107	h	h
109	Cyclopentyl	Phenylsulfonyl	96	2300	24	h	h
110	Cyclopentyl	Phenylamino	12.4	144	11.6	h	h
111	(R)-1-methyl-2-phenyl-ethyl	Amino	9.6	1530	159	h	h
112	(R)-1-methyl-2-phenyl-ethyl	Phenylthio	210	1000	4.7	h	h
113	(R)-1-methyl-2-phenyl-ethyl	Phenylamino	152	320	2.1	h	h

*Data sources are listed in Table 1.

5. 5'-Modified Adenosine Analogs

There are two major classes of 5'-modified adenosines, the hydroxy replacements, in which the 5'-hydroxyl group has been replaced with a group such as hydrogen, halogen, or thiol, and the uronic acid derivatives, where the 5'-hydroxymethyl group has been oxidized to an acid. 5'-Chloro-5'-deoxyadenosine (*114*) has A_1 and A_2 binding affinities of 19.5 and 151 nM (Table 6), which are probably close to those of adenosine itself (Bruns et al., 1986). The 5'-iodo derivative *115* is over fourfold weaker at both receptors, and 5'-methylthioadenosine (*116*) is twice as weak as *115* (280 and 1100 nM). 5'-Deoxyadenosine (*117*) has micromolar, relatively balanced affinity (1470 and 2600 nM).

Adenosine-5'-uronic acid *118* is an extremely weak agonist (most likely because of its charge at neutral pH), but the ethyl ester *119* is moderately potent (A_1 174 nM, A_2 390 nM) with strong biological activity (Prasad et al., 1976). Compound *119* has been tested in humans as a potential antianginal therapy (Irshad et al., 1977). Affinity is greatly enhanced by substituting an amide function for the ester in *119* (Prasad et al., 1980). The simple uronamide *120* is over twice as potent as *119* at both receptors. The *N*-methyl derivative *121* is of similar affinity, but seems to have slightly better A_2 selectivity than other members of the series (A_1 84 nM, A_2 67 nM, ratio 0.8). The optimum for A_1 and A_2 affinity occurs with the *N*-ethyl derivative *122*, which is the balanced agonist NECA (A_1 6.3 nM, A_2 10.3 nM). NECA is used in the A_2 binding assay, and is widely used by pharmacologists as a reference agonist, especially to identify A_2-mediated effects. Its extraordinary potency in many in vivo tests is not a simple reflection of its binding, since many N^6-modified agonists have similar or greater affinity at both receptors, but exhibit less in vivo activity. The high in vivo potency of NECA may be attributable to its much greater hydrophilicity compared to most other adenosine analogs. The *N*-cyclopropyl analog *123* is almost as potent as NECA, but with larger alkyl groups, the binding affinity and biological potency fall off rapidly, as illustrated by the *N*-cyclohexyl analog *124* (K_i values 640 and 4300 nM). The work of Prasad and colleagues (Prasad et al., 1976, 1980) has established that there is a binding site at the 5'-position that is optimally filled by an ethyl or cyclopropyl group.

6. N^6,5'-Disubstituted Adenosine Analogs

Doubly modified adenosines have been made largely in attempts to enhance the binding selectivity or biological potency of N^6-substituted agonists. In this section, we will survey two series of adenosines modified at both C5 and N^6 that illustrate the two major trends that have been observed to date. Several of the adenosines with 5'-hydroxyl replacements have A_1 receptor affinities similar to, or somewhat weaker than, those of the parent N^6-substituted adenosines, but the A_2 affinities are considerably decreased, resulting in improved A_1 selectivity. 5'-N-Ethyluronamides have essentially the same A_1 affinity as the parent N^6-substituted adenosine, but in some cases have enhanced A_2 affinity. Thus, 5'-hydroxyl-replaced compounds have been made to enhance the A_1 selectivity of agonists, and 5'-uronamides have been made in the search for A_2-selective agonists.

The enhancement of A_1 selectivity by 5'-chloro substitution was first reported for 5'-chloro-5'-deoxy-R-PIA (*125*), which had 1000-fold A_1 selectivity, compared to 100-fold for the parent R-PIA (Taylor et al., 1986). The 10-fold boost in A_1 selectivity was the result of a 10-fold worsening in A_2 affinity, coupled with an unchanged A_1 affinity. The selectivity-enhancing effects of 5'-halogenation extend to even more selective N^6-derivatives. N^6-Cyclopentyladenosine (CPA) *20* has 0.59 nM A_1 affinity and is 780-fold A_1-selective. 5'-Chloro-5'-deoxy-CPA *126* is almost as potent an A_1 agonist (K_i 0.71 nM), but is over three times as A_1 selective as CPA. 5'-Bromo-5'-deoxy-CPA *127* is over 5000-fold A_1 selective, but its A_1 affinity falls off to 3.7 nM. The 5'-deoxy and 5'-azido analogs *128* and *129* are also potent and more selective than CPA, whereas the 5'-methylthio analog *130* is about equal to CPA in selectivity. The most potent compound in this series is *131*, the 5'-chloro-5'-deoxy analog of the extremely A_1-selective N^6-(2-*endo*-norbornyl)adenosine (*24*). Compound *131* has the same A_1 selectivity as *127* (4900-fold), but is almost 10 times more potent (A_1 K_i 0.42 nM).

The N-ethyluronamide of CPA, *132*, is marginally more potent than CPA at the A_1 receptor, but with 134 nM affinity at the A_2 receptor is threefold less A_1 selective than CPA. The NECA analog of R-PIA (*133*) shows the same relative change in affinities at both receptors (A_1 1.37 nM, A_2 46 nM). These results suggest that N-ethyluronamide

Table 6

A_1 and A_2 Binding Affinities of 5'-Modified and N^6,5'-Doubly Modified Adenosines

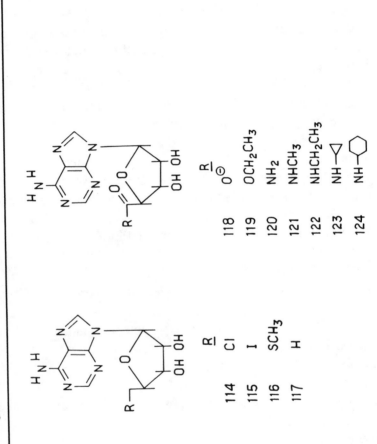

R
114	Cl
115	I
116	SCH₃
117	H

R
118	O⁻
119	OCH₂CH₃
120	NH₂
121	NHCH₃
122	NHCH₂CH₃
123	NH—△
124	NH—⬡

Compound	5'-modification	N^6-substituent	K_i, nM A_1	A_2	Ratio A_2/A_1	Data source* A_1 A_2
114	5'-chloro-5'-deoxy		19.5	151	7.7	h h
115	5'-deoxy-5'-iodo		93	640	6.9	e e
116	5'-methylthio		280	1100	3.9	e e
117	5'-deoxy		1470	2600	1.75	h h
118	5'-uronic acid					
119	5'-uronic acid ethyl ester		174	390	2.2	e e
120	5'-uronamide		73	120	1.66	e e
121	5'-methyluronamide		84	67	0.80	e e
122	5'-ethyluronamide (NECA)		6.3	10.3	1.64	e e
123	5'-cyclopropyluronamide		6.4	13.4	2.1	e e
124	5'-cyclohexyluronamide		640	4300	6.8	e e
	5'-modification	N^6-substituent				
125	Chloro	(R)-1-methyl-2-phenylethyl	1.60	1900	1200	i i
126	Chloro	Cyclopentyl	0.71	1870	2600	b b
27	Bromo	Cyclopentyl	3.7	20,000	5500	h h
128	Deoxy	Cyclopentyl	9.1	11,600	1270	h h
129	Azido	Cyclopentyl	2.9	4600	1580	h h
130	Methylthio	Cyclopentyl	11	10,700	970	h h
131	Chloro	2-endo-norbornyl	0.42	2100	4900	h h
132	NECA	Cyclopentyl	0.50	134	270	h h
133	NECA	(R)-1-methyl-2-phenylethyl	1.37	46	34	e e

*Data sources are listed in Table 1.

substitution at the 5'-position may provide a small enhancement of A_2 affinity for N^6 derivatives. Although Ukena et al. (1987a) saw a similar small enhancement of A_2 affinity in adenylate cyclase, Olsson et al. (1986) did not see any enhancement of coronary vasodilator activity. Greater albumin binding of the NECA derivatives might explain the latter finding.

7. Adenosine Analogs—Other Positions

Modification of adenosine at positions other than those already discussed (C2, N^6, and 5') has generally not resulted in potent or subtype-selective agonists. Only a few of these positions can be modified without drastic loss of affinity. In the purine ring backbone, a 2-aza modification retains at least A_2 activity (Born et al., 1965; Bruns, 1980), but 1-deaza (*134*), 3-deaza (*135*), 7-deaza (*136*), and 9-deaza (*137*) modifications diminish affinity strongly, with 1-deaza substitution being the least detrimental and 7- and 9-deaza being the most (Table 7). 1-Position modification can also spare activity, as illustrated by 1-methylisoguanosine (*138*) (however, 1-methyladenosine is inactive because it forces N^6 into an imino configuration; Bruns, 1980). Activity falls off rapidly with groups larger than methyl at C-1 (Baird-Lambert et al., 1980). 8-Bromoadenosine is inactive, possibly because it forces the purine ring into the *syn* conformation, but 8-amino- and 8-methylaminoadenosine retain A_2 activity in human fibroblasts (Bruns, 1980).

Replacement of the ribose oxygen bridge with methylene or thio has only slight effects on A_2 activity in human fibroblasts (Bruns, 1980). Removal (*139* and *140*) or inversion of the 2'- or 3'-hydroxyl group destroys affinity (Bruns, 1980; Taylor et al., 1986).

8. Xanthine Adenosine Antagonists

Adenosine antagonists possessing selectivity for adenosine receptor subtypes have only appeared within the last few years. This is mainly because of the fact that theophylline and caffeine (*141* and *142*; Table 8), the first adenosine antagonists to be reported (Sattin and Rall, 1970), do not distinguish between the different subtypes (Bruns et al., 1980, 1986).

Table 7
A_1 and A_2 Binding Affinities of Miscellaneous Adenosine Analogs

		K_i, nM		Ratio	Data source*	
Compound	Structure	A_1	A_2	A_2/A_1	A_1	A_2
134	1-deazaadenosine	370	–	–	i	–
135	3-deazaadenosine	7600	>100,000	–	h	h
136	7-deazaadenosine	>25,000	–	–	i	–
137	9-deazaadenosine	>10,000	>100,000	–	h	h
138	1-methyl-2-oxo-adenosine	150	3300	22	e	e
139	2'-deoxy-R-PIA	2300	13,100	5.8	i	i
140	3'-deoxy-R-PIA	36	6900	193	i	i

Data sources are listed in Table 1.

8.1. A_1-Selective Antagonists

8.1.1. 1- and 3-Position Modifications

Since the 1- and 3-positions are often modified together, it is necessary to discuss them together. From the fact that 3-propylxanthine (143) is very weak (A_1 K_i 29 μM, A_2 K_i 103 μM), whereas 1-methylxanthine (144) is only slightly less active than theophylline, it

Table 8
A_1 and A_2 Binding Affinities of Xanthine Adenosine Antagonists

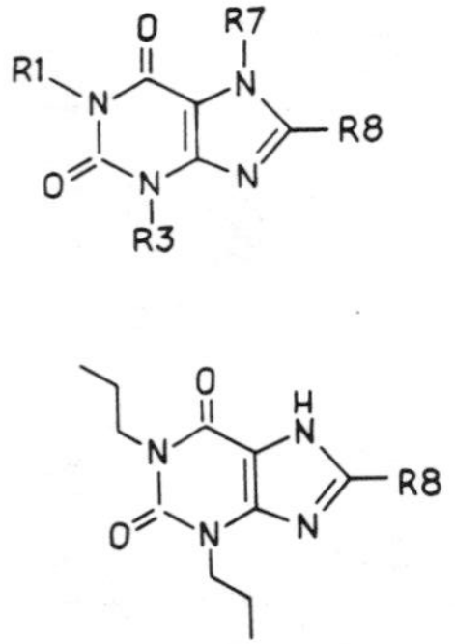

159-162, 173

159 R8 = 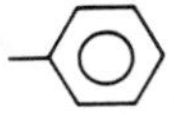

160 R8 = 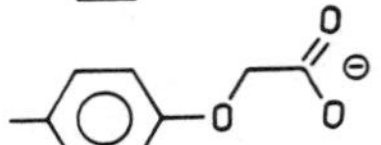

161 R8 = 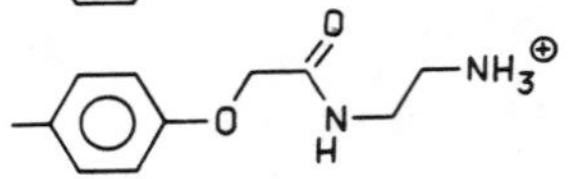

162 R8 = 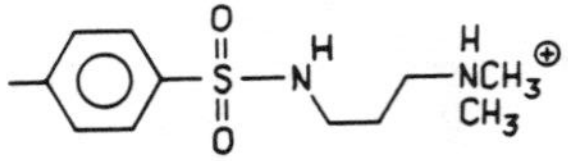

173 R8 = 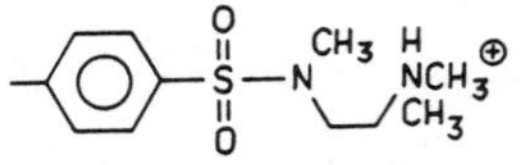

Compound	Structure	K_i, nM		Ratio	Data source*	
		A_1	A_2	A_2/A_1	A_1	A_2
141	1,3-dimethylxanthine	8500	25,000	3.0	*e*	*e*
142	1,3,7-trimethylxanthine	29,000	48,000	1.65	*e*	*e*
143	3-propylxanthine	29,000	103,000	3.6	*e*	*e*
144	1-methylxanthine	11,400	36,000	3.2	*e*	*e*
145	3-isobutyl-1-methylxanthine	2500	13,800	5.6	*e*	*e*
146	3,7-dimethyl-1-ethylxanthine	21,000	38,000	1.82	*e*	*e*

147	1-butyl-3,7-dimethylxanthine	13,000	36,000	2.8	e	e
148	1,3-dipropylxanthine	450	5200	11.5	e	e
149	7-benzyl-1,3-dimethylxanthine	6000	–	–	k	–
150	8-phenyltheophylline	86	850	9.9	e	e
151	8-phenylcaffeine	14,700	–	–	w	–
152	8-(p-chlorophenyl)theophylline	48	370	7.8	e	e
153	8-(m-dimethylaminophenyl)-theophylline	950	1530	1.62	h	h
154	8-(p-sulfophenyl)theophylline	2600	15,300	5.8	–	–
155	8-(o-methoxyphenyl)-theophylline	4000	11,500	2.9	h	h
156	8-(o-aminophenyl)theophylline	83	1430	17.2	e	e
157	8-(o-hydroxyphenyl)-theophylline	230	4100	17.8	h	h
158	8-(2-amino-4-chlorophenyl)-1,3-dipropylxanthine	2.5	92	37	e	e
159	1,3-dipropyl-8-phenylxanthine	10.2	180	17.8	e	e
160	XCC	58	–	–	p	–
161	XAC	0.86	27	31	h	h
162	PD 113,297	5.6	70	12.5	e	e
163	8-propyltheophylline	370	10,000	27	e	e
164	8-cyclopentyltheophylline	10.9	1440	133	e	e
165	8-cyclopentyl-1,3-dipropylxanthine	0.46	340	740	j	f
166	8-cyclopropyltheophylline	195	5900	30	h	h
167	8-cyclobutyltheophylline	64	3800	58	h	h
168	8-cyclohexyltheophylline	25	1480	59	h	h
169	1,3-diethylxanthine	2700	22,000	8.2	e	e
170	1,3-diethyl-8-phenylxanthine	44	860	19.4	e	e
171	3,7-dimethyl-1-propargylxanthine	45,000	–	–	x	–
172	1,3-dimethyl-7-propylxanthine	21,000	–	–	x	–
173	PD 115,199	13.9	15.5	1.11	g	g

*Data sources are listed in Table 1.

is apparent that the 1-methyl group (but not the 3-methyl) of theophylline is essential for receptor affinity. Because *143* (also known as enprofylline) retains theophylline's bronchodilator activity (Persson et al., 1986) without its adenosine antagonism, it has been used extensively to identify potential adenosine-independent actions of theophylline.

Although the 3-methyl group is not essential for adenosine antagonist activity, larger 3-position moieties can enhance A_1 receptor affinity, as illustrated by the threefold greater A_1 affinity of 3-isobutyl-1-methylxanthine (*145*, A_1 2.5 µM) compared to theophylline (*141*, A_1 8.5 µM). 1-Position extension to ethyl or butyl can also increase A_1 affinity (*146* and *147*). The most favorable pattern of 1,3-substitution is 1,3-dipropyl (*148*), which results in a 20-fold boost in A_1 affinity (K_i 450 nM) relative to theophylline; A_2 affinity is increased to a lesser extent (K_i 5200 nM), resulting in 11.5-fold A_1 selectivity for this compound.

8.1.2. 7-Position

7-Methyl substitution of theophylline (*141*) reduces affinity at both A_1 and A_2 receptors, resulting in little change in selectivity (*142*, caffeine, A_1 29 µM, A_2 48 µM). 7-Benzyl substitution (*149*) enhances A_1 binding affinity more than affinity for A_2 blockade in guinea pig cortical slices, with the result that A_1 selectivity is increased about threefold (Daly et al., 1986b). 7-Position substitution is known to abolish the affinity-enhancing effect of 8-phenyl substitution (compare *141* with *150* and *142* with *151*) (Bruns, 1981; Ukena et al., 1986a), possibly because of steric interference between the two adjacent positions. Perhaps for this reason, 7-position variations have not been studied extensively.

8.1.3. 8-Position

8-Position substitution can be used to attain substantial improvements in A_1 receptor affinity and selectivity of xanthine adenosine antagonists. A_1 affinity can be enhanced by phenyl, alkyl, and cyclo-alkyl groups; each of these chemical classes will be discussed separately below.

8-Phenyltheophylline (*150*) was the first adenosine antagonist with nanomolar affinity to be reported (Smellie et al., 1979a; Bruns, 1981), and extensive explorations of the structure–activity relationships of this series have been carried out. The 8-phenyl substitution *per se* enhances A_1 affinity more than A_2 affinity, resulting in about 10-fold selectivity of *150* for the A_1 receptor (A_1 86 n*M*, A_2 850 n*M*). Aromatic substitution of the 8-phenyl ring has been studied extensively (Bruns et al., 1983; Daly et al., 1986c). In bovine A_1 binding, *p*-substituents on the phenyl ring (*152*) increase affinity by up to four--fold, whereas *m*-substituents (*153*) consistently reduce affinity (Bruns et al., 1983). Modest affinity for the A_1 receptor is retained when the *para* position is substituted with a charged group (for instance, *154*, 8-PSPT) (Bruns et al., 1980). Substitution at the *o*-position has contradictory effects, depending on the size of the group. Groups that are medium-sized or larger (for instance methoxy, *155*) strongly reduce affinity, whereas small groups, such as amino and hydroxy (*156* and *157*), leave binding affinity unaltered. This effect may be the result of steric interference between the *o*-group and N^7 (Bruns, 1981). Although single aromatic substitutions produce only modest effects on A_1/A_2 balance, combinations have more marked effects, as illustrated by 8-(2-amino-4-chlorophenyl)-1,3-dipropylxanthine (PACPX, *158*), which possesses about 37-fold A_1 selectivity (A_1 2.5 n*M*, A_2 92 n*M*). However, most of this increase is because of 1,3-dipropyl substitution, since *158* is only about twice as A_1 selective as the unsubstituted phenyl derivative *159*. 1,3-Dipropyl-8-(2-hydroxy-4-methoxyphenyl)xanthine has also been reported to possess significant selectivity for the A_1 receptor (Daly et al., 1986c).

Jacobson and colleagues have prepared an extensive series of functionalized congeners of 1,3-dipropyl-8-phenylxanthine, in which the *p*-position is substituted with a spacer group ending in any of a variety of functional groups. Typical of the series are the simple carboxylate congener *160* and the amine congener *161* (Jacobson et al., 1985c). The latter has about 10-fold higher A_1 affinity (K_i 0.86 nM) than the parent *159*, but A_2 affinity is increased to almost the same degree, resulting in little increase in preference for the A_1 receptor compared to *159*. The distal portion of the receptor appears to accept quite large and complex functional groups, including amino acids (Jacobson et al., 1986), fluorescent probes (Jacobson et al., 1987a), spin labels (Jacobson et al., 1987a), and even peptides (Jacobson et al., 1987b).

Hamilton et al. (1985) also described a series of 8-phenylxanthines with hydrophilic "tail" groups attached at the *p*-position. A representative example of this series is PD 113,297 (*162*), which has an A_1 affinity of 5.6 nM and 12-fold A_1 selectivity.

Structure–activity relationships for the 8-alkylxanthines have not been characterized extensively. 8-Propyl substitution of theophylline enhances A_1 affinity more than A_2, resulting in an A_2/A_1 K_i ratio of 27 for *163*.

By far the most successful approach to A_1-selective adenosine antagonists has been in the 8-cycloalkylxanthine series. 8-Cyclopentyl substitution of theophylline (*164*) increases A_1 affinity by a factor of 800 (K_i 10.9 nM), while increasing A_2 affinity only 20-fold (K_i 1440 nM), resulting in 130-fold selectivity for the A_1 receptor (Bruns et al., 1984, 1986). A further boost in affinity and A_1 selectivity is seen with the corresponding 1,3-dipropyl derivative (CPX, *165*), whose A_1 affinity of 0.46 nM and selectivity of 700-fold are by far the greatest yet reported (Lee and Reddington, 1986; Ukena et al., 1986a; Bruns et al., 1987b; Martinson et al., 1987). In the 8-cycloalkyltheophylline series, the rank order of A_1 affinity and selectivity is cyclopentyl > cyclohexyl ≥ cyclobutyl > cyclopropyl (*164, 166–168*). The same rank order is seen for the 8-cycloalkyl-1,3-dipropylxanthine series (Martinson et al., 1987). 8-Cyclopentyltheophylline shows A_1-selective adenosine antagonism by the oral route in vivo (Bruns et al., 1988).

8.2. A₂-Selective Antagonists

Only moderate success has been forthcoming so far in the search for xanthines with preferential affinity for the A_2 receptor.

8.2.1. 1-, 3-, and 7-Positions

As previously mentioned, 1,3-dipropyl substitution increases both A_1 and A_2 affinity. Selectivity for the A_1 receptor is enhanced by this modification, because A_1 affinity is increased more than A_2. In contrast, 1,3-diethyl modification has been reported by Jacobson et al. (1987c) to enhance A_2 affinity (human platelet adenylate cyclase) without affecting A_1 affinity, resulting in improved selectivity for the A_2 receptor. These results are in agreement with previous results in an A_2-stimulated human fibroblast system, in which 1,3-diethylxanthine (*169*; Table 8) was almost as potent as 1,3-dipropylxanthine (*148*) (Bruns, 1981). However, this pattern is not seen in the rat brain A_2 binding assay, where the 1,3-diethyl derivatives are similar to the corresponding 1,3-dimethyl compounds (compare *141*, *169*, and *148*, as well as *150*, *159*, and *170*). The difference may reflect either a human–rat species difference (Ferkany et al., 1986; Stone et al., 1988) or the existence of subclasses of A_2 receptors.

3,7-Dimethyl-1-propargylxanthine (*171*) and 1,3-dimethyl-7-propylxanthine (*172*) have been reported to show approximately fourfold selectivity for the A_2 receptor when affinities in A_1 binding in rat brain and A_2-stimulated adenylate cyclase in rat PC12 cells are compared (Ukena et al., 1986b).

8.2.2. 8-Position

Although 8-substituted xanthines have generally shown some degree of A_1 selectivity, PD 115,199 (*173*) had equally high affinity in A_1 and A_2 binding (A_1 13.9 nM, A_2 15.5 nM) (Bruns et al., 1987c). Like PD 113,297 (*162*), *173* contains an amine side chain linked to the *p*-position of 1,3-dipropyl-8-phenylxanthine (Hamilton et al., 1985). Side-chain structure–activity relationships have not been explored in detail in this series and may be a fruitful area for future investigation.

9. Non-Xanthine Adenosine Antagonists

Although the nonxanthine adenosine antagonists have not been the subject of as much chemical effort as the more well-known xanthines, quite a few chemical series have been identified, and unusual patterns of subtype selectivity have been seen in several cases.

9.1. A_1-Selective Agents

9.1.1. 9-Methyladenine Series

Although 9-methyladenine (174; Table 9) was identified as a (rather weak) A_2 antagonist in 1981 (Bruns, 1981), the question of possible structure–activity parallels between this series and the adenine nucleosides was only investigated recently (Ukena et al., 1987b). As in the adenosine series, N^6-cyclopentyl substitution (175) resulted in a substantial increase in A_1 affinity (K_i 540 nM) and A_1 selectivity (based on affinity in A_2-stimulated adenylate cyclase). The 9-methyladenine counterparts of R- and S-PIA (176 and 177) showed stereoselectivity in the same direction as in the adenosine series, confirming the likelihood that the 9-methyladenine derivatives bind to the same N^6 receptor domain as adenosine.

9.1.2. [1,2,4]Triazolo[4,3-a]quinoxaline Series

The ability of an appropriately placed cyclopentyl group to enhance A_1 affinity and selectivity was also observed in a series of [1,2,4]triazolo[4,3-a]quinoxaline-4-amines. These compounds, which were originally reported in the patent literature as antidepressants (Sarges, 1985a,b), bore some structural resemblance to adenine and were found to have weak adenosine antagonist activity (Trivedi and Bruns, 1988). Replacement of the N-isopropyl group in 178 with a cyclopentyl moiety (179, CPEQ) resulted in a considerable improvement in A_1 affinity and selectivity, which could be further improved by replacement of the 1-ethyl group in *178* and *179* with a trifluoromethyl group (*180*, CPQ).

9.1.3. Pyrazolo[3,4-b]pyridine Series

Several pyrazolo[3,4-*b*]pyridines with anxiolytic and benzo-diazepine receptor-modulating activity were also found to possess adenosine antagonist activity (Psychoyos et al., 1982; Murphy and Snyder, 1981). Cartazolate (*181*) has A_1 affinity in the high nano-molar range and sixfold selectivity.

9.1.4. Pyrazolo[4,3-d]pyridine Series

Several 5-aryl-7-oxopyrazolo[4,3-*d*]pyrimidines were found to have significant affinity for adenosine A_1 receptors (Hamilton et al., 1987b). Quantitative structure–activity relationship analysis for this series indicated that changes in affinity caused by substitution of the 5-aryl group in this series were similar to the patterns seen in the 8-phenyltheophylline series, implying that the 6-membered ring of the pyrazolo[4,3-*d*]pyrimidine series may bind to the same receptor domain as the 5-membered ring of the xanthines. In support of this idea, the affinity of the phenyl derivative *182* (K_i 580 nM) is enhanced by 2-amino-4-chloro substitution (*183*) (K_i 176 nM) (compare with *150* and *158*).

9.1.5. Pyrazolo[3,4-d]pyrimidine Series

Davies et al. (1983) reported that several 4-thiopyrazolo[3,4-*d*]pyrimidines had significant affinity as adenosine antagonists. We independently became interested in this series because of the activity shown by the pyrazolo[4,3-*d*]pyrimidines. Several 4-aminopyra-zolo[3,4-*d*]pyrimidines that had been synthesized in the late 1950s by Edward C. Taylor (Princeton University) were found to possess con-siderable affinity in A_1 and A_2 binding. The most potent activity was shown by the 1,6-diphenyl derivative *184*, which had a balanced receptor profile with A_1 affinity of 23 nM and A_2 affinity of 35 nM (Bruns and Hamilton, 1987).

Table 9
A$_1$ and A$_2$ Binding Affinities of Nonxanthine Adenosine Antagonists

Compound	Structure	K_i, nM		Ratio	Data source*	
		A_1	A_2	A_2/A_1	A_1	A_2
174	9-methyladenine	57,000	66,000	1.14	h	h
175	N^6-cyclopentyl-9-methyl-adenine	540	—	—	y	—
176	9-methyl-N^6-[(R)-1-methyl-2-phenylethyl]adenine	2500	—	—	y	—
177	9-methyl-N^6-[(S)-1-methyl-2-phenylethyl]adenine	10,000	—	—	y	—
178	1-ethyl-N-(1-methylethyl)-[1,2,4]triazolo[4,3-a]-quinazoline	240	4300	17.7	y	y
179	CPEQ	24	3000	125	y	y
180	CPQ	7.3	1000	137	y	y
181	Cartazolate	360	2200	6.3	e	e
182	1,4-dihydro-1,3-dimethyl-5-phenyl-7H-pyrazolo[4,3-d]-pyrimidin-7-one	580	—	—	m	—
183	5-(2-amino-4-chlorophenyl)-1,4-dihydro-1,3-dimethyl-pyrazolo[4,3-d]pyrimidine-7-one	176	—	—	m	—
184	4-amino-1,6-diphenyl-pyrazolo[3,4-d]pyrimidine	23	35	1.50	d	d
185	Alloxazine	5200	2700	0.52	e	e
186†	CGS 15943	20	3.3	0.161	n	n
187	ADQZ	600	310	0.52	c	c
188	HTQZ	3000	124	0.042	c	c

*Data sources are listed in Table 1.

†Compound 186 is drawn as the amino tautomer to emphasize its similarity to adenine.

9.2. A_2-Selective Agents

9.2.1. Alloxazine

The benzo[*g*]pteridine alloxazine (*185*), identified as an adenosine antagonist in the human fibroblast A_2 system (Bruns, 1981), showed twofold A_2 selectivity in A_1 and A_2 binding, with affinity in the low micromolar range (A_2 K_i 2.7 μM).

9.2.2. [1,2,4]Triazolo[1,5-c]quinazoline Series

CGS 15943 (*186*) showed sevenfold A_2 selectivity and a very potent affinity of 3 nM in A_2 binding (Williams et al., 1987). This compound, which bears some resemblance to adenine, came from a series of benzodiazepine receptor antagonists.

9.2.3. Quinazolines

The simple quinazoline ADQZ (*187*) was discovered in screening of assorted quinazolines in A_1 and A_2 binding (Bruns and Coughenour, 1987). ADQZ is of interest because of its amine-containing side chain and its twofold A_2 selectivity (A_1 600 nM, A_2 310 nM).

9.2.4. Thiazolo[2,3-b]quinazoline Series

HTQZ (*188*) was also found in a broad screening program (Bruns and Coughenour, 1987). This compound stands out because of its very high (25-fold) A_2 selectivity (A_1 3000 nM, A_2 124 nM). However, the high A_2 selectivity of this compound is only seen in rat A_1/A_2 binding, with less selectivity observed in dog, cow, or human binding (unpublished results). Nevertheless, *188* may provide an important chemical lead towards A_2-selective antagonists.

10. Pharmacological Tools
for Adenosine Research

The ultimate objective of structure–activity analysis is of course to design new pharmacological tools and therapeutic agents. The utility of adenosine agonists and antagonists in vivo is governed not only by receptor affinity and selectivity, but also by such factors as intrinsic activity, solubility, and pharmacokinetics.

10.1. Use of Structure–Activity Relationships
to Define Adenosine Receptor Subtypes

Two prominent differences in structure–activity relationships have been used to define A_1 and A_2 receptors. At A_1 receptors, R-PIA is more potent than NECA, whereas at A_2 receptors, the reverse order of potency holds (Londos et al., 1980). In addition, the R diastereomer of PIA is 20–100 times more potent than the S diastereomer at A_1 receptors, but less than 10 times as potent as A_2 receptors (Smellie et al., 1979b). However, these differences have not been seen consistently in all A_1 and A_2 systems, leading to situations where the subtype classification can be ambiguous (Stone et al., 1988). Because of these problems, new structure–activity criteria for receptor subclassification have been proposed (Bruns et al., 1987a).

In several neurotransmitter release systems that are presumed to be A_1, NECA is more potent than R-PIA (McCabe and Scholfield, 1985). The reasons for this are unclear, but the explanation is likely to be related to either the much greater hydrophobicity of R-PIA compared to NECA or the very slow binding kinetics of the N^6-modified adenosines. An alternative pair of compounds that could be used to distinguish A_1 and A_2 receptors is CPA and N^6-benzyladenosine (Bruns et al., 1987a). Although the two compounds are from the same structural class and have identical log P values (log octanol:water partition coefficient), their relative affinities at A_1 and A_2 receptors differ markedly. In A_1 binding, CPA is about 300 times more potent than N^6-benzyladenosine, whereas in A_2 binding, N^6-benzyladenosine is slightly the more potent of the two. The two antagonists CPX and PD 115,199 could also be used to distinguish adenosine receptor subtypes: CPX is 30-fold more potent than PD 115,199 in A_1 binding, whereas a 20-fold difference in the opposite direction is seen in the A_2 binding assay (Bruns et al., 1987a).

10.2. A Pharmacological "Toolbox" for Adenosine Studies

It often happens that an investigator may wish to characterize an adenosine-responsive system, but because of time and budget limitations will be unable to examine more than a small number of compounds. Of the hundreds of adenosine agonists and antagonists that have been reported, which small subset will provide the most useful information to the investigator? The answer to this question is of course quite subjective, but our own suggested "toolbox" is given in Table 10. Since many agonists with good solubility and pharmacokinetics are available, the agonists in Table 10 were chosen to represent the widest possible assortment of structural classes and receptor selectivities. For the antagonists, the additional criteria of solubility and in vivo activity by different routes were also given high priority.

Acknowledgments

We thank Edward Badger, James Bristol, Harriet Hamilton, Wendy Kramer, Walter Moos, William Patt, and Deedee Szotek for providing compounds, and James Fergus and Gina Lu for providing affinities of compounds in A_1 and A_2 binding.

Table 10

Adenosine Agonists and Antagonists as Pharmacological Tools*

Compound	Structure	K_i, nM		Ratio	Features
		A_1	A_2	A_2/A_1	
Agonists					
88	3,5-dimethoxy-CI-936	30	6.1	0.20	Moderately A_2 selective
106	CV-1808	560	120	0.21	Moderately A_2 selective
44	N^6-(1-naphthylmethyl)ado	24	9.3	0.39	Moderately A_2 selective
122	NECA	6.3	10.3	1.64	Potent, balanced
100	2-chloroado	9.3	63	6.8	Proxy for adenosine
20	CPA	0.59	460	780	Highly A_1 selective
131	5'-chloro-5'-deoxy-N^6-(2-*endo*-norbornyl)ado	0.42	2100	4900	Extremely A_1 selective
Antagonists					
186	CGS 15493	20	3.3	0.161	Moderately A_2 selective, active p.o.
173	PD 115,199	14.0	16.0	1.1	Potent, moderately soluble
154	8-PSPT	2600	15,300	5.8	Soluble, does not cross membranes
162	PD 113,297	5.6	70	12.5	Potent, soluble, active iv
161	XAC	0.86	27	31	Potent, soluble, active ip
164	CPT	10.9	1440	133	A_1 selective, active p.o.
165	CPX	0.46	340	740	Highly A_1 selective, active ip

*Note added in proof: subsequent to the completion of this manuscript, several agonists with even greater selectivity were reported (Bridges et al., 1988; Hutchison, et al., 1989; Trivedi et al., 1989). These can be used in place of compound 88 and 131.

References

Baird-Lambert, J., Marwood, J. F., Davies, L. P., and Taylor, K. M. (1980) 1-Methylisoguanosine: An orally active marine natural product with skeletal muscle and cardiovascular effects. *Life Sci.* **26,** 1069–1077.

Born, G. V. R., Haslam, R. J., Goldman, M., and Lowe, R. D. (1965) Comparative effectiveness of adenosine analogues as inhibitors of blood-platelet aggregation and as vasodilators in man. *Nature* **205,** 678–680.

Bridges, A. J., Moos, W. H., Szotek, D. S., Trivedi, B. K., Bristol, J. A., Heffner, T. G., Bruns, R. F., and Downs, D. A. (1987) N^6-(2,2-diphenylethyl)adenosine, a novel adenosine receptor agonist with antipsychotic activity. *J. Med. Chem.* **30,** 1709–1711.

Bridges, A. J., Bruns, R. F., Ortwine, D. F., Priebe, S. R., Szotek, D. L., and Trivedi, B. K. (1988) N^6-[2-(3,5-Dimethoxyphenyl)-2-(2-methylphenyl)ethyl]adenosine and its uronamide derivatives. Novel adenosine agonists with both high affinity and high selectivity for the adenosine A_2 receptor. *J. Med. Chem.* **31,** 1282–1285.

Bristol, J. A., Bridges, A. J., Bruns, R. F., Downs, D. A., Heffner, T. G., Moos, W. H., Ortwine, D. F., Szotek, D. S., and Trivedi, B. K. (1988) The search for purine and ribose-substituted adenosine analogs with potential clinical application, in *Physiology and Pharmacology of Adenosine and Adenine Nucleotides* (Paton, D. M., ed.), Taylor and Francis, London, pp. 17–26.

Bruns, R. F. (1980) Adenosine receptor activation in human fibroblasts: Nucleoside agonists and antagonists. *Canad. J. Physiol. Pharmacol.* **58,** 673–691.

Bruns, R. F. (1981) Adenosine antagonism by purines, pteridines, and benzopteridines in human fibroblasts. *Biochem. Pharmacol.* **30,** 325–333.

Bruns, R. F. and Coughenour, L. L. (1987) New non-xanthine adenosine antagonists. *Pharmacologist* **29,** 146.

Bruns, R. F. and Hamilton, H. W. (1987) 6-Arylpyrazolo[3,4-d]pyrimidines with high affinity in adenosine A_1 and A_2 binding. *Fed. Proc.* **46,** 393.

Bruns, R. F., Daly, J. W., and Snyder, S. H. (1980) Adenosine receptors in brain membranes: Binding of N^6-cyclohexyl-[^{3}H]adenosine and 1,3-diethyl-8-[^{3}H]phenylxanthine. *Proc. Natl. Acad. Sci. USA* **77,** 5547–5551.

Bruns, R. F., Daly, J. W., and Snyder, S. H. (1983) Adenosine receptor binding. Structure–activity analysis generates extremely potent xanthine antagonists. *Proc. Natl. Acad. Sci. USA* **80,** 2077–2080.

Bruns, R. F., Lu, G. H., and Pugsley, T. A. (1984) Binding of ^{3}H-NECA to a subtype of A_2 adenosine receptor in rat striatal membranes. *Soc. Neurosci. Abstr.* **10,** 957.

Bruns, R. F., Lu, G. H., and Pugsley, T. A. (1986) Characterization of the A_2 adenosine receptor labeled by [^{3}H]NECA in rat striatal membranes. *Mol. Pharmacol.* **29,** 331–346.

Bruns, R. F., Lu, G. H., and Pugsley, T. A. (1987a) Adenosine receptor subtypes: Binding studies, in *Topics and Perspectives in Adenosine Research* (Gerlach, E. and Becker, B., eds.), Springer-Verlag, Berlin Heidelberg, pp. 59–73.

Bruns, R. F., Fergus, J. H., Badger, E. W., Bristol, J. A., Santay, L. A., Hartman, J. D., Hays, S. J., and Huang, C. C. (1987b) Binding of the A₁-selective adenosine antagonist 8-cyclopentyl-1,3-dipropylxanthine to rat brain membranes. *Naunyn-Schmiedebergs Arch. Pharmacol.* **335,** 59–63.

Bruns, R. F., Fergus, J. H., Badger, E. W., Bristol, J. A., Santay, L. A., and Hays, S. J. (1987c) PD 115,199: An antagonist ligand for adenosine A₂ receptors. *Naunyn-Schmiedebergs Arch. Pharmacol.* **335,** 64–69.

Bruns, R. F., Davis, R. E., Ninteman, F. W., Poschel, B. P. H., Wiley, J. N., and Heffner, T. G. (1988) Adenosine antagonists as pharmacologic tools, in *Physiology and Pharmacology of Adenosine and Adenine Nucleotides* (Paton, D. M., ed.) Taylor and Francis, London, pp. 39–49.

Cristalli, G., Grifantini, M., Vittori, S., Balduini, W., and Cattabeni, F. (1985) Adenosine and 2-chloroadenosine deaza-analogues as adenosine receptor agonists. *Nucleosides Nucleotides* **4,** 625–639.

Daly, J. W., Butts-Lamb, P., and Padgett, W. (1983) Subclasses of adenosine receptors in the central nervous system: Interaction with caffeine and related methylxanthines. *Cell. Mol. Neurobiol.* **3,** 69–80.

Daly, J. W., Padgett, W., Thompson, R. D., Kusachi, S., Bugni, W. J., and Olsson, R. A. (1986a) Structure–activity relationships for N⁶-substituted adenosines at a brain A₁-adenosine receptor with a comparison to an A₂-adenosine receptor regulating coronary blood flow. *Biochem. Pharmacol.* **35,** 2467–2481.

Daly, J. W., Padgett, W. L., and Shamim, M. T. (1986b) Analogues of caffeine and theophylline: Effect of structural alterations on affinity at adenosine receptors. *J. Med. Chem.* **29,** 1305–1308.

Daly, J. W., Padgett, W. L., and Shamim, M. T. (1986c) Analogues of 1,3-dipropyl-8-phenylxanthine: Enhancement of selectivity at A₁-adenosine receptors by aryl substituents. *J. Med. Chem.* **29,** 1520–1524.

Davies, L. P., Brown, D. J., Chen Chow, S., and Johnston, G. A. R. (1983) Pyrazolo[3,4-d]pyrimidines, a new class of adenosine antagonists. *Neurosci. Lett.* **41,** 189–193.

Dunwiddie, T. V. and Proctor, W. R. (1987) Mechanisms underlying physiological responses to adenosine in the central nervous system, in *Topics and Perspectives in Adenosine Research* (Gerlach, E. and Becker, B. F., eds.), Springer-Verlag, Berlin, pp. 499–508.

Einstein, R., Angus, J. A., Cobbin, L. B., and Maguire, M. H. (1972) Separation of vasodilator and negative chronotropic actions in analogues of adenosine. *Eur. J. Pharmacol.* **19,** 246–250.

Ferkany, J. W., Valentine, H. L., Stone, G. A., and Williams, M. (1986) Adenosine A1 receptors in mammalian brain: Species differences in their interactions with agonists and antagonists. *Drug Develop. Res.* **9,** 85–93.

Hamilton, H. W., Ortwine, D. F., Worth, D. F., Badger, E. W., Bristol, J. A., Bruns, R. F., Steffen, R. P., and Haleen, S. J. (1985) Synthesis of xanthines as adenosine antagonists, a practical quantitative structure–activity relationship application. *J. Med. Chem.* **28,** 1071–1079.

Hamilton, H. W., Taylor, M. D., Steffen, R. P., Haleen, S. J., and Bruns, R. F.

(1987a) Correlation of adenosine receptor affinities and cardiovascular activity. *Life Sci.* **41**, 2295–2302.

Hamilton, H. W., Ortwine, D. F., Worth, D. F., and Bristol, J. A. (1987b) Synthesis and structure–activity relationships of pyrazolo[4,3-*d*]pyrimidine-7-ones as adenosine receptor antagonists. *J. Med. Chem.* **30**, 91–96.

Hamprecht, B. and van Calker, D. (1985) Nomenclature of adenosine receptors. *Trends Pharmacol. Sci.* **6**, 153-154.

Hutchison, A. J., Webb, R. L., Oei, H. H., Ghai, G. R., Zimmerman, M. B., and Williams, M. (1989) CGS 21680C, an A_2 selective adenosine receptor agonist with preferential hypotensive activity. *J. Pharmacol. Exp. Ther.* **251**, 47–55.

Irshad, F., Dagenais, G. R., Fedor, E. J., and Cavanaugh, J. H. (1977) Trials with an adenosine analogue as antianginal medication. *Clin. Pharmacol. Ther.* **22**, 470–474.

Jacobson, K. A., Kirk, K. L., Padgett, W. L., and Daly, J. W. (1985a) Functionalized congeners of adenosine: Preparation of analogues with high affinity for A_1-adenosine receptors. *J. Med. Chem.* **28**, 1341–1346.

Jacobson, K. A., Kirk, K. L., Padgett, W., and Daly, J. W. (1985b) Probing the adenosine receptor with adenosine and xanthine biotin conjugates. *FEBS Lett.* **184**, 30–35.

Jacobson, K. A., Kirk, K. L., Padgett, W. L., and Daly, J. W. (1985c) Functionalized congeners of 1,3-Dialkylxanthines: Preparation of analogues with high affinity for adenosine receptors. *J. Med. Chem.* **28**, 1334–1340.

Jacobson, K. A., Kirk, K. L., Padgett, W. L., and Daly, J. W. (1986) A functionalized congener approach to adenosine receptor antagonists: Amino acid conjugates of 1,3-dipropylxanthine. *Mol. Pharmacol.* **29**, 126–133.

Jacobson, K. A., Ukena, D., Padgett, W., Kirk, K. L., and Daly, J. W. (1987a) Molecular probes for extracellular adenosine receptors. *Biochem. Pharmacol.* **36**, 1697–1707.

Jacobson, K. A., Lipkowski, A. W., Moody, T. W., Padgett, W., Pijl, E., Kirk, K. L., and Daly, J. W. (1987b) Binary drugs: Conjugates of purines and a peptide that bind to both adenosine and substance P receptors. *J. Med. Chem.* **30**, 1529–1532.

Jacobson, K. A., Ukena, D., Padgett, W., Daly, J. W., and Kirk, K. L. (1987c) Xanthine functionalized congeners as potent ligands at A_2-adenosine receptors. *J. Med. Chem.* **30**, 211–214.

Jahn, W. (1969) N^6-[Naphthyl-(1)]-methyl-adenosin, ein adenosin-derivat mit lang anhaltender coronarwirkung. *Arzneim.-Forsch.* **19**, 701–704.

Kusachi, S., Thompson, R. D., Bugni, W. J., Yamada, N., and Olsson, R. A. (1985) Dog coronary artery adenosine receptor: Structure of the N^6-alkyl subregion. *J. Med. Chem.* **28**, 1636–1643.

Kusachi, S., Thompson, R. D., Yamada, N., Daly, D. T., and Olsson, R. A. (1986) Dog coronary artery adenosine receptor: Structure of the N^6-aryl subregion. *J. Med. Chem.* **29**, 989–996.

Kwatra, M. M., Leung, E., Hosey, M. M., and Green, R. D. (1987) N^6-phenyladenosines: Pronounced effect of phenyl substituents on affinity for A_2 adenosine receptors. *J. Med. Chem.* **30**, 954–956.

Lee, K. S. and Reddington, M. (1986) 1,3-Dipropyl-8-cyclopentylxanthine (DPCPX) inhibition of [³H]*N*-ethylcarboxamidoadenosine (NECA) binding allows the visualization of putative non-A₁ adenosine receptors. *Brain Res.* **368,** 394–398.

Londos, C., Cooper, D. M. F., and Wolff, J. (1980) Subclasses of external adenosine receptors. *Proc. Natl. Acad. Sci. USA* **77,** 2551–2554.

Martinson, E. A., Johnson, R. A., and Wells, J. N. (1987) Potent adenosine receptor antagonists that are selective for the A₁ receptor subtype. *Mol. Pharmacol.* **31,** 247–252.

Marumoto, R., Shima, S., Omura, K., Tanabe, M., Fujiwara, S., and Furukawa, Y. (1985) Synthetic studies of 2-substituted adenosines. III. Coronary vasodilatory activity of 2-arylaminoadenosines. *J. Takeda Res. Lab.* **44,** 220–230.

Marumoto, R., Yoshioka, Y., Miyashita, O., Shima, S., Imai, K. I., Kawazoe, K., and Honjo, M. (1975) Synthesis and coronary vasodilating activity of 2-substituted adenosines. *Chem. Pharm. Bull.* **23,** 759–774.

McCabe, J. and Scholfield, C. N. (1985) Adenosine-induced depression of synaptic transmission in the isolated olfactory cortex: Receptor identification. *Pflügers Arch.* **403,** 141–145.

Moos, W. H., Szotek, D. S., and Bruns, R. F. (1985) Cycloalkyladenosines: Potent, A₁-selective adenosine antagonists. *J. Med. Chem.* **28,** 1383,1384.

Murphy, K. M. M. and Snyder, S. H. (1981) Adenosine receptors in rat testes: Labeling with ³H-cyclohexyladenosine. *Life Sci.* **28,** 917–920.

Olsson, R. A., Kusachi, S., Thompson, R. D., Ukena, D., Padgett, W., and Daly, J. W. (1986) *N*⁶-substituted *N*-alkyladenosine-5'-uronamides: Bifunctional ligands having recognition groups for A₁ and A₂ adenosine receptors. *J. Med. Chem.* **29,** 1683–1689.

Paton, D. M., Olsson, R. A., and Thompson, R. T. (1986) Nature of the *N*⁶ region of the adenosine receptor in guinea pig ileum and rat vas deferens. *Naunyn-Schmiedebergs Arch. Pharmacol.* **333,** 313–322.

Persson, C. G. A., Andersson, K. E., and Kjellin, G. (1986) Effects of enprofylline and theophylline may show the role of adenosine. *Life Sci.* **38,** 1057–1072.

Prasad, R. N., Fung, A., Tietje, K., Stein, H. H., and Brondyk, H. D. (1976) Modification of the 5' position of purine nucleosides. 1. Synthesis and biological properties of alkyl adenosine-5'-carboxylates. *J. Med. Chem.* **19,** 1180–1186.

Prasad, R. N., Bariana, D. S., Fung, A., Savic, M., Tietje, K., Stein, H. H., Brondyk, H., and Egan, R. S. (1980) Modification of the 5' position of purine nucleosides. 2. Synthesis and some cardiovascular properties of adenosine-5'-(*N*-substituted)carboxamides. *J. Med. Chem.* **23,** 313–319.

Psychoyos, S., Ford, C. J., and Phillipps, M. A. (1982) Inhibition by etazolate (SQ 20009) and cartazolate (SQ 65396) of adenosine-stimulated [³H]cAMP formation in [2-³H]adenine-prelabeled vesicles prepared from guinea pig cerebral cortex. *Biochem. Pharmacol.* **31,** 1441, 1442.

Sarges, R. *US Patent* 4,457,901, 1985a.

Sarges, R. *US Patent* 4,495,187, 1985b.

Sattin, A. and Rall, T. W. (1970) The effect of adenosine and adenine nucleotides on the cyclic adenosine 3',5'-phosphate content of guinea pig cerebral cortex slices. *Mol. Pharmacol.* **6**, 13–23.

Schwabe, U. and Trost, T. (1980) Characterization of adenosine receptors in rat brain by (-)[³H]N^6-phenylisopropyladenosine. *Naunyn-Schmiedebergs Arch. Pharmacol.* **313**, 179–187.

Scott, R. H. and Dolphin, A. C. (1987) Inhibition of calcium currents by an adenosine analogue 2-chloroadenosine, in *Topics and Perspectives in Adenosine Research* (Gerlach, E. and Becker, B. F., eds.), Springer-Verlag, Berlin, pp. 549–558.

Smellie, F. W., Davis, C. W., Daly, J. W., and Wells, J. N. (1979a) Alkylxanthines: Inhibition of adenosine-elicited accumulation of cyclic AMP in brain slices and of brain phosphodiesterase activity. *Life Sci.* **24**, 2475–2482.

Smellie, F. W., Daly, J. W., Dunwiddie, T. V., and Hoffer, B. J. (1979b) The dextro and levorotatory isomers of *N*-phenylisopropyladenosine: Stereospecific effects on cyclic AMP-formation and evoked synaptic responses in brain slices. *Life Sci.* **25**, 1739–1748.

Stone, G. A., Jarvis, M. F., Sills, M. A., Weeks, B., and Williams, M. (1988) Species differences in high affinity adenosine A_2 receptor binding in mammalian striatal membranes. *Drug Develop. Res.* **15**, 31–46.

Taylor, M. D., Moos, W. H., Hamilton, H. W., Szotek, D. S., Patt, W. C., Badger, E. W., Bristol, J. A., Bruns, R. F., Heffner, T. G., and Mertz, T. E. (1986) Ribose-modified adenosine analogs as adenosine receptor agonists. *J. Med. Chem.* **29**, 346–353.

Trivedi, B. K. (1988) Studies toward synthesis of C2-substituted adenosines: An efficient synthesis of 2-(phenylamino)adenosine (CV-1808). *Nucleosides Nucleotides* **7(3)**, 393–402.

Trivedi, B. K., Bristol, J. A., Bruns, R. F., Haleen, S. J., and Steffen, R. P. (1988a) N^6-Arylalkyladenosines: Identification of N^6-(9-fluorenylmethyl)adenosine as a highly potent agonist for the adenosine A_2 receptor. *J. Med. Chem.* **31**, 271–273.

Trivedi, B. K. and Bruns, R. F. (1988) [1,2,4]Triazolo[4,3-*a*]quinoxaline-4-amines: A new class of A_1 receptor selective adenosine antagonists. *J. Med. Chem.* **31**, 1011–1014.

Trivedi, B. K., Blankley, C. J., Bristol, J. A., Hamilton, H. W., Patt, W. C., Kramer, W. J., Bruns, R. F., Cohen, D. M., and Ryan, M. J. (1988b) N^6-substituted adenosine receptor agonists: Potential antihypertensive agents. Abst. 32 presented in the division of Medicinal Chemistry of the American Chemical Society at the Third Chemical Congress of North America in Toronto, Canada.

Trivedi, B. K., Bridges, A. J., Patt, W. C., Priebe, S. R., and Bruns, R. F. (1989) N^6-Bicycloalkyladenosines with unusually high potency and selectivity for the A1 adenosine receptor. *J. Med. Chem.* **32**, 8–11.

Ukena, D., Jacobson, K. A., Padgett, W. L., Ayala, C., Shamim, M. T., Kirk, K. L., Olsson, R. O., and Daly, J. W. (1986a) Species differences in structure–activity relationships of adenosine agonists and xanthine antagonists at brain A_1 adenosine receptors. *FEBS Lett.* **209**, 122–128.

Ukena, D., Shamim, M. T., Padgett, W., and Daly, J. W. (1986b) Analogs of caffeine: Antagonists with selectivity for A₂ adenosine receptors. *Life Sci.* **39,** 743–750.

Ukena, D., Olsson, R. A., and Daly, J. W. (1987a) Definition of subclasses of adenosine receptors associated with adenylate cyclase: Interaction of adenosine analogs with inhibitory A₁ receptors and stimulatory A₂ receptors. *Can. J. Physiol. Pharmacol.* **65,** 365–376.

Ukena, D., Padgett, W. L., Hong, O., Daly, J. W., and Olsson, R. A. (1987b) *N⁶*-substituted 9-methyladenines: A new class of adenosine receptor antagonists. *FEBS Lett.* **215,** 203–208.

van Calker, D., Muller, M., and Hamprecht, B. (1979) Adenosine regulates via two different types of receptors, the accumulation of cyclic AMP in cultured brain cells. *J. Neurochem.* **33,** 999–1005.

Williams, M. and Risley, E. (1980) Biochemical characterization of putative central purinergic receptors by using 2-chloro[³H]adenosine, a stable analog of adenosine. *Proc. Natl. Acad. Sci. USA* **77,** 6892–6896.

Williams, M., Francis, J., Ghai, G., Braunwalder, A., Psychoyos, S., Stone, G. A., and Cash, W. D. (1987) Biochemical characterization of the triazoloquinazoline, CGS 15943, a novel, nonxanthine adenosine antagonist. *J. Pharmacol. Exp. Ther.* **241,** 415–420.

Yeung, S. M. H. and Green, R. D. (1984) [³H]5'-*N*-ethylcarboxamide adenosine binds to both Rₐ and Rᵢ adenosine receptors in rat striatum. *Naunyn-Schmiedebergs Arch. Pharmacol.* **325,** 218–225.

CHAPTER 4

Signal Transduction Mechanisms for Adenosine

Dermot M. F. Cooper and Kevin K. Caldwell

1. Introduction

Significant advances have been made during the last 20 yr in understanding the mechanisms of adenosine receptor-mediated signal transduction at the level of the plasma membrane. The pioneering studies in the early 1970s by Rall, Daly, Schwabe, and Fain clearly established that adenosine and its analogs could modulate both cyclic AMP levels in intact cells and adenylate cyclase activity in broken cell preparations. Systematic studies by Londos and Wolff (1977) established that the effects of adenosine on the intact cell could be separated into those that could be considered to be receptor-mediated (by the so-called "R-site"), and thus, amenable to the conceptual and methodological strategies available for the study of such processes, and those that could be viewed as being mediated by the so-called "P-sites," which displayed few of the regulatory features exhibited by previously characterized extracellular receptors. Achieving this discrimination set the stage for the subsequent separation of the conventional ("R-site") receptor into two subtypes termed R_i and R_a^*, which inhib-

Adenosine and Adenosine Receptors Editor: Michael Williams ©1990 The Humana Press Inc.

ited and stimulated, with differing pharmacological profiles, respectively, adenylate cyclase activity in plasma membranes isolated from a wide variety of cells (Londos et al., 1980). At nearly the same time, Van Calker et al. (1979) proposed the same subclassification of the "R-site" based on studies of cyclic AMP production in response to a range of adenosine analogs in two cultured cell lines. Because there is now a wide variety of hormones, neurotransmitters and autocoids that have been shown to both stimulate and inhibit adenylate cyclase via distinct receptors that can be generically referred to as "R_i" and "R_a", respectively, the more specific convention of A_1 (inhibitory) and A_2 (stimulatory) has been generally adopted in referring to the subtypes of the adenosine receptor. The physiological relevance and molecular nature of the "P-site" remains in some doubt because of its utilization of rarely achieved concentrations of adenosine and apparently unconventional transduction mechanisms. Therefore, it will not be considered further in this chapter. (The reader is referred to the article by Wolff et al. [1981] for a detailed account of this "receptor.")

Both the regulatory features of the signal transduction processes linking A_1 and A_2 receptors to changes in adenylate cyclase activity, as well as the structural aspects of the A_1 receptor, have been studied extensively during the past 10 yr; however, several questions remain to be addressed and resolved. In addition, several recent reports link adenosine receptors to signal transduction mechanisms other than those associated with the regulation of adenylate cyclase activity. Consequently, the present chapter will deal in some detail with:

1. The regulatory and structural aspects of adenosine receptor-mediated regulation of adenylate cyclase; this will set the conceptual stage for the regulatory mechanisms by which adenosine receptors might be associated with other signaling systems.

---

*Abbreviations: R_a(A2), stimulatory receptor coupled to adenylate cyclase; R_i(A1), inhibitory receptor coupled to adenylate cyclase; G_i, inhibitory guanine nucleotide-binding protein that couples R_i to adenylate cyclase; G_s, stimulatory guanine nucleotide-binding protein that couples R_a to adenylate cyclase; PDE, cyclic nucleotide phosphodiestrerase; GTPγS, guanosine 5'-(γ-thio)triphosphate; PK-A, cyclic AMP-dependent protein kinase; PK-C, Ca^{2+}-phospholipid-dependent protein kinase; R-PIA, R-phenylisopropyladenosine; NECA, 5'-*N*-ethylcarboxamide adenosine, CHAPS, 3-[(3-cholamidopropyl)-dimethylammonio]-1-propane-sulfonate.

2. Evidence that adenosine receptors regulate signaling processes other than adenylate cyclase and

3. The strategies needed to probe the association of adenosine receptors with "novel" signaling systems.

2. Adenosine Receptor Regulation of Adenylate Cyclase

As mentioned in the introductory remarks, adenosine can both stimulate and inhibit adenylate cyclase activity via pharmacologically distinct receptors. The potency series R-PIA>adenosine> NECA is characteristic of A_1 (or R_i) receptors, whereas the reverse potency series is observed at A_2 (or R_a) receptors (Wolff et al., 1981). This precise potency difference is not always observed when comparing A_1 and A_2 receptors in different tissues, but whether this represents heterogeneity of adenosine receptor subtypes or technical issues (such as differences in drug accessibility to receptor sites) is unclear (Ferkany et al., 1986; Stone et al., 1988). Adenosine receptors are relatively ubiquitous in mammalian tissues and cells derived therefrom. The present chapter does not attempt to deal in any detail with receptor distribution among different tissues, since comprehensive reviews have appeared on this subject by other (e.g., Stiles, 1986; Ferkany et al., 1986; Stone et al., 1988). However, it appears that most mammalian tissues contain either an A_1 or A_2 receptor, or both. The A_1 receptor is well represented in neuronal tissue and appears to have been detected in every major neuronal structure. A_2 receptors are also present in the brain, particularly in the caudate nucleus, in addition to being prominent in liver and blood cells. No guiding physiological rationale appears to account for the widespread distribution of these receptors, although it might be argued that the A_1 subtype plays the more relevant physiological role, since these receptors are sensitive to concentrations of adenosine that are often achieved in the circulation. However, this argument can be tempered by considering the high local concentrations of adenosine that may be achieved in certain situations.

It is possible to generalize on the properties of the regulation of adenylate cyclase by adenosine from the large number of studies describing the effects of adenosine on adenylate cyclase activity in broken cell preparations and purified membranes. As has been found for all other receptors that are coupled to adenylate cyclase, exogenous GTP is absolutely required in order to observe stimulatory or inhibitory effects of adenosine on adenylate cyclase activity in purified membrane preparations (Cooper and Londos, 1979). This GTP requirement translates into the involvement of distinct GTP regulatory proteins (G_s and G_i) that mediate these effects (Rodbell, 1980). Interestingly, it was as a result of work on the adenosine A_1 receptor that some of the earliest evidence was gathered suggesting that a unique GTP regulatory complex (different from the previously identified G_s complex) mediated the inhibitory effects of receptors on adenylate cyclase. This evidence hinged on the fact that selective treatments could eliminate the inhibitory effects of adenosine analogs, without perturbing stimulatory effects of hormones, and vice versa (Cooper, et al., 1979; Schlegel et al., 1980). That distinct GTP regulatory-proteins were involved was confirmed by the use of two toxins, which selectively modify the two processes by ADP-ribosylation reactions. Cholera toxin, by modifying G_s, enhances or overrides the consequences of adenosine A_2 receptor occupancy (Lad et al., 1980), causing a lasting activation of adenylate cyclase. Pertussis toxin, on the other hand, eliminates G_i-mediated processes, and in so doing, abolishes the ability of adenosine A_1 receptors to inhibit adenylate cyclase (Hazeki and Ui, 1981).

There now is abundant evidence available on which to base the following description of a dually regulated adenylate cyclase system. Mammalian adenylate cyclase systems comprise five distinct functional elements: a catalytic unit, which has recently been purified from a number of sources and which displays a mol wt of 120,000 daltons (Smigel, 1986); distinct stimulatory and inhibitory receptors, which elevate or depress, respectively, the activity of the enzyme (familiar examples would be the β-adrenergic and the α-2-adrenergic receptors; Lefkowitz et al., 1981); and stimulatory and inhibitory GTP-regulatory elements (termed G_s or N_s, and G_i or N_i, respectively), which transduce the signal of hormone receptor occupancy into either an increase or decrease in the activity of the catalytic unit, as appropriate (Rodbell, 1980; Cooper, 1982). These GTP-regulatory elements have also been purified and are composed of three subunits, which are termed α, β, and γ. The β and γ subunits (mol wt 35,000 and 10,000 daltons, respectively) appear to be identical for both G_s and G_i, whereas the α subunits are unique to the particular G-protein. (However, it should be noted that Fong et al. (1987) and Gao et al. (1987) have recently demonstrated the existence of two distinct forms of the β subunit, based on deduced amino acid sequences of complementary DNA clones. Whether these two forms differ in their association with the α_s and α_i subunits is presently unclear.) α_s from G_s displays a mol wt of 45,000 daltons and is a substrate for *Vibrio cholerae* toxin, whereas α_i from G_i displays a mol wt of 41,000 daltons and is a substrate for *Bordetella pertussis* toxin (Gilman, 1984). Two distinct, though not necessarily mutually exclusive, mechanisms for the regulation of adenylate cyclase by these subunits have been proposed. In both models, stimulation is viewed to be based on the dissociation of the $\alpha\beta\gamma$ complex of G_s, resulting in the liberation of α_s, which directly activates the catalytic unit. The models differ in how inhibition is viewed to proceed. In the first, known as the "chelation" model, $\beta\gamma$ subunits are released upon the activation of G_i by hormone/neurotransmitter; these $\beta\gamma$ subunits then chelate α_s subunits as a result of simple mass action and thereby reverse activation. In the so-called "direct" model, the α_i, that is liberated from G_i, directly inhibits the catalytic unit. Both models can be supported, based largely on studies that have been performed in detergent solution. It seems likely that in many situations both models may apply (Gilman, 1984).

In neuronal tissues, the regulation of adenylate cyclase is somewhat more complex. Adenylate cyclase in every region of the brain is regulated to some extent by Ca^{2+}/calmodulin. In the striatum, this complex stimulates activity by only 40%, whereas in the hippocampus, five-fold increments are achieved (Girardot et al., 1983; Cooper et al., 1986). It is significant that this stimulation is totally dependent upon Ca^{2+} concentrations in the very low micromolar range. These concentrations $(0.6–2 \, \mu M)$ correspond closely to the intracellular Ca^{2+} concentrations that are achieved upon neuronal depolarization or upon mobilization of Ca^{2+} following the hydrolysis of phosphatidylinositol bisphosphate by Ca^{2+}-mobilizing neurotransmitters, such as acetylcholine and norepinephrine, acting through muscarinic cholinergic M_1 receptors and α-1 adrenergic receptors, respectively

(Berridge and Irvine, 1984; Rasmussen et al., 1984). The fact that a several-fold stimulation of adenylate cyclase activity may be achieved by elevations in intracellular Ca^{2+} is in contrast to the modest elevation in activity that can be elicited by hormones/neurotransmitters that are directly linked via G_s proteins to adenylate cyclase in neuronal tissue. These latter agents have rarely been reported to stimulate activity by more than 50% in studies in which basal activity has not been allowed to decay (thereby exaggerating the magnitude of the stimulation) (Cooper, et al., 1988). This regulation by Ca^{2+}/calmodulin has important implications for neurotransmitter regulation of the enzyme. Generally, it has been observed that, unless basal cyclase activity is stimulated by Ca^{2+}/calmodulin, inhibitory neurotransmitters, such as adenosine acting through A_1 receptors, cannot inhibit the enzyme (Girardot et al., 1983; Ahlijanian and Cooper, 1987). In invitro studies, other activators of adenylate cyclase, such as forskolin, will sustain inhibition, but within the physiological context, it appears that the stimulation by Ca^{2+}/calmodulin provides a feedback inhibitory loop through which neurotransmitters that are released into synapses in association with a Ca^{2+} influx, can inhibit the synthesis of cyclic AMP and thereby modulate their own release.

3. Structural Studies of Adenosine Receptors

In determining the molecular mechanisms of adenosine receptor coupling to adenylate cyclase or other signaling systems, an essential prerequisite is the purification and characterization of the receptors. Since several recent reports indicate that adenosine receptors are linked to signaling mechanisms other than adenylate cyclase, it would not be surprising if distinct molecular species (receptor subtypes) were involved in each case. It is therefore of interest to review some of the structural strategies that are currently being pursued in order to predict whether such strategies may assist in the resolution of this issue. Because the majority of the structural studies on adenosine receptors have focused on the A_1 subtype of the receptor, because of the greater technical ease with which this subtype can be investigated (*see* Cooper, 1988 for a discussion of this issue), this chapter will review only these studies.

3.1. Solubilization of A_1 Receptors

Several reports have described successful solubilization of adenosine A_1-receptors from various brain areas (Gavish et al., 1982; Klotz et al., 1986; Bruns et al., 1983; Stiles, 1985; Perez-Reyes et al., 1987; Nakata and Fujisawa, 1983). The use of brain tissue reflects the relative abundance of A_1 receptors in brain and the mass of material available. Relatively standard solubilization conditions have been used in these studies, although some uncertainties are present, as discussed in detail in Cooper (1988). (This article presents a detailed account of the technical considerations associated with structural aspects of adenosine receptor characterization, which fall outside the scope of the present chapter, but are essential to understanding the potential difficulties that may arise in studies of the adenosine receptor, e.g., the presence of endogenous phospholipid and the critical micellar concentrations of detergents.)

An apparently unusual feature of adenosine receptor solubilization is the ease with which adenosine receptors can be solubilized in a high affinity state that is regulated by guanine nucleotides. In most of the studies reported, guanine nucleotides regulate the binding of agonist ligands to the receptor in detergent solution. At first sight, this might appear unusual, compared with findings with other neurotransmitter receptors that inhibit (or stimulate) adenylate cyclase: generally, solubilization of high affinity forms of receptors, which can be regulated by guanine nucleotides, requires the prior binding of agonist (Smith and Limbird, 1981; Michel et al., 1981; Kilpatrick and Caron, 1983). The unusual nature of this finding with adenosine receptors has been commented on previously (Stiles, 1985). A possible cause for this effect is the presence of endogenous adenosine. In several studies of adenosine receptors, steps were taken to reduce the level of adenosine prior to solubilization (Gavish, et al., 1982; Stiles, 1985; Nakata and Fujisawa, 1983; Perez-Reyes et al., 1987). However, even in situations where attempts are made to reduce adenosine levels prior to solubilization, it must be recognized that, although it is relatively easy to eliminate adenosine from solution by treatment with adenosine deaminase, adenosine dissociates slowly from its receptors (the half-time for dissociation is in excess of 3 h at 24°C; Yeung et al.,

1987). Thus, under normal solubilization conditions (10–30 min at 4°C), adenosine would remain bound to its receptor and therefore, would stabilize high affinity forms of the receptor during solubilization.

The foregoing considerations suggest that the same general principles apply to the solubilization of adenosine A_1 receptors as to any other receptors that regulate adenylate cyclase, and consequently, similar tactics might be expected to be needed in their structural characterization, i.e., agonists stabilize high affinity, guanine nucleotide-regulated forms of the receptor; antagonists may be expected to stabilize low affinity, guanine nucleotide-resistant forms of the receptor. The latter form may be most easily purified (Florio and Sternweis, 1985).

3.2. Hydrodynamic Analysis

The solubilized A_1 receptor has been subjected to several independent hydrodynamic studies, as a means of obtaining some insights into its molecular size. Receptors that were solubilized following exposure to the slowly dissociating agonist [^{3}H]R-PIA tended to be recovered with 80–90% efficiency, whereas only ca. 4–10% of receptors that were occupied by the more rapidly dissociating, endogenous adenosine, and which were therefore intrinsically less stable, were recovered—as measured by agonist postlabeling (Yeung et al., 1987). This difference underlines the uncertain composition and potentially irrelevant sizes of these latter "complexes." Consequently, it might be suggested that, prior to the performance of any hydrodynamic analysis, receptors should be stabilized with either an agonist (to stabilize an R–G complex) or an antagonist (to stabilize a free receptor—noting the difficulties that may be encountered in the case of adenosine receptors, in actually obtaining conditions in which the antagonist can occupy a significant proportion of the receptor population, because of the presence of endogenous adenosine).

Hydrodynamic studies have been performed on solubilized adenosine receptors both below and above the critical micelle concentration (CMC) of the detergent during sedimentation analyses. Both in this laboratory and elsewhere, studies were performed below the CMC, which gave rise to complexes with apparent sedimentation coefficients of 12 to 16S (Svedberg) (or >500,000 daltons; Stiles, 1985, Cooper et al., 1985). When the concentration of cholate or digitonin was maintained above the CMC during such analyses, sedimentation coefficients of 7.7S and Stokes' radii of 7.2 nm were observed. These values yielded an approximate particle size of 250,000 daltons for agonist-binding, GTP-sensitive forms of the receptor (Nakata and Fujisawa, 1983; Yeung et al., 1987). It is likely that these latter studies, performed under conditions in which the solubility of the receptor was maintained (i.e., above the detergent CMC), convey the most information on the hydrodynamic properties of the adenosine receptor.

3.3. Photolabeling

Specific, high affinity, radio-iodinated photolabels have been developed for the adenosine A_1 receptor. Both agonist and antagonist iodo-azidoderivatives have been prepared. The availability of such covalent labels is a critical requirement in the ultimate purification of the receptor. These ligands label a protein(s) of apparent mol wt of 34,000–38,000 daltons in membrane preparations from a variety of cell types (Linden et al., 1987; Stiles, 1985). The consensus of the mol wt observed strongly suggests that this species represents a specific component of the receptor. The size of approximately 36,000 daltons is somewhat smaller than might be anticipated for a receptor coupled to adenylate cyclase. When other such receptors are considered, sizes greater than 60,000 daltons have generally been encountered; for instance, the adrenergic receptors are all approximately 58,000–63,000 daltons (Caron et al., 1985), whereas the muscarinic cholinergic receptor is 80,000 daltons (Venter, 1983). It is conceivable, therefore, that the photolabeled adenosine target protein is one subunit of a more complex receptor assembly. The A_2 receptor has also been photoaffinity labeled and found to have a size of 45,000 daltons (Barrington et al., 1989).

3.4. Target Size Analysis

The cerebral cortex adenosine A_1 receptor has been analyzed by irradiation inactivation. The major advantage of this technique is that the functional size for the receptor can be estimated in a native membrane environment, without disrupting the binding properties of the receptor. A disadvantage of this method is that only a size for a functional entity is provided; no information is provided on the composition of the target (Kempner and Schlegel, 1979). In addition, because receptors exist in both high and low affinity forms, it is necessary to perform detailed binding isotherms at each irradiation dose, so that the disappearance of each receptor state can be plotted. When such analysis is performed on the high affinity adenosine receptor from cerebral cortex, a linear decay is observed, which corresponds to a target size of 63,000 daltons (Frame et al., 1986). As an internal calibration for these studies, the muscarinic cholinergic receptor which is also present in these membranes, was used. The latter receptor displays a target size of 80,000 daltons, as well as an electrophoretic mobility of M_r 80,000 daltons (Venter, 1983). A similar size for the A_1 receptor was encountered by Reddington et al. (1987).

As is evident from the foregoing presentation, disparate sizes arise for the adenosine A_1 receptor from the various methods of target size analysis, hydrodynamic analysis, and photolabeling. However, these differences may be reconciled if the nature of the entity being measured in the various methods is considered. The size of the receptor that was obtained from target-size analysis may be very close to the size of the native receptor. The discrepancy between this value and that observed by photolabeling may indicate either that the size that is interpolated from the migration of standards on polyacrylamide gels may not reflect the size of the native protein, or that a subunit(s), additional to that which is photolabeled, comprises the native receptor. The particle size of 250,000 daltons that is obtained for the high affintiy agonist-binding receptor upon hydrodynamic analysis is compatible with a receptor size of 63,000 daltons, plus a bound GTP regulatory complex of approximately 100,000 daltons, with an additional quantity of bound detergent and lipid. Hydrodynamic analysis of antagonist-binding forms of the receptor may provide a particle size estimate for the free receptor.

3.5. Monoclonal Antibodies

A critical tool in the discrimination between different subclasses of adenosine receptors may rely on the development of specific monoclonal antibodies. Such antibodies will be useful in identifying differential distribution of the receptors among tissues, as well as providing potentially powerful screening devices for the detection of the expression of mRNAs encoding the receptor.

As a step in this direction, recently, in this laboratory, monoclonal antibodies were prepared that were directed against a crude sodium cholate-solubilized preparation of rat cortical A_1 receptors that was eluted from a TSK 3000 sizing column. Inhibition of reversible binding was chosen as the screen for the immunological identification of the receptor for two major reasons:

1. antibodies that would precipitate photolabeled receptor could be directed against a region of the receptor that was common to many membrane-inserted proteins and thus would not prove specific in later immunoblotting experiments, and
2. it was felt that inhibition of agonist binding was a functional criterion that was specific to adenosine receptors, which might prove quite useful in purification strategies that sought to retain receptor function.

In preliminary studies in this laboratory, two monoclonal antibodies that inhibited receptor binding were produced (Perez-Reyes et al., 1987). Upon the production of such an antibody, the first requirement is the demonstration of specificity for the adenosine receptor, since the possibility of generating nonselective antibodies is high. Three major possibilities for nonspecific inhibitory effects of an antibody on receptor binding were considered and evaluated: first, the antibody may be directed against the G_i component of the solubilized R_i–G_i complex, which may sustain the binding activity of the receptor in detergent solution; second, the antibody may be directed against the detergent; and third, it may be directed against a general membrane spanning domain that is common to many membrane-inserted proteins. In the present case, it was considered important to evaluate both an antagonist and an agonist-binding form of the receptor. The former would be believed to represent a free R_i complex, the latter an R_i–G_i complex. If the antibody inhibited binding to both forms of the receptor equally well, it would tend to eliminate the possibility that the antibody was directed against the G-protein. The antibodies tested inhibited binding of both agonist and antagonist forms equally well. This observation seemed to rule out the possibility that the antibodies were directed against G_i complexes. By preparing receptors in different detergents, it was possible to determine whether the antibodies were directed against the detergent. The antibodies were equally capable of inhibiting receptor binding in native membranes, sodium cholate- or CHAPS-solubilized preparations, which suggested that they were not directed against the detergent. Finally, a lack of inhibition of binding to any other type of receptor would suggest that the monoclonal antibody was not directed against a general, nonspecific membrane-spanning domain. Adenosine A_1, β-adrenergic, muscarinic cholinergic, and D_2-dopaminergic binding activities were assessed. In this latter test, the antibodies inhibited adenosine receptor binding, not β-adrenergic or muscarinic cholinergic binding, but they did inhibit dopamine D_2-receptor binding. This quasi-specificity may turn out to be interesting, when more is learned of the structure of adenosine A_1 receptors compared with dopamine D_2 receptors. However, these antibodies are not immediately useful, unless they are applied in a tissue where other cross-reacting receptors are not present.

4. Regulation of Adenosine Receptor Binding

In preparation for the development of strategies to be used in the study of the coupling of adenosine receptors to signaling systems other than adenylate cyclase, the regulation of adenosine A_1 receptor binding will be presented as a basis from which to compare the differences and similarities between receptors that are linked to different effector systems.

4.1. Role of GTP and Mg^{2+}

Mg^{2+} and GTP exert opposing effects on the regulation of binding to adenosine A1 receptors, as is the case with many other receptors that regulate adenylate cyclase activity. Thus, in the absence of regulators, adenosine receptors in plasma membrane preparations are distributed between high and low affinity states (Yeung and Green, 1983; Ukena et al., 1984; Frame et al., 1986). Mg^{2+} shifts this distribution predominantly towards high affinity forms, whereas GTP shifts the distribution predominantly towards low affinity forms. When Mg^{2+} is combined with GTP, the effect of the cation seems to predominate, i.e., the cation appears to reverse the effect of the guanine nucleotides (Goodman et al., 1982; Ukena et al., 1984; Frame et al., 1986). Such effects do not lend themselves to simple interpretation of their structural or molecular significance. Nevertheless, the state that is induced in the combined presence of Mg^{2+} and GTP is presumably a state that can be generated when adenosine receptors are coupled to the inhibition of adenylate cyclase.

Several studies on the fat cell A_1-adenosine receptor system have aimed to address the significance of this apparent interaction between Mg^{2+} and GTP. The alkylating agent, *N*-ethylmaleimide (NEM), had been introduced by Harden et al. (1982) to compromise the functioning of G_i-mediated processes in cardiac membranes. Conditions were devised that allowed an analogous mild treatment of fat cell membranes with this agent (Yeung et al., 1985a). Following such treatment, it was observed that the ability of Mg^{2+} to reverse the effect of GTP on binding was eliminated, whereas the ability of GTP to regulate binding was retained. The functional correlate to these observations was that the ability of adenosine agonists to inhibit adenylate cyclase was abolished. On the other hand, nonhydrolyzable guanine nucleotide analogs retained their ability to inhibit the enzyme following this treatment, which was taken to suggest that at least one aspect of G_i–C coupling was not impaired. In addition, the fact that GTP continued to regulate receptor binding suggested that at least one measure of R_i–G_i coupling was retained. Thus, the ability of the separate R_i–G_i and G_i–C coupling events to result in the regulation of catalytic activity by receptors had been eliminated. The fact that this "uncoupling" was associated with the elimination of the ability of Mg^{2+} and GTP to interact at the level of binding regulation led to the proposal that this effect of Mg^{2+} on binding was a monitor of an event that was central to the integration of inhibitory receptor occupancy with adenylate cyclase regulation (Cooper et al., 1984, 1985; Yeung et al., 1985a,b). A less mechanistic interpretation of these findings might simply be that nonhydrolyzable guanine nucleotide analogs should not necessarily be expected to mimic GTP in every circumstance.

Studies that were performed with detergent-solubilized preparations of cortical membranes (Perez-Reyes and Cooper, 1986) supported the contention that the interplay between Mg^{2+} and GTP was critical, and reinforced the conclusion on the importance of the role played by Mg^{2+} and the merit in identifying its site and mode of action (Yeung et al., 1985a,b; Cooper et al., 1985) Not all interactions between Mg^{2+} and GTP may be considered of regulatory significance. As discussed elsewhere (Cooper, 1988), Mg^{2+} can promote the hydrolysis of GTP by nonspecific nucleotide phosphohydrolases, leading to the elimination of GTP and its effects, unless steps are taken to protect the nucleotide.

Current views from studies of other receptors that are linked to adenylate cyclase suggest that low affinity receptors may represent free receptors, whereas high affinity forms may represent receptor–G-protein complexes (Smith and Limbird, 1981; Kilpatrick and Caron, 1983). Stated explicitly, it is believed that receptors may exist either free, or coupled to G-proteins; hormones bind preferentially to R–G forms (which exhibit high affinity for agonists). Mg^{2+} promotes the formation of these states, whereas GTP stimulates the liberation of free receptor (which exhibits low affinity for agonists) and G-proteins. These contentions are coupled with the proposition that hormones act to free G-proteins of prebound GDP and allow occupancy of G-proteins by GTP. GTP then occupies the G-protein and stimulates the dissociation of the hormone from its receptor. In so doing, it promotes low affinity binding states (Ross and Gilman, 1980; Rodbell, 1980; Caron et al., 1985). Most of the experimental data available from solubilization/reconstitution studies support this view, which provides a testable description of the system, even though problems are encountered when trying to extrapolate from studies that have been performed in detergent solution to the situation that pertains in unperturbed plasma membranes. For example, although weakened associations between receptors and G-proteins may be detected in detergent solution following the application of the regulator, such as GTP, such observations do not necessarily imply that actual association and dissociation of these components occurs in intact plasma membranes. It is possible that, in native membranes, strong interactions between components could be converted by GTP to weaker interactions, without leading to actual dissociation of the components. Nevertheless, aside from these reservations upon the interpretation of Mg^{2+}/GTP effects on binding, the experimental observations remain consistent features that may be important clues to the structural and regulatory events associated with the regulation of adenylate cyclase activity.

4.2. Effects of Pertussis Toxin Treatment

Since the regulation of A_1 receptor binding by GTP (and therefore the functional response mediated by A_1 receptors, i.e., inhibition of adenylate cyclase; Londos et al., 1980) is mediated through Gi, *Bordetella pertussis* toxin, which eliminates G_i-mediated processes, would be expected to modulate the effect of GTP on agonist binding (Kurose et al., 1983). This prediction has been fulfilled. Generally, the binding isotherms that are observed in the absence of regulators are converted to those that are observed in the presence of GTP following pertussis toxin treatment, i.e., low affinity binding is generated (Kurose et al., 1983).

When extensive binding isotherms were performed on the fat cell A1 receptor, following treatment of intact fat cells with the toxin, the distribution between high and low affinity states that was observed in control membranes was shifted towards predominantly low affinity states. As a result, the effect of GTP in promoting the formation of low affinity states was muted. This finding is the widely encountered effect of the toxin treatment. However, the ability of Mg^{2+} ions to induce high affinity binding states of the receptor was quite unperturbed (D. M. F. Cooper, unpublished observation). Thus, it might be concluded that pertussis toxin affected only the resting distribution of A_1 receptors between high and low affinity states, thereby compromising the magnitude of the GTP effect, without modifying the Mg^{2+} effect. The earlier findings with the fat cell A_1-receptor had led us to believe that the Mg^{2+} effect on binding was a sensitive monitor of the ability of the receptor occupancy signal to be translated into an inhibition of adenylate cyclase activity (Yeung et al., 1985a,b; Cooper et al., 1984, 1985; *see above*). Consequently, the toxin effect on binding is difficult to reconcile with its effect on receptor-mediated inhibition of adenylate cyclase activity. Treatment with the toxin does inhibit the GTPase activity of the G_i-protein, which seems essential for the inhibitory effects of hormones. Consequently, the inhibition of the GTPase activity, which correlates with the diminution of the effect of GTP on binding, suggests that the ability of GTP to transduce the signal of receptor occupancy to the catalytic unit of the cyclase is compromised. However, the lack of effect of the toxin on the regulation by Mg^{2+}, which seemed from the earlier studies to be a vital aspect of inhibitory regulation, stresses the need for caution in attempting to assign priorities to, or to integrate into regulatory cycles, partial reactions of that cycle.

5. Cyclic AMP-Independent Effector Systems Regulated by Adenosine Receptors

The study of neurotransmitters and hormones other than adenosine indicates that these substances are often coupled to more than one effector mechanism. This raises the posssibility that the activity of effector systems other than adenylate cyclase/cyclic AMP may be regulated by adenosine. In this section, preliminary evidence is presented indicating the coupling of adenosine receptors to several "novel" signaling systems.

5.1. Effects of Adenosine Analogs on Plasma Membrane Ion Channels

A growing body of evidence suggests that adenosine receptors are coupled to plasma membrane ion channels. Though this evidence is not always compelling, the available data are presented in this section in terms of the following most common means whereby receptor binding of an agonist may alter the activity of plasma membrane ion channels:

1. The receptor binding site and the ion channel are within the same protein complex, such that binding of the agonist results in changes in channel activity, probably via protein conformational changes, e.g., the nicotinic cholinergic receptor–cation channel (Changeux et al., 1984)

2. Receptor binding leads to the modification of the activity of an intramembrane intermediate that directly regulates ion channel activity, e.g., the atrial, inwardly rectifying, potassium channel that is coupled to adenosine and acetylcholine receptors via a guanine nucleotide-binding protein (*see below*) and

3. Receptor binding results in the production of an intracellular messenger (e.g., cyclic AMP, Ca^{2+}), which regulates ion channel activity, either directly—e.g., the calcium-dependent

potassium channels present in a variety of cells (Petersen and Maruyama, 1984)—or indirectly, e.g., a ventricular calcium channel that is coupled to β-adrenergic receptors via changes in intracellular cyclic AMP-dependent protein kinase activity, which via phosphorylation processes modifies channel acitivity (Kameyama et al., 1985).

Adenosine regulates ion channel activity by these latter two mechanisms in several cell types, with its actions on cardiac cells being probably the best characterized. In the ventricles, adenosine antagonizes the actions of β-adrenergic agonists on the slow calcium inward current via a cyclic AMP-dependent mechanism (Isenberg et al., 1987). Adenosine also increases potassium conductance in pacemaker cells and atrial muscel, but not in ventricular cells (West et al., 1987), apparently via a mechanism involving a GTP-regulatory protein (*see below* for discussion). The effects of adenosine on cardiac calcium and potassium currents account, at least in part, for the negative chronotropic, dromotropic, and inotropic actions of adenosine on the heart.

5.1.1. Effects on Cardiac Potassium Channels

As mentioned above, adenosine, and its derivatives, apparently regulate cardiac potassium ion (K^+) channel currents by a mechanism involving intramembrane intermediates, rather than intracellular messengers. The evidence for this mechanism derives from several studies that are discussed below.

Kurachi et al. (1986a,b) extensively studied the effects of adenosine and acetylcholine (ACh) on guinea pig atrial cell potassium channels. Measurements of potassium currents in the whole-cell and cell-attached membrane patch configurations revealed adenosine- and ACh-induced currents, which displayed the same relaxation times and current–voltage dependencies, indicating that the same or a kinetically similar current is activated by these two agents. Upon excision of the patch, producing an "inside-out" patch, adenosine- and ACh-responsive currents were not detectable, when GTP and Mg^{2+} were absent from the bathing solution to which the intracellular face of the membrane was exposed. Addition of either GTP or Mg^{2+} alone to the bath solution did not result in recovery of adenosine- or ACh-sensitivity. However, the combination of both agents restored

adenosine- and ACh-responsiveness. When the A subunit of preactivated pertussis toxin was added to the internal solution, the adenosine-
and ACh-induced currents gradually disappeared. GTPγS, in the
presence of Mg^{2+}, slowly (within 10–15 min) induced an outward
current with characteristics similar to those of the adenosine- and
ACh-induced current. The activation by GTPγS was irreversible and
persisted following the removal of GTPγS and Mg^{2+} from the bathing
solution. The presence of cyclic AMP in the internal solution did not
significantly alter channel opening, which indicated that the effects of
adenosine and ACh in the whole (intact) cell were not mediated by
changes in intracellular cyclic AMP concentration, and ultimately
PK-A activity. Since the calcium concentration in the internal solution was held constant throughout the procedure, the effects of adenosine and ACh on channel activity were apparently not mediated by
changes in intracellular calcium ion concentrations. The authors concluded that activation of potassium channels in guinea pig atria by
adenosine and ACh is mediated by the α subunit, rather than by the β
and/or γ subunit(s), of a Mg^{2+}-dependent guanine nucleotide-binding
protein. This conclusion is based on the observation that β and/or γ
subunits liberated by β-adrenergic receptor activation of G_s apparently do not regulate channel opening, as β-adrenergic agents have no
effect on this current.

     These results strongly indicate that adenosine and ACh receptors
are coupled to the same population of K^+ channels in the guinea-pig
atria via a pertussis toxin-sensitive G-protein, and that the transduction mechanism linking the receptor and the channel is cyclic AMP-
and Ca^{2+}-independent. The full identification of the molecular mechanism coupling adenosine receptors and atrial, inwardly rectifying,
potassium channels is not presently known. Recent studies, however,
have begun to elucidate the mechanism involved in coupling muscarinic receptors (and, thus, presumably adenosine receptors) to these
channels, but controversy remains as to the exact molecular species
that activates the channel.

     Yatani et al. (1987a,b) studied the effects of the addition of
purified G-proteins on potassium currents in "inside-out" membrane
patches from guinea pig atrial cells. These currents displayed the
same kinetic characteristics as the potassium currents that were induced by ACh in the same patches. In the presence of Mg^{2+}, addition

of preactivated (with guanine nucleotides) G_i (a pertussis toxin-sensitive, guanine nucleotide-binding protein purified from erythrocytes, having an α subunit of 40,000 dalton mol wt) to the bathing solution led to the opening of the potassium channels, whereas addition of unactivated G_i was without effect. Neither preactivated nor unactivated G_s (purified from human erythrocytes) affected the K^+ currents. GTPγS-activated G_o (a pertussis toxin-sensitive guanine nucleotide-binding protein purified from bovine brain; Sternweis and Robishaw, 1984) was less than 5% as effective as was GTPγS-activated G_i in opening the potassium channels. These investigators also reported that tetradecanoyl phorbol ester, a PK-C activator, did not affect the K^+ channel current. In addition, substitution of AMPP(NH)P (adenyl-5'yl imidodiphosphate) for ATP in the bath solution did not affect the activation of the K^+ currents by G_i. These results indicate that the action of the purified G_i is direct, i.e., it is independent of PK-C or PK-A activation and subsequent phosphorylation of a channel-associated protein.

Codina et al. (1987), employing the same procedure as Yatani et al. (1987a,b), investigated the effects of purified G-protein subunits on atrial potassium currents. Addition of GTPγS-complexed α subunits purified from erythrocyte G_i induced by preactivated G_i. In addition, the GTPγS-complexed α subunits were equipotent on a molar basis with preactivated G_i. Additions of the $\beta\gamma$ subunit dimer prepared from erythrocytes, in concentrations approximately $1000 \times$ greater than those required to observe an effect of the α subunit, were ineffective at eliciting channel opening. The authors concluded that the α subunit of G_i, rather than the $\beta\gamma$ subunits, regulates channel opening.

In contrast to the conclusions of Kurachi et al. (1986a,b), Yatani et al. (1987a,b), and Codina et al. (1987), Logothetis et al. (1987a,b) conclude that the $\beta\gamma$ subunits of G_i control the activity of the atrial, cholinergic-sensitive, K^+ channel. These investigators measured potassium channel currents in inside-out membrane patches excised from 14-d embryonic chick atrial cells. Addition of purified $\beta\gamma$ subunits to the medium bathing the intracellular face of the membrane increased the frequency of channel openings, whereas addition of purified "α_{41}" or "α_{39}" to the medium did not. The effect of the $\beta\gamma$ subunits was prevented by preincubation of the subunits with an excess of α subunits, indicating that complexed βγ subunits are incapable of activating the channel. Though these results appear in sharp contrast to those of Birnbaumer and colleagues, the issue may be reconciled if the stimulatory effect of the βγ subunits observed by Logothetis et al. was the result of βγ "chelation" of an inhibitory α subunit. That is, βγ subunits may act indirectly by "disinhibiting" the channel. If such an inhibitory G-protein exists, it could be activated by an endogenous inhibitory hormone or neurotransmitter (e.g., adenosine—*see 6.5. below*), which was not washed out by the bathing medium. Alternatively, the "purified" βγ subunits used by Logothetis et al. may have been contaminated with sufficient quantities of stimulatory α subunits to open the channel.

Thus, there is a compelling body of evidence suggesting that adenosine and ACh receptors in atrial cells directly regulate inwardly rectifying potassium channels by a pertussis toxin-sensitive, Mg^{2+}-dependent, guanine nucleotide-binding protein, which is the same as, or very similar to, the previously characterized "G_i"-protein that regulates adenylate cyclase.

5.1.2. Effects on Calcium Channels

In addition to its actions on potassium channel currents, adenosine also regulates calcium channel currents in some cells. However, the signal transduction mechanisms linking adenosine receptors to many of these channels remain uncharacterized.

In guinea pig atrial and ventricular myocytes, adenosine inhibits an inward calcium current. The observation of this effect in ventricular myocytes is dependent on the prestimulaton of the current by agents that elevate intracellular cyclic AMP levels (e.g., β-adrenergic drugs or forskolin), whereas in atrial myocytes the effect can be observed in the absence of these agents (Isenberg et al., 1987). Thus, the mechanisms by which adenosine regulates the activity of the current in the ventricle is apparently cyclic AMP-dependent, but the mechanism involved in the atrial cells is apparently cyclic AMP-independent. This conclusion is supported by the data of Bohm et al. (1984), who reported that adenosine not only antagonized the positive inotropic action of isoproterenol on the guinea pig left auricle, but also produced a strong negative inotropic effect in unstimulated auricles, while having little or no effect on basal or isoproterenol-stimulated cyclic AMP content of the tissue. However, it should be noted that adenosine did inhibit isoproterenol-stimulated adenylate cyclase activity in atrial particulate fractions.

Dolphin and colleagues (Dolphin and Scott, 1987; Scott and Dolphin, 1987) have studied the effects of 2-chloroadenosine (2-CADO) on whole-cell calcium currents in cultured rat dorsal root ganglia. These investigators found that 2-CADO acting via an A_1 receptor, inhibited both a low-threshold (activated between –40 and –30 mV) and a high-threshold (activated at about 0 mV) calcium current, with the high-threshold current appearing to be more sensitive to the action of adenosine. The same, or kinetically similar, calcium currents were also inhibited by the selective $GABA_B$ agonist, baclofen. Several results indicate that the effects of adenosine and baclofen are not the result of altered (decreased) intracellular cyclic AMP levels, but are the result of a direct action of a G-protein(s), which couples $GABA_B$ and A_1 receptors to the channel(s). Neither the presence of cyclic AMP in the patch pipet, or forskolin in the medium bathing the cells affected the amplitude of the maximal calcium current or its degree of inactivation during the voltage step. In addition, the effect of baclofen (and presumably adenosine) in control cells was not reduced by cyclic AMP or forskolin application. Inclusion of GTPγS in the pipet solution in equilibrium with the intracellular compartment increased the inhibitory response to 2-CADO compared to control (non-GTPγS-treated cells), whereas introduction of GDPβS into the intracellular space decreased the response to 2-CADO. The effect of GTPγS was prevented by treatment of the cells with pertussis toxin. GTPγS alone induced a largely noninactivating calcium current, similar in properties to the previously characterized "L" calcium current (Nowycky et al., 1985). The current(s) observed in the presence of GDPβS appeared to be analogous to the "N" and/or "T" calcium currents described by these workers. These results indicate that 2-CADO (and baclofen) regulated the activity of calcium channels (in particular "L" calcium currents) in these cells by a direct mechanism involving a guanine nucleotide-binding protein. However, they do not exclude the possibility that the effects of both agents are mediated by PK-C or by altered intracellular Ca^{2+} levels.

Adenosine has also been shown to inhibit calcium channels in other cell types. In hippocampal CA3 neurons (Madison et al., 1987), 2-CADO inhibited and "N-like" calcium channel without affecting "T-like" or "L"-like calcium currents. In GH_3 pituitary tumor cells, adenosine A_1 receptors have been demonstrated to inhibit a thyrotropin releasing hormone-stimulated calcium channel by a pertussis toxin-sensitive mechanism (Cooper et al., 1989). However, the exact molecular mechanisms coupling adenosine receptors to such channels is presently unclear.

5.2. Effects on Phosphatidylinositol Metabolism

The growing awareness of the importance of inositol phosphates in Ca^{2+}-mediated signaling has prompted an examination of the effects of adenosine analogs on the levels of these substances in neuronal preparations. In rat striatal minces, adenosine analogs have been found to inhibit the histamine-stimulated accumulation of inositol monophosphate (Petcoff and Cooper, 1987). A similar finding has been encountered in mouse cerebral cortical slices (Hill and Kendall, 1987a). By contrast, in guinea pig cerebral cortical slices, adenosine analogs augment histamine-stimulated inositol monophosphate accumulation (Hill and Kendall, 1987b). It is not believed that these results reflect a direct effect of the adenosine analogs on phospholipase C activity, but rather, that turnover of the phosphatidylinositol pool is inhibited or stimulated by the analogs, as the result of altered intracellular calcium concentrations that are mediated by adenosine effects on ion channels.

5.3. Cyclic AMP-Independent Effects of Adenosine on Adipocytes

The adipocyte has proven a fertile ground for the unraveling of important endocrine regulatory mechanisms. The activity of these cells is regulated by several hormones and neurotransmitters, including insulin, ACTH, adrenergic agonists, and adenosine. Until recently, adenosine was thought to regulate adipocyte functioning solely as the result of its inhibition of adenylate cyclase activity (and consequent decreases in the intracellular cyclic AMP concentration). However, studies by Londos and colleagues during the last few years have demonstrated that the effects of adenosine on the adipocyte appear also to be mediated by cyclic AMP-independent processes. These investigators developed an improved method for the isolation of metabolically active fat cells, in which the cyclic AMP dependence and independence of the effects of various agents can be reproducibly studied. The reader is referred to a recent chapter by Londos (1988) for a thorough discussion of this procedure and the understanding of the regulation of fat cell functioning that has been gained from studies in which it has been employed. In the present chapter, only a brief discussion of the actions of adenosine that have been demonstrated in these studies will be presented.

The effects of the adenosine A_1 agonist, R-PIA, on adipocyte cyclic AMP concentrations were determined by measuring changes in the $-cAMP/+cAMP$ activity ratios of PK-A. In these cells, the anti-lipolytic effects of R-PIA were fully described by changes in the activity of PK-A, i.e., they were fully cyclic AMP-dependent. However, R-PIA exerted several other effects on the fat cell that were apparently cyclic AMP-independent. Most notably, it increased the sensitivity of the cell to insulin, as measured by a decrease in the EC_{50} values for both inhibition of lipolysis and stimulation of plasma membrane glucose transport by insulin. Neither of these effects of R-PIA were related to changes in the activity of PK-A. Thus, the effects of adenosine A_1 agonists on the fat cell are both cyclic AMP-dependent and cyclic AMP-independent. The exact molecular mechanisms involved in the cyclic AMP-independent actions of PIA are unclear, but they appear to be mentioned by G_i-like, GTP-binding proteins.

5.4. Cyclic AMP-Independent Effects of Adenosine on Neurotransmitter Release

Studies by Fredholm et al., (1986), Dunwiddie and Proctor, (1987), and others have appeared recently indicating that neurotransmitter release in a variety of tissues is negatively regulated by adenosine derivatives in a cyclic AMP-independent fashion. Such an effect is quite consistent with current perceptions on the role of "cyclase inhibitory" receptors within the nervous system; i.e., it is generally believed that cyclase inhibitory receptors (e.g., α2-adrenergic) are located presynaptically and act as negative regulators of neurotransmitter release, possibly as the result of the direct inhibition of voltage-activated Ca-channels and/or activation of K^+ currents, both of which would lead to decreased Ca^{2+} influx and, consequently, decreased neurotransmitter release (Scholfield, 1986).

5.5. Other Apparently
Cyclic AMP-Independent Effects of Adenosine

Several other isolated reports have recently indicated that adenosine elicits various effects in an apparently cyclic AMP-independent manner, e.g., adenosine potentiates the release of mediators from rat mast cells (Church and Hughes, 1985; Church et al., 1986; Griffiths and Holgate, 1990). A thorough discussion of each of these reports is beyond the scope of the present chapter. The reader is referred to the comprehensive review by Ribeiro and Sebastiao (1986) for a discussion of several of these effects.

6. Strategies for Implicating and Characterizing the Involvement of a GTP-Regulatory Protein in a Transduction Mechanism

The increasing evidence of adenosine receptor coupling to a variety of cyclic AMP-independent signaling processes, combined with the increasing discovery of novel GTP-regulatory proteins (as judged by the discovery of cDNA-coding sequences that are analogous to, but different from, the already known GTP-regulatory proteins, the purification of novel GTP-binding proteins or pertussis toxin substrates, and the variety of both cholera and pertussis toxin substrates that can be seen on two-dimensional gels), raises the problem of the strategies that are available for identifying the GTP-regulatory proteins that mediate particular signal transduction processes. The tools currently available are crude, but it is conceivable that some of those that have proven useful in discovering the role of GTP-regulatory proteins in the regulation of adenylate cyclase may prove useful in determining the type of regulatory protein involved in other processes.

6.1. NEM Treatment

The amino acid sequence information that is currently emerging on novel, putative GTP-regulatory proteins indicates a sharp degree of conservation of structure. It is possible from these sequences, for instance, to predict whether the particular protein will possess an ADP-ribosylation site for cholera or pertussis toxin, or both (Sullivan et al., 1986). From the sequences that are currently reported, it appears that many putative GTP-regulatory proteins possess cysteine groups that appear to have been the target for alkylating agents, such as NEM, in studies that were performed on the regulation of cardiac and adipocyte membrane adenylate cyclase activity (e.g., Harden et al., 1982; Yeung et al., 1985a; *see* section 4.1. *above* for discussion). The possession of such groups makes it likely that such alkylating agents will continue to prove useful in probing the role of GTP-regulatory proteins in particular processes, even in situations where specific toxins are not available.

6.2. Toxin Treatments

Pertussis toxin ADP-ribosylates several proteins of varying apparent mol wt, including G_i, from a variety of cell types, yet the effector system(s) coupled to many of these proteins is unknown. Preliminary studies, however, indicate that at least one of these substrates is coupled to plasma membrane calcium channels. G_o (Sternweis and Robishaw, 1984) appears to couple D-Ala^2, D-Leu^5-enkephalin (DADLE) receptors to voltage-dependent calcium channels in neuroblastoma × glioma cells (Hescheler et al., 1987). The demonstration of this coupling was dependent on the restoration of DADLE-sensitivity to pertussis toxin-treated cells by the addition of the purified α subunit of G_o to the assay medium. This same strategy may be anticipated to be applied to the study of the regulation of several other pertussis toxin-sensitive processes. That is, the ability to demonstrate the reversal of the effects of pertussis toxin on the regulation of a process by a known agonist, following the addition of a purified pertussis toxin substrate to the assay medium, would be strong evidence for the potential involvement of that G-protein in the transduction of the signal in the intact, undisturbed cell. By a similar strategy, the effector systems regulated by identified cholera toxin substrates could be studied. Through rigorous studies of the effects of each of the identified pertussis toxin and cholera toxin substrates on a variety of toxin-sensitive mechanisms, it will become possible to generate relative activity ratios for the regulation of the processes by these proteins. Such strategies are among the few that are currently available for determining which particular toxin substrate regulates particular processes. It will be tempting to speculate that the most potent substrate is the physiological regulator of the process under study. However, the demonstration of lesser or greater affinities in such a study are not proof of specific associations in the intact cell. More sophisticated techniques will be required to provide less equivocal insights.

6.3. Regulation of Binding by GTP and Mg^{2+}

As discussed above, the regulation of the binding of adenosine analogs by GTP and Mg^{2+} has proven to be a useful tool in the study of the regulation of adenylate cyclase activity by adensoine. Similar strategies can be employed, noting the potential pitfalls discussed above, in the characterization of the coupling of adenosine receptors to effector systems other than adenylate cyclase. Studies of the effects of GTP and Mg^{2+} on the binding characteristics and physiologic effects of several recently developed, structurally diverse, adenosine analogs (*see* section 6.4. *below*) will aid in the definition of receptor subtypes and the elucidation of the subtle structural and functional differences, or similarities, between these subtypes, which may correlate with their abilities to couple to different, or the same, effector systems. Data obtained from NEM, toxin, and binding experiments can then be integrated into a working model of a cell's responses, in which different receptor subtypes may be coupled to various effector systems, while eliciting an integrated cellular response to a single extracellular messenger.

6.4. The Need for More Selective/Discriminating Ligands in Identifying Novel Subtypes of Adenosine Receptors

Recent reports indicating that adenosine receptors control processes other than adenylate cyclase activity raise the possibility that, like the adrenergic receptor system, adenosine receptors will be found to exhibit more subtypes than the two (i.e., A_1 and A_2) that are currently recognized. Indeed, numerous studies (*see* Ribeiro and Sebastiao, 1986; Bruns et al., 1987a) have reported adenosine analog and antagonist potency series for specific physiologic responses that seem incompatible with a simple two-receptor subtype system. However, it should be noted that caveats can be raised regarding pharmacodynamic differences between individual tissues that may influence the measurement of complex physiological events.

A prerequisite to addressing the hypothesis of the existence of multiple adenosine receptor subtypes is the development of more selective adenosine analogs (both agonists and antagonists) and more sensitive means to measure subtle differences in binding regulation and physiologic effects. Recently, Bruns and colleagues (Bruns et al., 1986; Bruns et al., 1987a,b,c) and Olsson and colleagues (Daly et al., 1986; Paton et al., 1987; Ukena et al., 1987) have synthesized a series of adenosine analogs and performed extensive characterization of their binding to plasma membranes. Comparison of the regulatory characteristics and analog preference series of the binding of these agents indicates that not all A_1 and A_2 receptors are the same, and that more subtypes of adenosine receptors may occur than are presently recognized (Brun et al., 1987a; Williams, 1990). These studies represent an important initial step in the development of discriminating agents for the characterization of adenosine receptor subtypes. Subsequent studies are needed on the effects of these agents on the activity of "novel" signaling systems, in order to define more clearly the potentially novel receptor subtypes that may be linked to these effectors.

6.5. Technical Difficulties in Studying Adenosine Receptor-Mediated Processes: The Role of Endogenous Adenosine

It is important to acknowledge that many of the physiological, structural, and regulatory studies that have been performed with adenosine may have been confounded by the presence of the nucleoside, which may either be present in the biological material being assayed or may be generated from nucleotide precursors in the assay. It is to be anticipated that these problems will continue to confound studies on the regulation by adenosine of novel processes. In several studies on adenosine receptor binding and regulation of adenylate cyclase activity, these problems have been tackled by two approaches:

1. The use of adenosine deaminase to metabolize adenosine to inosine, which is ineffective at adenosine receptors, or
2. The use of substrates other than ATP which avoids the generation of adenosine. These strategies are equally applicable to the study of new receptor subtypes linked to "novel" signaling devices, and thus will be discussed in some detail.

The inclusion of adenosine deaminase in adenylate cyclase assays (Londos and Preston, 1977) permitted the effects of nondeaminatable (e.g., N^6-substituted compounds) or deaminase-resistant (e.g., 5' -N-ethylcarboxamide adenosine) adenosine analogs to be observed. This is an adequate solution for most situations, although it should be borne in mind that some drugs may act to inhibit adenosine deaminase activity. This latter possibility, if not considered, could give rise to some dramatic misinterpretations of the mode of action of the compound. An addtional consideration when using adenosine deaminase is that some adenosine analogs may be competitive inhibitors of the enzyme. The effects of such compounds could be the result of both a direct (receptor) action and the accumulation of endogenous adenosine. The use of adenosine deaminase has also been introduced in binding studies, since most adenosine receptor ligands are poor substrates for adenosine deaminase (Schwabe and Trost, 1980). In order to study the effects of adenosine itself, adenosine deaminase cannot be used. In this situation, 2' deoxy ATP can be used as an adenylate cyclase substrate and 2' deoxy cyclic AMP used to protect newly formed [^{32}P-]2' deoxy cyclic AMP from the activity of PDE (Cooper and Londos, 1979). In contrast to the ability to overcome the problem of endogenous adenosine in physiological and regulatory studies, it continues to confound solubilization studies (*see* section 3.1.). A detailed account has been presented (Cooper, 1988) on the implications and difficulties posed by endogenous adenosine in solubilization studies. The reader is referred to that source for a presentation of those details. In the present context, only the regulatory implications of endogenous adenosine will be considered.

It is tempting to speculate that adenosine may be found to regulate previously "unresponsive" systems, if adequate steps are taken to eliminate endogenous adenosine from the experimental system. In many situations, multiple receptors occur on cells that are capable of regulating individual transduction processes. Generally, the combined effects of a range of hormones are nonadditive. Therefore, in the presence of endogenous adenosine, it is possible that the effects of other ligands are obscured. A further implication of the presence of endogenous adenosine is that systems may be permanently (either partially or fully) activated (or inhibited), thus rendering the interpretation of basal activity states difficult. Such chronic occupancy of adenosine receptors could also give rise to desensitization—not only of adenosine receptors, but also other receptors that are present in the same cell by heterologous mechanisms, as recently detailed by Sibley et al. (1987). It is also conceivable that adenosine receptor activation of processes could lead to an ambiguous interpretation of the regulation of the system. For instance, if GTP-regulatory proteins do exert their effects by chelation mechanisms, as well as by the direct interaction of their α subunits with target enzymes, then activation by endogenous adenosine could lead to an association of α subunits with a target enzyme. This would then give rise to the possibility that the addition of an exogenous $\beta\gamma$ subunit would show reversal of the stimulatory or inhibitory effects of the α subunits, i.e., "inhibition" or "activation" ("disinhibition"), respectively. If the possibility that such effects were at work were overlooked, then inappropriate interpretations could be placed on the findings that $\beta\gamma$ subunits directly activated or inhibited the process (e.g., *see* section 5.3. *above*).

7. Future Directions

The future of structural and functional studies on the adenosine receptors awaits their purification. If stable and functional receptors are to be purified, it is likely that they will be antagonist-stabilized forms. Hydrodynamic analysis of antagonist-binding forms of the purified receptor will provide key information on the minimal size of a functional receptor. If function is not the first goal following purification, then upon the unequivocal demonstration that a photolabeled protein is all or part of the receptor, purification of this protein will be a key development. The production of specific antibodies against native and denatured configurations of the receptor will assist in the performance of mechanistic, morphological, and developmental studies. Following the purification of the receptor, the problems that remain to be resolved are those of a more cell biological nature. (For example, how is the receptor configured in the cell to perform its physiological function?) Several receptors that are linked to adenylate cyclase have now been purified to homogeneity. Their cDNA sequences and implied amino acid sequences have been obtained. The GTP regulatory proteins that transduce their cellular signal have also been purified and cloned. The problems that remain to be resolved for these systems are the same as those for the adenosine receptor system. How is the association between the receptor and its regulatory proteins(s) maintained in the intact cell, how are these interdependent proteins expressed developmentally and genetically, and how are they configured in the plasma membrane to be at the point at which they are most effective in terms of receiving extracellular input? Clues available for the resolution of some of these ultrastructural issues may be provided from such functional interactions as those demonstrated in the interaction between Mg^{2+} and GTP in regulating receptor binding. Understanding the structural basis to these nuances of binding regulation may be a key element in understanding the significance of partial reactions in the regulation of adenylate cyclase activity along with the structural implications of these reactions. Some of what may seem currently to be extraneous regulatory subtleties may turn out to be sensitive determinants of the association of receptors with additional cofactors.

Another broad area that will need to be approached in the future is the full range of adenosine responses. How many processes does it actually regulate, how many receptor subtypes and G-proteins are involved, and are the receptor subtypes pluripotential? The recent reports of several cyclic AMP-independent actions of adenosine raise a multitude of questions about the mechanism by which information is integrated by the cell, e.g., how is the regulation of several signaling systems by a single extracellular messenger integrated into a coordinated cellular response? Integration of signaling must occur both within the plasma membrane, prior to the generation of intracellular messenger, and within the intracellular space, subsequent to the generation of these messengers. It is hoped that application of the strategies discussed above will be useful in addressing such issues.

Acknowledgments

The research described in this chapter emanated from D. M. F. Cooper's laboratory and was supported by NIH grants GM 32483 and NS 09199.

References

Ahlijanian, M. K. and Cooper, D. M. F. (1987) Calmodulin may play a pivotal role in neurotransmitter-mediated inhibition and stimulation of rat cerebellar adenylate cyclase. *Mol. Pharm.* **32,** 127–132.

Barrington, W. W., Jacobson, K. A., Hutchinson, A. J., Williams, M., and Stiles, G. L. (1989) Identification of the A_2 adenosine receptor binding subunit by photoaffinity crosslinking. *Proc. Natl. Acad. Sci. USA* **86,** 6572–6576.

Berridge, M. J. and Irvine, R. F. (1984) Inositol trisphosphate, a novel second messenger in cellular signal transduction. *Nature* **312,** 315–321.

Bohm, M., Bruckner, R., Hackbarth, I., Haubitz, B., Linhart, R., Meyer, W., Schmidt, B., Schmitz, W., and Scholz, H. (1984) Adenosine inhibition of catecholamine-induced increase in force of contraction in guinea-pig atrial and ventricular heart preparations. Evidence against a cyclic AMP- and cyclic GMP-dependent effect. *J. Pharmacol. Exp. Ther.* **230,** 483–492.

Bruns, R. F., Lawson-Wendling, K., and Pugsley, T. A. (1983) A rapid filtration assay for soluble receptors using polyethylimine-treated filters. *Anal. Biochem.* **132,** 74–81.

Bruns, R. F., Lu, G. H., and Pugsley, T. A. (1986) Characterization of the A_2 adenosine receptor labeled by [^{3}H]NECA in rat striatal membranes. *Mol. Pharmacol.* **29,** 331–346.

Bruns, R. F., Lu, G. H., and Pugsley, T. A. (1978a) Adenosine receptor subtypes: binding studies, in *Topics and Perspectives in Adenosine Research* (Gerlach, E. and Becker, B. F., eds.) Springer-Verlag, Berlin, pp. 59–73.

Bruns, R., F., Fergus, J. H., Badger, E. W., Bristol, J. A., Santay, L. A., Hartman, J. D., Hays, S. J., and Huang, C. C. (1987b) Binding of the A_1-selective adenosine antagonist 8-cyclopentyl-1,3-dipropylxanthine to rat brain membranes. *Naunyn-Schmiedebergs Arch. Pharmacol.* **335,** 59–63.

Bruns, R. F., Fergus, J. H., Badger, E. W., Bristol, J. A., Santagy, L. A., and Hays, S. J. (1987c) PD 115,199: An antagonist ligand for adenosine A_2 receptors. *Naunyn-Schmiedebergs Arch. Pharmacol.* **335,** 64–69.

Caron, M. G., Cerione, R. A., Benovic, J. L., Strulovici, B., Staniszewski, C., Lefkowitz, R. J., Codina-Salada, J., and Birnbaumer, L. (1985) Biochemical characterization of the adrenergic receptors: Affinity labeling, purification, and reconstitution studies, in *Advances in Cyclic Nucleotide and Protein Phosphorylation Research*, vol. 19 (Cooper, D. M. F. and Seamon, K. B., eds.), Raven, New York, pp. 1–12.

Changeux, J. P., Devillers-Thiery, A., and Chemouilli, P. (1984) Acetylcholine receptor: an allosteric protein. *Science* **225,** 1335–1345.

Church, M. K. and Hughes, P. J. (1985) Adenosine potentiates immunological histamine release from rat mast cells by a novel cyclic AMP-independent cell-surface action. *Br. J. Pharmacol.* **85**, 3–5.

Church, M. K., Hughes, P. H., and Vardey, C. J. (1986) Studies on the receptor mediating cyclic AMP-independent enhancement by adenosine of IgE-dependent mediator release from rat mast cells. *Br. J. Pharmacol.* **87**, 233–242.

Codina, J., Yatani, A., Grenet, D., Brown, A. M., and Birnbaumer, L. (1987) The α subunit of the GTP binding protein G_k opens atrial potassium channels. *Science* **236**, 442–445.

Cooper, D. M. F. (1982) Bimodal regulation of adenylate cyclase. *FEBS Lett.* **138**, 157–163.

Cooper, D. M. F. (1988) Structural studies of adenosine A_1 receptors, in *Receptor Biochemistry and Methodology*, vol. 11 (Cooper, D. M. F. and Londos, C., eds.) Alan Liss, New York, pp. 63–74.

Cooper, D. M. F. and Londos, C. (1979) Evaluation of the effects of adenosine on hepatic and adipocyte adenylate cyclase under conditions where adenosine is not generated endogenously. *J. Cyclic Nucleotide Res.* **5**, 289–302.

Cooper, D. M. F., Ahlijanian, M. K., and Perez-Reyes, E. (1988) Calmodulin plays a dominant role in determining neurotransmitter regulation of neuronal adenylate cyclase. *J. Cell. Biochem.* **36**, 417–427.

Cooper, D. M. F., Schlegel, W., Lin, M. C., and Rodbell, M. (1979) The fat cell adenylate cyclase system. Characterization and manipulation of its bimodal regulation by GTP. *J. Biol. Chem.* **254**, 8927–8931.

Cooper, D. M. F., Yeung, S.-M. H., Perez-Reyes, E., and Fossom, L. H. (1984) A central role for magnesium in the regulation of inhibitory adenosine receptors, in *Neurotransmitter Receptors. Mechanisms of Action and Regulation* (Kito, S., Segawa, T., Kuriyama, K., Yamamura, H. I., and Olsen, R. W., eds.), Plenum, New York, pp. 17–30.

Cooper, D. M. F., Bier-Laning, C. M., Halford, M. K., Ahlijanian, M. K., and Zahniser, N. R. (1986) Dopamine, acting through D_2 receptors, inhibits rat striatal adenylate cyclase by a GTP-dependent process. *Mol. Pharm.* **29**, 113–119.

Cooper, D. M. F., Yeung, S.-M. H., Perez-Reyes, E., Owens, J. R., Fossom, L. H., and Gill, D. L. (1985) Properties required of a functional N_i, the GTP regulatory complex that mediates the inhibitory actions of neurotransmitters on adenylate cyclase, in *Advances in Cyclic Nucleotide and Protein Phosphorylation Research*, vol. 19 (Cooper, D. M. F. and Seamon, K. B., eds.), Raven, New York, pp. 75–86.

Cooper, D. M. F., Caldwell, K. K., Boyajian, C. L., Petcoff, D., and Schlegel, W. (1989) Adenosine A_1 receptors inhibit adenylate cyclase activity and a TRH-activated calcium current by a pertussis-sensitive mechanism in GH_3 cells. *Cellular Signalling* **1**, 85–97.

Daly, J. W., Padgett, W., Thompson, R. D., Kusachi, S., Bugni, W. J., and Olsson, R. A. (1986) Structure–activity relationships for N^6-substituted adenosines at a brain A_1-adenosine receptor with a comparison to an A_2-adenosine receptor regulating coronary blood flow. *Biochem. Pharmacol.* **35**, 2467–2481.

Dolphin, A. C. and Scott, R. H. (1987) Calcium channel currents and their inhibition by (–)-baclofen in rat sensory neurones: modulation by guanine nucleotides. *J. Physiol.* **386,** 1–17.

Dunwiddie, T. V. and Proctor, W. R. (1987) Mechanisms underlying physiological responses to adenosine in the central nervous system, in *Topics and Perspectives in Adenosine Research* (Gerlach, E. and Becker, B. F., eds.), Springer-Verlag, Berlin, pp. 499–507.

Ferkany, J. W., Valentine, H. L., Stone, G. A., and Williams, M. (1986) Species differences in the binding of adenosine A_1 radioligands to mammalian brain tissue: Differences in agonist and antagonist affinity and implications for drug discovery. *Drug Develop. Res.* **9,** 85–93.

Florio, V. A. and Sternweis, P. C. (1985) Reconstitution of resolved muscarinic cholinergic receptors with purified GTP-binding proteins. *J. Biol. Chem.* **260,** 3477–3483.

Fong, H. K. W., Amatruda, T. T. III, Birren, B. W., and Simon, M. I. (1987) Distinct forms of the β subunit of GTP-binding regulatory proteins identified by molecular cloning. *Proc. Natl. Acad. Sci. USA* **84,** 3792–3796.

Frame, L. T., Yeung, S.-M. H., Venter, J. C., and Cooper, D. M. F. (1986) Target size of the adenosine R_i receptor. *Biochem. J.* **235,** 621–624.

Fredholm, B. B., Fastbom, J., and Lindgren, E. (1986) Effects of N-ethylmaleimide and forskolin on glutamate release from rat hippocampal slices. Evidence that prejunctional adenosine receptors are linked to N-proteins, but not to adenylate cyclase. *Acta. Physiol. Scand.* **127,** 381–386.

Gao, B., Gilman, A. G., and Robishaw, J. D. (1987) A second form of the β subunit of signal-transducing G proteins. *Proc. Natl. Acad. Sci. USA* **84,** 6122–6125.

Gavish, M., Goodman, R. R., and Snyder, S. H. (1982) Solubilized adenosine receptors in the brain: Regulation by guanine nucleotides. *Science* **215,** 1633–1635.

Gilman, A. G. (1984) G proteins and dual control of adenylate cyclase. *Cell* **36,** 577–579.

Girardot, J.-M., Kempf, J., and Cooper, D. M. F. (1983) Role of calmodulin in the effect of guanyl nucleotides on rat hippocampal adenylate cyclase: Involvement of adenosine and opiates. *J. Neurochem.* **41,** 848–859.

Goodman, R. R., Cooper, M. J., Gavish, M., and Snyder, S. H. (1982) Guanine nucleotide and cation regulation of the binding of [^{3}H]cyclohexyladenosine and [^{3}H]diethylphenylxanthine to adenosine A1 receptors in brain membranes. *Mol. Pharmacol.* **21,** 329–335.

Griffiths and Holgate, this volume.

Harden, T. K., Scheer, A. G., and Smith, M. M. (1982) Differential modification of the interaction of cardiac muscarinic cholinergic and beta-adrenergic receptors with a guanine nucleotide-binding component(s). *Mol. Pharmacol.* **21,** 570–580.

Hazeki, O. and Ui, M. (1981) Modification by istlet-activating protein of receptor-mediated regulation cyclic AMP accumulation in isolated rat heart cells. *J. Biol. Chem.* **256,** 2856–2862.

Hescheler, J., Rosenthal, W., Trautwein, W., and Schultz, G. (1987) The GTP-binding protein, G_o, regulates neuronal calcium channels. *Nature* **325,** 445–447.

Hill, S. J. and Kendall, D. A. (1987a) Adenosine inhibits histamine-induced inositol phospholipid hydrolysis in mouse cerebal cortex slices. *Br. J. Pharmacol.* **90,** 77.

Hill, S. J. and Kendall, D. A. (1987b) Studies on the adenosine-receptor mediating the augmentation of histamine-induced inositol phospholipid hydrolysis in guinea pig cerebral cortex. *Br. J. Pharmacol.* **91,** 661–669.

Isenberg, G., Cerbai, E., and Klockner, U. (1987) Ionic channels and adenosine in isolated heart cells, in *Topics and Perspectives in Adenosine Research* (Gerlach, E. and Becker, B. F., eds.) Springer-Verlag, Berlin, Heidelberg, pp. 323–334.

Kameyama, M., Hofmann, F., and Trautwein, W. (1985) On the mechanism of β-adrenergic regulation of the Ca channel in the guinea-pig heart. *Pflugers Arch.* **405,** 285–293.

Kempner, E. S. and Schlegel, W. (1979) Size determination of enzymes by radiation inactivation. *Anal. Biochem.* **92,** 2–10.

Kilpatrick, B. F. and Caron, M. G. (1983) Agonist binding promotes a guanine nucleotide reversible increase in the apparent size of the bovine anterior pituitary dopamine receptors. *J. Biol. Chem.* **258,** 13528–13534.

Klotz, K.-N., Lohse, M. J., and Schwabe, U. (1986) Characterization of the solubilized A_1 adenosine receptor from rat brain membranes. *J. Neurochem.* **46,** 1528–1534.

Kurachi, Y., Nakajima, T., and Sugimoto, T. (1986a) On the mechanism of activation of muscarinic K^+ channels by adenosine in isolated atrial cells: involvement of GTP-binding proteins. *Pflugers Arch.* **407,** 264–274.

Kurachi, Y., Nakajima, T., and Sugimoto, T. (1986b) Role of intracellular Mg^{2+} in the activation of muscarinic K^+ channel in cardiac atrial cell membrane. *Pflugers Arch.* **407,** 572–574.

Kurose, H., Katada, T., Asano, T., and Ui, M. (1983) Specific uncoupling by islet-activating protein, pertussis toxin, of negative signal transduction via α-adrenergic, cholinergic, and opiate receptors in neuroblastoma × glioma hybrid cells. *J. Biol. Chem.* **258,** 4870–4875.

Lad, P. M., Nielsen, T. B., Londos, C., Preston, M. S., and Rodbell, M. (1980) Independent mechanisms of adenosine activation and inhibition of the turkey erythrocyte adenylate cyclase system. *J. Biol. Chem.* **255,** 10841–10846.

Lefkowitz, R. J., De Lean, A., Hoffman, B. B., Stadel, J. M., Kent, R., Michel, T., and Limbird, L. (1981) Molecular pharmacology of adenylate cyclase-coupled α- and β-adrenergic receptors, in *Advances in Cyclic Nucleotide Research,* vol. 14 (Dumont, J. E., Greengard, P., and Robison, G. A., eds.), Raven, New York, pp. 145–161.

Linden, J., Earl, C. Q., Patel, A., Craig, R. H., and Daluge, S. M. (1987) Agonist and antagonist radioligands and photoaffinity labels for the adenosine A_1 receptor, in *Topics and Perspectives in Adenosine Research* (Gerlach, E. and Becker, B. F., eds.) Springer-Verlag, Berlin, Heidelberg, pp. 3–13.

Logothetis, D. E., Kurachi, Y., Galper, J., Neer, E. J., and Claphan, D. E. (1987a) The cardiac muscarinic K⁺ channel is regulated by the purified βγ subunits of GTP-binding proteins. *Biophys. J.* **51**, 412a.

Logothetis, D. E., Kurachi, Y., Galper, J., Neer, E. J., and Claphan, D. E. (1987b) The βγ subunits of GTP-binding proteins activate the muscarinic K⁺ channel in heart. *Nature* **325**, 321–326.

Londos, C. (1988) Receptor G_s and G_i complexes in rat adipocyte plasma membranes: Regulators of AMP-related and AMP-independent processes, in *Receptor Biochemistry and Methodology*, vol. 11 (Cooper, D. M. F. and Londos, C., eds.), Alan Liss, New York, pp. 75–86.

Londos, C. and Preston, M. S. (1977) Regulation by glucagon and divalent cations of inhibition of hepatic adenylate cyclase by adenosine. *J. Biol. Chem.* **252**, 5951–5956.

Londos, C. and Wolff, J. (1977) Two distinct adenosine-sensitive sites on adenylate cyclase. *Proc. Natl. Acad. Sci. USA* **74**, 5482–5486.

Londos, C., Cooper, D. M. F., and Wolff, J. (1980) Subclasses of external adenosine receptors. *Proc. Natl. Acad. Sci. USA* **77**, 2551–2554.

Madison, D. V., Fox, A. P., and Tsien, R. W. (1987) Adenosine reduces an inactivating component of calcium current in hippocampal CA3 neurons. *Biophys. J.* **51**, 30a.

Michel, T., Hoffman, B. B., Lefkowitz, R. J., and Caron, M. G. (1981) Different sedimentation properties of agonist- and antagonist-labeled platelet alpha$_2$ adrenergic receptors. *Biochem. Biophys. Res. Comm.* **100**, 1131–1136.

Nakata, H. and Fujisawa, H. (1983) Solubilization and partial characterization of adenosine binding sites from rat brainstem. *FEBS Lett.* **158**, 93–97.

Nowycky, M. C., Fox, A. P., and Tsien, R. W. (1985) Three types of neuronal calcium channel with different calcium agonist sensitivity. *Nature* **316**, 440–443.

Paton, D. M., Lockwood, P. A., Martlew, D. L., and Olsson, R. A. (1987) Potency of N^6-modified N-alkyl adenosine-5'-uronamides at presynaptic adenosine receptors in guinea-pig ileum. *Naunyn-Schmiedebergs Arch. Pharmacol.* **335**, 301–304.

Perez-Reyes, E. and Cooper, D. M. F. (1986) Interaction of the inhibitory GTP regulatory component with soluble cerebral cortical adenylate cyclase. *J. Neurochem.* **46**, 1508–1516.

Perez-Reyes, E., Yeung, S.-M. H., Lasher, R. L., and Cooper, D. M. F. (1987) Structural studies of adenosine receptors from rat cerebral cortex, in *Topics and Perspectives in Adenosine Research* (Gerlach, E. and Becker, B. F., eds.), Springer-Verlag, Berlin, Heidelberg, pp. 37–47.

Petcoff, D. W. and Cooper, D. M. F. (1987) Adenosine receptor agonists inhibit inositol phosphate accumulation in rat striatal slices. *Eur. J. Pharmacol.* **137**, 269–271.

Rasmussen, H., Kojima, I., Kojima, K., Zawalich, W., and Apfeldorf, W. (1984) Calcium as intracellular messenger: Sensitivity modulation, C-kinase pathway, and sustained cellular response, in *Advances in Cyclic Nucleotide and*

Protein Phosphorylation Research, vol. 18 (Greengard, P. and Robison, G. A., eds.), Raven, New York, pp. 159–193.

Reddington, M., Alexander, S. P., Erfurth, A., Lee, K. S., and Kreutzberg, G. W. (1987) Biochemical and autoradiographic approaches to the characterization of adenosine receptors in brain, in *Topics and Perspectives in Adenosine Research* (Gerlach, E. and Becker, B. F., eds.), Springer-Verlag, Berlin, Heidelberg, pp. 49–58.

Ribeiro, J. A. and Sebastiao, A. M. (1986) Adenosine receptors and calcium: basis for proposing a third (A_3) adenosine receptor. *Prog. Neurobiol.* **26,** 179–209.

Rodbell, M. (1980) The role of hormone receptors and GTP-regulatory proteins in membrane transduction. *Nature* **284,** 17–22.

Ross, E. M. and Gilman, A. G. (1980) Biochemical properties of hormone-sensitive adenylate cyclase. *Ann. Rev. Biochem.* **49,** 533–564.

Schlegel, W., Cooper, D. M. F., and Rodbell, M. (1980) Inhibition and activation of fat cell adenylate cyclase by GTP is mediated by structures of different size. *Arch. Biochem. Biophys.* **210,** 678–682.

Scholfield, C. N. (1986) Adenosine and transmitter release. *Trends Pharmcol. Sci.* **17,** 478.

Schwabe, U. and Trost, T. (1980) Characterization of adenosine receptors in rat brain by (–) [^{3}H]N^6-phenylisopropyladenosine. *Naunyn-Schmiedebergs Arch. Pharmacol.* **313,** 179–187.

Scott, R. H. and Dolphin, A. C. (1987) Inhibition of calcium currents by an adenosine analogue 2-chloroadenosine, in *Topics and Perspectives in Adenosine Research* (Gerlach E. and Becker, B. F., eds.), Springer-Verlag, Berlin, Heidelberg, pp. 549–558.

Sibley, D. R., Benovic, J. L., Caron, M. G., and Lefkowitz, R. J. (1987) Regulation of transmembrane signaling by receptor phosphorylation. *Cell* **48,** 913–922.

Smigel, M. D. (1986) Purification of the catalyst of adenylate cyclase. *J. Biol. Chem.* **261,** 1976–1982.

Smith, S. K. and Limbird, L. E. (1981) Solubilization of human platelet α-adrenergic receptors: Evidence that agonist occupancy of the receptor stabilizes receptor-effector interactions. *Proc. Natl. Acad. Sci. USA* **78,** 4026–4030.

Sternweis, P. C. and Robishaw, J. D. (1984) Isolation of two proteins with high affintiy for guanine nucleotides from membranes of bovine brain. *J. Biol. Chem.* **259,** 13806–13816.

Stiles, G. L. (1985) The A_1 adenosine receptor. Solubilization and characterization of a guanine nucleotide-sensitive form of the receptor. *J. Biol. Chem.* **260,** 6728–6732.

Stiles, G. L. (1986) Adenosine receptors: Structure, function, and regulation. *Trends Pharmacol. Sci.* **17,** 486–490.

Stone, G. A., Jarvis, M. F., Sills, M. A., Weeks, B., and Williams, M. (1988) Species differences in high affintiy adenosine A_2 receptor binding in mammalian atrial membranes. *Drug Dev. Res.,* in press.

Sullivan, K. A., Liao, Y.-C., Alborzi, A., Beiderman, B., Chang, F.-H., Masters, S. B., Levinson, A. D., and Bourne, H. R. (1986) Inhibitory and stimulatory

G proteins of adenylate cyclase: cDNA and amino acid sequences of the α chains. *Proc. Natl. Acad. Sci. USA* **83**, 6687–6691.

Ukena, D., Olsson, R. A., and Daly, J. W. (1987) Definition of subclasses of adenosine receptors associated with adenylate cyclase: Interaction of adenosine analogs with inhibitory A_1 receptors and stimulatory A_2 receptors. *Can. J. Physiol. Pharmacol.* **65**, 365–376.

Ukena, D., Poeschla, E., and Schwabe, U. (1984) Guanine nucleotide and cation regulation of radioligand binding to R_i adenosine receptors of rat fat cells. *Naunyn-Schmiedebergs Arch. Pharmacol.* **326**, 241–247.

van Calker, D., Muller, M., and Hamprecht, B. (1979) Adenosine regulates via two different types of receptors, the accumulation of cyclic AMP in cultured brain cells. *J. Neurochem.* **33**, 999–1005.

Venter, J. C. (1983) Muscarinic cholinergic receptor structure. Receptor size, membrane orientation, and absence of a major phylogenetic structural diversity. *J. Biol. Chem.* **258**, 4842–4848.

West, G. A., Giles, W., and Belardinelli, L. (1987) The negative chronotropic effect of adenosine in sinus node cells, in *Topics and Perspectives in Adenosine Research* (Gerlach, E. and Becker, B. F., eds.), Springer-Verlag, Berlin, Heidelberg, pp. 336–342.

Williams, M. (1990) Adenosine receptor: an historical perspective in *Adenosine and Adenosine Receptors* (Williams, M., ed.) Humana, Clifton, New Jersey, pp. 1–15.

Wolff, J., Londos, C., and Cooper, D. M. F. (1981) Adenosine receptors and the regulation of adenylate cyclase, in *Advances in Cyclic Nucleotide Research*, vol. 14 (Dumont, J. E., Greengard, P., and Robison, G. A., eds.), Raven, New York, pp. 199–214.

Yatani, A., Codina, J., Brown, A. M., and Birnbaumer, L. (1987a) Direct activation of a mammalian atrial K channel by a human erythrocyte pertussis toxin (PTX)-sensitive G protein, G_k. *Biophys. J.* **51**, 37a.

Yatani, A., Codina, J., Brown, A. M., and Birnbaumer, L. (1987b) Direct activation of mammalian atrial muscarinic potassium channels by GTP regulatory protein G_k. *Science* **235**, 207–211.

Yeung, S.-M. H. and Green, R. D. (1983) Agonist and antagonist affinities for inhibitory adenosine receptors are reciprocally affected by 5'-guanylylimidodiphosphate or N-ethylmaleimide. *J. Biol. Chem.* **258**, 2334–2339.

Yeung, S.-M. H., Perez-Reyes, E., and Cooper, D. M. F. (1987) Hydrodynamic properties of adenosine R_i receptors solubilized from rat cerebral cortical membranes. *Biochemical J.* **248**, 635–642.

Yeung, S.-M. H., Fossom, L. H., Gill, D. L., and Cooper, D. M. F. (1985a) Magnesium ion exerts a central role in the regulation of inhibitory adenosine receptors. *Biochem. J.* **229**, 91–100.

Yeung, S.-M. H., Frame, L. T., Venter, J. C., and Cooper, D. M. F. (1985b) Magnesium ions exert a central role in integrating adenosine receptor occupancy with the inhibition of adenylate cyclase, in *Purines, Pharmacology, and Physiological Roles* (Stone, T. W., ed.), VCH, Weinheim, London, pp. 203–214.

Chapter 5

Electrophysiological Aspects of Adenosine Receptor Function

Thomas V. Dunwiddie

1. Introduction

Within the past decade, the importance of adenosine as a regulator of cellular activity in a variety of physiological systems has become increasingly apparent. Although some of adenosine's actions were recognized over 50 years ago (Drury and Szent-Györgyi, 1929), it has only been relatively recently that the nature of these actions, the receptors that are involved, and the mechanisms by which receptor occupation is translated into physiological responses have been explicitly described. Although adenosine receptors have now been found in many different tissues, much of the early interest in this area focused around the central nervous system (CNS), in part because of the high concentrations of adenosine receptors in the brain, and because of the profound physiological actions of purines on the activity of the CNS. Historically, the observation by Sattin and Rall (1970) that adenosine stimulated the formation of cyclic adenosine 3', 5'-monophosphate in brain slices, and that methylxanthines, such as theophylline and caffeine, were competitive pharmacological antago-

Adenosine and Adenosine Receptors Editor: Michael Williams ©1990 The Humana Press Inc.

nists of this response, was instrumental in stimulating much of the recent interest in adenosine.

At the descriptive level, adenosine and adenosine analogs have a variety of effects upon neuronal activity, both in the central as well as the peripheral nervous system. Adenosine inhibits the spontaneous firing of neurons in many different brain regions, can modulate the efficacy of synaptic transmission by altering transmitter release, and in both neuronal and nonneuronal cells, has been shown to regulate the activity of ion channels that are selective for potassium and calcium ions. In the majority of cases, these responses appear to be mediated via extracellularly directed, highly specific, high affinity receptors for adenosine that have been identified on the basis of ligand binding studies (Williams and Jacobson, 1990). What remains to be determined in many of these cases is which adenosine receptors are involved in these responses, and to characterize the role played by potential second messengers, such as cyclic AMP.

The physiological role of adenosine as a regulator of cellular activity in the nervous system has been the subject of several reviews (Fredholm and Hedqvist, 1980; Phillis and Wu, 1981; Stone, 1981a; Daly 1982; Fredholm, 1982; Su, 1983; Dunwiddie, 1985; Williams, 1987). Some of the more general aspects of "purinergic" function have been the subject of several recent volumes (Berne et al., 1983; Stone, 1985; Stefanovich et al., 1985; Gerlach and Becker, 1987). This review will make no attempt to summarize this extensive literature, but rather, will provide a selective discussion of recent developments concerning the electrophysiological aspects of adenosine action, particularly with reference to the ionic mechanisms that underlie such responses. In this context, we would like to direct attention to both the similarities and differences that are found in the effects of adenosine on different physiological systems.

2. Adenosine Effects upon Cellular Physiology

2.1. General Aspects

Many of the early studies concerning the electrophysiological actions of adenosine on the nervous system used iontophoretic techniques to study the effects of local application of adenosine to neurons in many different brain regions. Phillis and colleagues (Phillis et al., 1974; Phillis and Kostopoulos, 1975; Kostopoulos and Phillis, 1977) demonstrated that spontaneous neuronal firing in cerebral cortex, hippocampus, thalamus, cerebellum, and superior colliculus was inhibited by the local iontophoretic application of adenosine, adenine nucleotides, such as 5'-AMP, ADP and ATP, and related purine compounds, and that such inhibitions were antagonized by concurrent application of the methylxanthine adenosine receptor antagonists caffeine and theophylline (Phillis et al., 1979; Perkins and Stone, 1980). However, it is generally not possible to determine from these types of studies the mechanisms that are responsible for the inhibition of cellular activity; adenosine could either inhibit firing via a direct action upon the neurons tested, most likely by activating a conductance that would hyperpolarize these cells, or could reduce the tonic excitatory drive that underlies "spontaneous" firing in some neurons. More recent experiments have indicated that both of these mechanisms may contribute; in in vitro experiments, where spontaneous firing can occur in the absence of synaptic transmission, adenosine can still inhibit spontaneous firing (Shefner and Chiu, 1986). However, in addition, adenosine can clearly interfere with excitatory synaptic transmission in a variety of brain regions, and the loss of this input may lead to a reduction in firing rate as well. The mechanisms that underlie the inhibitory effects of adenosine on neuronal firing have been discussed in detail in a previous review (Phillis and Wu, 1981).

A second and perhaps related action of adenosine that has attracted considerable attention is the purinergic depression of synaptic transmission. Although this phenomenon was originally noted in the peripheral nervous system, a similar action has now been described in the CNS in such regions as the olfactory cortex (Okada and Kuroda, 1975, 1980; Scholfield, 1978) cerebellum (Kocsis et al., 1984) and hippocampus (Schubert and Mitzdorf, 1979; Dunwiddie and Hoffer, 1980; Okada and Ozawa, 1980). Where the pharmacology has been characterized to a sufficient degree, the depression of synaptic transmission appears to be mediated via an adenosine receptor of the A_1 subtype (McCabe and Scholfield, 1985; Dunwiddie and Fredholm, 1984; Dunwiddie et al., 1986; Dunwiddie and Fredholm,

1989). Although depressant effects of adenosine occur in many synaptic systems, there is a considerable range in the maximal extent of the depression. In the Schaffer/commissural pathway in rat hippocampus, the maximal effect is a 95–100% depression of the postsynaptic response (Dunwiddie and Hoffer, 1980; Lee et al., 1983; Dunwiddie, 1984), in the olfactory cortex the maximal effect is a 70–80% depression (Scholfield, 1978; Okada and Kuroda, 1980), whereas the EPSP elicited in the superior colliculus by stimulation of the optic tract is unaffected by even high concentrations of adenosine (Okada and Saito, 1979). The mechanism by which adenosine exerts these depressant effects remains unclear. Because the pharmacology is consistent with receptors of the A_1 subtype, this would suggest that inhibition of adenylate cyclase might play a role, but direct tests of this hypothesis have been largely negative. The mechanism underlying the inhibition of transmission is discussed at length in section 2.4. below.

The third major action of adenosine on neural activity is an inhibition of convulsive activity. Dunwiddie (1980) originally described the adenosine-mediated inhibition of the rate of penicillin-induced interictal spiking in the rat hippocampus in vitro, and other groups have described similar effects (Lee et al., 1984; Haas et al., 1984; Schubert and Lee, 1986). Although this initially might appear to be yet another manifestation of an inhibition of synaptic transmission by adenosine, this is not the case. Either spontaneous or evoked repetitive bursting can be maintained in hippocampal slices in the absence of synaptically released transmitter substances, and even under these conditions, adenosine has marked anticonvulsant effects (Haas et al., 1984; Haas and Greene, 1988). It would appear that these effects reflect direct actions upon ionic conductances in the hippocampal pyramidal neurons. It is not clear whether the adenosine-mediated inhibition of burst rate reflects a direct action on the mechanism underlying the burst discharges (which may involve voltage-activated calcium currents) or the activation of an adenosine-sensitive potassium channel (section 2.2.), which would tend to limit burst activity.

These three major actions of adenosine—the inhibition of spontaneous neuronal activity, inhibition of synaptic transmission, and inhibition of convulsive electrical activity—are likely to reflect actions that are mediated via multiple cellular mechanisms. Two of the most clearly understood ionic mechanisms of adenosine action, which involve a reduction in a calcium conductance and an increase in potassium conductance, certainly could provide mechanistic explanations for these three phenomena. However, the specific ways in which these cellular mechanisms operate to produce these responses and the role played by other more complex cellular processes, such as cyclic AMP formation, the subsequent protein phosphorylation, and so on, remain to be determined.

2.2. Effects of Adenosine upon Potassium Channels

2.2.1. Direct Effects

Many of the early investigations of the actions of adenosine reported that it could hyperpolarize neurons, but the mechanism responsible was unclear, the effects were sometimes transient, and there was often a high degree of variability between preparations. One of the first observations in this regard was the study of Tomita and Watanabe (1973) of the guinea pig taenia coli, in which they observed a hyperpolarizing effect of purines that was insensitive to changes in chloride concentrations and was accompanied by a decrease in the membrane resistance, suggesting mediation via a potassium conductance. More recent in vitro studies of the CNS have reported that adenosine has no effect on the membrane potential or conductance (Scholfield, 1978), can elicit moderate and/or variable hyperpolarizations with similarly variable changes in membrane resistance (Edstrom and Phillis, 1976; Proctor and Dunwiddie, 1983; Siggins and Schubert, 1981; Greene and Haas, 1985), or can consistently hyperpolarize neurons with an associated decrease in the input resistance (Segal, 1982). These actions of adenosine are direct postsynaptic actions, because treatments that block neurotransmitter release have no effect on such responses (Siggins and Schubert, 1981; Segal, 1982; Proctor and Dunwiddie, 1983). This original hypothesis of a potassium channel regulated by adenosine receptor agonists is also supported by biochemical experiments. For example, 2-chloroadenosine (2-CADO) can be shown to increase the internal negative membrane potential of synaptic plasma membrane vesicles (as monitored by the accumulation of a radioactive organic cation), and this hyperpolarization is sensitive to K^+, but not Cl^- gradients (Michaelis and Michaelis, 1981).

Patch-clamp studies on cultured neurons have provided direct confirmation that adenosine can activate potassium channels in postsynaptic neurons and have also provided insights into the reasons for some of the variability in earlier studies. In studies on cultured striatal neurons, Trussell and Jackson (1985) have described the activation of a conductance by relatively low concentrations of adenosine that hyperpolarizes neurons concurrently with increases in membrane conductance; furthermore, in solutions with altered potassium concentrations, the reversal potential for this current was changed in a fashion consistent with mediation via a potassium conductance. This conductance was also observed to be highly voltage dependent, so that, at membrane potentials depolarized with respect to the membrane potential, the current was attenuated. Some of the variability attributed to this response in earlier studies might well reflect this underlying voltage dependence. At the normal resting potential, the hyperpolarizing responses to adenosine are usually rather small; if the cell is depolarized, either because of a poor cell impalement, or by current injection in an effort to increase the hyperpolarizing response, the conductance change would actually be reduced because of the voltage dependency of this response.

In another communication from this same group (Trussell and Jackson, 1987), several other very interesting aspects of this response were reported. The earlier study had noted that, when recorded in the whole cell patch mode, the response to adenosine rapidly decreased with time, particularly if an electrode with a large diameter tip was used. This suggested that some internal cell constituent was required in order for adenosine to activate the ion channel, and that with time this factor diffused out into the pipet and was lost; the more recent study identified this factor as GTP, which when included in the patch pipet prevented the loss of adenosine responsivity. Furthermore, since pretreatment with pertussis toxin blocked the subsequent adenosine response, this provided evidence for an interaction of the adenosine receptor with a guanine nucleotide regulatory protein (either G_i or G_o) as an obligatory step in the activation of the ion channel. Since neither forskolin, dibutyryl cyclic AMP, nor Walsh protein kinase inhibitor affected the response, it would appear that a guanine nucleotide regulatory protein couples the adenosine receptor to the ion channel in such a fashion that cyclic AMP does not play an intermediary role.

Andrade et al. (1986) reported that a potassium conductance start with similar properties is activated in hippocampal pyramidal neu-rons by serotonin, $GABA_B$ agonists, and by adenosine. As in the preceding study, these receptors were coupled to the ion channel via a G-type regulatory protein, the response was not sensitive to cyclic nucleotides, could be blocked by pertussis toxin treatment, and in addition, could be inhibited by treatment with phorbol esters. Although the mechanism that underlies the activation of this hippocampal potassium conductance has been well characterized, neither this study nor the studies by Trussell and Jackson has provided sufficient evidence to determine whether these responses are mediated via any of the known adenosine receptor subtypes. Clearly one of the most important issues that remains unresolved is whether A_1 and A_2 adenosine receptors are coupled to cellular effector mechanisms *only* through the G_i or G_s, respectively, act exclusively via cyclic nucleotide-related mechanisms, and a third, undefined receptor activates a potassium conductance through G_o, or alternatively, either the A_1 or A_2 receptor might be able to act both through G_s/G_i, and through G_o. If this latter situation is the case, one might envision adenosine regulating activity both in the short term, via changes in membrane potential, and in the longer term via changes in the phosphorylation state of proteins that are substrates for cyclic AMP-dependent protein kinase. These issues are discussed in greater detail in a recent review (Fredholm and Dunwiddie, 1988).

Although the preceding studies provide the most unequivocal evidence for a role of potassium conductance changes in physiological responses to adenosine, other studies using more indirect techniques have suggested similar conclusions. For example, 4-aminopyridine (a well-characterized potassium channel blocker) is able to reverse the adenosine-mediated inhibition of synaptic transmission, low calcium-induced bursting, and reduction in spontaneous neuronal activity (Perkins and Stone, 1980; Okada and Ozawa, 1980; Stone, 1981b; Schubert and Lee, 1986; Dunwiddie and Proctor, 1987; Scholfield and Steel, 1988).

Another study with direct relevance to the issue of potassium channels is that of Akasu et al. (1984), which characterized a synaptically evoked response in cat vesical parasympathetic ganglion that appeared to be mediated via adenosine receptors. This response was a slow hyperpolarization of the postsynaptic cells that could be antagonized either by competitive adenosine receptor antagonists, such as caffeine, or by adenosine deaminase, which would catabolize endogenously released adenosine to the relatively inactive metabolite inosine. The hyperpolarization was accompanied by a reduction in the input resistance, and was sensitive to the extracellular concentration of potassium, strongly suggesting that this response results from an increase in potassium conductance.

Although the primary focus of this review is upon neuronal adenosine receptors and their associated mechanisms of action, a complementary picture has developed from the study of adenosine actions in the heart. In the atrium in particular, adenosine can activate a potassium conductance that shares many of the characteristics of that observed in brain (*see also* Mullane and Williams, this vol.). Hartzell (1979) initially reported that adenosine activates a potassium conductance in the frog sinus venosus, and that this response could be antagonized by methylxanthine adenosine receptor antagonists. Other investigators have demonstrated similar effects in tortoise, rabbit, and guinea pig atrial tissue (*see* Table 1). In the heart, both adenosine and acetylcholine (the latter acting via muscarinic receptors) activate potassium channels; moreover, the conductances that are activated are nonadditive, suggesting that both adenosine and muscarinic cholinergic receptors act independently upon the same population of ion channels (Kurachi et al., 1986; West and Belardinelli, 1985). Other aspects of these potassium conductances also closely resemble what has been described in the brain; the voltage dependence of the responses is quite similar (Kurachi et al., 1986; Trussell and Jackson, 1985), both depend upon intracellular GTP (Kurachi et al., 1986; Trussell and Jackson, 1987), and can be blocked by pretreatment with pertussis toxin

Table 1
Comparison of Atrial and Neuronal Potassium Channels Activated by Adenosine

	Atrial potassium channel	Neuronal potassium channel
Basis for identification of potassium conductance	Voltage clamp[4] Flux measurements[4,5] Dependence upon extracellular potassium[6]	Voltage clamp[6] Dependence upon extracellular potassium[2]
Requirements Receptor antagonists	Intracellular GTP[2] Aminophylline[1] Theophylline[2]	Intracellular GTP[7] Theophylline[6]
Other antagonists	Pertussis toxin[2,3] TTX not effective[1]	Pertussis toxin[7,8] TTX does not block[6] Cobalt does not block[6]
Voltage dependence	Inactivates around -50 mV[2]	Begins to inactivate at -50 mV[6]
Potassium channel antagonists Effective		4-aminopyridine[10,11,12,13,14] 3,4 diaminopyridine, cesium[13] 4-aminopyridine[9,15] TEA,[10,13,15] barium,[13] rubidium,[13] cadmium[15]
Ineffective	Cesium (5 mM)[1]	Serotonin, GABA[8]
Common agonists	Muscarinic cholinergic[1,2]	

[1]West and Belardinelli, 1985 (rabbit atrium)
[2]Kurachi et al., 1986 (guinea pig atrium)
[3]Bohm et al., 1986 (guinea pig atrium)
[4]Jochem and Nawrath, 1983 (guinea pig atrium)
[5]Hutter and Rankin, 1984 (tortoise heart)
[6]Trussell and Jackson, 1985 (cultured striatal neurons)
[7]Trussell and Jackson, 1986 (cultured striatal neurons)
[8]Andrade et al., 1986 (hippocampus)
[9]Segal, 1982 (hippocampus)
[10]Schubert and Lee (convulsive bursting, hippocampus)
[11]Okada and Ozawa, 1980 (synaptic transmission, olfactory cortex)
[12]Dunwiddie and Proctor, 1987 (synaptic transmission, hippocampus)
[13]Scholfield and Steel, 1987 (synaptic transmission, olfactory cortex)
[14]Perkins and Stone, 1980 (spontaneous firing, cerebral cortex)
[15]Greene and Haas, 1985 (hippocampus)

(Kurachi et al., 1986; Bohm et al., 1986; Trussell and Jackson, 1987). Many of the known properties of these conductances are summarized in Table 1.

2.2.2. Effects upon Calcium-Dependent Potassium Conductances

The preceding studies have outlined the properties of quite similar adenosine- and voltage-sensitive potassium channels that are found in both brain and atrial tissue. In addition, however, there are several reports of actions of adenosine on potassium channels that can clearly be dissociated from the preceding responses. Haas and Greene (1984) have demonstrated that, in addition to hyperpolarizing hippocampal neurons, adenosine has a biphasic effect upon the calcium-dependent potassium conductance that is responsible for the long-duration afterhyperpolarization (AHP) that is observed in these cells, facilitating the AHP at low doses and, at high doses, reducing the AHP. At least part of the potentiation of the AHP was the result of reduction in the rate of decay, and it was suggested that the intracellular metabolism of calcium might be affected in some way by adenosine (Greene and Haas, 1985). In our own studies, using a range of adenosine concentrations, we have not observed significant potentiation of the AHP, although the reduction at higher adenosine concentrations can be readily observed (Dunwiddie and Proctor, 1987). Other studies in rat superior cervical ganglion (Henon and McAfee, 1983a), rat dorsal root ganglion (Dolphin et al., 1986), and mouse dorsal root ganglion cells in culture (Macdonald et al., 1986) have all reported that adenosine and a range of adenosine analogs either reduced or had no effect on calcium-dependent potassium conductances. A possible explanation for this apparent discrepancy is that the facilitation of the AHP by adenosine may not be direct, but may involve mediation by a secondary process that is in some way dependent upon the basal state of the cell.

One possible explanation is suggested by the well-characterized antagonistic relationship between adenosine and β-adrenergic receptor agonists in the heart. In the ventricle, for example, β-adrenergic receptors mediate an increase in calcium conductance that is clearly linked to the activation of adenylate cyclase and increases in cyclic AMP. Also, this response is attenuated by adenosine, even though adenosine by itself has little or no direct effect on ventricular physiology (*see* Isenberg and Belardinelli, 1984). This antagonistic relationship may reflect the ability of adenosine to inhibit adenylate cyclase activity via A_1 receptors and, hence, to antagonize the activation of adenylate cyclase by such agents as isoproterenol. In the hippocampus, β-adrenergic receptor activation results in a reduction in the calcium-dependent potassium conductance, and this response is clearly linked to increases in cyclic AMP (Madison and Nicoll, 1986). Thus, it is possible that adenosine, again acting via A_1 receptors, might reverse the effects of endogenous agents that increase cyclic AMP, such as histamine or norepinephrine, and hence produce an apparent facilitation of the AHP. The lack of a facilitory effect of adenosine on the AHP that we and others have observed might simply reflect a lower basal level of adenylate cyclase activity in our slices. Testing this hypothesis will require experiments in which responses to adenosine are examined under a variety of conditions where the background level of adenylate cyclase activation is systematically varied by isoproterenol or other agents that affect cyclic AMP levels.

2.2.3. Adenosine Receptors
Coupled to Potassium Channels in Xenopus Oocytes

Most of the effects of adenosine on potassium channels that have been described thus far are associated with an adenosine receptor that is pharmacologically most similar to the A_1 receptor. Based upon the effects of pertussis toxin, it would appear these are coupled to potassium channels via either the G_i or G_o regulatory protein. However, a very different situation occurs in the *Xenopus* oocyte. In this giant cell, adenosine activates a potassium conductance (Lotan et al., 1982), but there is compelling evidence that this is directly mediated via increases in intracellular cyclic AMP. The hyperpolarizing response to adenosine can be mimicked by intracellularly injected cyclic AMP, and is blocked by protein kinase inhibitors (Lotan et al., 1985). Theophylline has a dual effect; when applied extracellularly, it antagonizes the response to adenosine via a competitive interaction with the receptor, whereas when applied intracellularly, it facilitates the hyperpolarizing response, most likely because of an inhibition of phosphodiesterase. Forskolin in low concentrations facilitates the adenosine response, and mimics it in higher concentrations (Stinnakre and Van Renterghem, 1986). Finally, when adenosine is applied extracellularly, it increases the intracellular cyclic AMP concentration, and the magnitude of this increase is sufficient to account for the permeability changes that are observed (Lotan et al., 1985).

Another unique aspect of this response is that, unlike the situation in the heart, where adenosine antagonizes β-adrenergic responses, adenosine and the β-adrenergic agonist isoproterenol have a synergistic effect in the oocyte, as might be expected of two agents that increase cyclic AMP (Stinnakre and Van Renterghem, 1986). On the other hand, acetylcholine, which mimics the effect of adenosine in the heart, induces a long-lasting antagonism of the adenosine response of the oocyte; this effect is mediated via muscarinic receptors, is mimicked by phorbol esters, and appears to the result of the activation of protein kinase C by acetylcholine (Stinnakre and Van Renterghem, 1986; Dascal et al., 1985).

The adenosine-sensitive potassium channel found in oocytes appears to be quite unlike any that has been reported either in the brain or in the heart. Clearly, it will be important to determine whether A_2 receptors that are found in brain and elsewhere mediate similar responses, or whether the coupling of cyclic AMP to the activation of potassium conductances is unique to the oocyte. What is clear is that, in at least some regions of the CNS, increases in neuronal cyclic AMP are not coupled to increased cellular potassium permeability (e.g., Madison and Nicoll, 1986). However, until the cellular localization of A_2 receptors in the brain is better established, it is clearly premature to rule out the possibility of a cyclic AMP-activated potassium channel.

2.3. Effects of Adenosine upon Calcium Channels

Another mechanism of adenosine action that has attracted considerable interest is the possibility of an adenosine-mediated reduction in calcium currents. Such a hypothesis would provide a mechanistic explanation for the purinergic inhibition of transmitter release observed in many systems (*see* section 2.4.), but one that would be difficult to test electrophysiologically, because there is no direct way of measuring calcium conductances in the presynaptic terminals. One way in which presynaptic calcium currents can be examined is by measuring calcium influx into synaptosomal preparations, and some studies of this type have provided direct evidence for a reduction in calcium flux by adenosine (Ribeiro et al., 1979; Wu et al., 1982). In these experiments, the effect of adenosine on the uptake of labeled ^{45}Ca into synaptosomes during potassium depolarization was examined. Unfortunately, the data in support of this hypothesis are not entirely consistent. Ribeiro et al. (1979) found that adenosine actually facilitated ^{45}Ca influx at short intervals (15 s), and inhibited at longer intervals and higher concentrations. The primary response to adenosine in these studies was an inhibition of the steady-state ^{45}Ca concentration (measured after 1–2 min of depolarization), whereas the initial phase of calcium influx, which is most likely to reflect the activity of voltage-sensitive calcium channels responsible for transmitter release, was not studied. Thus, it is difficult to relate this study directly to functional effects on neurotransmission. Wu et al. (1982) did characterize a more rapid phase of calcium influx and reported that adenosine receptor agonists inhibited ^{45}Ca flux in very low concentrations. Subsequently, several groups have been unsuccessful in inhibiting the initial rapid phase of ^{45}Ca flux with adenosine (Barr et al., 1985; Michaelis et al., 1988; R. A. Harris, unpublished).

A less direct way of testing for possible effects of adenosine on calcium fluxes is to determine the calcium sensitivity of electrophysiological responses to adenosine. If adenosine inhibits calcium influx, then high concentrations of calcium might be expected to antagonize the depressant effect of adenosine. Kuroda et al. (1976) demonstrated that the depression of a field EPSP response in the olfactory cortex slice by adenosine is reversed in a dose-dependent fashion by increasing the extracellular concentration of calcium, and suggested that adenosine inhibits calcium flux into the nerve terminal. Similar effects have been reported by Dunwiddie (1984) in the rat hippocampus in vitro. However, in the latter study, the magnitude of the depressant response to adenosine was shown to be dependent upon several factors. In particular, the amplitude of the test EPSP prior to adenosine treatment was clearly an important variable; adenosine was found to reduce the amplitude of small test responses to a much greater degree than large responses. Because increasing the extracellular calcium concentration also increases the magnitude of the baseline test response, it is not possible to determine whether the reduced response to adenosine in high calcium medium results from a direct interaction between adenosine and calcium, or simply the effects of an increase in the baseline response. Taking into consideration the effects of calcium on the magnitude of the synaptic response, it is possible to account for the apparent calcium sensitivity of the response without invoking a specific interaction between adenosine and calcium. However, this clearly does not rule out such a possibility.

These observations are consistent with those of Silinsky (1984) at the frog neuromuscular junction. At this synapse, the response to adenosine (using quantal measurements of transmitter release) was not dependent upon the calcium/magnesium content of the medium. Furthermore, adenosine could inhibit transmitter release that was independent of the voltage-dependent influx of calcium into the nerve terminal, suggesting some type of action upon the release process *per se* (*see* section 2.4.). Again, these data do not rule out an effect of adenosine on calcium currents, but suggest that other mechanisms are operative as well.

Although it is not possible to record calcium currents from presynaptic nerve terminals, voltage-dependent calcium conductances are found in many types of neurons and, in some cases, are sensitive to adenosine. Henon and McAfee (1983a,b) have reported that calcium spikes are inhibited in rat superior cervical ganglion and similar effects have been observed in pyramidal neurons in the CA1 region of the rat hippocampus in vitro, where adenosine inhibits regenerative calcium-dependent action potentials (calcium spikes; Proctor and Dunwiddie, 1983). In both types of cells, inhibition of the voltage-dependent sodium conductance with tetrodotoxin uncovers a calcium spike that can be evoked by depolarizing current pulses, and these calcium spikes are inhibited by adenosine. Under these conditions, endogenous release of neurotransmitters is blocked, so the effect of adenosine is most likely to be a direct postsynaptic action. However, an important unresolved issue is whether these effects reflect a direct action upon the calcium channel or whether adenosine activates a potassium conductance that indirectly reduces the calcium flux (*see*, e.g., North and Williams, 1983). Henon and McAfee (1983b) suggest that the effect of adenosine on the calcium spike in the sympathetic ganglion is likely to be direct, since voltage-clamp measurements in the presence of the potassium channel blockers Cs^+ and tetraethylammonium demonstrated an adenosine-mediated inhibition of inward current.

Although these reports would suggest that adenosine may reduce calcium currents, Halliwell and Scholfield (1984) reported that adenosine did not appear to affect calcium currents in voltage-clamp experiments on hippocampal and olfactory cortex neurons, and concluded that the previously reported effects on calcium spikes (Proctor and Dunwiddie, 1983) were likely to be indirect. What complicates the situation is the fact that most neurons appear to have not one, but three different calcium conductances (Nowycky et al., 1985): a long-lasting conductance that requires strong depolarization to be activated (L channel), a transient current activated by relatively weak depolarizations from highly negative holding potentials (T channel), and another channel that shows an intermediate rate of inactivation and is activated by strong depolarization from highly negative holding potentials (N channel). Although the two currents that were reported by Halliwell and Scholfield to be unaffected by adenosine were not identified, they appear to correspond most closely to the L and T channels. The voltage-clamp protocols that they used would be unlikely to activate the N channel. However, a recent report by Madison et al. (1987) suggests that it is only the N channel that is affected by adenosine in hippocampal neurons, and this was not tested in the former study.

Studies in other brain regions and tissues have provided additional supportive evidence for an action of adenosine upon calcium conductances. Adenosine shortened the duration of calcium spikes in dorsal root ganglion cells in culture and reduced an inward calcium current studied under voltage-clamp conditions (Macdonald et al., 1986). The voltage dependency of this current was consistent with activation of an N channel. Two aspects of this study were somewhat unusual; first, adenosine and adenosine analogs did not appear to have any effect upon the passive membrane properties of these neurons. Thus, unlike most neurons that have been studied, adenosine receptors do not seem to be coupled to potassium channels in dorsal root ganglion cells, which makes it even more unlikely that the effects on the calcium currents are the indirect result of effects upon potassium conductances. The other unique aspect of this study was the pharmacology of the receptors that mediated this response. R-PIA was quite active (effective in the low nanomolar range), which would normally indicate mediation via an A_1 receptor subtype. However, S-PIA was also very potent (more so than cyclohexyladenosine), which is not characteristic of A_1 receptors. Furthermore, 5'-N-ethylcarbox-amidoadenosine (NECA), which usually has a moderate potency at A_1 as well as A_2 receptors, was virtually inactive, suggesting that neither A_1 nor A_2 receptors were involved. Although the relative potencies of these drugs were inconsistent with what has been described for either A_1 or A_2 receptors, the responses were antagonized by moderate concentrations of methylxanthines, suggesting that they still fit into the general category of P_1 adenosine receptors (A_1, A_2, and other unknown receptors with similar pharmacology).

Somewhat similar observations concerning the effects of 2-CADO on rat dorsal root ganglion cells were made by Dolphin et al. (1986). Adenosine reduced the duration of the calcium spike, and this effect was not reduced by potassium channel blockers, such as TEA, barium, and intracellular cesium. As in the previous study, 2-CADO did not appear to increase outward potassium currents, but markedly reduced (although it did not abolish) a calcium current that again appeared to have the voltage-dependence characteristic of the N-channel. However, 2-CADO appeared to be much more potent in rat dorsal root ganglion neurons (effective at 50 nM) than in mouse (threshold > 1 μM).

Thus, although there remains some controversy concerning the particulars, it appears that in some cells adenosine can reduce inward calcium currents. The reasons why such effects are not observed in some situations may have more to do with hetereogeneity in calcium channels than with variability in the response. Of the three types of calcium channels that have been proposed, only one (the N channel) has been reported to be sensitive to adenosine, and in cases where adenosine appears to have no effect, this may simply represent cases in which the other calcium channels are the primary contributors to the flux that is measured. However, until the effects of adenosine are determined on each calcium channel independently, the possibility that some of the effects of adenosine on calcium currents (e.g., calcium spikes in hippocampus) are indirectly mediated via potassium conductances cannot be ruled out.

2.4. Inhibition of Transmitter Release: Potential Mechanisms

Perhaps one of the best characterized physiological actions of adenosine is the inhibition of the release of neurotransmitters, an action that occurs at many peripheral and central synapses that use a variety of transmitters. This is a heterologous regulation of release, in that adenosine itself is not thought to be the major transmitter (or in most cases even a cotransmitter) at these synapses. The best characterized examples of purinergic inhibition of neurotransmitter release are found in the peripheral nervous system, where adenosine has been reported to inhibit the release of acetylcholine at the neuromuscular junction (Ginsborg and Hirst, 1972), and the release of NE at synapses in the sympathetic nervous system (Fredholm, 1976; Clanachan et al., 1977; Verhaege et al., 1977). Direct measurements of the efflux of radiolabeled or endogenous transmitter release from tissue, or electrophysiological measurements of the quantal release of transmitter (e.g., Silinsky, 1984) have confirmed that the primary action of adenosine is to inhibit transmitter release, rather than to reduce postsynaptic sensitivity. The inhibition of neurotransmitter release by adenosine, with particular emphasis on effects in the peripheral nervous system, has been the subject of several reviews (Fredholm and Hedqvist, 1980; Fredholm et al., 1983a).

Neurotransmission within the central nervous system also appears to be modulated to a significant extent by adenosine, and as in the periphery, the primary mechanism appears to be an inhibition of transmitter release. The release of acetylcholine (Vizi and Knoll, 1976; Jhamandas and Sawynok, 1976; Harms et al., 1979; Murray et al., 1982; Pedata et al., 1983), serotonin (Harms et al., 1979), norepinephrine (Harms et al., 1978; Ebstein and Daly, 1982; Fredholm et al., 1983b), dopamine (Michaelis et al., 1979; Harms et al., 1979;

Ebstein and Daly, 1982), GABA (Harms et al., 1979; Hollins and Stone, 1980), and glutamate (Dolphin and Archer, 1983; Burke and Nadler, 1988) all have been reported to be sensitive to adenosine. The majority of such experiments have been conducted using brain slices or synaptosomes, although in a few cases the effects of purines on the release of transmitters has been studied on directly superfused cortical tissue. However, with the exception of the synaptosomal experiments, these effects could result from an indirect depressant effect of adenosine. Because adenosine can inhibit neuronal firing, a decrease in transmitter efflux from a brain slice could result from a reduction in the number of times a nerve terminal fires, rather than a decrease in the amount of transmitter released per impulse.

Although the ability of adenosine to modulate the release process appears to be fairly ubiquitous, the mechanism that underlies this response is still unclear. Three basic hypotheses must be considered in this regard. First, if adenosine facilitates outward potassium currents in the presynaptic nerve terminal, this would tend to reduce calcium influx during nerve terminal depolarization and lead to a reduction in the release of neurotransmitter. Second, if adenosine directly affects calcium conductances in nerve terminals, this again would lead to a direct reduction in transmitter efflux. A third alternative has been suggested by Silinsky (1986), who has proposed that the reduction in transmitter release is a direct effect of adenosine upon the calcium sensitivity of the release process, rather than an effect mediated via changes in ion fluxes. A related issue that will be discussed in this section as well is whether changes in cyclic nucleotides play a role in inhibition of transmitter release, regardless of the proximate mechanism by which release is inhibited.

Evidence in support of the first two hypotheses (effects of adenosine on potassium and calcium fluxes) has largely been indirect. It has been argued that the postsynaptic cell can serve as a model for the presynaptic nerve terminal, and that the ionic conductances regulated by adenosine in neurons and dendrites are similarly affected in nerve terminals. However, in most cases, the validity of this assumption has yet to be demonstrated. Nevertheless, circumstantial evidence can be presented in support for roles for both potassium and calcium channels. For example, in the cases where the somatic calcium channel affected by adenosine can be identified, it appears to be the N-type channel that is sensitive to adenosine (Madison et al., 1987; Macdonald et al., 1986; Dolphin et al., 1986), and it is the N-channel that is thought to be responsible for the influx of calcium that results in neurotransmitter release (Perney et al., 1986; Miller, 1987).

In terms of the "potassium" hypothesis, several investigations have demonstrated that potassium channel blockers are able to antagonize partially or completely the inhibitory effects of adenosine on transmitter release (Okada and Ozawa, 1980; Stone, 1981b; Dunwiddie and Proctor, 1987; Scholfield and Steel, 1988). The latter study (Scholfield and Steel, 1988) is particularly interesting, in that these authors investigated the effects of a variety of potassium channel blockers on synaptic transmission in the olfactory cortex and related the ability of these agents to block adenosine-mediated changes in transmitter release to the direct effect of these drugs upon the presynaptic spike potential. Perfusion with cesium, 3,4-diaminopyridine, and 4-aminopyridine all prolonged the presynaptic action potential and reduced the effect of adenosine upon synaptic responses. Tetraethylammonium, barium, and rubidium had no apparent effect upon either of these measures, suggesting that there was a link between prolongation of the presynaptic response and the sensitivity to adenosine. A potential complication to all these studies arises from the fact that these potassium channel blockers have a direct effect of their own, in that they not only increase the presynaptic spike, but also increase the baseline synaptic response prior to adenosine perfusion. However, their ability to block adenosine responses does not appear to be related to this increase in the basal response, since treatment with increased magnesium or tetrodotoxin, both of which reduce the synaptic response, did not restore adenosine sensitivity (Scholfield and Steel, 1988).

A somewhat related mechanism may underlie the actions of Substance B, a partially characterized factor from the brain that blocks inhibitory neuromodulation by a number of agents, including adenosine. Opiate agonists, α-2-adrenergic agonists, and adenosine can all inhibit the efflux of [³H]ACh from ileal synaptosomes and likewise inhibit the contractions of the longitudinal muscle-myenteric plexus of the guinea pig. The effects of all of these agents upon release and evoked contractions have been shown to be inhibited by Substance B (Pearce et al., 1986; Benishin et al., 1986). We have examined the effects of Substance B in the hippocampus, and have found that, like 4-aminopyridine, it increases the amplitude of the presynaptic fiber spike and markedly reduces the effect of adenosine upon the synaptic response. These results suggest that Substance B may be an endogenous regulator of potassium channel activity, and that the antagonism of the responses to adenosine by both 4-aminopyridine and Substance B reflects the blockade of an adenosine-sensitive potassium channel. However, an alternative hypothesis that must be considered is that the apparent antagonism of adenosine responses by all of these agents may be indirect. If by prolonging the presynaptic nerve terminal action potential calcium flux is increased to the point where calcium-sensitive release sites are saturated, then it is possible that a direct inhibition of calcium flux by adenosine might not produce an appreciable diminution of transmitter release. Although the study of Scholfield and Steel (1988) would suggest that this is not the case, the extent to which decreases in calcium flux and/ or increases in potassium flux contribute to the modulation of transmission is not yet clear.

An alternative to the "potassium" and "calcium" hypotheses discussed above is that adenosine, in some manner, affects the release process itself, rather than ion fluxes in the nerve terminal. Specifically, it has been suggested that adenosine can reduce the affinity of the intracellular calcium-binding proteins that are involved in transmitter release for calcium (Silinsky, 1984, 1986). Several lines of evidence appear to support this hypothesis; in particular, adenosine can inhibit not only the normal release of neurotransmitter, but also release that is evoked in the absence of extracellular calcium (e.g., by calcium-filled liposomes). In addition, modeling experiments suggest that it is not possible to account for the observed effects of adenosine on calcium or strontium-induced transmitter release on any other basis than a reduction in the affinity of calcium for intracellular binding proteins. Although the evidence provided in support of this hypothesis is rather compelling, it is difficult at this point to determine the extent to which this proposed mechanism is solely responsible for the inhibition of transmitter release. In particular, if adenosine does act through some type of second-messenger system (*see below*), it is quite possible that multiple cellular mechanisms may underlie the inhibition of transmission. This hypothesis is supported to some extent by the effects of adenosine and 2-CADO at the neuromuscular junction; whereas adenosine inhibits both evoked as well as spontaneous release of transmitter, 2-CADO only inhibits evoked release (Silinsky, 1984). This would suggest that at least two different mechanisms can underlie adenosine-mediated inhibition of transmitter efflux.

In terms of the role of potential second-messenger systems, it is unclear at this point whether they are involved in the inhibitory effect of adenosine on synaptic transmission. In most cases, the receptor that mediates the inhibition of transmitter release has the pharmacological characteristics of the adenosine A_1 receptor (McCabe and Scholfield, 1985; Dunwiddie et al., 1986; Paton et al., 1986), which might indicate that inhibition of adenylate cyclase by adenosine either directly or indirectly inhibits the release process. However, direct tests of this hypothesis have been largely unsuccessful; for example, cAMP analogs appear unable to reverse the inhibition of transmission mediated by adenosine in the hippocampus (Dunwiddie and Fredholm, 1985; Dunwiddie and Proctor, 1987), although it has been reported that forskolin, which activates adenylate cyclase, can partially reverse adenosine-mediated depression of transmission in the olfactory cortex (McCabe and Scholfield, 1985). It has been observed that pertussis toxin, which can block the inhibition of adenylate cyclase by hormones and transmitters, can antagonize the ability of phenylisopropyladenosine to inhibit release of glutamate from cultured cerebellar neurons, concurrently with a blockade of the inhibitory effects of R-PIA on adenylate cyclase (Dolphin and Prestwich, 1985). However, as was discussed previously, it is now clear that pertussis toxin can inhibit responses mediated via several types of G proteins (at least G_i and G_o). Furthermore, pertussis toxin does not block the ability of adenosine to modulate the release of several neurotransmitters in the hippocampus (Fredholm et al., 1989), making it even less likely that an inhibition of adenylate cyclase is involved. Therefore, these experiments do not directly implicate adenylate cyclase or cAMP in the inhibition of transmitter release. CAMP and

its analogs have been reported to have a variety of effects upon transmitter release, including facilitation of release (e.g., Dolphin, 1983;
Kuba et al., 1981; Dolphin and Archer, 1983) and inhibition of release
(e.g., Silinsky and Hirsch, 1987; *see also* Dunwiddie and Hoffer,
1982, for review).

Resolving these issues is going to be a complex process, and at
this point, it is unlikely that a single hypothesis can account for all the
experimental observations. It is clear that adenosine has the ability to
depress the release of both excitatory and inhibitory neurotransmitters in the periphery and in the brain, and in most cases reflects a direct action on the nerve terminal. However, several points need to be
clarified before mechanisms of action can be reliably established.
Perhaps the most pressing need is to characterize the identity of the
receptors mediating these responses in greater detail. Even in very
similar preparations (e.g., the inhibition of excitatory transmission
in hippocampus and olfactory cortex), differences are observed; in
the hippocampus, the rank order of potency of adenosine analogs is
R-PIA ≥ CHA > NECA (Dunwiddie et al., 1986; Reddington et al.,
1982), whereas in the olfactory cortex, the order was NECA > CHA
> R-PIA. Other preparations show even more significant differ-
ences, both in rank order as well as absolute potencies of analogs,
and most studies have not tested enough analogs to characterize the
receptor adequately. Although some studies try to identify the receptor based upon the absolute potencies of a few analogs, this approach
can be very misleading, particularly if spare receptors are involved.
Highly selective antagonists that can be used for Schildtype analysis
of competitive receptor antagonism are needed for more definitive
receptor identification (e.g., Dunwiddie and Fredholm, 1989). Until
more such studies are carried out, it may be premature to attempt to
compare mechanisms in different preparations that might well involve different receptors.

3. General Conclusions

At this point, it is clear that the physiological basis of responses
to adenosine is far more complex than would have been suspected a
few years ago. There now seem to be several well-established models
in which adenosine can modulate distinct ionic conductances, re-

ducing calcium currents in some cells and activating potassium conductances in others. Three primary issues need to be resolved at this point. First, as was mentioned above, there clearly is a need for more sophisticated pharmacological approaches to the issue of receptor identification. It would appear at present that at least three and perhaps more receptors are required to account for the pharmacology observed in different systems; however, without more pharmacological data, this is difficult to establish. Secondly, the use of more definitive physiological approaches, such as voltage- and patch-clamping, coupled with a better understanding of the multiple types of ion channels, will presumably lead to a better identification of the ion upon which adenosine acts. Furthermore, these approaches should permit us to differentiate direct effects of adenosine upon ion channels from effects that reflect indirect modulation of voltage- or ion-sensitive channels. Finally, as we develop better tools for the study of second-messenger systems, such as pertussis toxin, intracellular perfusion, patch-clamping, and so on, it should be possible to determine the importance of potential second messengers, such as cAMP, and the permissive role of intracellular constituents, such as GTP. When we have identified well-characterized receptors that modify the activity of specific ion channels, and when the role of potential second messengers has been clearly established, it may then be possible to determine whether there are defined subclasses of adenosine receptors, each of which acts through the same mechanism in different tissues, or whether the same receptor can activate a variety of cellular responses, depending upon the cell type and perhaps the coupling proteins through which it acts.

References

Akasu, T., Shinnick-Gallagher, P., and Gallagher, J. P. (1984) Adenosine mediates a slow hyperpolarizing synaptic potential in autonomic neurones. *Nature* **311,** 62–65.

Andrade, R., Malenka, R. C., and Nicoll, R. A. (1986) A GTP binding protein may directly couple 5-HT1a and GABA-B receptors to potassium (K) channels in rat hippocampal pyramidal cells. *Society for Neuroscience Abstracts* **12,** 15.

Barr, E., Daniell, L. C., and Leslie, S. W. (1985) Synaptosomal calcium uptake unaltered by adenosine and 2-chloroadenosine. *Biochem. Pharmacol.* **34,** 713–715.

Benishin, C. G., Pearce, L. B., and Cooper, J. R. (1986) Isolation of a factor (substance B) that antagonizes presynaptic modulation: Pharmacological properties. *J. Pharmacol. Exp. Ther.* **239**, 185–191.

Berne, R. M., Rall, T. W., and Rubio, R. (eds.) (1983) *Regulatory Function of Adenosine* (Martinus Nijhoff Publishers, Boston, The Hague).

Bohm, M., Bruckner, R., Neumann, J., Schmitz, W., Scholz, H., and Starbatty, J. (1986) Role of guanine nucleotide-binding protein in the regulation by adenosine of cardiac potassium conductance and force of contraction. Evaluation with pertussis toxin. *Naunyn-Schmiedebergs Arch. Pharmacol.* **332(4)**, 403–405.

Burke, S. P. and Nadler, J. V. (1988) Regulation of glutamate and aspartate release from slices of the hippocampal CA2 area: Effects of adenosine and baclofen. *J. Neurochem.* **51**, 1541–1551.

Clanachan, A. S., Johns, A., and Paton, D. M. (1977) Presynaptic inhibitory actions of adenine nucleotides and adenosine on neurotransmission in the rat vas deferens. *Neuroscience* **2**, 597–602.

Daly, J. W. (1982) Adenosine receptors: Targets for future drugs. *J. Med. Chem.* **25**, 197–207.

Dascal, N., Lotan, I., Gillo, B., Lester, H. A., and Lass, Y. (1985) Acetylcholine and phorbol esters inhibit potassium currents evoked by adenosine and cAMP in Xenopus oocytes. *Proc. Natl. Acad. Sci. (USA)* **82**, 6001–6075.

Dolphin, A. C. (1983) The adenosine agonist 2-chloroadenosine inhibits the induction of long-term potentiation of the perforant path. *Neurosci. Lett.* **39**, 83–89.

Dolphin, A. C. and Archer, E. R. (1983) An adenosine agonist inhibits and a cyclic AMP analogue enhances the release of glutamate but not GABA from slices of rat dentate gyrus. *Neurosci. Lett.* **43**, 49–54.

Dolphin, A. C. and Prestwich, S. A. (1985) Pertussis toxin reverses adenosine inhibition of neuronal glutamate release. *Nature* **316**, 148–150.

Dolphin, A. C., Forda, S. R., and Scott, R. H. (1986) Calcium-dependent currents in cultured rat dorsal root ganglion neurons are inhibited by an adenosine analogue. *J. Physiol.* **373**, 47–61.

Drury, A. N. and Szent-Györgyi, A. (1929) The physiological activity of adenine compounds with especial reference to their action upon the mammalian heart. *J. Physiol.* **68**, 213–237.

Dunwiddie, T. V. (1980) Endogenously released adenosine regulates excitability in the in vitro hippocampus. *Epilepsia* **21**, 541–548.

Dunwiddie, T. V. (1984) Interactions between the effects of adenosine and calcium on synaptic responses in rat hippocampus in vitro. *J. Physiol.* **350**, 545–559.

Dunwiddie, T. V. (1985) Physiological role of adenosine in the nervous system. *Int. Rev. Neurobiol.* **27**, 63–139.

Dunwiddie, T. V. and Fredholm, B. B. (1984) Adenosine receptors mediating inhibitory electrophysiological responses in rat hippocampus are different from receptors mediating cyclic AMP accumulation. *Naunyn-Schmiedebergs Arch. Pharmacol.* **326**, 294–301.

Dunwiddie, T. V. and Fredholm, B. B. (1985) Adenosine modulation of synaptic responses in rat hippocampus: Possible role of inhibition or activation of adenylate cyclase, in *Advances in Cyclic Nucleotide and Protein Phosphorylation Research* (Cooper, D. M. F. and Seamon, K. B., eds.), vol 19, pp. 259–272.

Dunwiddie, T. V. and Fredholm, B. B. (1989) Adenosine A1 receptors inhibit adenylate cyclase activity and neurotransmitter release and hyperpolarize pyramidal neurons in rat hippocampus. *J. Pharmacol. Exp. Ther.* **249,** 31–37.

Dunwiddie, T. V. and Hoffer, B. J. (1980) Adenine nucleotides and synaptic transmission in the in vitro rat hippocampus. *Br. J. Pharmacol.* **69,** 59–68.

Dunwiddie, T. V. and Hoffer, B. J. (1982) The role of cyclic nucleotides in the nervous system, in *Handbook of Experimental Pharmacology* (Kebabian, J. W. and Nathanson, J. A., eds.), vol 58, pp. 389–463.

Dunwiddie, T. V. and Proctor, W. R. (1987) Mechanisms underlying physiological responses to adenosine in the central nervous system, in *Topics and Perspectives in Adenosine Research* (Gerlach, E. and Becker, B. F., eds.), Springer-Verlag, Berlin, pp. 499–508.

Dunwiddie, T. V., Worth, T. S., and Olsson, R. A. (1986) Adenosine analogs mediating depressant effects on synaptic transmission in rat hippocampus: Structure–activity relationships for the N6 Subregion. *Naunyn-Schmiedebergs Arch. Pharmacol.* **334,** 77–85.

Ebstein, R. P. and Daly, J. W. (1982) Release of norepinephrine and dopamine from brain vesicular preparations: Effects of adenosine analogues. *Cell. Mol. Neurobiol.* **2,** 193–204.

Edstrom, J. P. and Phillis, J. W. (1976) The effects of AMP on the potential of rat cerebral cortical neurons. *Can. J. Physiol. Pharmacol.* **54,** 787–790.

Fredholm, B. B. (1976) Release of adenosine-like material from isolated prefused dog adipose tissue following sympathetic nerve stimulation and its inhibition by adrenergic alpha-receptor blockade. *Acta Physiol. Scand.* **96,** 422–430.

Fredholm, B. B. and Hedqvist, P. (1980) Modulation of neurotransmission by purine nucleosides and nucleotides. *Biochem. Pharmacol.* **25,** 1583–1588.

Fredholm, B. B. (1982) Adenosine receptors. *Med. Biol.* **60,** 289–293.

Fredholm, B. B., Gustafsson, L., Hedqvist, P., and Sollevi, A. (1983a) Adenosine in the regulation of neurotransmitter release in the peripheral nervous system, in *Regulatory Function of Adenosine* (Berne, R., Rall, T., and Rubio, R., eds.), Martinus Nijhoff, The Hague, pp. 479–495.

Fredholm, B. B., Jonzon, B., and Lindgren, E. (1983b) Inhibition of noradrenaline release from hippocampal slices by a stable adenosine analogue. *Acta. Physiol. Scand.* **Suppl. 515,** 7–10.

Fredholm, B. B. and Dunwiddie, T. V. (1988) How does adenosine inhibit transmitter release? *Trends in Pharmacological Sciences,* **9,** 130–134.

Fredholm, B. B., Proctor, W., van der Ploeg, I., and Dunwiddie, T. V. (1989) In vivo pertussis toxin treatment attenuates some but not all adenosine A_1 effects in slices of rat hippocampus. *Eur. J. Pharmacol.,* in press.

Gerlach, E. and Becker, B. F. (1987) *Topics and Perspectives in Adenosine Research* (Springer-Verlag, Berlin).

Ginsborg, B. L. and Hirst, G. D. S. (1972) The effect of adenosine on the release of the transmitter from the phrenic nerve of the rat. *J. Physiol. (London)* **224**, 629–645.

Greene, R. W. and Haas, H. L. (1985) Adenosine actions on CA1 pyramidal neurones in rat hippocampal slices. *J. Physiol.* **366**, 119–127.

Haas, H. L. and Greene, R. W. (1984) Adenosine enhances afterhyperpolarization and accommodation in hippocampal pyramidal cells. *Pflugers Arch.* **402**, 244–247.

Haas, H. L. and Greene, R. W. (1988) Endogenous adenosine inhibits hippocampal CA1 neurones: Further evidence from extra- and intracellular recording. *Naunyn-Schmiedebergs Arch. Pharmacol.* **337**, 561–565.

Haas, H. L., Jeffreys, J. G., Slater, N. T., and Carpenter, D. O. (1984) Modulation of low calcium induced field bursts in the hippocampus by monoamines and cholinomimetics. *Pflugers Arch.* **400**, 28–33.

Halliwell, J. V. and Scholfield, C. N. (1984) Somatically recorded Ca-currents in guinea pig hippocampal and olfactory cortex neurones are resistant to adenosine action. *Neurosci. Lett.* **50**, 13–18.

Harms, H. H., Wardeh, G., and Mulder, A. H. (1978) Adenosine modulates depolarization-induced release of ^{3}H-noradrenaline from slices of rat brain neocortex. *Eur. J. Pharmacol.* **49**, 305–308.

Harms, H. H., Wardeh, G., and Mulder, A. H. (1979) Effect of adenosine on depolarization-induced release of various radiolabeled neurotransmitters from slices of rat corpus striatum. *Neuropharmacol.* **18**, 577–580.

Hartzell, H. C. (1979) Adenosine receptors in frog sinus venosus: Slow inhibitory potentials produced by adenine compounds and acetylcholine. *J. Physiol.* **293**, 23–49.

Henon, B. K. and McAfee, D. A. (1983a) The ionic basis of adenosine receptor actions on post-ganglionic neurones in the rat. *J. Physiol.* **336**, 607–620.

Henon, B. K. and McAfee, D. A. (1983b) Modulation of calcium currents by adenosine receptors on mammalian sympathetic neurons, in *Regulatory Function of Adenosine* (Berne, R., Rall, T., and Rubio, R., eds.), Martinus Nijhoff, The Hague, pp. 455–466.

Hollins, C. and Stone, T. W. (1980) Adenosine inhibition of gamma-aminobutyric acid release from slices of rat cerebral cortex. *Br. J. Pharmacol.* **69**, 107–112.

Hutter, O. F. and Rankin, A. C. (1984) Ionic basis of the hyperpolarizing action of adenyl compounds on sinus venosus of the tortoise heart. *J. Physiol.* **353**, 111–125.

Isenberg, G. and Belardinelli, L. (1984) Ionic basis for the antagonism between adenosine and isoproterenol on isolated mammalian ventricular myocytes. *Circ. Res.* **55**, 309–425.

Jhamandas, K. and Sawynok, J. (1976) Methylxanthine antagonism of opiate and purine effects on the release of acetylcholine, in *Opiates and Endogenous Opioid Peptides* (Kosterlitz, H. W., ed.), North-Holland Publishing Co., Amsterdam, pp. 161–168.

Kocsis, J. D., Eng, D. L., and Bhisitkul, R. B. (1984) Adenosine selectively blocks parallel-fiber-mediated synaptic potentials in rat cerebellar cortex. *Proc. Natl. Acad. Sci. (USA)* **81**, 6531–6534.

Kostopoulos, G. K. and Phillis, J. W. (1977) Purinergic depression of neurons in different areas of the rat brain. *Exp. Neurol.* **55**, 719–724.

Kuba, K., Kato, E., Kumamoto, E., Koketsu, K., and Hirai, K. (1981) Sustained potentiation of transmitter release by adrenaline and dibutyryl cyclic AMP in sympathetic ganglia. *Nature* **291**, 654–656.

Kurachi, Y., Nakajima, T., and Sugimoto, T. (1986) On the mechanism of activation of muscarinic K^+ channels by adenosine in isolated atrial cells: Involvement of GTP-binding proteins. *Pflugers Arch.* **407**, 264–274.

Kuroda, Y., Saito, M., and Kobayashi, K. (1976) High concentrations of calcium prevent the inhibition of postsynaptic potentials and the accumulation of cyclic AMP induced by adenosine in brain slices. *Proc. Japan Acad.* **52**, 86–89.

Lee, K. S., Reddington, M., Schubert, P., and Kreutzberg, G. (1983) Regulation of the strength of adenosine modulation in the hippocampus by a differential distribution of the density of A_1 receptors. *Brain Res.* **260**, 156–159.

Lee, K. S., Schubert, P., and Heinemann, U. (1984) The anticonvulsive action of adenosine: A postsynaptic, dendritic action by a possible endogenous anticonvulsant. *Brain Res.* **321**, 160–164.

Lotan, I., Dascal, N., Cohen, S., and Lass, Y. (1982) Adenosine-induced slow ionic currents in the Xenopus oocyte. *Nature* **298**, 564–572.

Lotan, I., Dascal, N., Oron, Y., Cohen, S., and Lass, Y. (1985) Adenosine-induced K^+ current in Xenopus oocyte and the role of adenosine 3', 5'-monophosphate. *Mol. Pharmacol.* **28**, 170–177.

Macdonald, R. L., Skerritt, J. H., and Werz, M. A. (1986) Adenosine agonists reduce voltage-dependent calcium conductance of mouse sensory neurones in cell culture. *J. Physiol.* **370**, 75–90.

Madison, D. V. and Nicoll, R. A. (1986) Cyclic adenosine 3', 5'-monophosphate mediates beta-receptor actions of noradrenaline in rat hippocampal pyramidal cells. *J. Physiol.* **372**, 245–259.

Madison, D. V., Fox, A. P., and Tsien, R. W. (1987) Adenosine reduces an inactivating component of calcium current in hippocampal CA3 Neurons. *Proc. Biophys. Soc.* **51**, 30.

McCabe, J. and Scholfield, C. N. (1985) Adenosine-induced depression of synaptic transmission in the isolated olfactory cortex: Receptor identification. *Pflugers Arch.* **403**, 141–145.

Michaelis, M. L., Johe, K. K., Moghadam, B., and Adams, R. N. (1988) Studies on the ionic mechanism for the neuromodulatory actions of adenosine in the brain. *Brain Res.* **473**, 249–260.

Michaelis, M. L. and Michaelis, E. K. (1981) Effects of 2-chloroadenosine on electrical potentials in brain synaptic membrane vesicles. *Biochim. Biophys. Acta* **648**, 55–62.

Michaelis, M. L., Michaelis, E. K., and Myers, S. L. (1979) Adenosine modulation of synaptosomal dopamine release. *Life Sci.* **24**, 2083–2092.

Miller, R. J. (1987) Multiple calcium channels and neuronal function. *Science* **235**, 46–52.

Mullane, K. M. and Williams, M. (1990), this volume.

Murray, T. F. (1982) Up-regulation of rat cortical adenosine receptors following chronic administration of theophylline. *Euro. J. Pharmacol.* **82**, 113–114.

Murray, T. F., Blaker, W. D., Cheney, D. L., and Costa, E. (1982) Inhibition of acetylcholine turnover rate in rat hippocampus and cortex by intraventricular injection of adenosine analogs. *J. Pharmacol. Exp. Ther.* **222**, 550–554.

North, R. A. and Williams, J. T. (1983) Opiate activation of potassium conductance inhibits calcium action potentials in rat locus coeruleus neurones. *Br. J. Pharmacol.* **80**, 225–228.

Nowycky, M. C., Fox, A. P., and Tsien, R. W. (1985) Three types of neuronal calcium channel with different calcium agonist sensitivity. *Nature* **316**, 440–443.

Okada, Y. and Kuroda, Y. (1975) Inhibitory action of adenosine and adenine nucleotides on the postsynaptic potential of olfactory slices of the guinea pig. *Proc. Jap. Acad.* **51**, 491–494.

Okada, Y. and Kuroda, Y. (1980) Inhibitory action of adenosine and adenosine analogs on neurotransmission in the olfactory cortex slice of guinea pig—structure–activity relationships. *Euro. J. Pharmacol.* **61**, 137–146.

Okada, Y. and Ozawa, S. (1980) Inhibitory action of adenosine on synaptic transmission in the hippocampus of the guinea pig in vitro. *Euro. J. Pharmacol.* **68**, 483–492.

Okada, Y. and Saito, M. (1979) Inhibitory action of adenosine, 5-HT (serotonin), and GABA (gamma-amino butyric acid) on the postsynaptic potential (PSP) of slices from olfactory cortex and superior colliculus in correlation to the level of cyclic AMP. *Brain Res.* **160**, 368–371.

Paton, D. M., Olsson, R. A., and Thompson, R. T. (1986) Nature of the N6 region of the adenosine receptor in guinea pig ileum and rat vas deferens. *Naunyn-Schmiedebergs Arch. Pharmacol.* **333**, 313–422.

Pearce, L. B., Benishin, C. G., and Cooper, J. R. (1986) Substance B: An endogenous brain factor that reverses presynaptic inhibition of acetylcholine release. *Proc. Natl. Acad. Sci. (USA)* **83**, 7979–7983.

Pedata, F., Antonelli, T., Lambertini, L., Beani, L., and Pepeu, G. (1983) Effect of adenosine, adenosine triphosphate, adenosine deaminase, dipyridamole, and aminophylline on acetylcholine release from electrically-stimulated brain slices. *Neuropharmacol.* **22**, 609–614.

Perkins, M. N. and Stone, T. W. (1980) 4-aminopyridine blockade of neuronal depressant responses to adenosine triphosphate. *Br. J. Pharmacol.* **70**, 425–428.

Perney, T. M., Hirning, L. D., Leeman, S. E., and Miller, R. J. (1986) Multiple calcium channels mediate neurotransmitter release from peripheral neurons. *Proc. Natl. Acad. Sci. (USA)* **83**, 6656–6659.

Phillis, J. W., Kostopoulos, G. K., and Limacher, J. J. (1974) Depression of corticospinal cells by various purines and pyrimidines. *Can. J. Physiol. Pharmacol.* **52**, 1226–1299.

Phillis, J. W. and Kostopoulos, G. K. (1975) Adenosine as a putative transmitter in the cerebral cortex. Studies with potentiators and inhibitors. *Life Sci.* **17**, 1085–1094.

Phillis, J. W., Edstrom, J. P., Kostopoulos, G. K., and Kirkpatrick, J. R. (1979) Effects of adenosine and adenine nucleotides on synaptic transmission in the cerebral cortex. *Can. J. Physiol. Pharmacol.* **57**, 1289–1312.

Proctor, W. R. and Dunwiddie, T. V. (1983) Adenosine inhibits calcium spikes in hippocampal pyramidal neurons in vitro. *Neurosci. Lett.* **35**, 197–201.

Reddington, M., Lee, K. S., and Schubert, P. (1982) An A1-adenosine receptor, characterized by [^{3}H] cyclohexyladenosine binding, mediates the depression of evoked potentials in a rat hippocampal slice preparation. *Neurosci. Lett.* **28**, 275–279.

Ribeiro, J. A., Sa-Almeida, A. M., and Namorado, J. M. (1979) Adenosine and adenosine triphosphate decrease 45Ca uptake by synaptosomes stimulated by potassium. *Biochem. Pharmacol.* **28**, 1297–1300.

Sattin, A. and Rall, T. W. (1970) The effect of adenosine and adenine nucleotides on the cyclic adenosine 3', 5'-monophosphate content of guinea pig cerebral cortex slices. *Mol. Pharmacol.* **6**, 13–23.

Scholfield, C. N. (1978) Depression of evoked potentials in brain slices by adenosine compounds. *Brit. J. Pharmacol.* **63**, 239–244.

Scholfield, C. N. and Steel, L. (1988) Presynaptic K-channel blockade counteracts the depressant effect of adenosine in olfactory cortex. *Neuroscience* **24**, 81–91.

Schubert, P. and Lee, K. S. (1986) Non-synaptic modulation of repetitive firing by adenosine is antagonized by 4-aminopyridine in a rat hippocampal slice. *Neurosci. Lett.* **67**, 334–338.

Schubert, P. and Mitzdorf, U. (1979) Analysis and quantitative evaluation of the depressant effect of adenosine on evoked potentials in hippocampal slices. *Brain Res.* **172**, 186–190.

Segal, M. (1982) Intracellular analysis of a postsynaptic action of adenosine in the rat hippocampus. *Eur. J. Pharmacol.* **79**, 193–199.

Shefner, S. A. and Chiu, T. H. (1986) Adenosine inhibits locus coeruleus neurons: an intracellular study in a rat brain slice preparation. *Brain Res.* **366**, 364–368.

Siggins, G. R. and Schubert, P. (1981) Adenosine depression of hippocampal neurons in vitro: An intracellular study of dose-dependent actions on synaptic and membrane potentials. *Neurosci. Lett.* **23**, 55–60.

Silinsky, E. M. (1984) On the mechanism by which adenosine receptor activation inhibits the release of acetylcholine from motor nerve endings. *J. Physiol.* **346**, 243–256.

Silinsky, E. M. (1986) Inhibition of transmitter release by adenosine: are Ca^{++} currents depressed or are the intracellular effects of Ca^{++} impaired? *Trends in Pharmacol. Sci.* **7**, 180–185.

Silinsky, E. M., Hirsh, J. K., and Vogel, S. M. (1987) Intracellular calcium mediating the actions of adenosine at neuromuscular junctions, in *Topics and Perspectives in Adenosine* (Berlach, E. and Becker, B. F., eds.), Springer-Verlag, Berlin, pp. 537–548.

Stefanovich, V., Rudolphi, K., and Schubert, P. (eds.) (1985) *Adenosine: Receptors and Modulation of Cell Function* (IRL Press, Oxford).

Stinnakre, J. and Van Renterghem, C. (1986) Cyclic adenosine monophosphate, calcium, acetylcholine and the current induced by adenosine in the Xenopus oocyte. *J. Physiol.* **374,** 551–569.

Stone, T. W. (1981a) Physiological role of adenosine and adenosine 5'-triphosphate in the nervous system. *Neurosci.* **6,** 391–398.

Stone, T. W. (1981b) The effects of 4-aminopyridine on the isolated vas deferens and its effects on the inhibitory properties of adenosine, morphine, noradrenaline, and gamma-aminobutyric acid. *Br. J. Pharmacol.* **73,** 791–796.

Stone, T. W. (ed.) (1985) *Purines: Pharmacology and Physiological Roles* (VCH Publishers, Weinheim, Germany).

Su, C. (1983) Purinergic neurotransmission and neuromodulation. *Ann. Rev. Pharmacol. Toxicol.* **23,** 397–411.

Tomita, T. and Watanabe, H. (1973) A comparison of the effects of adenosine triphosphate with noradrenaline and with the inhibitory potential of the guinea pig taenia coli. *J. Physiol.* **231,** 167–177.

Trussell, L. O. and Jackson, M. B. (1985) Adenosine-activated potassium conductance in cultured striatal neurons. *Proc. Natl. Acad. Sci. (USA)* **82,** 4857–4861.

Trussell, L. O. and Jackson, M. B. (1987) Dependence of an adenosine-activated potassium current on a GTP-binding protein in mammalian central neurons. *J. Neurosci.* **7,** 3306–3316.

Verhaege, R. H., Vanhoutte, P. M., and Shepherd, J. T. (1977) Inhibition of sympathetic neurotransmission in canine blood vessels by adenosine and adenine nucleotides. *Circ. Res.* **40,** 208–215.

Vizi, E. S. and Knoll, J. (1976) The inhibitory effect of adenosine and related nucleotides on the release of acetylcholine. *Neurosci.* **1,** 391–398.

West, G. A. and Belardinelli, L. (1985) Sinus slowing and pacemaker shift caused by adenosine in rabbit SA node. *Pflugers Arch.* **403,** 66–74.

Williams, M. (1987) Purinergic Receptors and CNS Function in *Psychopharmacology: The Third Generation of Progress* (Meltzer, H., ed.), Raven, New York, pp. 289–301.

Williams, M. and Jacobson, K. A. (1990), this volume.

Wu, P. H., Phillis, J. W., and Thierry, D. L. (1982) Adenosine receptor agonists inhibit K^+-evoked Ca^{++} uptake by rat brain cortical synaptosomes. *J. Neurochem.* **39,** 700–708.

CHAPTER 6

Adenosine Release

Trevor W. Stone, Andrew C. Newby, and Hilary G. E. Lloyd

1. Introduction

In contrast to the progress towards consensus regarding the sites and mechanisms of adenosine actions (Williams and Jacobson, 1990; Trivedi et al., 1990), the sites and mechanisms of adenosine formation and subsequent release are still the subjects of heated controversy. However, taken together with the demonstration of specific receptors for adenosine, the study of purine formation and release from tissues has contributed to at least two important hypotheses of tissue control and metabolic regulation. One is the purinergic nerve hypothesis (Burnstock, 1972, 1981), and the second is that adenosine might be an autonomous signal of cellular energy status (Berne, 1964; Lowenstein et al., 1983; Newby, 1984). In this chapter, the stimuli that give rise to adenosine release from various tissues will be reviewed in relation to the specific cell types involved. The biochemical mechanisms available for the formation of adenosine will then be discussed with emphasis given to the possibility of formation from both cytoplasmic and released nucleotides. Finally, some conclusions will be drawn as to the mechanisms of adenosine formation and release, and their physiological significance.

Adenosine and Adenosine Receptors Editor: Michael Williams ©1990 The Humana Press Inc.

2. Stimuli and Cellular Sources
of Adenosine Release

In the following paragraphs, the release of both adenosine and ATP will be discussed, since the relationship between the release of these purines is far from clear. For example, ATP released as a neurotransmitter or cotransmitter will rapidly give rise to adenosine as metabolism by ectoenzymes proceeds. In many situations, it is not certain whether physiologically released purines are designed primarily to bring ATP or its nucleoside metabolites into the vicinity of tissue receptors.

2.1. Neurons and Innervated Tissues

In those tissues that are innervated, the question of the origin of released purines has often been confined to a determination of whether the release occurs from neuronal (prejunctional) sites or from the postjunctional tissues themselves.

2.1.1. Neurons

Some of the earlier studies of purine release concentrated on neuronal tissues. Abood et al. (1962) and Kuperman et al. (1964) showed a depolarization-dependent release of radiolabeled purines from isolated axons and muscle, and similar observations have been made more recently by Maire et al. (1982, 1984) in which the purines released from desheathed vagus nerves were subjected to further analysis by HPLC. The preloaded radiolabeled efflux from the amphibian preparations used in the early studies was shown to include nucleotide components, especially adenine derivatives, although the amounts of nucleotide found on HPLC analysis were said to comprise less than 1% of the total purine release (Maire et al., 1982). Metabolic inhibition or reduced glucose availability enhanced the efflux of nucleoside, but not nucleotide fractions (Maire et al., 1984), and this axonal release was found to be independent of external calcium concentration.

The finding that ATP could stabilize axons made hyperexcitable by bathing in calcium-free medium further led to the suggestion that a calcium binding ATP complex in the axonal membrane was important for stability and, conversely, that destabilization of the membrane by depolarization might result in ATP being liberated from such a complex into the surrounding medium (Abood et al., 1962; Kuperman et al., 1964). This proposal was consistent with the observed correspondence between the release of ATP and calcium from stimulated tissues.

Other early experiments were performed by Holton and Holton (1954) and Holton (1959), who reported vasodilatation in the rabbit ear following antidromic stimulation of sensory nerves in this tissue. These workers also detected ATP in the perfusate from their preparation and proposed that the ATP released might be responsible for the vasodilatation. This remains one of the earliest suggestions for a functional and physiological role of ATP released from a specific population of nerves. In addition, an important conceptual point arises from the Holtons' work, namely that the ATP released from nonsynaptic regions of axonal membrane (and perhaps more quantitatively significant from the expansion of membrane surface area, which makes up the sensory ending arborizations) can be sufficient to produce measurable functional changes in a smooth muscle system. Burnstock (1976a), however, has pointed out that efferent axons have been demonstrated in dorsal roots raising the possibility that the ATP release seen by the Holtons could have been of orthodromic rather than antidromic origin.

More direct concern with the site of release began with the realization that ATP had profound motor effects, excitatory or inhibitory when applied to a wide range of smooth muscle tissues (Burnstock, 1972). Indeed ATP mimicked better than almost any other compound tested the effects of stimulation of those neurones that were not amenable to blockade by conventional cholinergic and adrenergic antagonists. This combination of factors resulted in the proposal that specific purinergic nerves might exist in such tissues (Burnstock, 1972), and a great deal of effort has been expended in attempting to define the specific sites of purine release in relation to these putatively purinergic neurones in a variety of tissues.

2.1.2. *Gastrointestinal Tract*
Among the first such investigations were those in which activation of nonadrenergic noncholinergic (NANC) inhibitory neurons lying within the stomach wall of toads or guinea pigs were activated

by stimulation of the vagus nerve (Burnstock et al., 1970; Satchell and Burnstock, 1971). The observed release was of adenosine and inosine, considered to be produced by the deamination of neurally released ATP. The authors argued that this purine release was probably not from sensory neurons stimulated antidromically (Burnstock, 1976b).

Preparations from the guinea pig alimentary tract also release purines following stimulation. Preloaded radiolabeled adenosine for example was released from the *taenia coli* in response to stimulation of the periarterial nerve supply (Su et al., 1971). More recent work on this same muscle has concentrated more on the detection specifically of ATP, since it is this compound rather than adenosine that has received most attention as the neurotransmitter of NANC nerves (Burnstock, 1972, 1981). The method of detection has usually been the firefly luciferin/luciferase assay.

However, although most groups have successfully demonstrated ATP release in this way, there is still disagreement about its origin. Su et al. (1971) reported that their radiolabeled purine release as well as of norepinephrine was inhibited by tetrodotoxin, implying a neuronal origin, and by guanethidine, an adrenergic neuron blocker. A later study by Rutherford and Burnstock (1978) reported that tetrodotoxin blocked only about one-third of the total purine release from the *taenia coli,* the tetrodotoxin-resistant component being attributable to release from directly stimulated muscle. Burnstock et al. (1978a) examined the release of endogenously derived ATP from the same tissue and observed a two- to sixfold enhancement of release even when changes of muscle length were prevented. Indeed direct stimulation of the muscle did not elicit any detectable release of ATP. The presumed neuronal release was prevented by tetrodotoxin though not in this case by guanethidine or 6-hydroxydopamine treatment. This latter finding clearly implies that the ATP is not originating from adrenergic neurons and stands in contrast to the report by Su et al. (1971).

White et al. (1981) later developed a technique of superfusion with luciferin/luciferase to detect ATP continuously as it was released from the tissue. Although this study supported the basic finding that ATP was released from the *taenia* and that this release was not the result of contraction of the muscle itself, neither the observed ATP release nor the muscle response was prevented by tetrodotoxin. This led the authors to conclude that the ATP was originating either from the muscle itself or from nerve terminals depolarized directly by the relatively high-intensity stimuli employed. The reasons for the discrepancy between this study and Rutherford and Burnstock (1978) are not clear, though it should be noted that changes of behavior of the *taenia* were seen by White et al. (1981) in the presence of luciferin/luciferase. This might imply an effect of the assay components upon the state of the preparation.

Using a preparation of myenteric plexus said to consist essentially of autonomic varicosities, White and Leslie (1982) demonstrated a calcium-dependent release of ATP upon potassium or veratridine depolarization. Since 6-hydroxydopamine depressed this release by about 50%, it was proposed that this fraction at least originated from adrenergic terminals (Alhumayyd and White, 1985).

2.1.3. Urinary Bladder

The urinary bladder has provided a similar lack of consensus. Burnstock et al. (1978b) found a tetrodotoxin-sensitive release of ATP from the guinea pig bladder, whereas the release seen by Chaudhry et al. (1984) was not sensitive to block by tetrodotoxin. Again the reasons for the difference are not clear, though a simple species difference could be a contributory factor.

2.1.4. Vas Deferens

Another favorite tissue for studies of purine function has been the vas deferens. Westfall et al. (1978) attempted to exclude any contribution of the smooth muscle contraction itself to purine release by bathing in a hypertonic medium. Under these conditions, the release of prelabeled norepinephrine and purines was increased by stimulation, and prevented by tetrodotoxin. When muscle contraction was permitted, however, a substantially greater release was obtained, indicating that unrestrained muscle contraction can contribute substantially to overall purine release.

White et al. (1981) were able to confirm the release of ATP from the guinea pig vas, but found that tetrodotoxin did not abolish the release, although it prevented contraction of the preparation presumably caused by the neuronal release of norepinephrine. It was therefore again suggested that the ATP release in this tissue was not associated with conducted action potentials.

A detailed analysis of purine release and ATP distribution in the rat vas deferens was later performed by Fredholm et al. (1982). Although transmural stimulation of the tissue caused an increased release of preloaded radiolabeled purines and norepinephrine, it was noted that the time course of release was significantly different. A 1-min period of electrical stimulation (5Hz) evoked a release of norepinephrine, which peaked within 1 min of ending the stimulus. Purine release, on the other hand, only reached a peak more than 2 min after ending the stimulation period. Furthermore, the α-adrenergic receptor blocking agent phentolamine enhanced norepinephrine outflow (by blocking presynaptic α-2 receptors), but reduced adenosine release. Adenosine, however, suppressed norepinephrine without affecting tritiated purine efflux. Dipyridamole greatly diminished purine, but not norepinephrine release. There are, therefore, several reasons for proposing that the greater part of norepinephrine and purine release from the rat vas arose from independent sources. The behavior of purine release with pharmacological manipulation, including the finding that norepinephrine would induce purine release, was consistent with a largely postjunctional site for purine release. Interestingly, analysis of the released purines indicated that <3% was present as ATP.

It is still not clear whether the profile of purine release from rat vas deferens is indeed different from that seen with the guinea pig preparation, possibly resulting from the smaller junctional gaps in the rat muscle, or whether other technical factors account for the differences observed. The use of vasa from castrated rats, for example, may have affected the preparation much more than simply altering the proportions of neuronal and muscular tissue (Fredholm et al., 1982). It is particularly unfortunate that these workers did not use tetrodotoxin or suppress muscle tone changes in some experiments.

The overall implication of this study then is that most, if not all, purine release is of postjunctional origin. Similar conclusions were drawn by White et al. (1981) as noted above, and other tissues have yielded the same result. The cat nictitating membrane exhibits a purine release that is stimulated by norepinephrine, acetylcholine, and tyramine to an extent that correlates well with the degree of muscle contraction. In all cases, the purine release can be prevented by α-adrenergic receptor blockade (Luchelli-Fortis et al., 1979). In both canine subcutaneous adipose tissue (Fredholm 1976; Fredholm and Hjemdahl, 1979) and the rabbit kidney (Fredholm and Hedqvist, 1978), the release of radiolabeled purines obtained by nerve stimulation can similarly be prevented by postjunctional α-receptor blockade.

2.1.5. Vascular Tissue

Several vascular preparations have also been used to examine purine release. Su (1975, 1978), for example, proposed that the ATP efflux from rabbit aorta and portal vein was coreleased with noradrenaline from adrenergic nerves, since it was prevented both by tetrodotoxin and by bretylium. Katsuragi and Su (1980, 1981) working with the rabbit pulmonary artery reported that labeled purine release was increased by elevated potassium levels, but depressed by calcium-free solutions or 6-hydroxydopamine treatments and not affected by postjunctional α-receptor block by phentolamine. This result therefore points to a presynaptic site for the purine origin (Su, 1983). The authors emphasize this point by testing strips of nerve-free aortic smooth muscle. Here potassium and clonidine produce contraction with no accompanying release of purines, although higher potassium concentrations induced a very small release of purine, which was not calcium dependent.

Burnstock et al. (1979) came to a similar conclusion that ATP release from the guinea pig portal vein was presynaptic in origin, since it was abolished by sympathectomy, whereas Levitt and Westfall (1982) attempted to quantify the relative contributions of pre- and postsynaptic sites to their observed purine release. By using the selective α-1-adrenergic receptor blocker prazosin to prevent release by activation of postjunctional sites without affecting α-2-receptors and therefore presynaptic release, these workers reported a reduction of approximately 20% of total purine release. The remaining 80% was presumed to originate presynaptically from a neuronal source, a suggestion supported by the parallel increase of norepinephrine and purine release seen on treatment with yohimbine and the decrease seen with clonidine. In addition, it was noted that release was reduced by approximately 55% following treatment with 6-hydroxydopamine, implying that a substantial fraction of the presynaptically released purines were coming from adrenergic neurones.

2.1.6. Cholinergic Systems

Studies by Meunier et al. (1975) and Israel et al. (1976) revealed a potassium-evoked release of ATP from preparations of Torpedo electroplaque, the release being enhanced by physostigmine and diminished by curare. This finding suggested a postsynaptic source for the released ATP, but subsequent careful analyses, including the examination of synaptosomes prepared from the electroplaque indicated that a small proportion of the nucleotide was originating from the presynaptic terminals together with acetylcholine (Morel and Meunier, 1981). This release was not detected by Michaelson (1978), although Zimmerman (1978), Zimmerman et al. (1979), and Zimmerman and Denston (1977) showed a corelease of acetylcholine and ATP from cholinergic vesicles.

In the mammalian neuromuscular system, Silinsky and Hubbard (1973) reported that stimulation of the rat phrenic nerve evoked a release of ATP sufficient to reach a concentration of about 100 μM in the extracellular fluid. This release was achieved in the presence of curare and was therefore presumed to be presynaptic in origin, a conclusion supported by the ability of hemicholinium to block the release and by the failure of carbachol to promote it.

Less work appears to have been performed on autonomic cholinergic systems, though botulinum toxin, a selective inhibitor of cholinergic neurones, depressed the NANC response of the guinea pig bladder (Mackenzie et al., 1982). Unfortunately, the effect of botulinum toxin directly upon ATP release has not been examined.

Depending on the tissue being studied, therefore, an argument can be made for a primarily pre- or postjunctional release from innervated tissues. One of the major difficulties with interpretation, however, is the uncertainty that surrounds the amount of ATP present in and available for release from presynaptic nerve terminals. Several authors have argued that a releaseable norepinephrine:ATP ratio of approximately 50:1 probably exists in sympathetic nerve terminals

(Smith 1977; Fredholm et al., 1982). This could easily result in pre-synaptically released purines escaping detection especially if swamped by a large excess of purine from postjunctional sources related to tissue metabolism and activity. Such a small presynaptic releaseable pool has been proposed as the explanation for the apparent absence of purine release following activation of sympathetic nerves to the spleen (Stjarne et al., 1970; Lagercrantz, 1976).

In a brief communication, Bencherif et al. (1986) reported that, in the frog paravertebral ganglion, prelabeled adenosine could be released by electrical stimulation of the preganglionic or postganglionic nerve trunks and by carbachol, the preganglionically evoked release being inhibited by cholinergic receptor antagonists. This study therefore indicates that the greater part of adenosine release originated from postsynaptic cell bodies, rather than the nerve terminals.

2.2. Central Nervous System

Interest in purine release from the central nervous system (CNS) tissue developed in many respects in parallel with interest in the concept of purinergic synaptic transmission, rather than resulting from it. This was because of the demonstration by Sattin and Rall (1970) and their colleagues that adenosine could activate adenylate cyclase in brain preparations. This action was seen not only in response to adenosine alone, but also as a potentiation of the effect of norepinephrine, both actions being mediated by membrane receptors and blocked by methylxanthines. The demonstration soon followed that adenosine, inosine, and hypoxanthine were released by depolarization from slices of guinea pig neocortex incubated with tritiated adenosine (Shimizu et al., 1970).

McIlwain (1972) in particular then developed this observation into the hypothesis that adenosine and possibly other purines might play a hormonal role within the CNS. It has since become clear that appreciable concentrations of adenosine may exist in cerebral extracellular fluid, values of around 1 μm being commonly recorded (Newman 1983; Newman and McIlwain, 1977; Rehncrona et al., 1978; Schrader et al., 1980; Winn et al., 1981; Zetterstrom et al., 1982). McIlwain and his colleagues (Pull and McIlwain, 1972, 1973, 1976) and Shimizu et al. (1970) showed that a variety of procedures causing depolarization of neuronal tissue, including electrical stimulation, high potassium, veratridine, hypoxia, and lowered glucose caused a calcium-dependent release of purines from cerebral preparations.

The ability of various manipulations that alter the metabolic status of the CNS to provoke an efflux of purines provides a strong argument that at least a proportion of the release can originate from sources that are independent of vesicular transmitter-related pools (McIlwain, 1972; Daval et al., 1980; Hollins and Stone, 1980; Hollins et al., 1980; Stone et al., 1981; Lloyd and Stone, 1980). The distinction is supported by the temporal contrast between the rapid release of transmitters, which is induced by high potassium depolarization and the relatively tardy efflux of purines (Daval et al., 1980; Lloyd and Stone, 1983; Fredholm and Vernet, 1978). A similar dissociation has been noted using the vas deferens (Fredholm et al., 1982). It is still unclear whether the slower efflux of purines compared with neurotransmitters results from a different site or mechanism of release, but it is interesting to note that other peculiar and as yet unexplained features of purine release have been reported. Most particularly, a chemical depolarizing stimulus delivered with a "square wave" profile causes a release of purine only when it is removed (Hollins and Stone, 1980; *see* Jonzon and Fredholm, 1985).

As with many of the peripheral tissues discussed earlier, however, interpretation of the site of purine release is often complicated by conflicting or confusing results. Pull and McIlwain (1973) reported a calcium-dependent release of prelabeled purines upon electrical stimulation of brain slices, but found that tetrodotoxin diminished the nucleoside efflux by 70–80% with little effect on the release of nucleotides. The potassium-induced release of purines was not dependent on calcium concentration, however, whether from brain slices (Pull and McIlwain, 1973) or from hypothalamic synaptosomes (Fredholm and Vernet, 1979), whereas the veratridine or electrically evoked release was calcium dependent. Fredholm and Vernet (1979) concluded their paper with the proposal that the calcium-dependent release of purines from neuronal tissues was the result of increased energy consumption or of diminished production.

A conflicting pattern of results was obtained by Bender et al. (1981) using cortical synaptosomes. In their hands, 60% of the radiolabeled purine content could be released by potassium stimulation, but this release was calcium dependent. Release by veratridine on the contrary was enhanced in calcium-free solutions.

One attempt to probe this confusion further has been made by Wu et al. (1984) using a mitochondrial P2 fraction from rat cortex, labeled with tritiated adenosine. In this system, the release induced by potassium, veratridine, or glutamate (which did not elicit release from hypothalamic synaptosomes) was not changed by removing calcium or adding calcium channel blocking agents. Indeed, calcium removal potentiated glutamate-induced release. Interestingly, the organic calcium channel blockers D600 and diltiazem selectively increased potassium-evoked purine release, but nifedipine enhanced selectively the release produced by veratridine. This fascinating finding may imply slightly different actions of the channel blockers or the existence of different releaseable pools of purines.

The technique of superfusion with components of the luciferin/luciferase ATP assay system was modified by White (1978) to reveal the release of ATP from depolarized hypothalamic synaptosomes. In a later study by Fredholm and Vernet (1979), no nucleotide release could be detected in response to potassium depolarization. ATP was obtained, however, with little loss because of metabolism, following lysis of the synaptosomes. Although not concerned primarily with ATP, Wyllie and Gilbert (1980) did detect ATP released from cortical synaptosomes, but this remained unmetabolized far longer than in the study by White (1978). The occurrence and metabolism of ATP released from central nerve terminals therefore is still something of a mystery, although the occurrence of a calcium-dependent, tetrodotoxin-insensitive release by potassium, and a low calcium-enhanced, tetrodotoxin-sensitive release by veratridine would be consistent with many studies of CNS release of more classically recognized neurotransmitters (White, 1978; White et al., 1985). A communication by Pollard and Pappas (1979), however, claimed that the veratridine-evoked release of ATP was calcium dependent.

Subsequent work by White's group has been directed at defining the source of released synaptosomal ATP in terms of specified populations of nerve terminal. There is some regional variation in the ability of brain samples to release ATP in response to potassium stimulation, with striatum and cortex being the best sources (Potter and White, 1980). The distribution of this release sensitivity correlates to some extent with the distribution of cholinergic neurons, but whereas botulinum toxin suppresses acetylcholine release, it does not alter ATP release (White et al., 1980).

Similarly, 6-hydroxydopamine pretreatment does not diminish ATP release (Potter and White, 1982). Unless, therefore, there is a very substantial metabolic or purely purinergic release of ATP sufficient to mask all other sources, it must be concluded that the release is not from cholinergic or adrenergic neuron terminals.

A different approach to the demonstration of purine release in the CNS was taken by Schubert's group, who injected radiolabeled adenosine into the entorhinal cortex of anesthetized rats and then examined the distribution of label after a suitable recovery period. The label, as expected, was taken up by local entorhinal neurons and some of the label was transferred, presumably by axoplasmic flow into the target areas of the entorhinal–hippocampal projections. Some of the tritium was clearly detectable in neurons linked synaptically with the injected neurons, leading to the proposal that some label had been transferred transneuronally. Interestingly, the amount of such transfer appeared to correlate with the amount of electrical stimulation delivered to the entorhinal neurons during the recovery period (Schubert et al., 1976). The sites of transfer also correlated with the localization of 5'nucleotidase (Schubert et al., 1979), an observation that was interpreted to indicate that the original tritiated adenosine had been taken up, converted to nucleotides, transported, and released as nucleotide at the neuron terminals, and the labeled hydrolysis product, adenosine, formed extracellularly by other neurons only when the relevant enzyme was present. Release of labeled adenosine by the relevant neuron terminals has been shown *ex vivo* by this group (Lee et al., 1982).

A release of purines has also been demonstrated from the surface of the cerebral cortex in vivo, including nucleoside (Perkins and Stone, 1983; Sulakhe and Phillis, 1975; Jhamandas and Dumbrille, 1980) and nucleotide (Wu and Phillis, 1978) components. Using a push–pull cannula, release can also be demonstrated in deeper structures of the brain (Barberis et al., 1984), although the recent trend is to use *in situ* electrochemical methods to detect the uric acid produced from adenosine.

Although most authors have speculated on the precise cellular origin of central purines, the problem has received little direct examination. Wojcik and Neff (1983) attempted an answer by lesioning components of the rat striatum. Although decortication and 6-hydroxydopamine lesions of afferent fibers had no effect on adenosine release, kainic acid, which destroys cell bodies within the area of administration, did decrease potassium-evoked release of endogenous adenosine. The release of adenosine in this case thus appears to originate entirely from elements within the body of the striatum, probably neuronal cell bodies.

Although the effects of several drugs have been examined on purine release from the CNS (Pull and McIlwain, 1976), opiates have proved particularly interesting. Morphine became of interest following the suggestion that purines might mediate some of the inhibitory effects of this drug on synaptic transmission in peripheral and central tissues (Sawynok and Jhamandas, 1976; Perkins and Stone, 1980; Stone and Perkins, 1979). Morphine has been shown to increase significantly the stimulated efflux of purines from brain slices in vitro (Fredholm and Vernet, 1978; Stone, 1981a; Wu et al., 1982) as well as from the brain in vivo after peripheral administration (Phillis et al., 1979). However, morphine applied directly to the cortical surface did not elicit any increase of purine release; rather, it suppressed glutamate-induced release (Jhamandas and Dumbrille, 1980). The endogenous opioid peptide, metenkephalin, was also found to enhance electrically evoked release of purines from rat cortex slices (Stone, 1981a), though in a study of the mouse vas deferens, Stone (1981b) reported that morphine did not increase purine release from this tissue. Since morphine potently depresses neurotransmission in this preparation, the generality of any relationship between opiate action and purine release is questionable.

Whereas the studies just discussed were concerned with the relationship between synaptic transmission and purine release, a different emphasis has been given by studies of the role of adenosine in cerebral blood flow. In such studies, other manipulations were employed aimed at increasing cerebral energy expenditure, or reducing its supply of substrates and oxygen. Increasing oxygen demand by inducing seizures with bicuculline in rats (Winn et al., 1980a) or cats (Schrader et al., 1980) rapidly increased tissue adenosine concentration. Ischemia also caused rapid accumulation of adenosine in excised brains (Deuticke and Gerlach, 1966). Hypoxia, hypercapnia, and hypotension as well as electrical stimulation increased adenosine concentration in brain tissue obtained by the "freeze-blow" technique (Rubio et al., 1975; Winn et al., 1979, 1980b, 1981). Sustained hypoglycemia also elevated brain adenosine concentration (Winn et al., 1983).

The cellular source of adenosine release in response to these stimuli has not been investigated systematically, though a recent investigation by Braas et al. (1986) may be relevant here. In these studies, an antibody raised to haptenized adenosine was used to localize the nucleoside cytochemically in fixed sections of brain tissue. A highly selective distribution of adenosine-containing neurons was revealed. Most importantly for this discussion, the intensity of staining was enhanced without changing its distribution by procedures that lengthened the ischemic time of the tissues prior to fixation. This suggests that the neurons localized may be those metabolically active in adenosine formation (Braas et al., 1986).

2.2.1. Retina

The accumulation into neurons and glia and the subsequent depolarization-evoked, calcium-dependent release of purines have also been demonstrated recently in preparations of the rabbit retina (Perez et al., 1986). Of special interest was the discovery that phenylisopropyladenosine, a compound not normally considered to be a substrate for the purine uptake systems, was also accumulated and released, though in this case the cells involved appeared to be only neurons (Perez and Ehinger, 1986). This finding may prove to have major implications for the many studies carried out using phenylisopropyladenosine as a stable analog of adenosine at purine receptors.

2.3. Nonneuronal Cells

2.3.1. The Heart

The earliest reports of purine release from isolated hearts (Berne, 1963; Gerlach et al., 1963; Imai et al., 1964; Richman and Wyborny, 1964; Katori and Berne, 1966) were contemporary with early studies in neuronal tissue. Attention focused on ischemia, hypoxia, or inhibition of energy production through oxidative phosphorylation as the triggers of adenosine formation. Taken together with the known coronary vasodilator action of adenosine, these observations suggested that adenosine might be the intrinsic agent responsible for autoregulation of coronary blood flow (Berne, 1964). Later experiments, including many with more intact preparations, such as the chronically instrumented, conscious dog, demonstrated that adenosine formation also accompanied an increase in myocardial energy demand (Miller et al., 1979; Foley et al., 1979; Saito et al., 1980; McKenzie et al., 1982; Knabb et al., 1983). Indeed, adenosine concentration was shown to be greater in the heart during systole than diastole (Thompson et al., 1980). Recent experiments by Bardenheuer and Schrader (1986) using isolated working guinea pig hearts have shown that the release of adenosine, which normally accompanies an increase in cardiac work, can be avoided if the supply of oxygen and metabolites is simultaneously increased by overperfusion. These data emphasize that it is an imbalance between energy supply and demand, rather than an increase in work rate *per se*, that is responsible for adenosine production.

The predominant purines released from the heart are adenosine, inosine, hypoxanthine, and uric acid, although low concentrations of adenine nucleotides have also been detected (Paddle and Burnstock, 1974; Clemens and Forrester, 1980; Schrader et al., 1982). Experiments performed in the presence of inhibitors of adenosine metabolism reveal that adenosine is the predominant purine released during myocardial ischemia (Achterberg et al., 1985; Newby et al., 1987).

The cellular source of adenosine production in the heart has been investigated. When guinea pig perfused hearts are prelabeled with low concentrations of [^{3}H]adenosine, perferential incorporation into vascular endothelium occurs (Nees et al., 1985a), leading to a higher relative specific activity in the endothelial nucleotide pool. Under normoxic conditions, released purines have high specific radioactivity (Schrader and Gerlach, 1977; Deussen et al., 1986), suggesting that they derive mainly from endothelium. This agrees with the conclusions of Achterberg et al. (1986), who showed that the major purine released from normoxic rat hearts is uric acid, which is formed by xanthine oxidase found exclusively in the microvascular endothelium. During hypoxic perfusion of prelabeled guinea pig hearts, release of [³H]adenosine increases, but its specific activity falls (Schrader and Gerlach, 1977; Deussen et al., 1986) indicating that adenosine release from endothelium is stimulated, but that release from cardiac myocytes is accelerated to an even greater extent. Whether adenosine formation from both pools occurs in direct response to an imbalance between energy supply and demand has yet to be established. An increase in cardiac work would be expected to lead to an increased energy demand primarily in the cardiac myocytes. How this results in adenosine release from endothelium remains to be investigated.

2.3.2. Skeletal Muscle

Early studies of globally ischemic rat skeletal muscles suggested that they were less active in adenosine formation than either heart or brain (Deuticke and Gerlach, 1966; Rubio et al., 1973). This was explained by the much higher activity of AMP-deaminase in skeletal muscle than in other tissues (*see* Ogasawara et al., 1978), which favored, therefore, deamination rather than dephosphorylation of AMP. Nonetheless, consistent reports of increased tissue adenosine concentration in response to ischemia or contraction have been presented with some skeletal muscle preparations (Dobson et al., 1971; Bockman et al., 1976; Phair and Sparks, 1979; Belloni et al., 1979), but not others (Phair and Sparks, 1979). This inconsistency may be explained by the dramatically greater activity of AMP deaminase in muscles of the glycolytic (fast twitch) rather than oxidative (slow twitch) types (Raggi et al., 1969; Bockman and McKenzie, 1983). Bockman and McKenzie (1983) showed that cat stimulated isolated soleus muscle, which contained a relatively low activity of AMP deaminase and a high activity of 5'-nucleotidase, did show adenosine accumulation, whereas cat gracilis muscle, which contained high AMP deaminase and low 5'-nucleotidase activity, did not. Dog gracilis muscle was biochemically similar to cat soleus muscle, whereas dog cardiac muscle had an even lower AMP deaminase to 5'-nucleotidase ratio. Examination of the relationship between the content of these enzyme activities and the capacity for adenosine formation in a wider range of muscles would, clearly, be valuable.

The cellular source of adenosine production in skeletal muscle does not appear to have been studied. In electrically stimulated muscle preparations, there is the possibility of both presynaptic and postsynaptic release of adenosine. Total tissue activities of AMP deaminase might be expected to have more impact on the production of adenosine postsynaptically. Release of ATP has also been observed from exercising skeletal muscle (Forrester, 1972). Its cellular origin is, likewise, unknown.

2.3.3. The Kidney

A role for adenosine has been proposed in the intrinsic control of glomerular filtration rate and in renin release (Spielman and Thompson, 1982; Osswald et al., 1982). Adenosine formation occurs in the kidney during global ischemia (Osswald et al., 1977) or during an increase in reabsorptive work caused by sodium loading (Osswald et al., 1980). The cellular site of adenosine formation is thought to be the epithelial cells responsible for sodium reabsorption.

2.3.4. Adipose Tissue

Adenosine inhibits lipolysis (Schwabe et al., 1973; Sollevi and Fredholm, 1981) and promotes insulin stimulation of glucose uptake (Schwabe et al., 1974; Martin and Bockman, 1986) in rat isolated adipocytes and dog adipose tissue *in situ*. Adenosine also increases dog adipose tissue blood flow (Sollevi and Fredholm, 1983). These effects appear to be directed toward preventing formation, toward reesterifying, or toward removing free fatty acid, which might otherwise uncouple adipocyte mitochondrial oxidative phosphorylation (Angel et al., 1971). Release of adenosine was observed from dog adipose tissue during stimulation of lipolysis, but only when this was accompanied by constriction of the local arteriole (Fredholm and Sollevi, 1981). According to these authors, this indicated that adenosine formation was not a direct consequence of neural activation or of lipolysis, but resulted from a deranged metabolic state in the adipocytes. The time course of adenosine formation, which lagged behind lipolysis, was also consistent with this proposal (Fredholm and Sollevi, 1981). Extensive studies of rat adipocytes in vitro also fail to show enhanced adenosine formation during stimulation of lipolysis (Schwabe et al., 1973; Fain, 1979). Indeed Fain (1979) found that, even when free fatty acid was allowed to accumulate such that adipocyte ATP concentrations were greatly reduced, AMP accumulated inside the fat cells and adenosine was not released. It is possible that dog adipocytes might behave differently. Nonetheless, other sources of adenosine formation in adipose tissue need to be considered. Fredholm and Sollevi (1981) reported that the specific activity of adenosine released from prelabeled adipose tissue was different when neural stimulation or noradrenaline infusion was used to stimulate lipolysis. This suggests the existence of multiple nucleotide pools that can release adenosine differentially.

2.3.5. Blood Platelets

Blood platelets release ATP, ADP, and other dense granule contents during the secondary phase of aggregation (for review, *see* Holmsen and Weiss, 1979; Huang and Detwiler, 1986). As much as 65% of the platelet adenine nucleotides may be in the intravesicular pool, which is not in rapid metabolic equilibrium with the cytosolic pool. Released ADP promotes further aggregation and is only slowly hydrolyzed in blood to form adenosine, which is capable of inhibiting platelet aggregation. Formation of adenosine from nucleotides is, however, greatly accelerated in blood perfusing vascular beds (Pearson, 1985). Platelet aggregation also leads to degradation of cytoplasmic ATP (Mills, 1973), although the release of purine from this pool has not been systematically studied.

2.3.6. Polymorphonuclear Leucocytes

Adenosine inhibits oxygen consumption and reactive oxygen metabolite production by stimulated human neutrophils (Cronstein et al., 1983; Roberts et al., 1985), suggesting that it might act as a physiological inhibitor of the inflammatory response. During phagocytosis, the ATP content of human neutrophils declines to about 50% (Borregaard and Herlin, 1982), and this degree of nucleotide catabolism results in adenosine formation in rat polymorphonuclear leucocytes (Newby and Holmquist, 1981). Adenosine production accounts for only 6% of total purine degradation, however, (Newby and Holmquist, 1981), so that it is possible that other sources, such as cells damaged by the action of reactive oxygen metabolites, might provide the regulatory pool of adenosine (Roberts et al., 1985).

2.3.7. Cells Isolated from Tissues by Enzyme Digestion

These experimental models are useful to define more precisely the biochemical mechanisms underlying adenosine formation, but they have some general limitations. First, immediately after isolation with proteolytic enzymes, cells tend to be extremely fragile and may release large quantities of nucleotides and/or cytosolic enzymes (Rodbell, 1966; Stanley et al., 1980; Pearson and Gordon, 1979; Jacobson and Piper, 1986). Such release may be exacerbated by experimental manipulations. Cells that remain viable in tissue culture subsequently show much greater stability (Meghji et al., 1985; Jacobson and Piper, 1986). Secondly, isolated cells show metabolic perturbation, such as loss of glycogen (Wagle et al., 1973), proteolysis of surface components, and increased membrane permeability to ions (Jacobson and Piper, 1986). Cells maintained in culture may cease to express or induce expression of cell surface components. Thirdly, isolated or cultured cells are often removed from the normal metabolic loads they sustain. This is particularly apparent in the case of myocytes, which although they may beat when isolated do not undergo loaded contractions. This has a profound effect on their rate of oxygen consumption compared to the tissues in vivo (Jacobson and Piper, 1986). For this reason, metabolic inhibitors, rather than more physiological stimuli, are required to elicit ATP breakdown and adenosine formation. A fourth drawback of unperfused, isolated cell systems is that released purines may be very efficiently taken up, and either deaminated or reconverted to nucleotides. Adenosine deaminase may be inhibited with 2'-deoxycoformycin and adenosine kinase with either 5'-iodotubercidin or 5'-deoxy-5'-amino adenosine (Newby, 1981). In the presence of these inhibitors, adenosine accumulates in static incubations of Ehrlich ascites tumor cells, polymorphonuclear leucocytes, or cultured heart cells (Lomax and Henderson, 1973; Newby and Holmquist, 1981; Meghji et al., 1985). These inhibitors may also be used, even with tissue preparations, to measure absolute rather than net rates of adenosine production. The relative importance of deamination vs dephosphorylation of AMP can also be assessed. Adenosine production in Ehrlich ascites tumor cells poisoned with 2-deoxyglucose accounted for 18% of the ATP broken down (Lomax and Henderson, 1973). Adenosine also accounted for 18% of ATP breakdown in cultured neonatal rat heart cells poisoned with oligomycin and 2-deoxyglucose (Meghji et al., 1985). This compared with values of 6% for polymorphonuclear leucocytes poisoned with 2-deoxyglucose (Newby and Holmquist, 1981) and 65%

for globally ischemic rat hearts (Newby et al., 1987). More cell types need to be analyzed in this way before the significance can be assessed of the very high propensity of ischemic hearts to produce adenosine.

Net formation of adenosine has been measured in cultured chick heart cells during hypoxia (Mustafa, 1979) and in freshly isolated adult cardiocytes poisoned with dinitrophenol or iodoacetate (Bukoski and Sparks, 1986). Net production of adenosine has also been measured in isolated hepatocytes under basal conditions (Bontemps et al., 1983) during incubation with glycerol or fructose (Des Rosiers et al., 1982) and during hypoxia (Belloni et al., 1985). Several preliminary reports have appeared regarding net formation of adenosine from vascular endothelium during hypoxia, hypercapnia, or treatment with metabolic inhibitors (Nees and Gerlach, 1983; Nees et al., 1985; Pearson and Gordon, 1985). The principal purine normally released from small vessel endothelium appears, however, to be uric acid (Nees et al., 1985a; Achterberg et al., 1986).

Release of ATP occurs from pig aortic endothelium during treatment with trypsin and EDTA (Pearson and Gordon, 1979). Nucleotide release was evoked also by mechanical agitation, collagenase treatment, and perhaps most significantly, after treatment with thrombin. Similar release did not, apparently, occur from guinea pig microvascular endothelial cells (Nees and Gerlach, 1983). Release of ^{32}P-labeled material also occurred from a number of cultured cell lines exposed to ATP (Trams, 1974). This raised the possibility that ATP release might propagate further release of ATP from neighboring cells (Trams, 1974).

These studies establish the capacity of certain cell types to release either nucleotides, nucleosides, or both. Different experimental approaches are required, however, to establish whether such release occurs in an organized tissue under any physiological stimulus.

3. Metabolic Sources
and Mechanisms of Adenosine Release

From the preceding discussion, two different physiological situations are apparent in which purines, including adenosine, are produced. Neuronal firing and platelet aggregation release nucleotides from a well-defined intravesicular store. On the other hand, purines are released both from these cells and from others that do not contain an obvious intravesicular nucleotide pool in response to an imbalance between energy supply and demand. These two processes are linked in tissues with innervation, since exocytosis is, in itself, energy requiring and neurotransmitter action often provokes energy expenditure postsynaptically. The following sections describe the known biochemical mechanisms that may contribute to adenosine formation and review the evidence regarding their relative contribution.

3.1. Adenosine Formation from Extracellular Nucleotides

This subject has been reviewed recently (Pearson, 1985; Pearson and Gordon, 1985; Gordon, 1986). Hydrolysis of released ATP occurs by the sequential action of distinct ATPase, ADPase, and 5'-nucleotidase enzymes (Pearson et al., 1980), all of which have their active sites directed to the extracellular space (Pearson, 1985). The ecto-ATPase is present on a large variety of mammalian cell types (Pearson, 1985), it has a K_m below 500 μM, and in endothelial cells, hydrolyzes the β,γ-bidentate complex of ATP-Mg^{2+} (Cusack et al., 1983). The ADPase has been sought in fewer tissues (Pearson, 1985), but it does occur together with ecto-ATPase in neutrophils, some lymphocytes, vascular smooth muscle cells, and endothelium. The enzyme from endothelium has a K_m of 160 μM and hydrolyzes the α,β-bidentate ADP-Mg^{2+} complex (Cusack et al., 1983). The ecto-5'-nucleotidase is the most thoroughly studied ectonucleotidase, and it is often thought to be distributed ubiquitously on nucleated mammalian cells. Its activity in different tissues and its location within tissues show, however, some remarkable variations between species (Nakatsu and Drummond, 1972; Lee et al., 1986). Table 1 and Fig. 1 show the total activities and cytochemical distribution of ecto-5'-nucleotidase in hearts from five species. Whereas in rat and guinea pig hearts there is an abundance of 5'-nucleotidase apparently distributed on all cell types, in rabbit heart the much lower activity appears confined to vascular smooth muscle cells. In the pigeon heart, the total activity of 5'-nucleotidase is extremely low, and residual activity does not appear concentrated in any visible structure.

Table 1
Activities of AMPases in Hearts of Different Species*

	Enzyme activity (μmol/min/g wet wt)		
Species	Ecto-5'-nucleotidase	Nonspecific phosphatase	Cytosolic-5'-nucleotidase
Rat	4900 ± 300	870 ± 150	410 ± 40
Guinea Pig	3600 ± 800	490 ± 240	300 ± 100
Turtle	510 ± 50	170 ± 20	116 ± 9
Rabbit	210 ± 40	78 ± 10	280 ± 40
Human (n = 3)	1000 ± 300	150 ± 50	150 ± 20
Pigeon	16 ± 4	34 ± 5	590 ± 40

*Ventricular, tissue (n = 6) was homogenized at 4°C in 9 vol of 20 mM dimethylglutarate, pH 7.0, 1 mM EDTA-dithiothreitol. Ecto-5'-nucleotidase activity was measured in the homogenate with 0.2 mM AMP as substrate (Newby et al., 1975) with 10 mM β-glycerophosphate present. Nonspecific phosphatase was estimated as the increase in AMPase activity when β-glycerophosphate was omitted (Newby, 1980). Cytosolic 5'-nucleotidase activity was measured in a 100,000 x g, 60-min supernatant with 3 mM IMP as substrate and with 5 mM ATP present (Worku and Newby, 1983). Values are mean ± SEM.

Similar variations between species were found in the regional distribution of 5'-nucleotidase within the hippocampal area of the brain (Lee et al., 1986).

Nucleotide monophosphates could also be hydrolyzed by an ecto-(nonspecific) phosphatase (Pearson, 1985). This enzyme appears less widely distributed than 5'-nucleotides, but is present together with ecto-ATPase on polymorphonuclear leucocytes (Pearson, 1985). In our study of nucleotidase activities in hearts from different species (Table 1), 5'-nucleotidase and nonspecific phosphatase activity varied roughly in parallel. It does not seem likely, therefore, that the nonspecific enzyme substitutes for 5'-nucleotidase in tissues where this is lacking.

Three principal means have been used to determine the contribution of hydrolysis of released nucleotides to adenosine production. First, as detailed in the sections relating to each tissue, release of nucleotides, particularly ATP, has been observed directly. The proportion of total purines released as ATP, has, however, been small and arguments that such values can be corrected up by one or two orders of magnitude so as to allow for ectonucleotidase activity

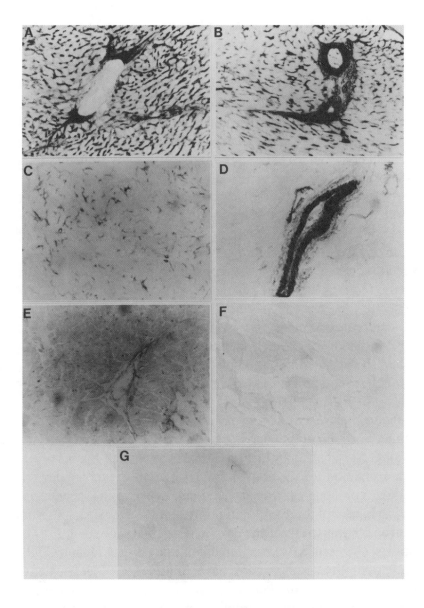

Fig. 1. Cytochemical localization of ecto-5'-nucleotidase in hearts. Ecto-5'-nucleotidase was localized by the method of Wachstein and Meisel (1957). Briefly, frozen sections (20μm) of (A) rat, (B) guinea pig, (C) turtle, (D) rabbit, (E) human, and (F) pigeon ventricular myocardium were air-dried, fixed, and incubated with AMP in the presence of lead nitrate. The lead phosphate precipitate was visualized with ammonium sulfite. Panel G shows the staining obtained with rat myocardium when AMP was omitted. Micrographs were obtained under bright field illumination at a magnification of 100x (Meghji, et al., 1988).

(Forrester, 1981) are provocative, but are no substitute for experimental observations. Unfortunately, there are no known potent inhibitors of ecto-ATPase, although inhibitors of ecto-ADPase (e.g., adenylylimidodiphosphate) and ecto-5'-nucleotidase (e.g., α,β-methylene-ADP, antibodies, and lectins) are available (Pearson, 1985). Inhibition of ecto-5'-nucleotidase should trap all released nucleotides as AMP. Promising preliminary data have been presented by Imai et al. (1986) using this approach to study ischemia-induced adenosine release from guinea pig hearts.

One reason why this approach has not been more widely adopted is the relative difficulty of measuring AMP concentration with the firefly luciferase assay. This assay method is both more sensitive than alternative chromatographic methods (Pearson, 1985) and can be used to measure release continuously. Application of the luciferase method in the presence of exogenous adenylate kinase, an ATP-regenerating system and an inhibitor of 5'-nucleotidase, may be a feasible approach to measuring total adenine nucleotide release.

A marker of tissue disruption is needed in any such study to establish the specificity of nucleotide release. Cell injury may pass through several thresholds (Morgan et al., 1986; Patel and Campbell, 1987). In the first, there is increased membrane permeability to ions, in the second, release of low molecular weight cellular contents, and in the third, disruption accompanied by release of cytoplasmic enzymes. No satisfactory marker has been applied to studies of adenosine formation for the nonspecific release of low molecular weight cellular components. For isolated-cell experiments, penetration of trypan blue or of the dye propidinium iodide, which becomes fluorescent on binding to DNA, can be used (Patel and Campbell, 1987). For tissues, simultaneous measurement of release of ATP and GTP may be a possibility. Many studies have, however, used lactate dehydrogenase release as a marker of complete cell disruption. Conclusions drawn from those studies in which no marker of cell integrity was employed should be viewed with scepticism.

A second method to decide whether adenosine arises from released nucleotides is to determine the influence of inhibitors of ecto-5'-nucleotidase on the rate of formation of adenosine. A competitive inhibitor, α,β-methylene ADP (Burger and Lowenstein, 1970) abolished the hypoxia-induced increase in tissue adenosine concentration in the isolated perfused rat kidney and reduced by 62% release of adenosine into the venous effluent (Ramos-Salazar and Baines, 1986). Although the inhibitor appeared to penetrate to all compartments of the kidney (Ramos-Salazar and Baines, 1986), this should not affect the interpretation of these experiments, since the cytosolic 5'-nucleotidases are not inhibited by α,β-methylene ADP (Worku and Newby, 1983).

Adenosine release into the perfusate of rat or guinea pig hearts was not blocked with α,β-methylene ADP (Frick and Lowenstein, 1976; Schrader, 1983), although the small component released into the lymphatic drainage may be prevented (Imai et al., 1986). Adenosine formation from cultured neonatal rat heart cells that remained >99% intact was also not impaired by an antiserum that inhibited ecto-5'-nucleotidase (Meghji et al., 1985). On the other hand, net release of adenosine from freshly isolated adult rat cardiocytes (Bukoski and Sparks, 1986) was partially inhibited by α,β-methylene ADP, although the degree of cellular disruption was not measured. The evidence of cellular disruption was not measured. The evidence for adenosine production from extracellular nucleotides in the heart is, therefore, unconvincing.

Adenosine production from superfused electrically stimulated synaptosome beds was also not prevented by α,β-methylene ADP (Pons et al., 1980; Daval and Barberis, 1981), despite evidence from other studies that ATP was released (*see above*). This suggests that production of adenosine was overwhelmingly from other sources. Recent studies with a highly purified preparation of cholinergic synaptosomes (Richardson et al., 1987) show that inhibition of 5'-nucleotidase can block both adenosine formation from endogenous nucleotides and its effect to reduce acetylcholine release. These experiments in which cytoplasmic sources of production were absent illustrate the point that local production of small quantities of adenosine in the synapses of selected neurons might still provide regulatory concentrations.

It is most disappointing that the function of endothelial ecto-enzymes has not been tested directly in relation to platelet–vessel wall interactions. The kinetic properties of the enzymes suggest that platelet aggregation in the vicinity of endothelium should result in a burst of proaggregatory ADP production and then a delayed burst of antiaggregatory adenosine formation (Gordon et al., 1986). Despite this, enhanced aggregation of platelets at the surface of a damaged vessel wall has yet to be demonstrated during inhibition of ecto-5'-nucleotidase.

The absolute rate of adenosine formation from intact polymorphonuclear leucocytes undergoing ATP breakdown was not at all reduced by inhibition of ecto-5'-nucleotidase with antibodies (Newby and Holmquist, 1981). Net formation of adenosine from resting or dinitrophenol poisoned hepatocytes were likewise unaffected by methylene ADP (Bontemps et al., 1983; Belloni et al., 1985). These studies provide further evidence for a pathway of adenosine formation that does not involve the ecto-5'-nucleotidase.

The distribution of ecto-nucleotidases has also been used as a third method for arguing in favor or against adenosine formation from extracellular nucleotides. Ecto-ATPase has been found, for example, in synaptosome fractions from the brain (Sorensen and Mahler, 1982; Nagy et al., 1983) and from the electric organ of *Torpedo marmorata* (Keller and Zimmerman, 1983). Ecto-5'-nucleotidase has, however, a largely extrasynaptic location according to cytochemical methods (review by Kreutzberg et al., 1986) or immunological methods (Richardson, 1983), although enzyme activity may be associated with some limited synaptosomal subpopulations (Kreutzberg et al., 1986; Richardson and Brown, 1987). The absence of a close association between 5'-nucleotidase and A_1-adenosine receptor localization in the hippocampus (Lee et al., 1986) and retina (Braas et al., 1987) of various mammals also argues against a general role for catabolism of released nucleotides in providing the adenosine for action at these receptors.

The virtual absence of ecto-5'-nucleotidase from the pigeon ventricle was recently exploited (Newby and Meghji, 1986) to test whether the enzyme was essential for ischemia-induced adenosine formation. Adenosine formation proceeded at a rate of 410 ± 4 nmol/min/g wet wt during the first 2 min of ischemia, despite the presence of only 16 ± 4 nmol/min/g wt of ecto-5'-nucleotidase. Mechanisms other than ecto- 5'-nucleotidase are, therefore, responsible for adenosine formation in this tissue.

3.2. Adenosine Formation from Intracellular Nucleotides

Tissue concentrations of adenosine are frequently in excess of those in tissue effluents, and this difference persists during ATP catabolism (reviewed by Arch and Newsholme, 1978). These data provided the first suggestion that adenosine might be produced by an intracellular pathway. This conclusion was challenged, however (Berne and Rubio, 1974), on the grounds that adenosine-metaboliz-ing enzymes are active in the cytosol and that tissue adenosine might be present instead in a poorly perfused interstitial compartment. A large proportion of tissue adenosine was later shown to be in a metabolically inactive complex with S-adenosylhomocysteine hy-drolase (Ueland and Saebo, 1979).

Intracellular formation of adenosine was first demonstrated conclusively in intact polymorphonuclear leucocytes (Newby and Holmquist, 1981) by simple separation of cells and medium. Simil-ar experiments were later conducted with neonatal rat heart cells in culture (Meghji et al., 1985). An ingeneous strategy was used to demonstrate cytosolic production of adenosine in the perfused guinea pig heart. Schrader et al. (1981) and Schutz et al. (1981) ex-ploited the reversibility of the cytosolic enzyme S-adenosylhomo-cysteine hydrolase to trap cytoplasmic adenosine as S-adenosyl-homocysteine by adding exogenous L-homocysteine. Accumulation of S-adenosylhomocysteine was enhanced in the simultaneous pres-ence of a nucleoside transport inhibitor. This ruled out the possibil-ity that homocysteine combined with adenosine originally released into the interstitial fluid and then was taken up again via the nucleoside transporter. Trapping of cytosolic adenosine with L-homocysteine has been used recently to demonstrate adenosine formation in the cytosol of brain tissue (McIlwain and Poll, 1986).

Nucleoside transport inhibitors have been shown to block the release of adenosine from neonatal heart cells (Meghji et al., 1985), from adult cardiocytes (Bukoski and Sparks, 1986), and from hepa-tocytes (Belloni et al., 1985). This provides further evidence for cytoplasmic production of adenosine and identifies the symmetric nucleoside transporter as the mechanism of adenosine release. It does, however, pose the question as to how nucleoside transport

inhibitors can act as coronary vasodilators. If they prevent release
of adenosine from cardiac myocytes, how can they increase the
concentration of adenosine at smooth muscle cell surface receptor
sites? This question has been addressed by a mathematical modeling study (Newby, 1986). The model predicted that the quantitative
effect of nucleoside transport inhibitors was always less profound on
adenosine release from cardiomyocytes than on uptake and inactivation by smooth muscle cells and endothelium. The potentiation of
adenosine action by nucleoside transport inhibitors was, thereby,
explained. The model reemphasized in addition the need for a sufficiently active nucleoside transporter in the putative adenosine-
forming cells so as to allow export rather than rephosphorylation or
deamination of cytosolic adenosine. This problem is particularly
acute with respect to adenosine kinase, which has a K_m value below 1
μM (Arch and Newsholme, 1978). Arch and Newsholme (1978)
suggested that the rate of adenosine formation might be sufficiently
rapid to saturate the adenosine kinase even under basal conditions.
Acceleration of adenosine formation would then lead to a dispro-
portionate rise in adenosine concentration owing to the operation of
a substrate cycle. Experimental tests of this hypothesis do reveal
formation and rephosphorylation of adenosine under basal conditions. They suggest, however, that the kinase is far from saturated
(Newby et al., 1983; Bontemps et al., 1983; Achterberg et al., 1986).

It is likely that an increase in adenosine concentration is brought
about largely by increasing its rate of formation. Acceleration of
adenosine formation can be observed even when adenosine metabolism is inhibited (Newby and Holmquist, 1981; Newby et al., 1983;
Meghji et al., 1985). Inhibition of the adenosine kinase has been
observed during extreme ATP-depletion (Newby et al., 1983), but
other evidence that it contributes significantly to increasing adenosine release is lacking. No physiological inhibitors of adenosine
deaminase are known.

3.2.1. Adenosine Formation by Concerted Action
of Adenylate Kinase and Cytosolic 5'-Nucleotidase

The pathways that might contribute to cytoplasmic adenosine
formation are illustrated in Fig. 2. The enzyme adenylate kinase is
highly active in mammalian cells, and therefore, catalyzes a thermo-

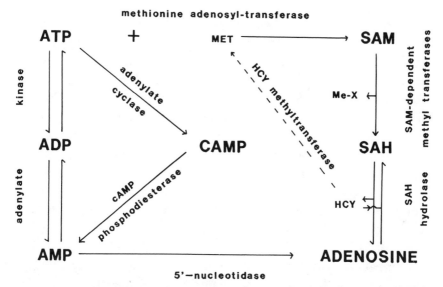

Fig. 2. Pathways of adenosine formation from cytosolic ATP. Abbreviations used: HCY, homocysteine; MET, methionine; SAH, S-adenosylhomocysteine; SAM, S-adenosylmethionine; HCY methyltransferase, Vitamin B12-dependent methyltetrahydrofolate-homocysteine methyltransferase.

dynamic equilibrium between the cytosolic concentrations of ATP, ADP, and AMP (Newsholme and Start, 1973). Since the ATP concentration in resting cells may be up to 1000x the AMP concentration, a very small percentage fall in ATP concentration can give rise to a large-fold increase in cytoplasmic AMP concentration (Newsholme and Start, 1973; Nishiki et al., 1978; Bunger and Soboll, 1986). Thus, the myokinase equilibrium provides an exquisite sensor of the balance between the rates of formation and hydrolysis of ATP (Lowenstein et al., 1983). Of the two products of ATP, ADP appears to control the rate of oxidative phosphorylation and AMP the rate of glycolysis (Lowenstein et al., 1983). It has been tempting, therefore, to suggest that hydrolysis of AMP to adenosine by a suitable 5'-nucleotidase provides the link between net ATP breakdown and adenosine release (Rubio et al., 1974; Lowenstein et al., 1983; Newby, 1984).

Early hypotheses concerning the mechanism of adenosine formation assumed that the plasma membrane 5'-nucleotidase was responsible for hydrolysis cytosolic AMP (Arch and Newsholme, 1978; Berne, 1980). The discovery that the enzyme is an ecto-enzyme (*see above*) forced a reappraisal of its role. Experiments by Frick and Lowenstein (1978) and Dornand et al. (1979) showed that radio-labeled AMP could be incorporated into cellular nucleotides more rapidly than adenosine. This suggested that 5'-nucleotidase might function as a "transmembrane hydrolase" taking AMP from the external membrane face and producing adenosine at the cytoplasmic side. The enzyme might then operate, physiologically, in the opposite direction. This attractive hypothesis appears, however, to be incorrect. First, incorporation of AMP into cellular nucleotides requires both 5'-nucleotidase and a separate adenosine transport protein (Fleit et al., 1975; Sasaki et al., 1983). Secondly, ATP and ADP are potent competitive inhibitors (Burger and Lowenstein, 1970; Pearson, 1985), suggesting that the enzyme would be virtually inactive if present at the cytoplasmic side of the membrane. Thirdly, inhibition of the enzyme fails to block adenosine or inosine formation from intact cells under conditions where the cytoplasmic concentrations of AMP and IMP are greatly elevated (Newby, 1980; Newby and Holmquist, 1981; Belloni et al., 1985; Meghji et al., 1985). Fourthly, if cells in which the ecto-enzyme is inhibited with antiserum at $4°C$ (to prevent endocytosis of antibody) are homogenized gently, no new activity is revealed (Stanley et al., 1980). Activity does increase, however, if such homogenates are treated with detergent to expose activity on the extracytoplasmic face of endocytic vesicles (Luzio et al., 1986). Lastly, the ecto-5'-nucleotidase appears to have a very small cytoplasmic domain (Baron et al., 1986).

A soluble 5'-nucleotidase is widely distributed in rat tissues (Newby et al., 1987) and in the hearts of mammals (Table 1). It has been purified to homogeneity from the livers and hearts of species as diverse as chickens and rats (Naito and Tsushima, 1976; Itoh, 1981a; Itoh and Oka, 1985; Itoh et al., 1986). The purified enzyme is an allosteric protein (Itoh, 1982) with four subunits. The preferred substrate of the enzyme is IMP ($K_m = 0.2$ mM). The K_m for AMP depends on the presence of activators and inhibitors, but minimum values fall in the range of 2.6–10 mM (Worku and Newby, 1983; Itoh, 1981a,b; Van den Berghe et al., 1977; Itoh et al., 1986). ATP activates the enzyme's activity towards AMP with a K_a of approximately 0.3 mM and a Hill coefficient close to 4. Inorganic phosphate inhibits the enzyme with a K_i of approximately 6 mM and a Hill coefficient close to 1.

Both the soluble and plasma membrane 5'-nucleotidases contain active site histidine (Worku et al., 1984), but they are otherwise different. The enzymes may be distinguished pharmacologically, since competitive inhibitors and antisera that block the ecto-enzyme activity have no effect on the cytosolic 5'-nucleotidase (Newby, 1980; Worku and Newby, 1983). Likewise, product nucleosides that are noncompetitive inhibitors of the soluble 5'-nucleotidase are only weak competitive inhibitors of the ecto-enzyme (Newby et al., 1975; Worku and Newby, 1982).

There is agreement that the cytosolic 5'-nucleotidase, by hydrolyzing IMP, controls the total purine nucleotide concentration of cells (Van den Berghe et al., 1977). The high K_m for AMP of the enzyme casts doubt, however, on its physiological role in adenosine formation (Van den Berghe et al., 1977; Itoh, 1981a). Its allosteric properties, on the other hand, suggest that the enzyme may be active towards AMP only during the early phases of nucleotide catabolism (Itoh, 1981b). This would allow the enzyme to fulfill a signal-generating capacity without further depleting an already-compromised nucleotide pool. Studies of intact polymorphonuclear leucocytes poisoned with 2-deoxyglucose (Worku and Newby, 1983) did, indeed, demonstrate an initial activation and subsequent inhibition of adenosine formation during ATP breakdown. Parallel studies were conducted with the purified soluble 5'-nucleotidase from rat liver using concentrations of ATP, ADP, AMP, IMP, and P_i, which mimicked those in the intact cell experiments. Both the maximum rate of adenosine formation and its biphasic nature could be accounted for by the purified enzyme (Worku and Newby, 1983). Similar studies using rat erythrocytes which contain an unusually low activity of soluble 5'-nucleotidase demonstrate a correspondingly low rate of adenosine production (Newby et al., 1987). This remains the only experimental evidence that the soluble 5'-nucleotidase is responsible for adenosine formation. A potent selective inhibitor of the enzyme would be invaluable to probe further its role.

Ischemia induces a very high rate of adenosine formation in the heart (Deuticke and Gerlach, 1966; Newby et al., 1987). These rates appear too great to be accounted for by the known kinetic properties of the purified soluble 5'-nucleotidase (Newby et al., 1987). There is a suggestion, however, that the enzyme may become modified during purification (Lowenstein et al., 1986). The 5'-nucleotidase activities were, therefore, reinvestigated in unpurified extracts of the pigeon ventricle, a tissue that lacks ecto-5'-nucleotidase (Meghji et al., 1988). The assay system contained concentrations of ATP, ADP, and AMP chosen to be in equilibrium with adenylate kinase. The pigeon ventricle extract catalyzed AMP hydrolysis sufficiently rapidly to explain ischemia-induced adenosine formation (Meghji et al., 1988). Further characterization is now needed of the enzyme or enzymes responsible.

Isolated mitochondria have been reported to generate adenosine (Bukoski et al., 1983; Asimakis et al., 1985; Bukoski et al., 1986), and this observation might explain the proposed correlation between the capacity of tissues for oxidative metabolism and for adenosine formation (*see above*). Mitochondria appear, however, to serve solely as a source of AMP; a separate 5'-nucleotidase enzyme is still needed to catalyze adenosine formation (Asimakis et al., 1985; Bukoski et al., 1986).

3.2.2. Adenosine Formation from Hydrolysis
of Adenosine 3.5'-Phosphate (Cyclic AMP)

It is possible that cyclic AMP may act as a source of adenosine either intracellularly following its metabolism to 5'AMP, or extracellularly after being lost from the cytosol (Cramer, 1977; Doore et al., 1975; Pull and McIlwain, 1977). This hypothesis has received little direct examination, however, although phosphodiesterase inhibitors produce small reduction of adenosine release from neuronal tissue consistent with an origin as cyclic AMP (Stone et al., 1981). It has been noted elsewhere (Stone, 1981c) that this idea would also be consistent with the observed peaking of adenosine concentration in stimulated neural tissue later than the concentration of cyclic AMP.

3.2.3. Adenosine Formation from the Transmethylation Pathway

In neuronal tissue, S-adenosylmethionine-dependent methyltransferases are involved in the biosynthesis and/or degradation of the biogenic amines—dopamine, epinephrine, norepinephrine, histamine, and 5-hydroxytryptamine (Borchardt, 1980), the synthesis of choline (Blusztajn et al., 1982), methylation of membrane phospholipids (Hirata and Axelrod, 1980), and protein carboxymethylation (*see* Paik and Kim, 1980). Transmethylation, which describes the transfer of the methyl group from S-adenosylmethionine (SAM) to a variety of methyl acceptors (as mentioned above), forms S-adenosylhomocysteine (SAH) as a product of the reaction. SAH, if allowed to accumulate, inhibits transmethylation (*see* Usdin et al., 1979), but normally is metabolized further by SAH-hydrolase, which catalyzes its reversible hydrolysis to adenosine and homocysteine (*see* Fig. 2). The hydrolysis of SAH is the only known metabolic pathway for the formation of homocysteine (Ueland, 1982). In addition, this reaction forms adenosine.

Direct experimental evidence for continuous turnover of SAM in mammalian brain was provided by the work of Spector et al. (1980), who demonstrated that ^{35}S-labeled methionine, injected iv into rats or intraventricularly into rabbits, is recycled via homocysteine (Fig. 2). Furthermore, the observation that inclusion of L-homocysteine thiolactone and ^{14}C-adenosine with incubated rat hippocampal slices leads to SAH formation (Reddington and Pusch, 1983) demonstrated that SAH-hydrolase is active in this preparation. L-homocysteine thiolactone also decreased the evoked release of adenosine from guinea pig cortical slices via formation of SAH (McIlwain, 1985; McIlwain and Poll, 1986). Incubation of hippocampal brain slices with ^{14}C-adenosine in the absence of homocysteine was found, however, not to result in significant labeling of SAH, which led Reddington and Pusch (1983) to conclude that accumulation of SAH is unlikely to be important in mediating any biological effects of adenosine in the CNS. The possibility that SAH may be important as a metabolic source of adenosine has not been studied in the brain.

The production of adenosine has generally been assumed to be principally from adenine nucleotides via the dephosphorylation of 5'-AMP (*see above*). Recently, however, the contribution of the transmethylation pathway to adenosine formation in the isolated, perfused, and nonworking guinea pig heart has been investigated (Lloyd and Schrader, 1986). The cellular transmethylation rate was estimated, in this preparation, by measuring the rate of dilution of a prelabeled SAH pool. This occurred as a result of SAH synthesis from unlabeled SAM. During steady-state conditions, the rate of SAH synthesis from SAM is equivalent to the transmethylation rate, which in turn reflects the net adenosine production rate from this pathway. In a separate series of experiments, the total rate of adenosine production in the isolated guinea pig heart was estimated by measuring adenosine release rate in the presence of adenosine deaminase and adenosine kinase inhibitors. From a comparison of transmethylation rate with total adenosine production rate, it was calculated that hydrolysis of SAH contributed more than 90% of the total amount of adenosine formed by the heart during normoxic perfusion (95% O_2). During hypoxic perfusion (30% O_2), this fell to less than 20%, the primary source of adenosine presumably then being 5'-AMP. Although the transmethylation pathway has not yet been quantified in neuronal tissues, it is possible that, as in the heart, it produced a significant and continuous supply of adenosine.

4. Conclusions and Implications for the Physiological Role of Adenosine

Release of adenine nucleotides has been demonstrated from neurons and blood platelets that contain vesicle-bound nucleotides and in small quantities from nonneuronal sources. Released nucleotides are broken down to adenosine by ecto-nucleotidases. The contribution of released nucleotides to production of adenosine may, except in particular locations, be small in comparison to release from cytoplasmic sources. Cytoplasmic production of adenosine has been demonstrated in polymorphonuclear leucocytes, heart, liver and brain. It is not catalyzed by the ecto-5'-nucleotidase. A soluble 5'-nucleotidase acting in concert with adenylate kinase may produce adenosine in response to an imbalance between ATP generation and utilization. Adenosine formation from cyclic AMP may occur, additionally, during activation of adenylate cyclase. Adenosine formation from the transmethylation pathway may make a substantial contribution especially to the basal rate of adenosine formation and particularly in tissues with high activities of methyl transferases.

In no case is the release of adenosine compatible with a conventional neurotransmitter function, as suggested by Sattin and Rall (1970). Adenosine is poorly released by potassium from preparations in comparison with transmitters, and more effectively released by ouabain and veratridine. Indeed, potassium (depolarization) induced release is often not dependent on calcium ions, and a simultaneous influx of sodium and calcium has been proposed as the relevant stimulus (Hollins and Stone, 1980). The differences in time course of release also represent an important distinction from neurotransmitters. The mechanism of adenosine production is also unlike that of any hormone (Newby, 1984), despite similarities between adenosine and hormones in their mechanism of action.

The difference in possible functional significance between adenosine produced from extracellular and from cytoplasmic nucleotides is illustrated in Fig. 3. In the left-hand panel, a stimulus provokes ATP release with or without a hormone or cotransmitter. ATP may exert a direct effect through purinoceptors, but is then degraded to adenosine by ecto-enzymes. Adenosine then acts on distinct purine receptors to bring about feedback or feedforward inhibition. Release of adenosine is a direct consequence of stimulation, and is similar therefore to that of a hormone or neurotransmitter. Its formation may, however, be delayed or occur at a remote site. Variations in its concentration might be damped, giving rise to the possibility of hysteresis. Adenosine may, therefore, be regarded as a modulator rather than a hormone or neurotransmitter.

The production of adenosine from cytosolic ATP is illustrated in the right-hand panel of Fig. 3. In this case, the response elicited by stimulation may be mechanical, electrical, or metabolic as well as secretion. There need be no fixed relationship between the magnitude of the stimulus and the rate of adenosine formation as well as demonstrated in the studies of Fredholm and Sollevi (1981) and Bardenheuer and Schrader (1986). This is because the production of adenosine depends also on the supply of oxygen and of exogenous and endogenous substrates, as well as on the energy-demands from simultaneously applied stimuli. Adenosine is not, therefore, a response to stimulation, but an autonomous stimulus generated by the target cell as a consequence of the intrinsic balance of energy supply and de-

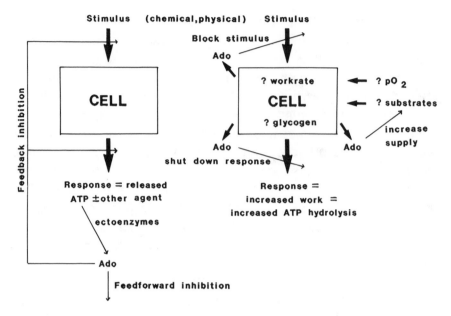

Fig. 3. Possible physiological roles of adenosine.

mand. The original stimulus thus provokes a new and opposing stimulus (adenosine), giving rise to the description "retaliatory" metabolite for adenosine (Newby, 1984).

The diverse aspects of adenosine's properties allow it to be viewed in many different ways, and to be classified variously as hormone, neuromodulator, protective, or retaliatory metabolite (Stone, 1981c; Newby, 1984; Snyder, 1985). The overall concept must be that of an ubiquitous, arguably primitive molecule importantly involved in homeostasis and the control of tissue function and integration. It is a compound for which further consideration of the sites and circumstances of its production and release must be undertaken, in parallel with an appreciation of its actions and receptors, if a complete understanding of its biological role is to be achieved.

References

Abood, L. G., Koketsu, K., and Miyamoto, S. (1962) Outflux of various phosphates during membrane depolarisation of excitable tissues. *Am. J. Physiol.* **202**, 469–474.

Achterberg, P. W., Harmsen, E., and De Jong, J. W. (1985) Adenosine deaminase inhibition and myocardial purine release during normoxia and ischemia. *Cardiovasc. Res.* **19,** 593–598.

Achterberg, P. W., Stroeve, R. J., and De Jong, J. W. (1986) Myocardial adenosine cycling rates during normoxia and under conditions of stimulated purine release. *Biochem. J.* **235,** 13–17.

Alhumayyd, M. and White, T. D. (1985) Adrenergic and possible nonadrenergic sources of ATP release from nerve varicosities isolated from ileal myenteric plexus. *J. Pharmacol. Exp. Ther.* **233,** 796–800.

Angel, A., Desai, K. S., and Halperin, H. L. (1971) Reduction in adipocyte ATP by lipolytic agents: relation to intracellular free fatty acid accumulation. *J. Lipid Res.* **12,** 203–213.

Arch, J. R. S. and Newsholme, E. A. (1978) The control of the metabolism and the hormonal role of adenosine. *Essays Biochem.* **14,** 82–123.

Asimakis, G. K., Wilson, D. E., and Conti, V. R. (1985) Release of adenosine and AMP from rat heart mitochondria. *Life Sci.* **37,** 2373–2380.

Barberis, C., Guibert, B., Daudet, F., Charriere, B., and Leviel, V. (1984) *In vivo* release of adenosine from cat basal ganglia—studies with a push pull cannula. *Neurochem. Int.* **6,** 545–551.

Bardenheuer, H. and Schrader, J. (1986) Supply-to-demand ratio for oxygen determines formation of adenosine by the heart. *Am. J. Physiol.* **250,** H173–H180.

Baron, M. D., Pope, B., and Luzio, J. P. (1986) The membrane topography of 5'-nucleotidase in rat hepatocytes. *Biochem. J.* **236,** 495–502.

Belloni, F. L., Elkin, P. L., and Giannotto, B. (1985) The mechanism of adenosine release from hypoxic rat liver cells. *Brit. J. Pharmac.* **85,** 441–446.

Belloni, F. L., Phair, R. D., and Sparks, H. V. (1979) The role of adenosine in prolonged vasodilation following flow-restricted exercise of canine skeletal muscle. *Circ. Res.* **44,** 759–766.

Bencherif, M., Berne, R. M., and Rubio, R. (1986) The release of purines by the frog sympathetic ganglion is the result of activation of postsynaptic elements. *J. Physiol.* **371,** 274P.

Bender, A. S., Wu, P. H., and Phillis, J. W. (1981) The rapid uptake and release of [^{3}H]adenosine by rat cerebral cortical synaptosomes. *J. Neurochem.* **36,** 651–660.

Berne, R. M. (1963) Cardiac nucleotides in hypoxia: possible role in regulation of coronary blood flow. *Am. J. Physiol.* **204,** 317–322.

Berne, R. M. (1964) Regulation of coronary blood flow. *Annu. Rev. Physiol.* **44,** 1–29.

Berne, R. M. (1980) The role of adenosine in the regulation of coronary blood flow. *Circ. Res.* **47,** 807–813.

Berne, R. M. and Rubio, R. (1974) Adenine and nucleotide metabolism in the heart. *Circ. Res.* **34** and **35** suppl. 3, 109–118.

Blusztajn, J. K., Zeisel, S. H., and Wurtman, R. J. (1982) Phospholipid methylation and cholinergic neurons, in *Biochemistry of S-Adenosylmethionine and Related Compounds* (Usdin, E., Borchardt, R. T., and Craveling, C. R., eds.), Macmillan, London, pp. 155–164.

Bockman, E. L. and McKenzie, J. E. (1983) Tissue adenosine content in active soleus and gracilis muscle of cats. *Am. J. Physiol.* **244,** H552–H559.

Bockman, E. L., Berne, R. M., and Rubio, R. (1976) Adenosine and active hyperaemia in dog skeletal muscle. *Am. J. Physiol.* **230,** 1531–1537.

Bontemps, F., Van den Berghe, G., and Hers, H. G. (1983) Evidence for a substrate cycle between AMP and adenosine in isolated hepatocytes. *Proc. Natl. Acad. Sci. USA* **80,** 2829–2833.

Borchardt, R. T. (1980) *N*- and *O*-methylation, in *Enzymatic Basis of Detoxification* (Jakoby, W. B., ed.), Academic, New York, pp. 43–62.

Borregaard, N. and Herlin, T. (1982) Energy metabolism of human neutrophils during phagocytosis. *J. Clin. Invest.* **70,** 550–557.

Braas, K. M., Zarbin, M. A., and Snyder, S. H. (1987) Endogenous adenosine and adenosine receptors localized to ganglian cells of the retina. *Proc. Natl. Acad. Sci. USA* **84,** 3906–3910.

Braas, K. M., Newby, A. C., Wilson, V. S., and Snyder, S. H. (1986) Adenosine-containing neurones in the brain localized by immunocytochemistry. *J. Neurosci.* **6,** 1952–1961.

Bukoski, R. D. and Sparks, H. V. (1986) Adenosine production and release by adult rat cardiocytes. *J. Mol. Cell. Cardiol.* **18,** 595–605.

Bukoski, R. D., Sparks, H. V., and Mela, L. M. (1983) Rat heart mitochondria release adenosine. *Biochem. Biophys. Res. Commun.* **113,** 990–995.

Bukoski, R. D., Sparks, H. V., and Mela-Riker, L. M. (1986) Mechanism of adenosine production by rat heart mitochondria. *Biochim. Biophys. Acta.* **884,** 25–30.

Bunger, R. and Soboll, S. (1986) Cytosolic adenylates and adenosine release in perfused working heart. *Eur. J. Biochem.* **159,** 203–213.

Burger, R. M. and Lowenstein, J. M. (1970) Preparation and properties of 5'-nucleotidase from smooth muscle of small intestine. *J. Biol. Chem.* **245,** 6274–6280.

Burnstock, G. (1972) Purinergic nerves. *Pharmacol. Revs.* **24,** 509–581.

Burnstock, G. (1976a) Do some nerve cells release more than one transmitter? *Neurosci.* **1,** 239–248.

Burnstock, G. (1976b) Purine nucleotides. *Adv. Biochem. Psychopharm.* **15,** 225–235.

Burnstock, G. (1981) Neurotransmitters and trophic factors in the autonomic nervous system. *J. Physiol.* **313,** 1–35.

Burnstock, G., Crowe, R., and Wong, H. K. (1979) Comparative pharmacological and histochemical evidence for purinergic inhibitory innervation of the portal vein of the rabbit but not guinea pig. *Br. J. Pharmacol.* **65,** 377–388.

Burnstock, G., Campbell, G., Satchell, D. G., and Smythe, A. (1970) Evidence that ATP or a related nucleotide is the transmitter substance released by nonadrenergic inhibitory nerves in the gut. *Brit. J. Pharmacol.* **40,** 668–688.

Burnstock, G., Cocks, T., Kasakov, L., and Wong, H. K. (1978a) Direct evidence for ATP release from nonadrenergic, noncholinergic (purinergic) nerves in the guinea-pig taenia coli and bladder. *Eur. J. Pharmacol.* **49,** 145–149.

Burnstock, G., Cocks, T., Crowe, R., and Kasakov, L. (1978b) Purinergic innervation of the guinea pig urinary bladder. *Br. J. Pharmacol.* **63,** 125–138.

Chaudhry, A., Downie, J. W., and White, T. D. (1984) Tetrodotoxin resistant release of ATP from superfused rabbit detrusor muscle during electrical field stimulation in the presence of luciferin-luciferase. *Can. J. Physiol. Pharmacol.* **62,** 153–156.

Clemens, M. G. and Forrester, T. (1980) Appearance of ATP in the coronary sinus effluent from isolated working rat heart in response to hypoxia. *J. Physiol.* **312,** 143–158.

Cramer, H. (1977) Cyclic 3'5'-nucleotides in extracellular fluids of neural systems. *J. Neurosci. Res.* **3,** 241–246.

Cronstein, B. N., Kramer, S. B., Weissman, G., and Hirschhorn, R. (1983) Adenosine: a physiological modulator of superoxide anion generation by human neutrophils. *J. Exp. Med.* **158,** 1160–1177.

Cusack, N. J., Pearson, J. D., and Gordon, J. L. (1983) Stereoselectivity of ectonucleotidases on vascular endothelial cells. *Biochem. J.* **214,** 975–981.

Daval, J. L. and Barberis, C. (1981) Release of radiolabeled adenosine from perfused synaptosome beds. *Biochem. Pharmacol.* **30,** 2559–2567.

Daval, J. L., Barberis, C., and Gayet, J. (1980) Release of adenosine derivatives from superfused synaptosome preparations: effects of depolarising agents and metabolic inhibitors. *Brain Res.* **181,** 161–174.

Des Rosiers, C., Lalanne, H., and Willemot, J. (1982) Glycerol-induced adenine nucleotide catabolism in rat liver cells. *Can. J. Biochem.* **60,** 1101–1108.

Deussen, A., Moser, G., and Schrader, J. (1986) Contribution of coronary endothelial cells to cardiac adenosine production. *Pflugers Arch.* **406,** 608–614.

Deuticke, B. and Gerlach, E. (1966) Abbau der freier Nucleotide in Herz, Skeletmuskel, Gehir und Leben der Ratter bei Sauerstoffmangel. *Pflugers Arch.* **292,** 239–254.

Dobson, J. G., Rubio, R., and Berne, R. M. (1971) Role of adenine nucleotides and inorganic phosphate in the regulation of skeletal muscle blood flow. *Circ. Res.* **29,** 375–384.

Doore, B. J., Basher, M. M., Spitzer, N., Mawe, R. C., and Saier, M. N. (1975) Cyclic AMP output from rat glioma cultures. *J. Biol. Chem.* **250,** 4371, 4372.

Dornand, J., Bonnafous, J. C., Gavach, C., and Mani, J. C. (1979) 5'-nucleotidase facilitated adenosine transport by mouse lymphocytes. *Biochemie* **61,** 973–977.

Fain, J. N. (1979) Effect of lipolytic agents on adenosine and AMP formation by fat-cells. *Biochim. Biophys. Acta.* **573,** 510–520.

Fleit, H., Conklyn, M., Stebbins, R. D., and Silber, R. (1975) Function of 5'-nucleotidase in the uptake of adenosine from AMP by human lymphocytes. *J. Biol. Chem.* **250,** 8889–8892.

Foley, D. H., Miller, W. L., Rubio, R., and Berne, R. M. (1979) Transmural distribution of myocardial adenosine content during coronary constriction. *Am. J. Physiol.* **226,** H833–H838.

Forrester, T. (1972) A quantitative estimation of ATP released from human forearm muscle during sustained exercise. *J. Physiol. (London)* **221,** 26P, 27P.

Forrester, T. (1981) Adenosine or ATP, in *Vasodilatation* (Vanhoutte, P. M. and Leusen, I., eds.), Raven, New York, pp. 205–229.

Fredholm, B. B. (1976) Release of adenosine like material from isolated perfused dog adipose tissue following sympathetic nerve stimulation and its inhibition by adrenergic alpha receptor blockade. *Acta. Physiol. Scand.* **96**, 422–430.

Fredholm, B. B. and Hedqvist, P. (1978) Release of [³H]purines from [³H]adenine labeled rat kidney following sympathetic nerve stimulation and its inhibition by alpha adrenoreceptor blockade. *Br. J. Pharmacol.* **64**, 239–246.

Fredholm, B. B. and Hjemdahl, P. (1979) Uptake and release of adenosine in isolated rat fat cells. *Acta. Physiol. Scand.* **105**, 257–267.

Fredholm, B. B. and Sollevi, A. (1981) The release of adenosine and inosine from canine subcutaneous adipose tissue by nerve stimulation and norepinephrine. *J. Physiol.* **313**, 351–367.

Fredholm, B. B. and Vernet, L. (1978) Morphine increases depolarisation induced purine release from hypothalamic synaptosomes. *Acta. Physiol. Scand.* **104**, 502–504.

Fredholm, B. B. and Vernet, L. (1979) Release of [³H]nucleotides from [³H]adenine labeled hypothalamic synaptosomes. *Acta. Physiol. Scand.* **106**, 97–107.

Fredholm, B. B., Fried, G., and Hedqvist, P. (1982) Origin of adenosine released from rat vas deferens by nerve stimulation. *Eur. J. Pharmacol.* **79**, 233–243.

Frick, G. P. and Lowenstein, J. M. (1976) Studies of 5'-nucleotidase in the perfused rat heart. *J. Biol. Chem.* **251**, 6372–6378.

Frick, G. P. and Lowenstein, J. M. (1978) Vectorial production of adenosine by 5'-nucleotidase in the perfused heart. *J. Biol. Chem.* **253**, 1240–1244.

Gerlach, E., Deuticke, B., and Dreisbach, R. H. (1963) Der Nucleotid Abbau in Hertzmuskel bei Sauerstoffmangel und seine Mogliche Bedeutung fur die Coronardurchblutung. *Naturwissenschaften* **50**, 228, 229.

Gordon, J. L. (1986) Extracellular ATP: Effects, sources, and fate. *Biochem. J.* **233**, 309–319.

Gordon, J. L., Pearson, J. D., and Slakey, L. L. (1986) Hydrolysis of extracellular adenine nucleotides by cultured endothelial cells from pig aorta: feed forward inhibition of adenosine production at the cell surface. *J. Biol. Chem.* **261**, 15496–15504.

Hirata, H. and Axelrod, J. (1980) Phospholipid methylation and biological signal transmission. *Science* **209**, 1082–1090.

Hollins, C. and Stone, T. W. (1980) Characteristics of the release of adenosine from slices of rat cerebral cortex. *J. Physiol.* **303**, 73–82.

Hollins, C., Stone, T. W., and Lloyd, H. (1980) Neuronal (Na, K)-ATPase and the release of purines from mouse and rat cerebral cortex. *Neurosci. Lett.* **20**, 217–221.

Holmsen, H. and Weiss, H. J. (1979) Secretable storage pools in platelets. *Annu. Rev. Med.* **30**, 119–134.

Holton, P. (1959) The liberation of ATP on antidromic stimulation of sensory nerves. *J. Physiol.* **145**, 494–504.

Holton, F. A. and Holton, P. (1954) The capillary dilator substances in dry powders of spinal roots: a possible role of ATP in chemical transmission from nerve endings. *J. Physiol.* **126**, 124–140.

Huang, E. M. and Detwiler, T. C. (1986) Stimulus-response coupling mechanisms, in *Biochemistry of Platelets* (Philips, D. R. and Schuman, M. A., eds.), Academic, London, pp. 1–68.

Imai, S., Imai, H., and Jin, H. (1986) Myocardial tissue fluid adenosine and the hyperemic responses. *Pflugers Archiv.* **407** suppl. 1, S17.

Imai, S., Riley, A. L., and Berne, R. M. (1964) Effect of ischemia on adenine nucleotides in cardiac and skeletal muscle. *Circ. Res.* **15**, 443–450.

Israel, M., Lesbats, B., Meunier, F. M., and Stinnakre, J. (1976) Post synaptic release of ATP induced by single impulse transmitter action. *Proc. Roy. Soc. B.* **193**, 461–468.

Itoh, R. (1981a) Purification and some properties of cytosol 5'-nucleotidase from rat liver. *Biochim. Biophys. Acta.* **657**, 402–410.

Itoh, R. (1981b) Regulation of cytosol 5'-nucleotidase by adenylate energy charge. *Biochim. Biophys. Acta.* **659**, 34–37.

Itoh, R. (1982) Studies on some molecular properties of cytosol 5'-nucleotidase from rat liver. *Biochim. Biophys. Acta.* **716**, 110–113.

Itoh, R. and Oka, J. (1985) Evidence for existence of a cytosol 5'-nucleotidase in chicken heart: comparison of some properties of heart and liver enzymes. *Comp. Biochem. Physiol.* **81B**, 159–163.

Itoh, R., Oka, J., and Ozasa, H. (1986) Regulation of heat cytosolic 5'-nucleotidase by adenylate energy charge. *Biochem. J.* **235**, 847–851.

Jacobson, S. L. and Piper, H. M. (1986) Cell cultures of adult cardiomyocytes as models of the myocardium. *J. Mol. Cell. Cardiol.* **18**, 661–678.

Jhamandas, K. and Dumbrille, A. (1980) Regional release of [³H]adenosine derivatives from rat brain *in vivo*: effects of excitatory amino acids, opiate agonists and benzodiazepines. *Can. J. Physiol. Pharmacol.* **58**, 1262–1278.

Jonzon, B. and Fredholm, B. B. (1985) Release of purines, noradrenaline and GABA from rat hippocampal slices by field stimulation. *J. Neurochem.* **44**, 217–224.

Katori, M. and Berne, R. M. (1966) Release of adenosine from anoxic hearts: relationship to coronary blood flow. *Circ. Res.* **19**, 420–425.

Katsuragi, T. and Su, C. (1980) Purine release from vascular adrenergic nerves by high potassium and a calcium ionophore A23187. *J. Pharmacol. Exp. Ther.* **215**, 685–690.

Katsuragi, T. and Su, C. (1981) Facilitation by clonidine of high KCl induced purine release from the rabbit pulmonary artery. *Br. J. Pharmacol.* **74**, 709–713.

Keller, F. and Zimmerman, H. (1983) Ecto-ATPase activity at the cholinergic nerve endings of the Torpedo electric organ. *Life Sci.* **33**, 2635–2641.

Knabb, R. M., Ely, S. W., Bacchus, A. N., Rubio, R., and Berne, R. M. (1983) Consistent parallel relationships among myocardial oxygen consumption, coronary blood flow and pericardial infusate adenosine concentration with various interventions and beta blockade in the dog. *Circ. Res.* **53**, 33–41.

Kreutzberg, G. W., Heymann, D., and Reddington, M. (1986) 5'-nucleotidase in the nervous system, in *Cellular Biology of Ecto-Enzymes* (Kreutzberg, G. W., Reddington, M., and Zimmerman, H., eds.), Springer-Verlag, Berlin, pp. 148–175.

Kuperman, A. S., Volpert, W. A., and Okamoto, M. (1964) Release of adenine nucleotides from nerve axons. *Nature* **204,** 1000,1001.

Lagercrantz, H. (1976) On the composition and function of large dense cored vesicles in sympathetic nerves. *Neurosci.* **1,** 81–92.

Lee, K. S., Schubert, P., Reddington, M., and Kreutzberg, G. W. (1986) The distribution of adenosine A_1 receptors and 5'-nucleotidase in the hippocampal formations of several mammalian species. *J. Comp. Neurol.* **246,** 427–434.

Lee, K., Schubert, P., Gribkoff, V., Sherman, B., and Lynch, G. (1982) A combined in vivo/in vitro study of the presynaptic release of adenosine derivatives in the hippocampus. *J. Neurochem.* **38,** 80–83.

Levitt, B. and Westfall, D. P. (1982) Factors influencing the release of purines and norepinephrine in the rabbit portal vein. *Blood Vessels* **19,** 30–40.

Lloyd, H. G. E. and Schrader, J. (1986) The importance of the transmethylation pathway in the production of adenosine. *Pflug. Arch.* **407** Suppl. 1, S21.

Lloyd, H. G. E. and Stone, T. W. (1980) Factors effecting the release of purines from mouse cerebral cortex: potassium removal and metabolic inhibitors. *Biochem. Pharmacol.* **30,** 1239–1243.

Lloyd, H. G. E. and Stone, T. W. (1983) A different time course of purine release from rat brain slices and synaptosomes. *J. Physiol.* **340,** 57P, 58P.

Lomax, C. A. and Henderson, J. F. (1973) Adenosine formation and metabolism during ATP catabolism in Ehrlich ascites tumour cells. *Cancer Res.* **33,** 2825–2829.

Lowenstein, J. M., Naito, Y., and Collinson, A. R. (1986) Regulatory properties of intracellular and ecto-5'-nucleotidases and their possible role in production of adenosine. *Pflugers Arch.* **407** suppl. 1, S9.

Lowenstein, J. M., Yu, M. K., and Naito, Y. (1983) Regulation of adenosine metabolism by 5'-nucleotidase, in *Regulatory Function of Adenosine* (Berne, R. M., Rall, T. W., and Rubio, R., eds.), Martinus Nijhoff, The Hague, pp. 117–131.

Luchelli-Fortis, M. A., Fredholm, B. B., and Langer, S. Z. (1979) Release of radioactive purines from cat nictitating membrane labeled [³H]adenine. *Eur. J. Pharmacol.* **58,** 389–398.

Luzio, J. P., Bailyes, E. M., Baron, M., Siddle, K., Mullock, B. M., Geuze, H. J., and Stanley, K. K. (1986) The properties, structure, function, intracellular localization and movement of hepatic 5'-nucleotidase, in *Cellular Biology of Ectoenzymes* (Kreutzberg, G. W., Reddington, M., and Zimmerman, H., eds.), Springer-Verlag, Berlin, pp. 89–116.

McIlwain, H. (1972) Regulatory significance of the release and action of adenine derivatives in cerebral systems. *Biochem. Soc. Symp.* **36,** 69–85.

McIlwain, H. (1985) The endogenously formed adenosine of the brain: its status as a regulator signal appraised in relation to actions of homocysteine, in *Purines Pharmacology and Physiological Roles* (Stone, T. W., ed.), Macmillan, London, pp. 215–220.

McIlwain, H. and Poll, J. D. (1986) Adenosine in cerebral homeostatic role: appraisal through actions of homocysteine, colchicine and dipyridamole. *J. Neurobiol.* **17,** 39–49.

Mackenzie, I., Burnstock, G., and Dolly, J. D. (1982) The effects of purified botulinum neurotoxin type A on cholinergic, adrenergic and nonadrenergic atropine resistant autonomic neuromuscular transmission. *Neurosci.* **7,** 997–1006.

McKenzie, J. E., Steffan, R. P., and Haddy, F. J. (1982) Relationship between adenosine and coronary resistance in conscious exercising dogs. *Am. J. Physiol.* **242,** H24–H29.

Maire, J. C., Medilanski, J., and Straub, R. W. (1982) Uptake and release of adenosine derivatives in mammalian non-myelinated nerve fibre at rest and during activity. *J. Physiol.* **323,** 589–602.

Maire, J. C., Medilanski, J., and Straub, R. W. (1984) Release of adenosine, inosine and hypoxanthine from rabbit non-myelinated nerve fibres at rest and during activity. *J. Physiol.* **357,** 67–78.

Mann, J. S., Renwick, A. G., and Holgate, S. T. (1986) Release of adenosine and its metabolites from activated human leucocytes. *Clin. Sci.* **70,** 461–468.

Martin, S. E. and Bockman, E. L. (1986) Adenosine regulates blood flow and glucose uptake in adipose tissue of dogs. *Am. J. Physiol.* **250,** H1127–H1135.

Meghji, P., Holmquist, C. A., and Newby, A. C. (1985) Adenosine formation and release from neonatal-rat heart cells in culture. *Biochem. J.* **229,** 799–805.

Meghji, P., Middleton, K. H., and Newby, A. C. (1988) Absolute rates of adenosine formation during ischaemia in rat and pigeon hearts. *Biochem. J.* **249,** 695–703.

Meunier, F. M., Israel, M., and Lesbats, B. (1975) Release of ATP from stimulated nerve electroplaque junctions. *Nature* **257,** 407, 408.

Michaelson, D. M. (1978) Is presynaptic acetylcholine release accompanied by the secretion of the synaptic vesicle contents? *FEBS Lett.* **89,** 51–53.

Miller, W. L., Belardinelli, L., Bacchus, A., Foley, D. H., Rubio, R., and Berne, R. M. (1979) Canine myocardial adenosine and lactate production, oxygen consumption and coronary blood flow during stellate ganglia stimulation. *Circ. Res.* **45,** 708–718.

Mills, D. C. B. (1973) Changes in adenylate energy charge in human blood platelets induced by adenosine diphosphate. *Nature* **243,** 220–222.

Morel, N. and Meunier, F. M. (1981) Simultaneous release of acetylcholine and ATP from stimulated cholinergic synaptosomes. *J. Neurochem.* **36,** 1766–1773.

Morgan, B. P., Luzio, J. P., and Campbell, A. K. (1986) Intracellular Ca^{2+} and cell injury: a paradoxical role of Ca^{2+} in complement membrane attack. *Cell. Calcium* **7,** 399–411.

Mustafa, S. J. (1979) Effects of coronary vasodilator drugs on the uptake and release of adenosine in cardiac cells. *Biochem. Pharmacol.* **28,** 2617–2624.

Nagy, A., Schuster, T. A., and Rosenberg, M. D. (1983) Adenosine triphosphate activity at the external surface of chick brain synaptosomes. *J. Neurochem.* **40,** 226–234.

Naito, Y. and Tsushima, K. (1976) 5'-nucleotidase from chicken liver. Purification and some properties. *Biochim. Biphys. Acta.* **438,** 159–168.

Nakatsu, K. and Drummond, G. I. (1972) Adenylate metabolism and adenosine formation in the heart. *Am. J. Physiol.* **223**, 1119–1127.

Nees, S. and Gerlach, E. (1983) Adenine nucleotide and adenosine metabolism in coronary endothelial cells, in *Regulatory Function of Adenosine* (Berne, R. M., Rall, T. W., and Rubio, R., eds.), Martinus Nijhoff, The Hague, pp. 347–360.

Nees, S., Bock, M., Herzog, V., Becker, B. F., Des Rosiers, C., and Gerlach, E. (1985a) The adenine nucleotide metabolism of the coronary endothelium: implications for the regulation of coronary flow by adenosine, in *Adenosine: Receptors and Modulation of Cell Function* (Staphanovic, V., Rudolphi, K., and Schubert, P., eds.), IRL, Oxford, pp. 419–436.

Nees, S., Herzog, V., Becker, B. F., Bock, M., Des Rosiers, C. H., and Gerlach, E. (1985b) The coronary endothelium: a highly active metabolic barrier for adenosine. *Basic Res. Cardiol.* **80**, 515–529.

Newby, A. C. (1980) Role of adenosine deaminase, ecto-5'-nucleotidase and ecto-(non-specific phosphatase) in cyanide-induced AMP catabolism in rat polymorphonuclear leucocytes. *Biochem. J.* **186**, 907–918.

Newby, A. C. (1981) The interaction of inhibitors with adenosine metabolising enzymes in intact isolated cells. *Biochem. Pharm.* **30**, 2611–2615.

Newby, A. C. (1984) Adenosine and the concept of retaliatory metabolites. *Trends Biochem. Sci.* **9**, 42–44.

Newby, A. C. (1986) How does dipyridamole elevate extracellular adenosine concentration? Predictions from a three compartment model of adenosine formation and inactivation. *Biochem. J.* **237**, 845–851.

Newby, A. C. and Holmquist, C. A. (1981) Adenosine production inside rat polymorphonuclear leucocytes. *Biochem. J.* **200**, 399–403.

Newby, A. C. and Meghji, P. (1986) The mechanism of adenosine formation in the heart. *Biochem. Soc. Trans.* **14**, 1110, 1111.

Newby, A. C., Luzio, J. P., and Hales, C. N. (1975) The properties and extracellular location of 5'-nucleotidase of the rat fat-cell plasma membrane. *Biochem. J.* **146**, 625–633.

Newby, A. C., Worku, Y., and Meghji, P. (1987) Critical evaluation of the role of ecto- and cytosolic 5'-nucleotidase in adenosine formation, in *Topics and Perspectives in Adenosine Research* (Gerlach, E. and Becker, B. F., eds.), Springer-Verlag, Berlin, pp. 155–170.

Newman, M. E. (1983) Adenosine binding sites in brain; relationship to endogenous levels of adenosine and to its physiological and regulatory roles. *Neurochem. Int.* **5**, 21–25.

Newman, M. E. and McIlwain, H. (1977) Adenosine as a constituent of the brain and of isolated cerebral tissues and its relationship to the generation of cyclic AMP. *Biochem. J.* **164**, 131–137.

Newsholme, E. A. and Start, C. (1973) Regulation of carbohydrate metabolism in muscle. *Regulation in Metabolism* (Wiley, London), pp. 111–113.

Nishiki, K., Erecinska, M., and Wilson, D. F. (1978) Energy relationships between cytosolic metabolism and mitochondrial respiration in rat heart. *Am. J. Physiol.* **234**, C73–C81.

Ogasawara, N., Goto, H., Yamada, Y., and Watanabe, T. (1978) Distribution of AMP deaminase isoenzymes in rat tissues. *Eur. J. Biochem.* **87,** 297–304.

Osswald, H., Hermes, H. H., and Nabakowski, G. (1982) Role of adenosine in signal transmission of tubuloglomerular feedback. *Kidney Int.* **22** suppl 12, S136–S142.

Osswald, H., Nabakowski, G., and Hermes, H. (1980) Adenosine as a possible mediator of metabolic control of glomerular filtration rate. *Int. J. Biochem.* **12,** 263–267.

Osswald, H., Schmitz, H. J., and Kemper, R. (1977) Tissue content of adenosine, inosine and hypoxanthine in the rat kidney after ischaemia and post ischaemic recirculation. *Pflugers Arch.* **371,** 45–49.

Paddle, B. M. and Burnstock, G. (1974) Release of ATP from perfused heart during coronary vasodilation. *Blood Vessels* **11,** 110–119.

Paik, W. K. and Kim, S. (1980) *Protein Methylation* (Wiley-Interscience, New York).

Patel, A. K. and Campbell, A. K. (1987) The membrane attack complex of complement induces permeability changes via thresholds in individual cells. *Immunol.* **60,** 135–140.

Pearson, J. D. (1985) Ectonucleotidases. Measurement of activities and use of inhibitors. *Methods in Pharm.* **6,** 83–108.

Pearson, J. D. and Gordon, J. L. (1979) Vascular endothelium and smooth muscle cells selectively release adenine nucleotides. *Nature* **281,** 384–386.

Pearson, J. D. and Gordon, J. L. (1985) Nucleotide metabolism by endothelium. *Annu. Rev. Physiol.* **47,** 617–627.

Pearson, J. D., Carleton, J. S., and Gordon, J. L. (1980) Metabolism of adenine nucleotides by ecto-enzyme of vascular endothelial and smooth muscle cells in culture. *Biochem. J.* **190,** 421–429.

Perez, M. T. R. and Ehinger, B. (1986) Adenosine uptake and release in the rabbit retina, in *Retina Signal Systems, Degenerations and Transplants* (Agardh, E. and Ehinger, B., eds.), Elsevier, Amsterdam, pp. 105–121.

Perez, M. T. R., Ehinger, B. E., Linstrom, K., and Fredholm, B. B. (1986) Release of endogenous and radioactive purines from the rabbit retina. *Brain Res.* **398,** 106–112.

Perkins, M. N. and Stone, T. W. (1980) Blockade of striatal neuron responses to morphine by aminophylline: evidence for adenosine mediation of opiate action. *Br. J. Pharmacol.* **69,** 131–137.

Perkins, M. N. and Stone, T. W. (1983) *In vivo* release of [^{3}H]purines by quinolinic acid and related compounds. *Brit. J. Pharmacol.* **80,** 263–267.

Phair, R. D. and Sparks, H. V. (1979) Adenosine content of skeletal muscle during active hyperemia and ischemic contraction. *Am. J. Physiol.* **237,** H1–H9.

Phillis, J. W., Jiang, Z. G., Chelack, B. J., and Wu, P. H. (1979) Morphine enhances adenosine release from the in vivo rat cerebral cortex. *Eur. J. Pharmacol.* **65,** 97–100.

Pollard, H. B. and Pappas, G. D. (1979) Veratridine activated release of ATP from synaptosomes: evidence for calcium dependence and blockade by tetrodotoxin. *Biochem. Biophys. Res. Comm.* **88,** 1315–1321.

Pons, F., Bruns, R. F., and Daly, J. W. (1980) Depolarization-evoked accumulation of cAMP in brain slices: the requisite intermediate adenosine is not derived from hydrolysis of released ATP. *J. Neurochem.* **34,** 1319–1323.

Potter, P. and White, T. D. (1980) Release of adenosine 5'-triphosphate from synaptosomes from different regions of rat brain. *Neurosci.* **5,** 1351–1356.

Potter, P. and White, T. D. (1982) Lack of effect of 6-hydroxydopamine pretreatment on depolarisation induced release of ATP from rat brain synaptosomes. *Eur. J. Pharmacol.* **80,** 143–147.

Pull, I. and McIlwain, H. (1972) Adenine derivatives as neurohumoral agents in the brain. The quantities liberated on excitation of superfused cerebral tissues. *Biochem. J.* **130,** 975–981.

Pull, I. and McIlwain, H. (1973) Output of ^{14}C adenine nucleotides and their derivatives from cerebral tissues. *Biochem. J.* **136,** 893–901.

Pull, I. and McIlwain, H. (1976) Centrally active drugs and related compounds examined for action on output of adenine derivatives from superfused tissues of the brain. *Biochem. Pharmacol.* **25,** 293–298.

Pull, I. and McIlwain, H. (1977) Adenine mononucleotides and their metabolites liberated from and applied to isolated tissue of the mammalian brain. *Neurochem. Res.* **2,** 203–216.

Raggi, A., Ronca-Testoni, S., and Ronca, G. (1969) Distribution of AMP aminohydrolase, myokinase and creatine kinase activities in skeletal muscle. *Biochim. Biophys. Acta.* **178,** 619–622.

Ramos-Salazar, A. and Baines, A. D. (1986) Role of 5'-nucleotidase in adenosine-mediated renal vasoconstriction during hypoxia. *J. Pharm. Exp. Ther.* **236,** 494–499.

Reddington, M. and Pusch, R. (1983) Adenosine metabolism in rat hippocampal slice preparation: incorporation into S-adenosylhomocysteine. *J. Neurochem.* **40,** 285–290.

Rehncrona, S., Siesjo, B. K., and Westerberg, E. (1978) Adenosine and cyclic AMP in cerebral cortex of rats in hypoxia, status epilepticus and hypercapnia. *Acta. Physiol. Scand.* **104,** 453–463.

Richardson, P. J. (1983) Presynaptic distribution of the cholinergic specific antigen chol-1 and 5'-nucleotidase in rat brain as determined by complement-mediated release of neurotransmitters. *J. Neurochem.* **41,** 640–648.

Richardson, P. J. and Brown, S. J. (1987) ATP release from affinity purified cholinergic nerve terminals. *J. Neurochem.* **48,** 622–630.

Richardson, P. J., Brown, S. J., Bailyes, E. M., and Luzio, J. P. (1987) Ecto-enzymes control adenosine modulation of immunoisolated cholinergic synapses. *Nature* **327,** 232–234.

Richman, H. G. and Wyborny, L. (1964) Adenine nucleotide degradation in the rabbit heart. *Am. J. Physiol.* **207,** 1139–1145.

Roberts, P. A., Newby, A. C., Hallet, M. B., and Campbell, A. K. (1985) Inhibition by adenosine of reactive oxygen metabolite production by human polymorphonuclear leucocytes. *Biochem. J.* **227,** 669–674.

Rodbell, M. (1966) The metabolism of isolated fat cells. *J. Biol. Chem.* **241,** 3909–3917.

Rubio, R., Berne, R. M., and Dobson, J. G. (1973) Sites of adenosine production in cardiac and skeletal muscle. *Am. J. Physiol.* **225**, 938–953.

Rubio, R., Wiedmeier, V. T., and Berne, R. M. (1974) Relationship between coronary flow and adenosine production and release. *J. Mol. Cell. Cardiol.* **6**, 561–566.

Rubio, R., Berne, R. M., Bockman, E. L., and Curnish, R. R. (1975) Relationship between adenosine concentration and oxygen supply in rat brain. *Am. J. Physiol.* **228**, 1896–1902.

Rutherford, A. and Burnstock, G. (1978) Neuronal and non-neuronal compartments in the overflow of labeled adenyl compounds from guinea-pig taenia coli. *Eur. J. Pharmacol.* **48**, 195–202.

Saito, D., Nixon, D. G., Vomacka, R. B., and Olsson, R. A. (1980) Relationship of cardiac oxygen usage adenosine content and coronary resistance in dogs. *Circ. Res.* **47**, 875–882.

Sasaki, T., Abe, A., and Sakagami, T. (1983) Ecto-5'-nucleotidase does not catalyse vectorial production of adenosine in perfused rat liver. *J. Biol. Chem.* **258**, 6947–6951.

Satchell, D. G. and Burnstock, G. (1971) Quantitative studies of the release of purine compounds following stimulation of non-adrenergic inhibitory nerves in the stomach. *Biochem. Pharmacol.* **20**, 1694–1697.

Sattin, A. and Rall, T. W. (1970) The effect of adenosine and adenine nucleotides on the cyclic AMP content of guinea-pig cerebral cortex slices. *Mol. Pharmacol.* **6**, 13–23.

Sawynok, J. and Jhamandas, K. H. (1976) Inhibition of acetylcholine released from cholinergic nerves by adenosine, adenine nucleotides and morphine: antagonism by theophylline. *J. Pharmacol. Exp. Ther.* **197**, 379–390.

Schrader, J. (1983) Metabolism of adenosine and sites of production in the heart, in *Regulatory Function of Adenosine* (Berne, R. M., Rall, T. W., and Rubio, R., eds.), Martinus Nijhoff, Boston, The Hague, pp. 133–156.

Schrader, J. and Gerlach, E. (1977) Compartmentation of cardiac adenine nucleotides and formation of adenosine. *Pflugers Arch.* **367**, 129–135.

Schrader, J., Schutz, W., and Bardenheuer, J. (1981) Role of S-adenosylhomocysteine hydrolase in adenosine metabolism in the mammalian heart. *Biochem. J.* **196**, 65–70.

Schrader, J., Thompson, C. I., Hiendlmayer, G., and Gerlach, E. (1982) Role of purines in acetylcholine-induced coronary vasodilation. *J. Mol. Cell. Cardiol.* **14**, 427–430.

Schrader, J., Wahl, M., Kuschinsky, W., and Kreutzberg, G. W. (1980) Increase of adenosine content in cerebral cortex of the cat during bicuculline-induced seizure. *Pflugers Arch.* **387**, 245–251.

Schubert, P., Komp, W., and Kreutzberg, G. W. (1979) Correlation of 5'-nucleotidase activity and selective transneuronal transfer of adenosine in the hippocampus. *Brain Res.* **168**, 419–424.

Schubert, P., Lee, K., West, M., Deadwyler, S., and Lynch, G. (1976) Stimulation dependent release of [³H] adenosine derivatives from the central axon terminals to target neurones. *Nature* **260**, 541, 542.

Schutz, W., Schrader, J., and Gerlach, E. (1981) Different sites of adenosine formation in the heart. *Am. J. Physiol.* **240**, H963–H970.

Schwabe, U., Ebert, R., and Erbler, H. C. (1973) Adenosine release from fat cells and its significance for the effects on cAMP levels and lipolysis. *Naunyn-Schmiederbergs Arch. Pharmacol.* **276**, 133–148.

Schwabe, U., Schonhofer, P. S., and Ebert, R. (1974) Facilitation by adenosine of the action of insulin on the accumulation of cAMP, lipolysis and glucose oxidation in isolated fat-cells. *Eur. J. Biochem.* **46**, 537–545.

Shimizu, H., Creveling, C. R., and Daly, J. (1970) Stimulated formation of cyclic AMP in cerebral cortex: synergism between electrical activity and biogenic amines. *Proc. Natl. Acad. Sci.* **65**, 1033–1044.

Silinsky, E. M. and Hubbard, J. I. (1973) Release of ATP from rat motor nerve terminals. *Nature* **243**, 404, 405.

Smith, A. D. (1977) Dale's principle today: adrenergic tissues, in *Neurone Concepts Today* (Szentagothai, J., Hamori, J., and Vizi, E. S., eds.), Akad. Kiado, Budapest, pp. 49–61.

Snyder, S. H. (1985) Adenosine as a neuromodulator. *Ann. Rev. Neurosci.* **8**, 103–124.

Sollevi, A. and Fredholm, B. B. (1981) The antilipolytic effect of endogenous and exogenous adenosine in canine adipose tissue in situ. *Acta. Physiol. Scand.* **113**, 53–60.

Sollevi, A. and Fredholm, B. B. (1983) Influence of adenosine on the vascular responses to sympathetic nerve stimulation in the canine subcutaneous adipose tissue. *Acta. Physiol. Scand.* **119**, 15–24.

Sorenson, R. G. and Mahler, H. R. (1982) Localisation of endogenous ATPases at the nerve terminal. *J. Bioenerg. Biomembr.* **14**, 527–547.

Spector, R., Coakley, G., and Blakely, R. (1980) Methionine recycling in brain: a role for folates and vitamin B-12. *J. Neurochem.* **34**, 132–137.

Spielman, W. S. and Thompson, C. I. (1982) A proposed role for adenosine in the regulation of renal hemodynamics and renin release. *Am. J. Physiol.* **242**, F423–F435.

Stanley, K. K., Edwards, M. R., and Luzio, J. P. (1980) Subcellular distribution and movement of 5'-nucleotidase. *Biochem. J.* **186**, 59–69.

Stjarne, L., Hedqvist, P., and Lagercrantz, H. (1970) Catecholamines and adenine nucleotide material in effluent from stimulated adrenal medulla and spleen: a study of the exocytosis hypothesis for hormone secretion and neurotransmitter release. *Biochem. Pharmacol.* **19**, 1147–1158.

Stone, T. W. (1981a) The effects of morphine and methionine-enkephalin on the release of purines from cerebral cortex slices of rats and mice. *Br. J. Pharmacol.* **74**, 171–176.

Stone, T. W. (1981b) Actions of adenine dinucleotides on the vas deferens, guinea-pig taenia aeci and bladder. *Eur. J. Pharmacol.* **75**, 93–102.

Stone, T. W. (1981c) Physiological roles for adenosine and ATP in the nervous system. *Neurosci.* **6**, 523–555.

Stone, T. W. and Perkins, M. N. (1979) Is adenosine the mediator of opiate action on neuronal firing rates. *Nature* **281**, 227, 228.

Stone, T. W., Hollins, C., and Lloyd, H. (1981) Methylxanthines modulate adenosine release from slices of cerebral cortex. *Brain Res.* **207**, 421–431.

Su, C. (1975) Neurogenic release of purine compounds in blood vessels. *J. Pharmacol. Exp. Ther.* **195**, 159–166.

Su, C. (1978) Purinergic inhibition of adrenergic transmission in rabbit blood vessels. *J. Pharmacol. Exp. Ther.* **204**, 351–361.

Su, C. (1983) Purinergic neurotransmission and neuromodulation. *Ann. Rev. Pharmacol. Toxicol.* **23**, 397–411.

Su, C., Bevan, J., and Burnstock, G. (1971) [³H]adenosine release during stimulation of enteric nerves. *Science* **173**, 337–339.

Sulakhe, P. V. and Phillis, J. W. (1975) The release of [³H]adenosine and its derivatives from cat sensorimotor cortex. *Life Sci.* **17**, 551–556.

Thompson, C. I., Rubio, R., and Berne, R. M. (1980) Changes in adenosine and glycogen phosphorylase activity during the cardiac cycle. *Am. J. Physiol.* **238**, H389–H398.

Trams, E. G. (1974) Evidence for ATP action on the cell surface. *Nature* **252**, 480–482.

Trivedi, B. K., Bridges, A. J., and Bruns, R. F. (1990) Structure-activity relationships of adenonine A_1 and A_2 receptors, in *Adenonine and Adenonine Receptors* chapter.

Ueland, P. M. (1982) Pharmacological and biochemical aspects of S-adenosylhomocysteine and S-adenosylhomocyteine hydrolase. *Pharmacol. Rev.* **34**, 223–253.

Ueland, P. M. and Saebo, J. (1979) Sequestration of adenosine in crude extracts from mouse liver and other tissues. *Biochem. Biophys. Acta.* **587**, 341–352.

Usdin, E., Borchardt, R. T., and Crevelling, C. R. (eds.) (1979) *Transmethylation* (Elsevier/North Holland, New York).

Van den Berghe, G., Van Pottlesberghe, C., and Hers, H. G. (1977) A kinetic study of the soluble 5'-nucleotidase of rat liver. *Biochem. J.* **162**, 611–616.

Wachstein, H. and Meisel, E. (1957) Histochemistry of hepatic phosphatases at physiological pH. *Am. J. Clin. Pathol.* **27**, 13–23.

Wagle, S. R., Ingebretsen, W. R., and Sampson, L. (1973) Studies on the *in vitro* effects of insulin on glycogen synthesis and ultrastructure in isolated rat liver hepatocytes. *Biochem. Biophys. Res. Commun.* **53**, 937–943.

Westfall, D. P., Stitzel, R. E., and Rowe, J. N. (1978) Post-junctional effects and neural release of purine compounds in guinea-pig vas deferens. *Eur. J. Pharmacol.* **50**, 27–38.

White, T. D. (1978) Release of ATP from a synaptosomal preparation by elevated extracellular potassium and by veratradine. *J. Neurochem.* **30**, 329–336.

White, T. D. and Leslie, R. A. (1982) Depolarisation-induced release of adenosine 5'-triphosphate from isolated varicosities derived from the myenteric plexus of the guinea-pig small intestine. *J. Neurosci.* **2**, 206–215.

White, T. D., Downie, J. W., and Leslie, R. A. (1985) Characteristics of potassium and veratradine induced release of ATP from synaptosomes prepared from dorsal and ventral spinal cord. *Brain Res.* **334**, 372–374.

White, T. D., Potter, P., and Wonnacott, S. (1980) Depolarisation induced release of ATP from cortical synaptosomes is not associated with acetylcholine release. *J. Neurochem.* **34**, 1109–1112.

White, T. D., Potter, P., Moody, C., and Burnstock, G. (1981) Tetrodotoxic-resistant release of ATP from guinea-pig taenia coli and vas deferens during electrical field stimulation in the presence of luciferin-luciferase. *Can. J. Physiol. Pharmacol.* **59**, 1094–1100.

Williams, M. and Jacobson, K. A. (1990) this volume.

Winn, H. R., Rubio, R., and Berne, R. M. (1979) Brain adenosine production during 60 seconds of ischaemia. *Circ. Res.* **45**, 486–492.

Winn, H. R., Rubio, R., and Berne, R. M. (1980a) Changes in brain adenosine during bicuculline-induced seizures in rats: Effects of hypoxia and altered systemic blood pressure. *Circ. Res.* **47**, 481–491.

Winn, H. R., Welsh, J. E., Rubio, R., and Berne, R. M. (1980b) Brain adenosine production in rats during sustained alteration in systemic blood pressure. *Am. J. Physiol.* **239**, H636–H641.

Winn, H. R., Rubio, R., and Berne, R. M. (1981) Brain adenosine concentration during hypoxia in rats. *Am. J. Physiol.* **241**, H235–H242.

Winn, H. R., Morii, S., Weaver, D. D., Reed, J. C., Ngai, A. C., and Berne, R. M. (1983) Changes in brain adenosine concentration during hypoglycemia and posthypoxic hyperemia. *J. Cereb. Blood Flow Metabol.* **3** suppl 1, S449, S450.

Worku, Y. and Newby, A. C. (1982) Nucleoside exchange catalysed by the cytoplasmic 5'-nucleotidase. *Biochem. J.* **205**, 503–510.

Worku, Y. and Newby, A. C. (1983) The mechanism of adenosine production in rat polymorphonuclear leucocytes. *Biochem. J.* **214**, 325–330.

Worku, Y., Luzio, J. P., and Newby, A. C. (1984) Identification of histidyl and cysteinyl residues essential for catalysis by 5'-nucleotidase. *FEBS Lett.* **167**, 235–240.

Wojcik, W. J. and Neff, N. H. (1983) Location of adenosine release and adenosine A2 receptors to rat striatal neurons. *Life Sci.* **33**, 755–763.

Wu, P. H. and Phillis, J. W. (1978) Distribution and release of adenosine triphosphate in rat brain. *Neurochem. Res.* **3**, 563–571.

Wu, P. H., Moron, M., and Barraco, R. (1984) Organic calcium channel blockers enhance [^{3}H]purine release from rat brain cortical synaptosomes. *Neurochem. Res.* **9**, 1019–1031.

Wu, P. H., Phillis, J. W., and Yuen, H. (1982) Morphine enhances the release of ^{3}H-purines from rat brain cerebral cortical prisms. *Pharmacol. Biochem. Behav.* **17**, 749–755.

Wyllie, M. G. and Gilbert, J. C. (1980) Exocytotic release of noradrenaline from synaptosomes. *Biochem. Pharmacol.* **29**, 1302, 1303.

Zetterstrom, T., Vernet, L., Ungerstedt, U., Tossman, U., Jonzon, B., and Fredholm, B. B. (1982) Purines levels in the intact rat brain. Studies with an implanted perfused hollow fibre. *Neurosci. Letts.* **29**, 111–115.

Zimmerman, H. (1978) Turnover of adenine nucleotides in cholinergic synaptical vesicles of the Torpedo electric organ. *Neurosci.* **3**, 827–836.

Zimmerman, H. and Denston, C. R. (1977) Recycling of synaptic vesicles in the cholinergic synapses of the Torpedo electric organ during induced transmitter release. *Neurosci.* **2,** 695–714.

Zimmerman, H., Dowdall, M. J., and Lane, D. A. (1979) Purine salvage at the cholinergic nerve endings of the Torpedo electric organ: the central role of adenosine. *Neurosci.* **4,** 979–994.

CHAPTER 7

Adenosine Deaminase and [³H] Nitrobenzylthioinosine as Markers of Adenosine Metabolism and Transport in Central Purinergic Systems

Jonathan D. Geiger
and James I. Nagy

1. Introduction

During the past decade, there has been a flood of information on the neuroregulatory actions of adenosine and its phosphorylated derivatives in the peripheral (PNS) and central nervous system (CNS) (for reviews, *see* Phillis and Wu, 1981; Dunwiddie, 1985). Many investigators now commonly refer to these substances as neurotransmitters or neuromodulators. The use of these terms tends to imply that purines are released from, accumulated by, and have actions on neurons through mechanisms analogous to other small molecular weight neuroactive agents. Since none of the currently identified transmitter or putative modulators are found ubiquitously distributed in neurons, it is conceivable that only particular neuronal populations in the CNS

Adenosine and Adenosine Receptors Editor: Michael Williams ©1990 The Humana Press Inc.

or PNS have modulatory capabilities utilizing purines. Thus, although no strict definitions or rules have been formulated to allow classification of neuroregulatory substances according to their actions, initial considerations of the possible scope of purine actions have relied on precedents set by other substances designated as putative neuromodulators. However, the examination of intracellular biochemical processes on which the extracellular roles of purines could ultimately depend may reveal some fundamentally unique features of purine neuromodulation that cannot be accommodated by current concepts of neuronal interactions (*see,* for example, Newby, 1984). Given the complexities involved in considering yet a new mode of neuroregulation, it seemed reasonable to establish first whether more traditional modes of intercellular communication apply to purines and whether specific neural systems in the CNS express neurochemical characteristics that might be used to classify them as purinergic. For the purposes of this chapter, the term purinergic will be used in discussing systems that may utilize adenosine and/or ATP as neuroregulatory substances.

Recognition of a neuromodulatory capacity has evolved from findings that:

1. Adenosine receptors are present in neurons of the CNS;
2. Adenosine is released upon depolarization of CNS tissues in a calcium-dependent manner; and
3. Enzymes for the synthesis and degradation of adenosine are present in nerve terminals.

Again, by analogy with more classical transmitter systems, the physiological activities of adenosine in vivo are likely determined not only by adenosine receptor subtypes, the cellular locations of these, and their mode of coupling to secondary effector systems, such as cAMP and ion channels, but also by mechanisms governing the availability of the purine. The latter, in turn, would be determined by the sum of such processes as release, reuptake into cells, and the capacity of tissues to generate and degrade adenosine. In an attempt to identify possible purinergic neural systems, we have documented the localization and activity of the degradative enzyme adenosine deaminase (ADA) and have investigated adenosine transport using [^{3}H]nitrobenzylthioinosine ([³H]NBI), a putative ligand for nucleoside transport binding sites. This chapter will focus on the potential functional significance of ADA and sites labeled by [³H]NBI.

2. Sites and Control of Adenosine's Actions

The identification and discrete localization of adenosine receptors in the CNS provide some justification to suspect the existence of distinct central purinergic systems. Adenosine receptors have been categorized pharmacologically as A_1 or A_2 based on the rank order of potencies with which adenosine agonists compete for these two receptor subtypes and, at least in certain tissues, on the basis of their ability, once activated, to inhibit (A_1) or stimulate (A_2) adenylate cyclase (Stone, 1985; Williams, 1984). More recently, the existence of other adenosine receptor subtypes has been proposed (Ribeiro and Sebastiao, 1986; Reddington et al., 1986; Bruns et al., 1986). In addition to these presumably cell surface embedded receptors, an intracellular receptor, designated the P site, has been identified (Londos et al., 1979) and recently localized to the catalytic subunit of the calmodulin-sensitive adenylate cyclase (Yeager et al., 1986). The A_1 and A_2 receptors are heterogeneously and differentially distributed within the CNS (Lee and Reddington, 1986), and have been localized on neurons in striatum (Wojcik and Neff, 1983a,c; Geiger, 1986), hippocampus (Murray and Cheney, 1982), and dorsal spinal cord (Geiger et al., 1984b), as well as on cerebellar granule cells (Wojcik and Neff, 1983b) and retinal projections to the superior colliculus (Goodman et al., 1983; Geiger, 1986). That receptor location specifies sites of adenosine action was suggested by the close correlation between adenosine receptor density and the ability of adenosine analogs to alter evoked neuronal potentials in rat hippocampus (Lee et al., 1983). However, neuronal firing rates in superior colliculus, an area with very high levels of mainly adenosine A_1 receptors, were found to be unaffected by adenosine applied directly into this structure (Okada and Saito, 1979). This observation may be partly explained by the very high levels of nucleoside transport sites as labeled by [³H]NBI (Geiger and Nagy, 1984) and ADA activity (Nagy et al., 1985; Geiger and Nagy, 1986) in the superior colliculus, but raises the issue of whether the presence and quantities of adenosine receptors on neurons are in themselves sufficient indicators of anatomical loci at which the actions of adenosine are expressed; perhaps not, since, in the absence of exogenously applicated adenosine, inhibitors of adenosine uptake and degradation depress (Phillis et al., 1979; Dunwiddie and Hoffer, 1980), whereas adenosine receptor antagonists increase (Dunwiddie and Hoffer, 1980) neuronal firing. These findings suggest not only that cortical neurons are under the continuing depressant influence of endogenous adenosine (Phillis and Wu, 1981), but also that the processes of uptake and intracellular metabolism contribute substantially to this depressant influence. Since it appears that an understanding of how adenosine levels are regulated may be pivotal to the formulation of concepts concerning its utilization as a neuroregulatory substance or "retaliatory metabolite" (Newby, 1984; Stone et al., 1988), some potentially key factors regarding enzymes responsible for adenosine production and catabolism need to be considered.

3. Adenosine Metabolism

3.1. Production

Some of the major biochemical pathways contributing to adenosine metabolism are illustrated in Fig. 1. In mammalian tissues, adenosine can be formed through *de novo* synthesis or from a variety of substrates. *De novo* synthesis involves formation of 5'-IMP and its subsequent conversion to 5'-AMP through the actions of adenylosuccinate synthetase (EC 6.3.4.4.) and lyase (EC 4.3.2.2.) (Schultz and Lowenstein, 1976). Since the levels of enzymes responsible for *de novo* synthesis of purines are very low in the adult brain (Allsop and Watts, 1983), this source of adenosine is generally believed to be of minor importance. Other sources of adenosine that are believed to contribute little to the functional "pool" of adenosine include: synthesis from adenine by the enzyme adenine phosphoribosyl transferase (EC 2.4.2.7.), a reaction limited mainly by substrate (adenine) availability in mammalian CNS; formation fom 3'-AMP derived from the breakdown of RNA; the condensation of adenine with ribose-1-phosphate catalyzed by purine nucleoside phosphorylase (PNP, EC 2.4.2.1.); and breakdown of NAD (Burnstock and Hoyle, 1985) or cAMP (Phillis and Wu, 1981; Burnstock and Hoyle, 1985). In addition, adenosine

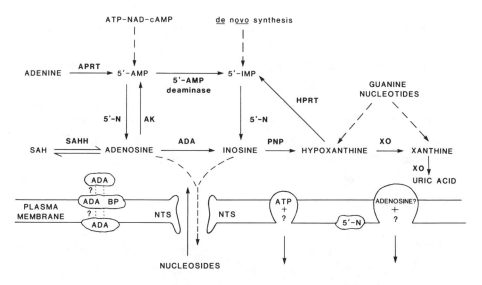

Fig. 1. Schematic representation of some major metabolic pathways involved in the production and disposition of adenosine. Question marks are shown where details are as yet unclear. ADA has been shown to be located intracellularly as well as on membranes in association with a binding protein (ADA-BP). The nucleoside transport system (NTS) is indicated as being capable of translocating various nucleosides, including adenosine and inosine, bidirectionally across membranes. The sites and mechanisms whereby ATP and adenosine may be released from cells are illustrated. The enzymes indicated include: adenine phosphoribosyltransferase (ARPT), 5'-nucleotidase (5'-N), adenosine kinase (AK), adenosine deaminase (ADA), S-adenosylhomocysteine hydrolase (SAHH), purine nucleoside phosphorylase (PNP), hypoxanthine phosphoribosyltransferase (HPRT), and xanthine oxidase (XO).

can be formed by nonspecific alkaline phosphatase (EC 3.1.3.1.), nonspecific acid phosphatase (EC 3.1.3.2.), and 2'-nucleotidase (Nagata et al., 1984). A potentially more important source is the hydrolysis of S-adenosylhomocysteine (SAH) by SAH hydrolase (EC 3.3.1.1). However, the low and tightly controlled levels of SAH limit its availability (Snyder, 1985). From a consideration of these various pathways, it appears that the major source of cellular adenosine is the breakdown of ATP and the subsequent dephosphorylation of 5'-AMP by 5'-nucleotidase (EC 3.1.3.5), and that adenosine production is subject to factors controlling ATP and 5'-AMP levels (Achterberg et al., 1986).

Intracellularly, ATP is contained in a variety of "pools" (Miller and Horowitz, 1986), and it may be released extracellularly from vesicular (presumably copackaged with other transmitter substances) and nonvesicular compartments (Tauc, 1982). It is noteworthy here that some enzymes catabolizing ATP are located both intracellularly and extracellularly, and that ecto-ATPases (EC 3.6.1.3) are capable of hydrolyzing ATP to ADP, as well as ADP to 5'-AMP (Nagy et al., 1986a). Thus, if ATP release is prevalent in CNS tissues, it may be particularly significant that adenosine can be formed not only intracellularly through the breakdown of 5'-AMP by endo-5'-nucleotidase, but also extracellularly by ecto-ATPases and ecto-5'-nucleotidase (Lowenstein et al., 1986).

3.2. Disposition

The examination of enzymes and processes involved in adenosine disposition rather than production represents an alternative strategy for the identification of possible purinergic systems. Various routes of adenosine disposition include its incorporation into SAH under certain conditions (Reddington and Pusch, 1983), deamination by ADA (adenosine aminohydrolase, EC 3.5.4.4) to inosine, or phosphorylation by adenosine kinase (AK) (EC 2.7.1.20) to 5'-AMP. The first of these, which will be discussed in the next section (3.3.), is not thought to be an important pathway for adenosine removal. This leaves principally ADA and AK as major contributors towards establishing physiological levels of adenosine. Controversy as to the relative importance of these enzymes has centered on analyses of their kinetic parameters and estimates of intracellular adenosine levels. Based primarily on findings that the K_m values of AK for adenosine are about 0.3 μM in rabbit brain microvessels (Mistry and Drummond, 1986) and range from 0.2 (Yamada et al., 1980) to 2.0 μM in rat brain (Arch and Newsholme, 1978; Phillips and Newsholme, 1979), AK has been traditionally considered to be the dominant pathway for adenosine metabolism; the K_m values of ADA for adenosine range from 17–100 μM (for references, *see* Geiger and Nagy, 1986). However, AK appears to be a lower capacity enzyme, since its V_{max} compared with that of ADA was found to be about 177× lower in rabbit brain microvessels (Mistry and Drummond, 1986) and 2–22× lower in human brain, depending on the region studied (Phillips and Newsholme, 1979). Although these two enzymes have roughly comparable activity in rat whole brain (Phillips and Newsholme, 1979; Arch and Newsholme, 1978), the activity of ADA in particular brain regions was found to be as much as ninefold higher than that of AK (Davies and Hambley, 1986; Geiger and Nagy, 1986).

With respect to adenosine levels, intracellular concentrations have been reported to be about 0.5 μM under normal physiological conditions (Winn et al., 1980; Sollevi, 1986). In contrast, hypoxia, tissue depolarization, and seizure activity are known to increase dramatically the production of adenosine (Kleihues et al., 1974; Berne et al., 1974) and that of its metabolites in vivo (Schultz and Lowenstein, 1978; Lewin and Bleck, 1981) and in vitro (Heller and McIlwain, 1973; Daval and Barberis, 1981). It should be noted that the very rapid breakdown of adenine nucleotides makes difficult the precise quantification of endogenous adenosine concentrations. For example, 30–50-fold differences in adenosine levels were observed in striatum of animals sacrificed by decapitation as compared with focused microwave irradiation (Wojcik and Neff, 1983a). Recently, however, the utilization of microdialysis sampling techniques has demonstrated adenosine, inosine, and hypoxanthine levels in rat cerebral cortical interstitial fluid to be 1.3, 3.3, and 7.2 μM, respectively. Furthermore, adenosine levels in the dialysate were increased 10-fold during hypoxia/hypotension, 30-fold during cerebral anoxia/ischemia, and fourfold after infusion of 80 mM potassium through the sampling cannula (Van Wylen et al., 1986).

The above findings suggest that AK is nearly saturated at physiological concentrations of adenosine and certainly saturated at higher concentrations that can be attained under certain conditions (Arch and Newsholme, 1978; Yamada et al., 1980; Fisher and Newsholme, 1984). Moreover, in brain and cultured cells, its activity is subject to potent substrate inhibition by relatively low adenosine concentrations (i.e., >0.5 μM). Therefore, although substantial AK activity is found rather uniformly distributed in at least rat and guinea pig brain (Phillips and Newsholme, 1979; Davies and Hambley, 1986), it remains questionable whether AK itself is entirely responsible for maintaining appropriate levels of adenosine under all, or even normal, physiological conditions in neurons. It may be reasonable to consider that, since ADA activity is heterogeneously distributed throughout the body (Arch and Newsholme, 1978; Geiger and Nagy, 1986), the relative rates of adenosine deamination and phosphorylation may differ among brain regions and various peripheral tissues. For example, depending on enzyme and endogenous adenosine levels, four metabolic patterns as described by Henderson (1979) are possible:

1. ADA activity could predominate at all adenosine concentrations
2. AK activity could predominate at all adenosine concentrations, as observed in ADA-deficient tissues
3. AK activity could predominate at low adenosine concentrations and ADA activity at high adenosine concentrations and
4. ADA activity could predominate at low adenosine concentrations and AK activity at high adenosine concentrations.

Pattern 4 has not been observed in any tissues, whereas pattern 1 is rare (Henderon, 1979). It may be argued that pattern 2 is the most vital pathway in all CNS neurons, and further, that ADA inhibition would not influence adenosine utilization and disposition. If, however, pattern 3 were feasible, particularly in putative purinergic neurons, then it may be argued that inhibition of ADA activity will elevate adenosine levels, reduce the levels of its metabolites, and augment its neuroregulatory effects not only as a consequence of this inhibition, but also through substrate inhibition of AK.

There are numerous examples of the functions of these pathways according to pattern 3. In the heart, phosphorylation and deamination of intracellular adenosine were about equal under normal conditions, and deamination predominated during ischemic conditions (Kohn and Garfinkel, 1977). Following uptake of [^{3}H] adenosine by microvessels isolated from bovine cerebral cortex, over 70% of the accumulated radioactivity was recovered as [^{3}H] inosine and about 20% as [^{3}H] nucleotides, thus demonstrating substantial ADA activity and lesser amounts of AK activity in these vessels (Stefanovich, 1983). In synaptosomes, appreciable quantities of inosine were produced following adenosine loading, and the levels of adenosine were increased and those of inosine decreased by erythro-9-(2-hydroxy-3-nonyl-adenine (ENHA), an ADA inhibitor (Daval and Barberis, 1981). Additionally, ischemia, electroshock treatment of rats, and increased neuronal activity were accompanied by increased levels of adenosine and its deaminated products within and emanating from CNS tissue (Kleihues et al., 1974; Schultz and Lowenstein, 1978; Daval and Barberis, 1981; Lewin and Bleck, 1981). Since inosine and hypoxanthine accumulation in CNS tissues of rats exposed to hypoxic conditions was reduced by EHNA administration (Zetterstrom et al., 1982), it is likely that these metabolites originated from the action of ADA. In *Aplysia,* the relative rates of inosine and 5'-AMP formation appear to be determined by the relative activities of ADA and AK in neuronal perikarya and their connective tissue sheath (McCaman, 1986). Taken together, these findings clearly show that adenosine is not only metabolized by AK, but also by ADA in nervous tissues. It remains to be determined whether distinct metabolic patterns for adenosine occur in certain cell types where adenosine might be involved in both normal intermediary metabolism and intercellular communication.

It may be of some significance, given the production of deaminated adenosine metabolites in the brain, that CNS is among those mammalian tissues having the highest activity of hypoxanthine phosphoribosyltransferase (HPRT, EC 2.4.2.8) and purine nucleoside phosphorylase (PNP, EC 2.4.2.1), which are enzymes principally involved in salvaging purines (Robins et al., 1953; Zimmerman et al., 1971; Gutensohn and Guroff, 1972). The HPRT salvage pathway, following the actions of ADA and PNP, may represent an important mechanism for the recapture of inosine and synthesis of 5'-AMP. On the other hand, it appears that the breakdown of ATP during ischemia leads not only to the formation of adenosine, inosine, and hypoxanthine, but also xanthine (Hagberg et al., 1986). This indicates that HPRT, although having high activity in the brain, is not capable of completely salvaging hypoxanthine. The theory that the production of xanthine in the brain is not insignificant derives from recent evidence showing the formation of uric acid via the degradation of xanthine by xanthine oxidase (EC 1.1.3.22) in CNS tissue (Betz, 1985; Kanemitsu et al., 1986; Honegger et al., 1986). Furthermore, uric acid is a prominent metabolite of adenosine in Aplysia neurons (McCaman, 1986). It has been proposed that uric acid, in addition to being simply an end product of purine metabolism, may function as an important antioxidant through its capacity to scavenge free radicals (Davies et al., 1986). Thus, the relationship of HPRT to neurons having the capacity to generate its substrate hypoxanthine, i.e., those containing ADA and PNP, remains an unclear, but potentially important issue. It may be informative, for example, to compare the anatomical and cellular localization of HPRT and ADA.

A possibility, only rarely considered (Phillis and Wu, 1981), is that a significant proportion of inosine and hypoxanthine may be produced via 5'-AMP deaminase not only under physiological conditions, as shown in rat brain (Schultz and Lowenstein, 1976) and human erythrocytes (Bontemps et al., 1986), but also during excessive neuronal electrical activity. For example, Schultz and Lowenstein (1978) suggested that adenylate deaminase activity was responsible for increased ammonia production in brains of rats subjected to electric shock. Among a variety of tissue examined, the level of 5'-AMP deaminase activity in the brain was second only to the level in muscle (Conway and Cooke, 1939). Thus, some proportion of inosine and hypoxanthine in CNS may be generated by degradation of 5'-IMP by 5'-nucleotidase. However, adenylate deaminase activity may not prevail under cellular conditions of low energy charge, such as hypoxia or ischemia, since allosteric activation of the enzyme occurs only at high levels of ATP (Schultz and Lowenstein, 1976). Nevertheless, the contribution of 5'-AMP deaminase to the formation of purine metabolites in the brain deserves further investigation.

3.3. Adenosine, 2'-Deoxyadenosine, and S-Adenosylhomocysteine Toxicity

Excessively high levels of adenosine are cytotoxic in several cell culture systems (Archer et al., 1985; Fox and Kelley, 1978; Henderson, 1983). One proposed biochemical mechanism for this toxicity that has gained considerable support involves the formation of SAH from S-adenosylmethionine (SAM) following essential cellular transmethylation reactions. SAH is a potent inhibitor of such reactions, and the relative levels of SAH and SAM appear to determine the extent of methyltransferase inhibition (de la Haba et al., 1986). Therefore, the degradation of SAH by SAH hydrolase (SAHH) to adenosine and homocysteine may be considered cyto-protective (Chiang, 1985). The reaction catalyzed by SAHH is, however, reversible and, in fact, favors SAH synthesis. To prevent or minimize the synthetic direction of this reaction adenosine and homocysteine, once formed from SAH or accumulated from other sources, must be metabolized and their concentrations kept low to favor SAH breakdown. Indeed, homocysteine levels appear to be sufficiently low such that little SAH was found in rat hippocampal slices exposed to relatively high concentrations of adenosine (Reddington and Pusch, 1983).

It has also been demonstrated that 2'-deoxyadenosine at relatively low concentrations is toxic to cells (Henderson et al., 1980; Parsons et al., 1986). The mechanism proposed for this is similar to that involving SAH in that 2'-deoxyadenosine appears to be a tight binding suicide inhibitor of SAHH (Hershfield, 1979) and prevents the reoxidation of enzyme-bound NADH (de la Haba et al., 1986). Administration of 2'-deoxycoformycin (DCF), a very potent inhibitor of ADA, to mice has been found to increase 2'-deoxyadenosine levels, inhibit SAHH activity, and produce lymphospecific toxicity (Ratech et al., 1981). Low ADA levels or ADA inhibition may result in a large overproduction of dATP and 2'-deoxyadenosine (Helland et al., 1983; Hershfield, 1979; Sylwestrowicz et al., 1982) through the ribonucleoside diphosphate reductase pathway (Henderson et al., 1980). These observations suggest the involvement of multiple mechanisms whereby ADA may prevent toxic reactions, since ADA is almost as effective in deaminating 2'-deoxyadenosine as it is adenosine (Simon et al., 1970; Parsons and Hayward, 1986). Moreover, ADA may maintain adenosine levels within a range where AK can perform effectively.

Other recent findings that suggest a link between ADA deficiency and cellular toxicity include observations of large losses of cerebellar Purkinje cells in patients lacking ADA and suffering from severe combined immunodeficiency disease (SCID) (Ratech et al., 1985a). Many of these patients exhibit neurological abnormalities (Hirschhorn et al., 1980; Daddona et al., 1983). Purkinje cell loss has also been observed in mice receiving relatively large doses of DCF (Ratech et al., 1985b).

4. Adenosine Deaminase (ADA)

4.1. Possible Role in CNS

It has been generally assumed that, because of its involvement in intermediary metabolism and its possible role in maintaining subtoxic levels of purine cellular constituents, ADA would be ubiquitous in mammalian cells. This was also thought to be the case in the CNS, despite the biochemical heterogeneity and morphological diversity of cells in this tissue. However, it has been known for a long time that some tissues, for reasons presently unclear, have exceptionally high ADA activity (Van der Weyden and Kelley, 1976; Arch and Newsholme, 1978). With respect to the CNS, it was hypothesized (Nagy et al., 1984b) that some neurons, namely those having an added metabolic burden imposed by their utilization of purines in intercellular communication, may also express relatively greater quantities of ADA. Using immunohistochemical methods, intense neuronal immunostaining for ADA in restricted regions of the posterior hypothalamus and several other structures in the rat brain was reported (Nagy et al., 1984b). Thus, although part of the hypothesis was confirmed, the notion that ADA-immunoreactive neurons are engaged in some form of purinergic transmission remains an open question.

There is now substantial evidence for the presence of high concentrations of not only transmitter synthetic, but also catabolic enzymes in a variety of well-characterized transmitter systems. Examples of these include monoamine oxidase in monoaminergic neurons, acetylcholinesterase in cholinergic neurons, and γ-aminobutyric acid (GABA) transaminase in GABAergic neurons. Inhibition of the activity of these degradative enzymes is known to alter profoundly the synaptic efficacy of their neurotransmitter substrates. It may be argued that, if ADA serves for purinergic neurons a function similar to that of classical transmitter catabolic enzymes, then perturbation of adenosine metabolism through ADA inhibition may influence purinergic neuromodulation. It has already been shown that rats and mice treated acutely with doses of potent ADA inhibitors sufficient to abolish ADA activity almost totally in the brain exhibit decreased spontaneous motor activity, altered sleep behavior, and decreased cardiac function (Radulovacki et al., 1983; Mendelson et al., 1983; Helland et al., 1983; Geiger et al., 1987; Szentmiklosi et al., 1982). Although inhibition of ADA clearly augments the actions of adenosine (Fredholm and Hedqvist, 1980) and potentiates physiological responses thought to be mediated by adenosine (Phillis et al., 1985, 1986), it is uncertain whether this potentiation results directly from increased quantities of releasable intraneuronal stores of adenosine or indirectly from secondary effects on purine metabolism. However, at least one study indicates that ADA inhibition can cause increased adenosine release (Zetterstrom et al., 1982).

The deaminated product of adenosine, inosine, is generally thought to be physiologically inactive, since its affinity for adenosine receptors is very low (Bruns et al., 1980). Interestingly, however, some of the actions of adenosine in guinea pig aorta were interpreted to be the result of its metabolism to inosine, which was found to be a fairly potent mediator of muscle contractility (Collis et al., 1986). In some cases, therefore, it may be worth considering the possibility that inosine could affect adenosine-mediated processes by, for example, acting as a competitive inhibitor of adenosine uptake through the nucleoside transport system. It appears premature to conclude that the exclusive role of ADA is to "inactivate" adenosine. The enzyme may provide deaminated products having biochemically or physiologically relevant functions.

4.2. *ADA Inhibition*

Various inhibitors of ADA include EHNA, coformycin, and DCF. Among these, DCF may have the greatest utility as a biochemical tool, since it is a transition-state noncompetitive inhibitor and exhibits high specificity and potency ($K_i = 10^{-11}M$) for the enzyme. In certain peripheral tissues, its nonspecific effects, e.g., inhibition of 5'-AMP deaminase, begin to occur only at concentrations many orders of magnitude higher than that required to inhibit ADA (Agarwal, 1982; Buc et al., 1986; Agarwal and Parks, 1977; Fishbein et al., 1981; Holland, 1986). Although many studies have been conducted on the effects of DCF in peripheral tissues (McConnell et al., 1978; Chassin et al., 1979; Ratech et al., 1981), very little is known regarding the ability of DCF to penetrate the blood–brain barrier and its dose–inhibition relationships in the CNS in vivo. In rat brain, DCF was found to be accumulated to a greater extent in the hypothalamus, the area with the highest levels of ADA activity, than three other brain regions examined (Geiger et al., 1987). This is consistent with reports that peripheral tissues with high ADA activity preferentially accumulate DCF (McConnell et al., 1978; Chassin et al, 1979). The rapid entry of DCF into the brain was followed by processes of elimination that appeared to have both fast and slow components. The t   values for the fast component ranged from 0.8 h for hypothalamus to 5.5 h for cortex. These values corresponded to the times at which peak DCF

levels were observed in CSF (2–3 h) following its peripheral administration (Chassin et al., 1979), and confirm findings of the rapid
distribution of DCF between blood and tissues. The t values of the
slow component varied little among brain regions, and the values of
about 50 h may reflect the formation of a stable complex between DCF
and ADA and the slow dissociation rate of this complex (Agarwal,
1982). The inhibition of brain ADA activity was most pronounced in
rats given the highest DCF dose tested of 5.0 mg/kg (18.6 μmol/kg).
The degree of inhibition was greater and the recovery rate slower in
brain compared with small intestine. This probably reflects differences in the turnover rates of the enzyme in these two tissues. It may be
especially significant that ADA activity in the brain was still only 66%
of control levels 50 d after a single ip injection of DCF (5.0 mg/kg).
This indicates long-term effects, perhaps involving neuronal degeneration. Whether such degeneration, if present, may be limited to ADA-
containing cells remains to be determined.

Clinically, ADA inhibitors are used as antileukemic, lymphocy-
topenic, and immunosuppressive agents, and are administered conco-
mittantly with compounds that are also antimetabolites, but are effective substrates for ADA (Agarwal, 1982). In addition, DCF is undergoing clinical trials as a primary therapeutic agent for treatment of
certain types of leukemia (Johnston et al., 1986). However, patients
given even low doses of DCF exhibit minor neurological side effects
(J. Johnston, personal communications), the neurochemical bases of
which are currently unknown.

4.3. Cellular and Ultrastructural Localization

In addition to reports on the regional (*see* section 4.5.) and subcellular distribution of ADA activity, some information is available
regarding its localization to particular cell types. ADA activity has
been measured directly in neuronal and glial enriched brain fractions
(Subrahmanyam et al., 1984), cultured cells of neuronal and glial origin—including mouse neuroblastoma (N-18), neonatal hamster astrocytes (NN), human astrocytomas (Cox-Clare), and oligodendro-
glioma (HOL) (Trams and Lauter, 1975)—and in primary cultures of
mouse astrocytes (Hertz, 1978). ADA has also been localized immu-

nohistochemically in specific cell types in thymus and various other peripheral tissues (Chechik and Sengupta, 1981; Chechik et al., 1981, 1983) and histochemically in macrophages (Tritsch et al., 1985).

The ultrastructural localization of ADA in the CNS is a virtually unexplored and perhaps far more complex issue than may be currently appreciated. For example, ADA is believed to be composed of protein aggregates ranging in mol wt from 35,000–298,000 (Van der Weyden and Kelley, 1976). This enzyme appears to be capable of associating with what has been termed an ADA-binding protein (BP), which converts ADA from a low to a high molecular weight form. BP has been purified and localized immunohistochemically to cytoplasmic membrane (Schrader and West, 1985) and, where examined, to the external surface of cells (Andy and Kornfeld, 1982). Although exteriorized ADA–BP complexes have been postulated to control adenosine concentrations near adenosine receptors (Schrader et al., 1983), their functions with respect to both ADA and the actions of adenosine are unknown; to our knowledge, the biochemical or anatomical relationship of BP to ADA in the brain has not been investigated. A curious finding is that, although BP has been detected in a variety of peripheral tissues (Schrader and West, 1985; Schrader et al., 1983; Trotta et al., 1979; Schrader and Stacy, 1979; Schrader and Pollara, 1978; Schrader and Bryer, 1982), there is enormous variation in the percent occupancy of BP by ADA among these tissues (Schrader and Stacey, 1979). Moreover, it is important to note that BP is expressed in mouse, guinea pig, rabbit, and human tissues, whereas rat tissues appear to lack it (P. E. Daddona, personal communications). Nevertheless, in the apparent absence of BP, ADA has been localized histochemically to the cytoplasmic membrane of rat erythrocytes (Bielat and Tritsch, 1986).

4.4. Subcellular Localization

The results of several reports on the subcellular distribution of ADA in the brain indicate that this enzyme is recovered predominantly in soluble fractions. Where examined, a significant fraction of the total cellular ADA activity was also associated with particulate fractions (Jordan et al., 1959; Pull and McIlwain, 1974; Van der Weyden and Kelley, 1976) and with synaptosomes (Subrahmanyam et al., 1984; Franco et al., 1986; Phillips and Newsholme, 1979) and mitochondria (Mustafa and Tewari, 1970). In rat brain, ADA was differentially distributed in subcellular fractions prepared from various brain regions (Yamamoto et al., 1987). In rat cortex, hypothalamus, and cerebellum, ADA activity in the P_2 (crude mitochondrial/synaptosomal) fraction was about 2–3× higher than that in the P_1 (nuclear) or P_3 (microsomal) fractions. In rat hippocampus, however, P_2 and P_3 preparations contained over 2× the activity found in similar fractions obtained from the other brain regions examined. Moreover, the activity in hippocampal P_2 fractions was much higher than that found in the P_1 and P_3 fractions of this region, an observation in agreement with immunohistochemical findings that ADA in the hippocampus is seen exclusively in fibers, whereas perikarya, from which more soluble ADA may be derived, as well as fibers immunoreactive for ADA, are seen in cortex, whole hypothalamus, and cerebellum. Consistent with demonstrations that anterior and midlevels of the hypothalamus are densely innervated by ADA-immunoreactive fibers and posterior hypothalamus contains both immunoreactive fibers and perikarya (Staines et al., 1987a,b) ADA activity was about 2× greater in particulate fractions (P_2 and P_3) of anterior compared with posterior hypothalamic regions. More detailed analysis of anterior hypothalamic tissue showed that about 50% of the activity in P_2 fractions was found in synaptosomes and about 30% in mitochondria. The localization of some proportion of ADA in synaptosomes suggests that this enzyme is present in nerve terminals where it could influence the levels of adenosine available for release.

4.5. Distribution of ADA Activity

A sensitive HPLC assay has been used to document the activity of ADA in 66 CNS regions and in a number of peripheral tissues of rat (Geiger and Nagy, 1986). Together with immunohistochemical studies, this approach has provided basic information that may be used in studies of adenosine neuromodulation in the CNS and possibly in ADA-containing neural systems in rat. The major outcome of these investigations has been that ADA is not uniformly distributed in rat CNS, and that the brain regions containing perikarya and fibers immunoreactive for ADA have the highest levels of ADA activity. This was especially evident in rat hypothalamus where the tubero-mammillary nucleus contained large numbers of ADA-immunoreactive perikarya and the greatest ADA activity. Similarly, superficial layers of superior colliculus, habenula, and choroid plexus exhibited relatively greater levels of ADA activity and density of neural elements immunoreactive for ADA. Although no ADA-immunostaining of perikarya has been observed in rat hippocampus, some immunostained fibers were seen in the temporal pole where ADA activity was about twice that of the septal pole. Largely because of difficulties inherent in microdissection and assay of very small nuclei and subnuclear regions, the levels of ADA in some of the areas studied likely represent underestimates of "true" subregional activity. However, this limitation does not detract from the close correspondence found between the relative density of immunoreactive elements and ADA activity in many brain regions.

4.5.1. ADA Immunohistochemistry

Despite uncertainties surrounding the relevancy of ADA to adenosine's neuromodulatory actions and the nature and functional significance of [³H]NBI binding sites (*see* section 6), the close correspondence between the distribution of these sites and ADA-immunoreactive systems in rat brain provided some of the impetus for continuing work on the anatomical localization of ADA in the entire rat brain, a detailed account of which is currently in preparation. With respect to specific systems we have described, ADA-immunoreactive neuronal cell bodies and/or fibers have been described in dorsal root ganglia and spinal cord (Nagy et al., 1984a, 1985; Nagy and Daddona, 1985), retina (Senba et al., 1986a), brainstem and spinal parasympathetic nuclei (Senba et al., 1987d), the mesencephalic nucleus of the trigeminal nerve in the rat (Nagy et al., 1986b) and have conducted more limited studies on hamster, guinea pig, and mouse brain (Yamamato et al., 1987). More detailed analyses of ADA-positive neurons in rat posterior hypo-thalamus (Senba et al., 1985; Staines et al., 1986a, 1987d) and the distribution of their associated fibers within the hypothalamus (Staines et al., 1987a,b) have been conducted. Finally, the pattern of immunostaining for ADA in the developing rat CNS (Senba et al., 1987a,b,c) 1987a,b) has been investigated.

4.5.1.1. PARASYMPATHETIC NUCLEI. Of particular interest was the observation of ADA-immunoreactivity in preganglionic parasympathetic neurons. In both spinal-cord and brainstem nuclei, immunostaining for ADA was found in a subpopulation of neurons having efferents to parasympathetic ganglia. In the brainstem, for example, ADA was localized to only those parasympathetic neurons projecting to sphenopalatine ganglia. Similarly, ADA was contained in only a subclass of spinal parasympathetic neurons innervating the pelvic ganglia. With respect to the proposal that ADA may be expressed by neurons having a purinergic component, these observations in rat may be particularly relevant given the electrophysiological demonstration by Akasu et al. (1984), albeit in cat, that stimulation of preganglionic parasympathetic nerves produced an adenosine-mediated slow hyperpolarizing potential in postganglionic cells of the vesical ganglion. It may now be worthwhile to conduct similar electrophysiological experiments on the sphenopalatine ganglia of the rat and, for comparison, on submandibular ganglia, which appear to receive non-ADA-containing preganglionic afferents. A separate, more speculative issue concerns the observation by Hara et al. (1985) that in rat the sphenopalatine ganglia is a rich source of cholinergic fibers innervating some of the major cerebral arteries. If purinergic transmission occurs in the sphenopalatine ganglion and given the dramatic effects of adenosine analogs on cerebral blood flow in a variety of species (Forrester et al., 1979; Heistad et al., 1981; Sollevi, 1986; Morii et al., 1986), the possibility exists that some of the cerebral vascular actions of these analogs may be exerted at the level of this ganglion.

4.5.1.2. RETINA AND SUPERIOR COLLICULUS. Relatively little is known about the actions of adenosine in the retina. Nevertheless, adenosine receptors linked to adenylate cyclase (Blazynski et al., 1986; Paes de Carvalho and de Mello, 1985), and nucleoside uptake (Schaeffer and Anderson, 1981) and release (Perez et al., 1986) in this structure have been described. It is therefore likely that purines have some modulatory role in retinal tissue. As in the CNS, ADA was detected immunohistochemically in only a subpopulation of cells in rat retina. Some ADA-immunoreactive cells, probably amacrine, were observed in the inner nuclear layer, and the processes of these were distributed in specific sublayers of the inner plexiform layer. In addition, small, sparsely distributed immunostained cells were seen in the ganglion cell layer. These were of particular interest given the retinal ganglion cell origin of the the majority of adenosine receptors in the superior colliculus (Goodman et al., 1983; Geiger, 1986). However, there appeared to be no relationship between ganglion cells bearing adenosine receptors on their terminals in the colliculus and ADA-containing cells in the ganglion cell layer, since more detailed studies showed that the latter cells were nonprojecting, probably displaced amacrine cells, and that ADA activity, unlike adenosine receptors, in SC were unaffected by enucleations (Geiger and Nagy, unpublished observations).

The distribution of ADA-containing neurons in the superior colliculus is somewhat more complex. It is clear that there is a differential distribution of several morphological types of ADA-positive neurons in the superficial collicular layers and that some of these project to the lateral posterior nucleus of the thalamus. An important issue is the possible synaptic relationship between ADA-positive collicular neurons and those collicular afferents from the retina on which adenosine receptors are localized. This may be technically difficult to determine, since it is not known whether all such afferents contain adenosine receptors. Still, the possibility that adenosine receptors are localized on terminals predominately presynaptic to ADA-immunoreactive structures is worth considering. Although highly speculative, morphological evidence for such a synaptic relationship would support the association of ADA with neural elements from which adenosine release may occur.

4.5.1.3. CEREBELLUM. The localization of adenosine receptors on cerebellar granule cells (Wojcik and Neff, 1983b; Goodman et al., 1983), the connection of these through parallel fibers with Purkinje cells, and the electrophysiological observation of presynaptic effects of adenosine on parallel fibers (Kocsis et al., 1984) suggest that, in analogy with the arguments posed above for superior colliculus, cerebellar Purkinje cells may release adenosine and would exhibit immunoreactivity for ADA. Some difficulties have been encountered, however. Of the roughly half-dozen antisera to ADA employed in immunohistochemical studies, all produce qualitatively the same immunostaining pattern for ADA in rat brain except one. This one exception produces the same pattern as the others in all respects, but unlike the rest, also gives a consistently positive reaction in cerebellar Purkinje cell bodies, dendrites, and axons over a wide range of tissue fixation conditions (Nagy et al., 1988). Absorption controls indicate that immunostaining with this antiserum is specific. There appears to be no reasonable explanation for this result. However, since the staining appears to be specific, tentative conclusions we can draw are that the anomalous antiserum reacts with ADA as well as with an ADA-like protein produced only by Purkinje cells, or alternatively that, since all the antisera we employ are polyclonal, Purkinje cells produce an altered form of ADA recognized only by a particular antibody species present in the anomalous antiserum. Clearly, some biochemical comparisons of ADA in cerebellum with that in other brain areas are required. However, if Purkinje cells express high levels of some form of ADA, then the degeneration of these cells seen in humans lacking ADA (Ratech et al., 1985a) or in mice after treatment with ADA inhibitors (Ratech et al., 1985b) would be consistent with the requirement of this enzyme for the prevention of cytotoxicity induced by adenosine or its derivatives in certain cell types in vitro.

4.5.1.4. HYPOTHALAMUS. In the hypothalamus, ADA-positive neurons have been described in what Bleir et al. (1979) had designated the tuberal, caudal, and postmammillary caudal magnocellular nuclei, but which we termed collectively as the posterior hypothalamic magnocellular nuclei. Subsequently, it was recognized that these nuclei were a single entity coextensive with what had originally been termed the tuberomammillary nucleus (TM) (*see* Staines et al., 1987b). ADA-immunoreactive neurons in TM have also been described recently by Patel and Tudball (1986). Irrespective of the metabolic function of ADA, immunostaining for this enzyme in neurons of TM has provided the opportunity to conduct detailed anatomical studies of this nucleus, which in the past has received little attention. As seen by immunostaining for ADA (Nagy et al., 1984b; Staines et al., 1987a) and on the basis of other anatomical evidence, cells in TM give rise to widespread projections throughout the brain. However, the area most richly innervated by these fibers is the hypothalamus itself (Staines et al., 1987b).

In view of the diverse brain regions to which TM projects, it is likely that this nucleus is able to influence an equal diversity of CNS functions. Thus, it may be worth keeping in mind that, if some manner of intercellular communication in the CNS is governed partly by the metabolic action of ADA, then inhibition of ADA activity (as discussed earlier) may have widespread consequences insofar as the function of TM is concerned. Although some suggestions have been made with regard to the function of TM (Nagy et al., 1986b), at present very little is known about this nucleus.

4.5.1.5. COEXISTENCE. In several neural systems studied by dual immunostaining procedures in the rat, it is evident that ADA is co-localized with transmitter synthetic and/or degradative enzymes, or with neuroactive peptides. In sensory ganglia, ADA-immunoreactivity was found exclusively in a subpopulation of neurons immunoreactive for the peptide somatostatin (Nagy and Daddona, 1985). In preganglionic parasympathetic neurons and some motor neurons (discussed below), which of course are known to be cholinergic, ADA was colocalized with choline acetyltransferase (Senba et al., 1987a,b). In TM, ADA was found in virtually all neurons containing glutamic acid decarboxylase (GAD) and histidine decarboxylase (Senba et al., 1985; Patel et al., 1986). In addition, subpopulations of ADA-immunoreactive TM neurons were found to contain the enzyme monoamine oxidase or the peptide galanin (Staines et al., 1986b). In striatum, ADA-immunoreactive neurons constitute a subpopulation of those we suspect may contain GAD. With further scrutiny, it is likely that ADA in other central neurons will be found to coexist with other transmitters and/or neuromodulators. Thus, if ADA in rat and possibly other species proves to be a valid marker for sites of purinergic modulation, it would appear that purines may be utilized to some extent as coneuromodulators with some of the better established transmitters or neuropeptides.

4.5.1.6. SENSORY GANGLIA AND SPINAL CORD. Although no transmitters of sensory ganglia neurons have been firmly established, a number of candidate substances have been suggested as neurotransmitters and/or neuromodulators at primary afferent endings. Among these are amino acids and peptides. On the basis of early biochemical work (Holton, 1959) and more recent behavioral and electrophysiological studies (Post, 1984; Holmgren et al., 1986; DeLander and Hopkins, 1986; Sawynok et al., 1986; Fyffe and Perl, 1984; Jahr and Jessel, 1983; Salt and Hill, 1983; Salter and Henry, 1985), purines have also been implicated as neuromodulators in the spinal cord or transmitters of sensory afferents. In this context, it is noteworthy that adenosine and its analogs have antinociceptive actions when administered directly into the spinal cord by the intrathecal route (Post, 1984; Sawynok et al., 1986; DeLander and Hopkins, 1986). Adenosine receptors in the dorsal horn have the characteristics of A_1 receptors described in the brain and are highly concentrated in the substantia gelatinosa (Geiger et al., 1984b). On the other hand, ATP, but not adenosine, was found to have excitatory actions on spinal-cord dorsal horn neurons, and it was suggested that this nucleotide or a related substance may mediate transmission by low-threshold primary afferents associated with mechanoreceptors (Salter and Henry, 1985; Jahr and Jessel, 1983). There is, however, some controversy regarding the nature of the afferent fibers potentially involved; that is, whether these are small-diameter or large-caliber fibers. This is pertinent, as elaborated below, to the possible relationship between ADA and purinergic transmission.

In sensory ganglia, immunostaining for ADA has been observed in a subpopulation of small type B neurons, which are believed to give rise to fine-diameter and unmyelinated C fibers. In the spinal cord, immunostained fibers were located in the substantia gelatinosa. Accordingly, ADA activity was about 50% higher in dorsal than in ventral spinal cord (Geiger and Nagy, 1986). In order to determine the localization of ADA in spinal cord and related sensory structures, 2-d-old rat pups were treated with capsaicin, a C-fiber neurotoxin that causes a lifelong depletion of unmyelinated primary sensory afferents in animals allowed to mature to adulthood (Nagy 1982; Nagy et al., 1983). Significant depletions of immunostained fibers in dorsal spinal cord and immunoreactive neurons in dorsal root ganglia were found in these capsaicin-treated animals (Nagy et al., 1984c). In similarly treated rats, significant reductions of ADA activity were observed in sciatic nerve, dorsal horn, and trigeminal ganglia (Table 1). Although not statistically significant, ADA levels in dorsal roots from capsaicin-treated rats were about 20% lower than those in control animals. No differences were found in ventral roots or ventral horn of the spinal

Table 1
Adenosine Deaminase Activity in Spinal-Cord Tissues
of Adult Rats Treated Neonatally with Capsaicin*

| Tissue | Activity, nmol/mg protein/30 min | | | |
	Control	Capsaicin		
Sciatic nerve	169 ± 12 (8)	132 ± 11 (8)	22%;	$P < 0.05$
Dorsal root ganglia	447 ± 27 (9)	439 ± 29 (11)		
Dorsal horn	86 ± 2 (8)	64 ± 3 (8)	26%;	$P < 0.001$
Ventral horn	49 ± 5 (8)	53 ± 3 (8)		
Trigeminal ganglia	301 ± 11 (8)	225 ± 7 (8)	25%;	$P < 0.001$
Dorsal roots	139 ± 13 (8)	116 ± 5 (8)		
Ventral roots	102 ± 9 (8)	110 ± 12 (8)		

*Two-day-old Sprague-Dawley rat pups were treated with 25 mg/kg capsaicin. Control littermates were given vehicle alone and the animals were allowed to mature to adulthood. ADA assays were conducted as previously described (Geiger and Nagy, 1986). Values represent means ± SEM of the number of determinations indicated in parentheses. Statistical analysis of data was performed using a two-tailed Student's t-test. Statistical significance was considered at the $P < 0.05$ level.

cord. ADA activity was significantly reduced by 25% in trigeminal ganglia, whereas its activity was unchanged in dorsal root ganglia. This difference between cranial and spinal ganglia may be the result of the presence of ADA-immunoreactive satellite cells in sensory ganglia and the apparently larger numbers of ADA-containing cells in trigeminal compared with spinal ganglia.

The above observations indicate that a significant proportion of ADA is contained in primary afferents having unmyelinated or fine-diameter fibers. Fyffe and Perl (1984) suggested that synaptic excitation in the spinal cord by precisely these fibers was mediated by ATP or an ATP-like agent. Thus, the anatomical localization of ADA in sensory ganglia may have some functional relevance to the transmission of particular sensory modalities. In view of uncertainties in this area, however, it may be revealing to determine whether ATP release by primary sensory neurons, as originally demonstrated by Holton (1959), could be reduced by procedures of capsaicin treatment that cause either the degeneration or inactivation of unmyelinated primary afferents. A further point in need of clarification is how the antinociceptive actions of adenosine in the spinal cord may be related to the proposal of ATP-mediated transmission by low-threshold primary afferents. In this regard, a major issue not dealt with in detail here, but possibly pertinent to the spinal cord and elswehere, is whether adenosine and ATP have actions at separate anatomical loci or whether the actions of the former follow the extracellular breakdown of the latter.

4.6. ADA During Development

Observations concerning the distribution pattern and restricted neuronal localization of ADA in adult rat CNS prompted investigation of whether a similar pattern existed in this species at various stages of development. The profile of ADA activity in whole brain was examined from the 15th day of gestation to adulthood and in nine brain regions from the first postnatal day to adulthood (Geiger and Nagy, 1987). In 1-d-old rat pups, ADA activity was highest in olfactory bulbs and lowest in pons; the difference between the two regions was about 4.5-fold. A more heterogeneous distribution was found in adult animals where activity in olfactory bulbs was about eightfold higher than in the hippocampus, in which activity was the lowest. ADA activity decreased during development by as much as fivefold in most brain regions examined, including whole brain, cortex, cerebellum, superior colliculus, hippocampus, olfactory bulbs, and olfactory nucleus. The levels in pons and remaining subcortical tissues remained relatively constant throughout this same developmental period. Hypothalamus was a striking exception in that activity increased by about twofold between the ages of 1 and 50 d. Analysis of whole brain showed that developmental changes in ADA activity were the result of differences in V_{max} rather than altered affinity (K_m) for its substrate. It may be noted that the developmental profiles of [^{3}H]NBI sites paralleled very closely those of ADA (Geiger, 1987). In contrast, adenosine receptor numbers are low in newborns and increase with development (Geiger et al., 1984a).

The different developmental patterns observed in ADA activity among rat brain regions were reflected by similar complex changes in immunostaining for ADA. In general, the immunohistochemical results indicated the very early appearance of ADA in several structures that exhibit immunostaining in adults (Senba et al., 1987c). For example, in the superior colliculus and hypothalamus, the intensity of perikaryal ADA-immunostaining progressively increased throughout development. However, in several other structures, such as some olfactory-related systems, immunostaining increased perinatally, but then gradually decreased and disappeared in young adult animals (Senba et al., 1987b). A striking example of this phenomenon was observed in facial and hypoglossal motoneurons, which lack ADA-immunoreactivity in adult animals, but exhibit intense transient immunostaining from the 18th gestational day to about 15 d of age. Interestingly, in the entire facial motor nucleus, only those motoneurons innervating the perioral facial muscles showed this transient change. Similarly, in the hypoglossal nucleus, only motoneurons in the dorsal subnucleus, which innervate tongue retractor muscles, exhibited ADA-immunoreactivity at late prenatal and early postnatal times. It was speculated that the requirement for ADA in these motoneurons may be related to the functional demands placed on perioral muscles during early feeding behavior and/or the demonstrated neuromodulatory actions of adenosine at the neuromuscular junction (Silinsky, 1984; Ribeiro and Sebastiao, 1985).

The generally high levels of ADA activity in neonates suggests a possible role of ADA in neural development. However, the disparate developmental patterns among brain regions did not appear to correspond to any one particular process, such as periods of rapid cell proliferation, cell death, synaptogenesis, or myelination. This lack of correspondence perhaps reinforces the notion that ADA and, by inference, adenosine are involved in several processes, including intermediary metabolism as well as neuroregulation and metabolic control of cellular activity through adenosine receptors (Newby, 1984; Magistretti et al., 1986).

4.7. Species Differences

As an aid to establishing the functional significance of putative markers for sites of adenosine action, numerous reports have appeared comparing the neurochemical characteristics and distribution patterns of possible markers across species. In this regard, considerable interspecies variations have been reported in the molecular weights of A_1 receptors (Klotz and Lohse, 1986) and the distribution of adenosine receptors in cerebral cortex (Ukena et al., 1986), hippocampus (Lee et al., 1986), and other brain regions (Murray and Cheney, 1982;

Fastbom et al., 1986, 1987). Moreover, interspecies differences have been found regarding the distribution of 5'-nucleotidase activity in hippocampus (Lee et al., 1986), the kinetic interactions of [³H]NBI with nucleoside transport sites, the displacement of [³H]NBI from its binding sites by nucleosides and nucleoside transport inhibitors (Hammond and Clanachan, 1985; Bisserbe et al., 1986), the distribution of [³H]NBI sites in the CNS (Hammond and Clanachan, 1983; Marangos, 1984; Verma and Marangos, 1985), adenosine-induced increases in cerebral blood flow (see Sollevi, 1986), and as mentioned earlier, the expression of ADA-binding protein (Schrader et al., 1983; Andy and Kornfeld, 1982). These results seem to suggest that species differences in biochemical processes related to adenosine are the rule rather than the exception. In view of this, the immunohistochemical localization and activity of ADA in mouse, rat, guinea pig, hamster, and rabbit were compared (Yamamoto et al., 1987).

In general, the distribution of ADA-immunoreactivity in mouse was similar to that in rat, particularly in the hypothalamus and spinal-cord substantia gelatinosa. However, certain differences are worthy of note. For example, immunostained neurons were virtually absent in mouse superior colliculus, yet the hippocampus, which in rat contained only sparsely distributed immunoreactive fibers, contained densely packed immunoreactive perikarya in the subiculum and granular layers of the denate gyrus. Unlike rat, prominent immuno-staining of glial cells was observed throughout the mouse CNS. The distribution of ADA-immunoreactive cells and fibers in hamster and guinea pig was less remarkable. A few immunostained neurons were found in hamster hypothalamus, whereas fibers were prominent in the habenular commissure and substantia gelatinosa of hamster and less so in the substantia gelatinosa of guinea pig. No glial cell staining was found in hamster, although extensive staining of glial cells and endo-thelial cells of blood capillaries was observed in guinea pig. Attempts to demonstrate ADA-immunoreactive structures in rabbit CNS tissues produced only nonspecific staining.

The distribution of ADA activity in 11 brain regions of mouse, rat, guinea pig, and rabbit is illustrated in Fig. 2. Among the four species, the differences between the areas with the highest and lowest activities were about 12-fold for rat, 4.2-fold for mouse, 2.4-fold for guinea pig, and threefold for rabbit. The more homogeneous distribution of ADA activity in mouse, guinea pig, and rabbit compared with rat is probably related, certainly in the cases of mouse and guinea pig, to the presence of ADA in glial cells, which are thought to be homogeneously and ubiquitously distributed in the CNS. The distribution of ADA in human brain was reportedly similar to (Phillips and Newsholme, 1979) and distinct from (Norstrand et al., 1984; Norstrand, 1985) that found in rat. In contrast to rat, high levels of ADA were associated with white matter and glial cells in human CNS. The significance of these and the other species differences cited above are likely to remain perplexing until their functional consequences to both adenosine metabolism and neuroregulation are determined.

5. Adenosine Uptake and Transport

It is now well established that synaptic levels, and hence receptor-mediated actions, of most neurotransmitters are regulated, in part, by their selective uptake into nerve terminals and cells adjacent to synapses. Two noteable exceptions are histaminergic and cholinergic neurons, which do not possess uptake systems for histamine or acetylcholine, respectively, but rather selectively accumulate the precursors of these substances (Kuhar, 1973; Hosli and Hosli, 1984). Since uptake processes are specifically expressed by neurons according to the transmitters they utilize, the study of uptake mechanisms has aided the identification of transmitters in neural systems. That reuptake may be significantly involved in regulating the actions of adenosine is suggested by findings that adenosine uptake inhibitors enhance the ability of adenosine to depress neuronal excitability (Motley and Collins, 1983), alter cAMP levels (Huang and Daly, 1974; Nimit et al., 1981), produce antinociception (Yarbrough and McGuffin-Clineschmidt, 1981), affect cardiovascular function (Stafford, 1966; Berne, 1980; Sollevi, 1986), and decrease animal locomotor activity (Crawley et al., 1983; Phillis et al., 1986). In discussing adenosine neuromodulation, it is therefore necessary to consider the membrane localized mechanisms responsible for governing the entry of adenosine into cells.

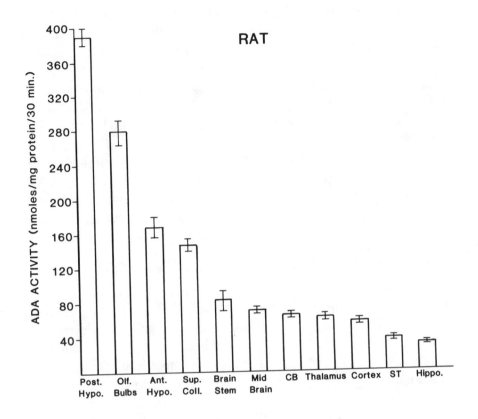

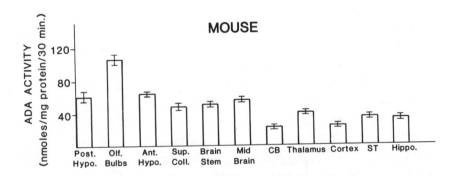

Fig. 2. Regional distribution of adenosine deaminase activity in rat, mouse, guinea pig, and rabbit brain. Abbreviations: posterior hypothalamus (Post. Hypo.), olfactory bulbs (Olf. Bulbs), anterior hypothalamus (Ant. Hypo.), superior colliculus (Sup. Coll.), cerebellum (CB), striatum (ST), hippocampus (Hippo.).

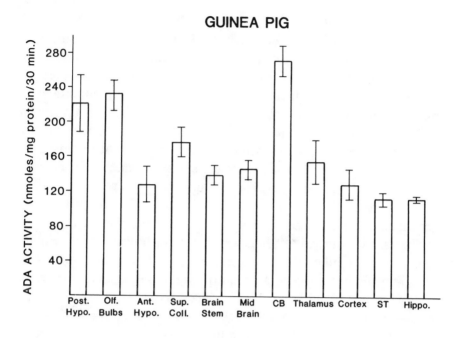

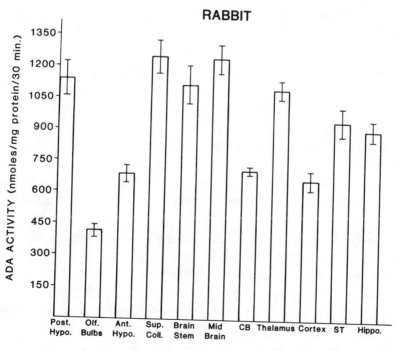

5.1. Defining Uptake and Transport Processes

Adenosine crosses cell membranes via two different mechanisms, namely, by passive diffusion and by what has classically been thought of as facilitated diffusion through a nucleoside transport system (NTS). Passive diffusion is slow and of minor importance at physiologically relevant concentrations of nucleosides. Facilitated diffusion, on the other hand, is a high affinity, high capacity process through which nucleosides are transferred bidirectionally across cytoplasmic membranes. These processes have been extensively reviewed elsewhere (Plagemann and Wohlhueter, 1980; Paterson et al., 1981; Young and Jarvis, 1983; Wu and Phillis, 1984). More recently, a third mechanism for transporting nucleosides has been identified in such tissues as choroid plexus, hepatocytes, renal brush border vesicles, and cultured epithelial cells (Spector, 1985; Paterson et al., 1987). This appears to be an energy-requiring process that is driven by the sodium gradient and is capable of concentrative transport. In certain tissues having this capability, nucleosides may be actively "pumped" into cells, but "leaked" back out through the NTS, thereby sharing characteristics with that of some glucose transporters. Findings from earlier studies that [^{3}H]adenosine uptake into mouse cerebral cortical slices (Banay-Schwartz et al., 1980) and rat brain synaptosomes (Bender et al., 1980) was reduced by 50 and 25%, respectively, in the absence of sodium suggests and indeed we have recently demonstrated that a sodium-dependent uptake process also occurs in the CNS (Johnston and Geiger, 1989). On the basis of these and other recent developments, it appears that nucleoside transport is more intricate than originally thought, and little is known regarding its complexities in CNS tissues. The term nucleoside transport will be used here to describe adenosine translocation by cells, be this facilitated or active.

The NTS-mediated permeation of adenosine across membranes is extremely rapid. This, in conjunction with the abundance of enzymes available to metabolize intracellular nucleosides, creates certain problems for determining the precise kinetics and characteristics of adenosine transport. For example, phosphorylation of adenosine by AK leads to the formation and accumulaton of 5'-AMP intracellularly, thereby trapping the permeated adenosine in the form of nucleotides. Under certain experimental conditions the rate-limiting step for adenosine accumulation may be misinterpreted as the K_m of AK rather than the K_m of the NTS for adenosine. *Transport* then is defined as the transfer of native substrate from one side of a membrane to another and differs from *uptake*, which refers to the accumulation of substrate without regard to its possible metabolism. According to current criteria, successful measurements of adenosine transport have only been accomplished using either very short incubation times, nonmetabolized substrates for the transporter, or tissues deficient in the enzymes responsible for metabolizing the transported substrate. These methodological considerations are discussed in detail elsewhere (Paterson et al., 1985).

5.2. Adenosine Uptake by CNS Tissues

The bulk of studies to date on adenosine accumulation into CNS tissues have used methodology that more likely reflected the measurement of uptake rather than transport. For example, Bender et al. (1981a,b) found that incubation of synaptosomes from rat cerebral cortex with [^{3}H]adenosine at 37°C for 1 min resulted in metabolic conversion of approximately 50% of the accumulated adenosine. The contribution to total uptake by particular CNS cell types further complicates the issue, since uptake processes have been localized on neuronal, glial, and endothelial cells (Thampy and Barnes, 1983a,b; Hertz, 1978; Bender and Hertz, 1986; Beck et al., 1983). With these caveats in mind, K_m values for [^{3}H]adenosine uptake were found to be 0.9 and 21 μM for rat and guinea pig cortical synaptosomes, respectively (Bender et al., 1981a; Barberis et al., 1981), and ranged from 9.6–140 μM for mouse and guinea pig brain slices (Banay-Schwartz et al., 1980; Shimizu et al., 1972; Davies and Hambley, 1986) and from 3.4–6.5 μM for primary cultures of mouse astrocytes and neurons (Hertz, 1978; Bender and Hertz, 1986).

5.3. Adenosine Transport by CNS Tissues

In CNS tissues, adenosine transport *per se* appears only to have been accomplished using dissociated brain cells from adult rat (Geiger et al., 1988) and primary cultures of chick neuronal and glial cells depleted of ATP. Dissociated brain cells expressed a high affinity (K_m

0.8 μM) system selective for the nucleoside adenosine as well as a lower affinity (K_m 259 μM) system. Neuronal cells contained a higher affinity transport system for adenosine (K_m 13 μM) than did glial cells (K_m 370 μM; or Thampy and Barnes, 1983a,b). In addition, neurons showed a greater selectivity than glial cells for transporting adenosine as compared with other nucleosides. This later finding agrees with previous reports that, although a variety of nucleosides are apparently able to pass through the NTS, this system in at least certain tissues appears to exhibit some selectivity for adenosine (Plagemann and Wohlhueter, 1980; Bender et al., 1980). On the other hand, it is important to note that glial cells have a greater capacity to accumulate [^{3}H]adenosine than do neurons (Ehinger and Perez, 1984; Bender and Hertz, 1986). It is currently not possible to attribute the different rates of neuronal or glial [^{3}H]adenosine uptake to differences in transport rates or purine metabolism.

5.4. Possible Relationship
Between Transport and Release

Although detailed discussion of adenosine release from nervous tissues is beyond the scope of this chapter (*see* reviews by Stone, 1981; White, 1985) a brief mention is warranted in the context of recent evidence suggesting a close relationship between this process and adenosine transport. Adenosine outflow from various central and peripheral tissues has been shown to occur as a consequence of depolarizing stimuli and hypoxia. However, at present it is not possible to attribute adenosine release to any particular cell type or neural system in the CNS. It is also unclear whether adenosine output, which has been demonstrated in synaptosomal (Sun et al., 1976), brain slice (Wojcik and Neff, 1983a), and in vivo (Barbaris et al., 1984) preparations, occurs as released nucleoside or whether extracellular adenosine appears secondarily to the efflux of adenine nucleotides and their subsequent metabolism by ecto-nucleotidases. Indeed, evidence has been presented in favor of (MacDonald and White, 1985) and against (Pons et al., 1980) the proposal that released adenine nucleotides represent the major source of extracellular adenosine. In addition to better understood modes, novel mechanisms for release of adenosine as well as classical transmitters have recently been observed (Fillenz,

1984). In the case of adenosine, sufficiently high intracellular levels may cause it to flow down its concentration gradient and pass out of cells through the NTS. This latter phenomenon was convincingly demonstrated in some peripheral tissues exposed to hypoxic challenge (Belloni et al., 1985; Bukoski and Sparks, 1986). The efflux of adenosine through the NTS in CNS tissues is suggested by reports that dipyridamole, a nucleoside transport inhibitor, decreased the evoked release of [³H]purines (Jonzon and Fredholm, 1985; Daval and Barbaris, 1981). The existence of such a "release" mechanism may explain some apparent discrepancies, namely, that adenosine can be extruded from cells in a fashion similar to (Barberis et al., 1984) or different from (Jonzon and Fredholm, 1985) that of traditional transmitter substances. Therefore, as the biochemical characteristics and locations of adenosine transport and release sites becomes clearer, so too will the possible functional relationships between them.

6. Radioligand Labeling of Adenosine Transport Sites

Resulting in large part from the difficulties in measuring adenosine transport in CNS tissue, equilibrium radioligand binding assays using ligands for transporter binding sites represents a promising approach to the study of membrane-associated mechanisms involved in nucleoside transport. Two putative ligands for the nucleoside transporter are [³H]NBI (Marangos et al., 1982b; Hammond and Clanachan, 1984b; Wu and Phillis, 1982; Geiger et al., 1985) and [³H]dipyridamole (Bisserbe et al., 1985, 1986; Deckert et al., 1987; Marangos and Deckert, 1987). The utility of [³H]NBI as a label for NTS sites is supported by findings that, in some peripheral tissues and cultured cells, occupancy of transport sites by [³H]NBI correlates with its ability to inhibit nucleoside transport and that the density of functional transport sites correlates on a sites/cell basis with the V_{max} of the NTS (Cass et al., 1974; Jarvis et al., 1982). However, for most tissues, the precise relationship of [³H]NBI sites to the transporter is less clear and is complicated by observations in some tissues where NTS systems are:

1. Sensitive to inhibition by NBI and contain high levels of [³H]NBI binding sites;

2. Less sensitive to NBI and contain few [³H]NBI binding sites; and

3. Less sensitive to NBI, but retain high numbers of [³H]NBI binding sites.

Furthermore, in some tissues, two types of NTS systems have been found where one is sensitive and the other insensitive to the transport blocking actions of NBI (Paterson et al., 1987). Since it appears that more than one site may be labeled by [³H]NBI, the association of [³H]NBI sites with the NTS must be determined for each tissue in which these sites are studied.

In the CNS, uncertainty surrounding the nature of [³H]NBI binding sites originated from the report that the concentrations of NBI necessary to inhibit adenosine accumulation into synaptosomal preparations were orders of magnitude higher than the affinity of the NTS for [³H]NBI (Phillis and Wu, 1983). However, in this study, at least a 20% reduction in accumulation was observed at nanomolar concentrations of NBI. Moreover, it is likely that uptake and not transport was examined. Interpretations of these results include the possibilities that adenosine uptake in these preparations is relatively insensitive to the blocking effects of such inhibitors as NBI and dipyridamole or, alternatively, that as discussed earlier two populations of NTS sites exist—one sensitive and another relatively insensitive to the effects of NBI (Jarvis, 1986). However, it was recently reported that adenosine uptake in rat brain synaptosomes under very brief incubation conditions (5 s) was inhibited by dipyridamole with an IC_{50} of 76 nM (Morgan and Stone, 1986), a value comparable to that observed for synaptosomal preparations from guinea pig, a species more sensitive than rat to the effects of nucleoside transport inhibitors (Hammond and Clanachan, 1985). Thus, part of the problem in establishing the nature of [³H]NBI binding sites originates from comparisons of ligand binding affinities with inhibition of adenosine uptake rather than adenosine transport.

6.1. Reversible Binding of [³H] Nitrobenzylthioinosine

Although a single class of high affinity [³H]NBI binding sites has been observed in most studies of CNS tissues, two reports have appeared describing the presence of both high and low affinity sites. In rabbit, the K_d and B_{max} values were 0.32 nM and 79 fmol/mg protein for the high affinity site and 4.1 nM and 520 fmol/mg protein for the lower affinity site, respectively (Hammond and Clanachan, 1984a). For rat synaptosomal membranes, the two classes of sites had K_d and B_{max} values of 0.05 nM and 113 fmol/mg protein and 190 μM and 1 nmol/mg protein, respectively (Wu and Phillis, 1982). The kinetic characteristics and physiological significance of the low affinity site are poorly understood at present; their analysis is made difficult by the high concentrations of radioligand required during tissue incubation. Where one class of [³H]NBI binding sites has been found, the reported kinetic values are summarized in Table 2. For rat brain, K_d values ranged from 0.05 (Wu and Phillis, 1982; Geiger et al;., 1985) to 0.18 nM (Marangos et al., 1982b), and B_{max} values ranged from 113 (Wu and Phillis, 1982) to 172 fmol/mg protein (Hammond and Clanachan, 1984a). Similar values were found for mouse brain (Hammond and Clanachan, 1984a; Verma and Marangos, 1985). The K_d and B_{max} values for human brain were 0.16 nM and 142 fmol/mg protein (Verma and Marangos, 1985) and 0.20 nM and 115 fmol/mg protein (Marangos, 1984), respectively. Although the K_d and B_{max} values for guinea-pig were generally similar to those of rat, mouse, and human, the variability among studies of guinea pig CNS tissues was much greater; K_d values ranged from 0.02 nM (Hammond, 1985) to 0.35 nM (Marangos, 1984), and B_{max} values from 115 (Hammond, 1985) to 433 fmol/mg protein (Hammond and Clanachan, 1984a). In membrane preparations of dog brain, [³H]NBI binding was of slightly lower affinity (K_d values ranged from 0.44–4.9 nM) and somewhat greater numbers of sites were found (B_{max} ranged from 174–638 fmol/mg protein) (Hammond and Clanachan, 1984a; Verma and Marangos, 1985).

From the above data, it appears that the binding characteristics of [³H]NBI are fairly consistent across species. In contrast, certain differences among species have been identified in studies where the competition for [³H]NBI binding sites by adenosine uptake inhibitors and nucleosides were examined. For uptake inhibitors, the K_i values for competition by NBI and nitrobenzylthioguanosine varied less than fourfold between human, dog, guinea pig, mouse, and rat CNS tissues. However, dilazep, hexobendine, and dipyridamole were far more potent in competing for [³H]NBI binding sites in human and dog than in rat,

Table 2

Pharmacological Characteristics of [³H]NBI Binding in Mammalian CNS

Tissue preparation	K_d, nM	B_{max}, fmol/mg protein	K_i, nM DPR	K_i, nM NBI	K_i, μM Adenosine	Ref.
Rat						
Whole brain	0.18	135	283	0.23	33	Marangos et al., 1982
Whole brain	0.15	120	1200	–	–	Marangos, 1984
Whole brain	0.06	147	16	0.05	3	Geiger et al., 1985
Brain sections	0.16	126	–	–	–	Nagy et al., 1985
Sol. brain*	0.12	133	1084	0.6	92	Verma et al., 1985
Cortex†	0.05	113	–	–	9	Wu and Phillis, 1982
Cortex†	0.11	172	711	–	–	Hammond and Clanachan, 1984b
Cortex†	0.18	170	1135	4.2	–	Verma and Marangos, 1985
Striatum	0.06	278	–	–	–	Geiger, 1986
Dorsal cord	0.42	110	–	–	–	
Ventral cord	0.54	74	–	–	–	
Dorsal roots	0.07	98	–	–	–	
Ventral roots	0.09	23	–	–	–	Geiger and Nagy, 1985
Guinea Pig						
Whole brain	0.35	150	6.5	–	–	Marangos, 1984
Whole brain	0.075	240	4.9	–	77	Jarvis and Ng, 1985
Cortex†	0.020	115	2.5	–	–	Hammond, 1985
Cortex†	0.16	300	11	0.22	123	Hammond and Clanachan, 1984a\
Cortex†	0.31	433	22	–	–	Hammond and Clanachan, 1984b
Cortex†	0.26	128	6.3	3.9	175	Verma and Marangos, 1985
Sol. brain*	0.13	150	36	1.9	88	Verma et al., 1985
Brain regions	0.15–0.38	201–644	–	–	–	Hammond and Clanachan, 1983

Mouse						
Cortex[†]	0.11	210	—	—	—	Hammond and Clanachan, 1984b
Cortex[†]	0.17	196	82.9	3.1	150	Verma and Marangos, 1985
Dog						
Cortex[†]	4.9	638	—	—	—	Hammond and Clanachan, 1984b
Cortex[†]	0.44	174	1.1	2.8	150	Verma and Marangos, 1985
Human						
Brain	0.20	115	2.9	—	—	Marangos, 1984
Cortex[†]	0.16	142	2.7	1.1	—	Verma and Marangos, 1985

*Sol.—solubilized.

[†]Cortex refers to cerebral cortex.

the species least sensitive to the effects of these agents (Verma and Marangos, 1985). For example, based on K_i values, dipyridamole was 1013× more potent in dog compared with rat tissues (Verma and Marangos, 1985). Similarly, studies of dipyridamole competition for [³H]NBI binding sites in the three other species showed the rank order of sensitivity to be rabbit > guinea pig > rat (Hammond and Clanachan, 1985). As summarized in Table 2, K_i values for dipyridamole ranged from 16 (Geiger et al., 1985) to 1200 nM (Marangos, 1984) for rat and from 2.5 (Hammond, 1985) to 36 nM (Verma et al., 1985) for guinea pig. Although a repeatable finding at the time, we have no explanation for the much lower K_i values obtained by us for DPR (Geiger et al., 1985) compared to those reported by others. A further difference among species regarding dipyridamole competition for [³H]NBI sites includes the observation of biphasic competition curves for guinea pig, rabbit, and dog, but not for rat or mouse (Hammond and Clanachan, 1985). The proportions of high and low affinity binding sites indicated by these biphasic curves were different among the species in which both types of sites were found. Thus, based on [³H]NBI binding data alone, it might be expected that rat compared with other species would be less susceptible to uptake inhibitor enhancement of adenosine's effects; this, in fact, is supported by findings from physiological studies (*see* Table 4 in Hammond and Clanachan, 1985).

Of critical importance with respect to the nature of sites labeled by [³H]NBI is their affinity and specificity for nucleosides. The K_i values of adenosine competition for [³H]NBI binding sites in rat brain have been reported to be 3 µM (Geiger et al., 1985), 9µM (Wu and Phillis, 1982), 33 µM (Marangos et al., 1982b), and 92 µM (Verma et al., 1985). As for specificity, adenosine was clearly more potent in competing for [³H]NBI sites than were other nucleosides (Wu and Phillis, 1982; Geiger et al., 1985). It remains an enigma, however, why low micromolar affinities for adenosine were found by some investigators and not others. One possible explanation for this may be the difficulty associated with the measurement of adenosine's competition for [³H]NBI binding sites in the presence of excessive endogenous levels of adenosine or alternatively metabolism of the competing nucleoside. Indeed, Hammond and Clanachan (1984b) found a somewhat reduced K_i value of adenosine competition for [³H]NBI binding to membranes from guinea pig brain in the presence of an adenosine deaminase inhibitor. Despite the species differences described above and the need for further studies validating this substance as a marker for the NTS in some tissues, [³H]NBI remains one of the most promising ligands for nucleoside transporters identified to date.

6.2. Covalent Labeling of Transport Sites

In addition to reversible interactions of [³H]NBI with nucleoside transport sites, this ligand is able to covalently label transporters in the presence of UV irradiation (Marangos et al., 1982a), a property that has been exploited to purify and characterize these sites partially. This approach has shown that nucleoside transporters labeled by [³H]NBI have similar molecular weights in rat and guinea pig lung (Shi et al., 1984), cultured mouse lymphoma cells (Young et al., 1984), human erythrocytes (Jammohamed et al., 1985), and guinea pig brain (Jarvis and Ng, 1985). The partially purified transporter protein from human erythrocytes, designated as a band 4.5 polypeptide, has been effectively reconstituted into phospholipid liposomes (Tse et al., 1985). Photoaffinity labeling with [³H]NBI may ultimately lead to clarification of the discrepancies and uncertainties as to the function and molecular nature of binding sites.

One brief report has appeared utilizing 2-azidoadenosine as a photoaffinity label for adenosine transport sites in the CNS (Calvalla and Neff, 1985). Further information regarding the binding characteristics of this ligand and its distribution in the CNS are not currently available.

6.3. Binding of [³H]Dipyridamole

An alternative ligand for the NTS is [³H]dipyridamole ([³H]DPR), which has recently been synthesized with very high specific activity (Marangos et al., 1985). In membranes of guinea pig brain, [³H]DPR exhibited high affinity binding with values of K_d ranging from 3.5–7.6 nM, and B_{max} ranging from about 320–530 fmol/mg protein (Marangos et al., 1985; Marangos and Deckert, 1987). When [³H]DPR binding was examined autoradiographically in sections of guinea pig brain, the K_d and B_{max} values obtained were 10 nM and 650 fmol/mg protein, respectively. Thus, the affinity of [³H]DPR for binding sites in the brain were found to be at least 10× lower and the number of sites 2×

higher compared with those labeled by [³H]NBI. In membrane preparations from guinea pig lung, the pharmacological characteristics of [³H]DPR were similar to those of [³H]NBI binding in this tissue (Shi and Young, 1986), whereas K_d values were about 10× higher than that observed for [³H]NBI. The greater numbers of [³H]DPR binding sites may reflect methodological difficulties arising from the highly lipophilic nature of dipyridamole, or the labeling of NTS sites possibly different from those labeled by [³H]NBI. In membranes of guinea pig brain, one principal difference between the characteristics of [³H]DPR and [³H]NBI labeled sites is that, although subnanomolar concentrations of NBI were inhibited the binding of both ligands, micromolar concentrations of NBI required to reach its IC_{50} value for [³H]DPR binding (Marangos and Deckert, 1987; Deckert et al., 1987). It has further been shown that the potency of NBI to compete for [³H]DPR binding varies significantly among brain regions. For example, NBI (500 n*M*) inhibited [³H]DPR binding by only about 10% in ependyma and cerebellar Purkinje cells, but by greater than 65% in anterior hypothalamus, nucleus tractus solitarius, area postrema, and brainstem arteries (Deckert et al., 1987). In contrast to guinea pig, specific binding of [³H]DPR to membranes of brain and lung from rat could not be measured apparently because of the low affinity of nucleoside transporters for this ligand (Marangos and Deckert, 1987; Shi and Young, 1986). It seems clear, therefore, that different types of [³H]DPR binding sites are expressed in the CNS of various species.

6.4. Distribution in the CNS

The CNS distribution of sites labeled by [³H]NBI in rat has been examined by autoradiographic and membrane binding methods (Geiger and Nagy, 1984, 1985; Bisserbe et al., 1985; Geiger, 1986), and the results of these studies are in close agreement. High levels of [³H]NBI sites were found in thalamic structures, midbrain, superficial layers of the superior colliculus, area postrema, choroid plexus of lateral, third, and fourth ventricles, hypothalamus, striatum, accumbens, nucleus of the solitary tract, pyriform cortex, and dorsal spinal cord. In contrast, very low levels were found in hippocampus, cerebellum, and white matter. Not withstanding the limitations of membrane binding and microdissection techniques, the distribution of these sites was very similar qualitatively to that observed autoradiographically (Geiger and Nagy, 1984). Of note was the nonuniformity in the density of [³H]NBI sites within particular brain regions. This is clearly illustrated by quantitative autoradiography (Table 3) where measurements of optical density (OD) revealed highly heterogeneous patterns: in thalamic nuclei, the OD was twofold higher in the mediodorsal nucleus compared with the paracentral nucleus and reuniens nucleus, in hypothalamus, the OD in TM was twice that found in more anterior portions, in superior colliculus, the OD in superficial layers was about twice that observed in deeper layers, and in the solitary nucleus, the OD was about twofold greater in dorsal compared with ventral portions of this structure.

Autoradiographic examination of [³H]DPR binding sites in guinea pig brain showed patterns both similar to and different from that observed for [³H]NBI sites in rat brain. As with [³H]NBI, high levels of [³H]DPR binding were observed in superficial layers of the superior colliculus, supraoptic nucleus, nucleus of the solitary tract, ependyma, choroid plexus, and hypothalamus. In contrast with [³H] NBI, high levels of [³H]DPR-labeled sites were found in the superficial layers of cerebral cortex, the interpeduncular nucleus, the central gray matter, and on brainstem arteries (Deckert et al., 1987). A dramatic difference between [³H]NBI and [³H]DPR was found in cerebellum that contained very low levels of [³H]NBI sites in rat, but where dense [³H]DPR binding was observed in the Purkinje cell layer in guinea pig (Bisserbe et al., 1985; Geiger and Nagy, 1984; Deckert et al., 1987). Because specific binding of [³H]DPR to rat brain was not measurable and the distribution of [³H]NBI sites in guinea pig brain has not yet been reported in detail (Deckert et al., 1986), it is not clear whether differences in the distribution of sites labeled by these two agents reflect species variations or labeling of different populations of binding sites associated with the nucleoside transporter.

7. Comparative Distributions
of Putative Purinergic Markers

Comparison of the distribution of ADA and [³H]NBI binding sites in rat brain with those of other putative markers for adenosine-utilizing neural systems are made difficult by the paucity of detailed information regarding these markers in rat. Nevertheless, in an attempt to ascribe functional significance to the heterogeneous distribu-

Table 3
Regional Distribution of [³H]NBI Binding Sites in Rat Brain as
Determined by Quantitative Autoradiography*

Brain area	OD	Brain area	OD
Cortex		Striatum	0.07
Frontal	0.05	Substantia nigra	0.07
Motor	0.03	Septum	0.04
Visual	0.04	Corpus collosum	0.02
Hypothalamus		Anterior commisure	0.02
Posterior	0.07	Thalamus	
Anterior	0.06	Whole	0.08
TM	0.10	Thalamic nuclei	
Preoptic	0.06	Reuniens	0.09
Superior colliculus		Anteromedial	0.09
Superficial	0.11	Mediodorsal	0.04
Deep	0.06	Paracentral	0.09
Inferior colliculus	0.05	Paratenial	0.09
Accumbens	0.04	Pulvinar	0.08
Hippocampus	0.03	Gelatinosus	0.05
Mammillary body	0.03	Parafascicular	0.08
Habenula		Ventromedial	0.08
Lateral	0.03	Zona Incerta	0.07
Medial	0.08	Ventroposterior	0.08
Solitary nucleus		Paraventricular	0.08
Whole	0.12	Anterior	0.08
Dorsal	0.18	Posterior	0.08
Ventral	0.11	Peri-aquaductal gray	
Reticular formation	0.02	Whole	0.06
Area postrema	0.22	Dorsal	0.08
Choroid plexus	0.10	Ventral	0.06
Cerebellum	0.02	Background	0.01

*Incubation conditions and qualitative descriptions of the distribution of [³H]NBI binding were reported previously (Geiger and Nagy, 1984). OD readings were performed using a computer-assisted image analysis system. Values represent means from at least two separate determinations of grain density on different sections. Nonspecific binding was subtracted from the OD measurements of individual brain regions.

tion patterns of ADA and sites labeled by [³H]NBI in rat brain, such comparisons may be instructive. These markers include adenosine A_1 and non-A_1 receptors, adenosine-like immunoreactivity, 5'-nucleotidase activity, ADA activity and immunoreactivity, and sites of adenosine and ATP release. A worthwhile comparison, of course, would be the distribution of adenosine transporter binding sites with that of regional adenosine transport capabilities. However, as mentioned earlier, very little quantitative information is available for nucleoside transport in CNS tissues. Where comparisons are possible, examples exist for similarities and disparities in distribution patterns, and only selected examples will be described.

7.1. Receptors

The distribution of adenosine receptors labeled with [³H] cyclohexyladenosine ([³H]CHA) (Goodman and Snyder, 1982), a relatively selective ligand for A_1 sites, was found to coincide to some degree with the distribution of [³H]NBI sites; high levels of both sites were found in superficial layers of superior colliculus, certain thalamic nuclei, striatum, and spinal cord substantia gelatinosa. Comparison of the cellular localization of [³H]CHA and [³H]NBI-labeled sites has revealed that, in striatum, [³H]CHA sites and to a lesser degree [³H]NBI sites were located on neuronal elements originating from perikarya within this structure (Geiger, 1986). In superior colliculus, adenosine receptors were located to a large extent (>50%) on terminals of retinal afferents; [³H]NBI sites were not detected on these afferents (Goodman et al., 1983; Geiger, 1986). However, in marked contrast to the very high levels of [³H]CHA sites present in cerebellum and hippocampus, these structures were among the areas containing the lowest levels of [³H]NBI sites. One possibility for some of the striking mismatches between the localization of adenosine receptors and [³H]NBI sites is that [³H]CHA is not labeling all receptor subtypes. The autoradiographic distribution of what were considered A_2 receptors was in fact recently reported to be significantly different from that of A_1 sites (Lee and Reddington, 1986; Jarvis et al., 1989). Nevertheless, disparities between the locations of adenosine receptors and [³H]NBI sites were still evident. If current markers for adenosine receptors are proven to be valid indicators of all sites of adenosine action, then other factors for the differential distribution patterns of these receptors and [³H]NBI binding sites deserve consideration. Foremost among these, of course, is whether there is a necessary anatomical and neurochemical relationship between nucleoside transport capability and all adenosine receptor-mediated events. This issue is likely to remain unresolved until some of the complexities surrounding adenosine transport sites and their labeling with radiolabeled transport inhibitors are better understood.

7.2. 5'-Nucleotidase

Heterogeneous distribution patterns of 5'-nucleotidase activity have been found in the CNS of mouse (Scott, 1967; Hess and Hess, 1986), rat (Nagata et al., 1984; Cammer et al., 1980; Schubert et al., 1979), and human (Lee et al., 1986). High levels were observed in hippocampus, cerebellum, and medulla oblongata of rat and mouse (Nagata et al., 1984) and in globus pallidus, striatum, substantia gelatinosa, olfactory cortex, and retina of mouse (Scott, 1967). More detailed species comparisons of 5'-nucleotidase localization in hippocampus revealed marked differences in enzyme patterns that did not correspond to the localization of [³H]CHA-labeled adenosine receptors (Lee et al., 1986). This lack of correspondence may partly reflect the association of 5'-nucleotidase with plasma membranes of both glial and neuronal cells (Bernstein et al., 1978; Kreutzberg et al., 1978). In view of this, it is perhaps not surprising that we have not observed any distinct colocalization between it and either [³H]NBI binding sites or ADA. In any case, 5'-nucleotidase remains a potentially important enzyme for producing adenosine intra- and/or extracellularly, and therefore, its relevance to adenosine neuromodulation warrants continued assessment.

7.3. Adenosine

Recently, antisera directed against a levulinic acid derivative of adenosine has been used to localize adenosine immunoreactivity in rat brain (Braas et al., 1986). Dense immunostaining of neuronal perikarya was found in hypothalamus, hippocampus, primary olfactory cortex, and motor nuclei. In hypothalamus, neurons apparently corresponding to those in TM that we observe to be immunoreactive for

ADA were reported to be intensely stained for adenosine. In addition, adenosine-immunoreactive fibers and terminals were found to be prominent in the spinal nucleus of the trigeminal nerve and moderate in the spinal-cord substantia gelatinosa. Although some similarities are apparent in the distribution of adenosine and ADA immunoreactivity, there are also some marked disparities. In general, immunostaining for adenosine was found to be much more widely distributed than that for ADA. Similarly, examples were found where the distribution pattern of adenosine-immunoreactivity either corresponded to, or differed from, other putative markers of adenosine systems (*see* Braas et al., 1986).

7.4. Adenosine and ATP Release

Without becoming embroiled in controversies regarding mechanisms of transmitter release, a brief outline of information available on the release of adenosine and its possible extracellular precursor, ATP, from CNS tissues is worth considering within the context of this chapter. Typically, depolarization-induced release of purines from tissue preparations in vitro has been evoked with high concentrations of potassium or veratridine. In synaptosomes derived from various brain regions and corrected for yield by measurements of occluded LDH levels, White and Leslie (1982) found the rank order of propensity for adenosine release ellicited by potassium to be amygdala >> cortex > striatum > hippocampus > thalamus > hypothalamus > cerebellum > pons-medulla. A similar pattern was observed for veratidine-induced release (MacDonald, 1988). Previously, the rank order of potassium-induced ATP release from synaptosomes, again according to specific activity of release, was found to be striatum > cortex > medulla > hypothalamus > cerebellum. A different pattern for veratridine-induced ATP release was observed: medulla > striatum > hypothalamus > cortex > cerebellum (Potter and White, 1980). Since ATP release in response to veratridine actually increased in the absence of calcium, which is in contrast to that induced by potassium, it was proposed that at least part of veratridine's actions may be on glial cells. This could partially explain the different regional patterns observed with veratridine and potassium at least for ATP release. There appears to be some quantitative correspondence of ATP and adenosine release among brain regions from which release was evoked by potassium. Meaningful interpretation of these release patterns in relation to other markers must await futher information on mechanisms whereby purines are released, the neuronal identity of synaptosomes from which release originates, and the impact of cellular metabolism on the quantities of purines available for release.

7.5. ADA and [³H]NBI

Comparisons of the distribution of ADA in rat brain with that of [³H]NBI binding sites have revealed some rather interesting trends. Qualitatively, the localization of ADA-immunoreactivity coincides remarkably well with the density of [³H]NBI sites identified autoradiographically (Nagy et al., 1985). This appears to be true regardless of the ADA-immunostained elements involved within particular brain regions, i.e., neuronal cell bodies, axons, or terminal fields. Examples of this include the high density of [³H]NBI binding sites in medial thalamic nuclei, which are richly innervated by ADA immunoreactive fibers, and in TM, which exhibits intense ADA-immunostaining of neuronal parikarya. Although these results do not demonstrate that [³H]NBI sites are located on ADA-immunoreactive elements, this is strongly suggested by the appearance of patterns in, for example, TM, where the distribution of autoradiographic grains closely matched that of ADA-positive cell bodies. Simultaneous localization of [³H] NBI sites and ADA-immunostaining in the same section at the cellular level may reveal a more exact correspondence.

In quantitative comparisons of ADA activity in homogenates with the levels of [³H]NBI sites in membrane preparations of various brain regions, the results are somewhat less clear, but equally interesting. As shown in Fig. 3, the brain areas examined fall roughly into two categories: those areas in which there appears to be a correlation between the relatively low and uniform levels of ADA activity and the broad range in the levels of [³H]NBI binding sites (correlation coefficient 0.64, $p < 0.01$, lower regression line), and those areas in which there appeared to be a strong relationship between [³H]NBI sites and a wide range of ADA values (correlation coefficient 0.90, $p < 0.01$, upper regression line). These latter areas included the septum, bed nucleus of the stria terminalis, anterior and whole hypothalamus,

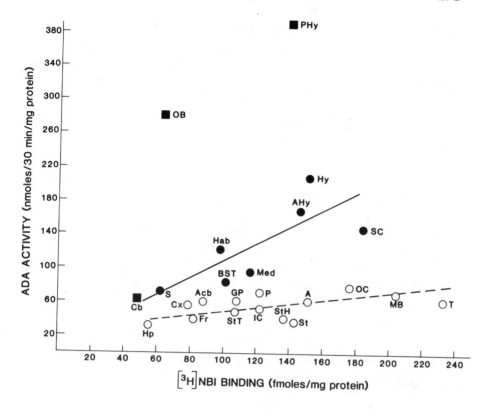

Fig. 3. Plot of ADA activity (Geiger and Nagy, 1986) in homogenates against [³H]NBI binding (Geiger and Nagy, 1984) in membrane preparations of rat brain regions. Areas containing a moderate to high density of ADA immunoreactive fibers (●)include: septum (S), bed nucleus of stria terminalis (BST), habenula (Hab), medulla (Med), anterior hypothalamus (AHy), whole hypothalamus (Hy), and superior colliculus (SC); these were taken for regression analysis separately from those containing a much lower density of fibers immunoreactive for ADA (○). These latter areas include: hippocampus (Hp), cerebral cortex (Cx), frontal cortex (Fr), accumbens (Acb), tail of striatum (StT), globus pallidus (GP), pons (P), inferior colliculus (IC), head of striatum (StH), whole striatum (St), amygdala (A), olfactory cortex (OC), midbrain (MB), and thalamus (T). Areas not included in these regression analyses (■) were cerebellum (Cb), olfactory bulbs (OB), and posterior hypothalamus (PHy).

habenula, medulla, and superior colliculus. Consistent with the above qualitative observations, a distinguishing feature of these regions is their dense concentration of ADA-immunoreactive fibers. On the other hand, there are some outlying points; posterior hypothalamus and olfactory bulb contained relatively higher levels of ADA than [^{3}H]NBI sites. Although the high content of ADA in large numbers of magnocellular neurons in TM may explain this result in posterior hypothalamus, the reason for the aberrant nature of olfactory bulb is less obvious. The cerebellum, although appearing to lie on the upper regression line, was excluded from regression analyses given the anomolous behavior of immunostaining for Purkinje cells (*see* section 4.5.1.3).

Taken together, the above results for rat brain may be interpreted as indicating that either:

1. [^{3}H]NBI labels two types of sites, one being highly concentrated in ADA-containing neural systems and the other being more uniformly distributed;
2. ADA labels one site, which is abundantly expressed by ADA-containing neurons and less so by other neurons and perhaps glia; or
3. [^{3}H]NBI labels only one site and the resolution of analyses has not as yet been sufficient to reveal total correspondence with ADA containing systems.

Whichever of these is correct, it is compelling to postulate a neuroregulatory role for purines in some neurons that appear to exhibit simultaneously at least two biochemical characteristics thought to be associated with adenosine disposition and metabolism.

Acknowledgments

The authors' work cited here has been supported by grants from the Medical Research Council of Canada (MRCC), Manitoba Health Research Council, University of Manitoba Faculty Fund, and Health Sciences Centre Research Foundation. JIN is a Scientist and JDG is a Scholar of the MRCC. We wish to thank Mark Whittaker, Kosala Sivananthan, Cheryl Baraniuk, Barb Hunt, and Lyn Polson for technical assistance and help in preparing this manuscript. We are grateful to Drs. P. Marangos and T. White for allowing us to read some of their manuscripts which at that time were unpublished.

References

Achterberg, P. W., Stroeve, R. J., and DeJong, J. W. (1986) Myocardial adenosine cycling rates during normoxia and under conditions of stimulated purine release. *Biochem. J.* **235**, 13–17.

Agarwal, R. P. (1982) Inhibitors of adenosine deaminase. *Pharmacol. Ther.* **17**, 399–429.

Agarwal, R. P. and Parks, R. E. Jr. (1977) Potent inhibiton of muscle 5'-AMP deaminase by the nucleoside antibiotics coformycin and deoxycoformycin. *Biochem. Pharmacol.* **26**, 663–666.

Akasu, T. Shinnick-Gallagher, P., and Gallagher, J. P. (1984) Adenosine mediates a slow hyperpolarizing synaptic potential in automatic neurons. *Nature* **311**, 62–64.

Allsop, J. and Watts, R. W. E. (1983) Purine *de novo* synthesis in liver and developing rat brain, and the effect of some inhibitors of purine nucleotide interconversion. *Enzyme* **30**, 172–180.

Andy, R. J. and Kornfeld, R. (1982) The adenosine deaminase binding protein of human skin fibroblasts is located on the cell surface. *J. Biol. Chem.* **257**, 7922–7928.

Arch, J. R. S. and Newsholme, E. A. (1978) Activities and some properties of 5'-nucleotidase, adenosine kinase and adenosine deaminase in tissues from vertebrates and invertebrates in relation to the control of the concentration and the physiological role of adenosine. Biochem. J. 174, 965–977.

Archer, S., Juranka, P. F., Ho, J. H., and Chan, V. L. (1985) An analysis of multiple mechanisms of adenosine toxicity in baby hamster kidney cells.. *J. Cell. Physiol.* **124**, 226–232.

Banay-Schwartz, M., deGuzman, T., and Lajtha, A. (1980) Nucleoside uptake by slices of mouse brain. *J. Neurochem.* **35**, 544–551.

Barberis, C., Minn, A., and Gayet, J. (1981) Adenosine transport into guinea pig synaptosomes. *J. Neurochem.* **36**, 347–354.

Barberis, C., Guibert, B., Daudet, F., Charriere, B., and Leviel, V. (1984) *In vivo* release of adenosine from cat basal ganglia: Studies with a push pull cannula. *Neurochem. Int.* **6**, 545–551.

Beck, D. W., Vinters, V. H., Hart, M. S., Henn, F. A., and Cancilla, P. A. (1983) Uptake of adenosine into cultured cerebral endothelium. *Brain Res.* **271**, 180–183.

Belloni, F. L., Elkin, P. L., and Giannotto, B. (1985) The mechanism of adenosine release from hypoxic rat liver cells. *Br. J. Pharmacol.* **85**, 441–446.

Bender, A. S. and Hertz, L. (1986) Similarities of adenosine uptake systems in astrocytes and neurons in primary cultures. *Neurochem. Res.* **11**, 1507–1524.

Bender, A. S., Wu, P. H., and Phillis, J. W. (1980) The characterization of [³H]adenosine uptake into rat cerebral cortical synaptosomes. *J. Neurochem.* **35**, 629–640.

Bender, A. S., Wu, P. H., and Phillis, J. W. (1981a) The rapid uptake and release of ³H-adenosine by cerebral cortical synaptosomes. *J. Neurochem.* **36**, 651–660.

Bender, A. S., Wu, P. H., and Phillis, J. W. (1981b) Some biochemical properties of the rapid adenosine uptake system in rat brain synaptosomes. *J. Neurochem.* **37**, 1282–1290.

Berne, R. M. (1980) The role of adenosine with regulation of coronary blood flow. *Circ. Res.* **46**, 807–813.

Berne, R. M., Rubio, R., and Curnish, R. R. (1974) Release of adenosine from ischemic brain. Effect of cerebral vascular resistance and incorporation into cerebral adenine nucleotides. *Circ. Res.* **35**, 262–271.

Bernstein, H., Weiss, J., and Luppa, H. (1978) Cytochemical investigations on the localization of 5'-nucleotidase in the rat hippocampus with special reference to synaptic regions. *Histochemistry* **55**, 261–267.

Betz, A. L. (1985) Identification of hypoxanthine transport and xanthine oxidase activity in brain capillaries. *J. Neurochem.* **44**, 574–579.

Bielat, K. and Tritsch, G. L. (1986) Adenosine deaminase activity localization at the erythrocyte cell membrane by electron microscopy. *Fed. Proc.* **45**, 360.

Bisserbe, J. C., Deckert, J., and Marangos, P. (1986) Autoradiographic localization of adenosine uptake sites in guinea pig brain using [³H]dipyridamole. *Neurosci. Lett.* **66**, 344, 345.

Bisserbe, J. C., Patel, J., and Marangos, P. J. (1985) Autoradiogrphic localization of adenosine uptake sites in rat brain using [³H]nitrobenzylthioinosine. *J. Neurosci.* **5**, 544–550.

Blazynski, C., Kinscherf, D. A., Geary, K. M., and Ferrendelli, J. A. (1986) Adenosine-mediated regulation of cyclic AMP levels in isolated incubated retinas. *Brain Res.* **366**, 224–229.

Bleire, R., Cohn, P., and Siggelkov, I. K. (1979) A cytoarchitectonic atlas of the hypothalamus and hypothalamic third ventricle of rat, in *Handbook of the Hypothalamus*, vol I, Anatomy of the Hypothalamus (Morgane, P. J. and Parksepp, J., eds.), Marcel Dekker, New York, pp. 137–220.

Bontemps, F., Van den Berghe, G., and Hers, H. G. (1986) Pathways of adenine nucleotide catabolism in erythrocytes. *J. Clin. Invest.* **77**, 824–830.

Braas, K. M., Newby, A. C., Wilson, V. S., and Snyder, S. H. (1986) Adenosine-containing neurons in the brain localized by immunocytochemistry. *J. Neurosci.* **6**, 1952–1961.

Bruns, R. F., Lu, G. H., and Pugsley, T. A. (1986) Characterization of the A_2 adenosine receptor labeled by [³H]NECA in rat striatal membranes. *Mol. Pharamacol.* **29**, 331–346.

Bruns, R. F., Daly, J. W., and Snyder, S. H. (1980) Adenosine receptors in brain membranes: Binding of N^6 cyclohexyl[³H]adenosine and 1,3-diethyl-8-[³H]phenylxanthine. *Proc. Nat. Acad. Sci. USA* **77**, 5547–5551.

Buc, H. A., Thiullur, L., Hamet, M., Garreau, F., Moncion, A., and Perignon, J. L. (1986) Energy metabolism in adenosine deaminase-inhibited human erythrocytes. *Clin. Chim. Acta* **156**, 61–70.

Bukoski, R. D. and Sparks, H. V., Jr. (1986) Adenosine production and release by adult rat cardiocytes. *J. Mol. Cell. Cardiol.* **18**, 595–605.

Burnstock, G. and Hoyle, C. H. V. (1985) Actions of adenosine dinucleotides in the guinea pig taenia coli: NAD acts indirectly on P_1-purinoceptors; NADP acts like a P_2-purinoceptor agonist. *Br. J. Pharmacol.* **84**, 825–831.

Calvalla, D. and Neff, N. H. (1985) 2-Azidoadenosine, a photoaffinity label for the CNS adenosine transporter. *Trans. Amer. Soc. Neurochem.* **16,** 173.

Cammer, W., Sirota, S. R., Zimmerman, T. R., Jr., and Norton, W. T. (1980) 5'-Nucleotidase in rat brain myelin. *J. Neurochem.* **35,** 367–373.

Cass, C. E., Gaudette, L. A., and Paterson, R. P. (1974) Mediated transport of nucleosides in human erythrocytes. Specific binding of the inhibitor nitrobenzylthioinosine to nucleoside transport sites in the erythrocyte membrane. *Biochem. Biophys. Acta* **345,** 1–10.

Chassin, M. M., Adamson, R. H., Zaharevitz, D. W., and Johns, D. G. (1979) Enzyme inhibition filtration assay for 2'-deoxycoformycin and its application to the study of the relationship between drug concentration and tissue adenosine deaminase in dogs and rats. *Biochem. Pharamcol.* **28,** 1849–1855.

Chechik, B. E. and Sengupta, S. (1981) Detection of human, rat and mouse adenosine deaminase by immunochemical and immunomorphologic methods using antiserum to calf enzyme. *J. Immunological Methods* **45,** 165–176.

Chechik, B. E., Baumal, R., and Sengupta, S. (1983) Localization and identity of adenosine deaminase-positive cells in tissues of the young rat and calf. *Histochemical Journal* **15,** 373–387.

Chechik, B. E., Schrader, W. P., and Minowada, J. (1981) An immunomorphologic study of adenosine deaminase distribution in human thymus tissue, normal lymphocytes, and hematopoietic cell lines. *J. Immunol.* **126,** 1003–1007.

Chiang, P. K. (1985) S-adenosylhomocysteine hydrolase. Measurement of activity and use of inhibitors, in *Methods in Pharamcology,* vol 6 (Paton, D. M., ed.), Plenum, New York, pp. 127–145.

Collis, M. G., Palmer, D. B., and Baxter, G. S. (1986) Evidence that the intracellular effects of adenosine in the guinea pig aorta are mediated by inosine. *Eur. Pharmacol.* **121,** 141–145.

Conway, E. J. and Cooke, R. (1939) The deaminase of adenosine and adenylic acid in blood and tissues. *Biochem. J.* **33,** 479–492.

Crawley, J. N., Patel, J., and Marangos, P. J. (1983) Adenosine uptake inhibitors potentiate the sedative effects of adenosine. *Neurosci. Lett.* **36,** 169–174.

Daddona, P. E., Mitchell, B. S., Meuwissen, H. J., Davidson, B. L., Wilson, J. M., and Killer, C. A. (1983) Adenosine deaminase deficiency with normal immune function. *J. Clin. Invest.* **72,** 483–492.

Daval, J.-L. and Barberis, C. (1981) Release of radiolabelled adenosine derivatives from superfused synaptosome beds. *Biochem. Pharmacol.* **30,** 2559–2567.

Davies, L. P. and Hambley, J. W. (1986) Regional distribution of adenosine uptake in guinea pig brain slices and the effect of some inhibitors: Evidence for nitrobenzylthioinosine-sensitive and insensitive sites? *Neurochem. Int.* **8,** 103–108.

Davies, K. J. A., Sevanian, A., Muakkasseh-Kelley, S. F., and Hochstein, P. (1986) Uric acid-iron ion complexes. A new aspect of the antioxidant functions of uric acid. *Biochem. J.* **235,** 747–754.

de la Haba, G., Agostini, S., Biozzi, A., Merta, A., Unson, C., and Cantoni, G. L. (1986) S-Adenosylhomocysteinase: Mechanism of reversible and irreversible inactivation by ATP, cAMP, and 2'-deoxyadenosine. *Biochem.* **25,** 8337–8342.

Deckert, J., Bisserbe, J. C., and Marangos, P. J. (1986) [³H]Dipyridamole and [³H]nitrobenzylthioinosine binding sites in guinea pig brain: A comparison. *Pflugers Archiv.* **407,** S29.

Deckert, J., Bisserbe, J. C., and Marangos, P. J. (1987) Heterogeneity of adenosine transporters in guinea pig brain as visualized by quantitative [³H] dipyridamole autoradiography. *Naunyn Schmiedebergs Arch. Pharmacol.* **335,** 660–666.

DeLander, G. E., and Hopkins, C. J. (1986) Spinal adenosine modulates descending antinociceptive pathways stimulated by morphine. *J. Pharmacol Exp. Therap.* **239,** 88–93.

Dunwiddie, T. V. (1985) The physiological role of adenosine in the central nervous system. *Intern. Rev. Neurobiol.* **27,** 63–139.

Dunwiddie, T. V. and Hoffer, B. J. (1980) Adenine nucleotides and synaptic transmission in the *in vitro* rat hippocampus. *Br. J. Pharmacol.* **69,** 59–68.

Ehinger, B. and Perez, M. T. R. (1984) Autoradiography of nucleoside uptake into the retina. *Neurochem. Int.* **6,** 369–381.

Fastbom, J., Pazos, A., Probst, A., and Palacios, J. M. (1986) Adenosine A₁-receptors in human brain: Characterization and autoradiographic visualization. *Neurosci. Lett.* **65,** 127–132.

Fastbom, J., Pazos, A., and Palacios, J. M. (1987) The distribution of adenosine A₁ receptors and 5'-nucleotidase in the brain of some commonly used experimental animals. *Neurosci.* **22,** 813–826.

Fillenz, M. (1984) Norepinephrine, in *Handbook of Neurochemistry*, 2nd Ed. vol 6, *Receptors in the Nervous System*, Plenum, New York, pp. 61–69.

Fishbein, W. N., Davis, J. I., Winkert, J. W., and Strong, D. M. (1981) Levels of adenosine deaminase, AMP deaminase, and adenylate kinase in cultured human lymphoblast lines: Exquisite sensitivity of AMP deaminase inhibitors. *Biochem. Med.* **26,** 377–386.

Fisher, M. N. and Newsholme, E. A. (1984) Properties of rat heart adenosine kinase. *Biochem. J.* **221,** 521–528.

Forrester, T., Harper, A. M., McKenzie, E. T., and Thomsen, E. M. (1979) Effect of adenosine triphosphate and some derivatives on cerebral blood flow and metabolism. *J. Physiol.* (Lond.) **296,** 343–355.

Fox, I. H. and Kelley, W. N. (1978) The role of adenosine and 2'-deoxyadenosine in mammalian cells. *Ann. Rev. Biochem.* **47,** 655–686.

Franco, R., Canela, E. I., and Bozal, J. (1986) Heterogeneous localization of some purine enzymes in subcellular fractions of rat brain and cerebellum. *Neurochem. Res.* **11,** 423–435.

Fredholm, B. B. and Hedqvist, P. (1980) Modulation of neurotransmission by purine nucleotides and nucleosides. *Biochem. Pharmacol.* **29,** 1635–1643.

Fyffe, R. E. W. and Perl, E. R. (1984) Is ATP a central synaptic mediator for certain primary afferent fibers from mammalian skin? *Proc. Natl. Acad. Sci. USA* **81,** 6890–6893.

Geiger, J. D. (1986) Localization of [³H]cyclohexyladenosine and [³H]nitrobenzyl-thioinosine binding sites in rat striatum and superior colliculus. *Brain Res.* **363**, 404–408.

Geiger, J. D. (1987) Adenosine uptake and [³H]nitrobenzylthioinosine binding in developing rat brain. *Brain Res.* **436**, 265–272.

Geiger, J. D., Johnston, M. E., and Yago, V. (1988) Pharmacological characterization of rapidly accumulated adenosine by dissociated brain cells from adult rat. *J. Neurochem.* **51**, 283–291.

Geiger, J. D., and Nagy, J. I. (1984) Heterogeneous distribution of adenosine transport sites labeled by [³H]nitrobenzylthioinosine in rat brain: An autoradiographic and membrane binding study. *Brain Res. Bull.* **13**, 657–666.

Geiger, J. D. and Nagy, J. I. (1985) Localization of [³H]nitrobenzylthioinosine binding sites in rat spinal cord and primary afferent neurons. *Brain Res.* **347**, 321–327.

Geiger, J. D. and Nagy, J. I. (1986) Distribution of adenosine deaminase activity in rat brain and spinal cord. *J. Neurosci.* **6**, 2707–2714.

Geiger, J. D. and Nagy, J. I. (1987) Ontogenesis of adenosine deaminase activity in rat brain. *J. Neurochem.* **48**, 147–153.

Geiger, J. D., LaBella, F. S., and Nagy, J. I. (1984a) Ontogenesis of adenosine receptors in the central nervous system of the rat. *Develop. Brain Res.* **13**, 97–104.

Geiger, J. D., LaBella, F. S., and Nagy, J. I. (1984b) Characterization and localization of adenosine receptors in rat spinal cord. *J. Neurosci.* **4**, 2303–2310.

Geiger, J. D., LaBella, F. S., and Nagy, J. I. (1985) Characterization of nitrobenzyl-thioinosine binding to nucleoside transport sites selective for adenosine in rat brain. *J. Neurosci.* **5**, 735–740.

Geiger, J. D., Lewis, J. L., MacIntyre, C. J., and Nagy, J. I. (1987) Pharmacokinetics of 2'-deoxycoformycin, an inhibitor of adenosine deaminase, in rat. *Neuropharmacol.* **26**, 1383–1387.

Goodman, R. R., Kuhar, M. J., Hester, L. L., and Snyder, S. H. (1983) Adenosine receptors: Autoradiographic evidence for their localization on axon terminals of excitatory neurons. *Science* **220**, 967–969.

Goodman, R. R. and Snyder, S. H. (1982) Autoradiographic localization of adenosine receptors in rat brain using [³H]cyclohyxyladenosine. *J. Neurosci.* **2**, 1230–1241.

Gutensohn, W. and Guroff, G. (1972) Hypoxanthine-guanine phosphoribosyltransferase from rat brain: Purification, kinetic properties, development and distribution. *J. Neurochem.* **19**, 2139–2150.

Hagberg, H., Anderson, P., Butchek, S., Sandberg, M., Lehmann, A., and Hamberger, A. (1986) Blockade of N-methyl-D-aspartate-sensitive acidic amino acid receptors inhibits ischemia-induced accumulation of purine catabolites in the rat striatum. *Neurosci. Lett.* **68**, 311–316.

Hammond, J. R. (1985) Photoaffinity labeling of benzodiazepine receptors. Lack of effect on ligand binding to the nucleoside transport system. *J. Neurochem.* **45**, 1327–1330.

Hammond, J. R. and Clanachan, A. S. (1983) Distribution of nucleoside transport sites in guinea-pig brain. *J. Pharm. Pharmacol.* **35,** 117,118.

Hammond, R. R. and Clanachan, A. S. (1984a) Heterogeneity of high affinity nitrobenzylthioinosine binding sites in mammalian cortical membranes: Multiple forms of CNS nucleoside transporters. *Can. J. Physiol. Pharmacol.* **62,** 961–963.

Hammond, J. R. and Clanachan, A. S. (1984b) [^{3}H]Nitrobenzylthionosine binding to the guinea pig CNS nucleoside transport system: A pharmacological characterization. *J. Neurochem.* **43,** 1582–1592.

Hammond, J. R. and Clanachan, A. S. (1985) Species differences in the binding of [^{3}H]nitrobenzylthioinosine to the nucleoside transport system in mammalian central nervous system membranes: Evidence for interconvertible conformations of the binding site/transporter complex. *J. Neurochem.* **45,** 527–535.

Hara, H., Hamill, G. S., and Jacobowitz, D. M. (1985) Origin of cholinergic nerves to the rat major cerebral arteries: Coexistence with vasoactive intestinal polypeptide. *Brain Res. Bull.* **14,** 179–188.

Heistad, D. D., Marcus, M. L., Courley, J. K., and Busija, D. W. (1981) Effect of adenosine and dipyridamole on cerebral blood flow. *Am. J. Physiol.* **240,** H775–H780.

Helland, S., Broch, O. J., and Ueland, P. M. (1983) Neurotoxicity of deoxycoformycin: Effect of constant infusion on adenosine deaminase, adenosine, 2'-deoxyadenosine and monoamines in the mouse brain. *Neuropharmacol.* **7,** 915–917.

Heller, I. H. and McIlwain, H. (1973) Release of [^{14}C]adenine derivatives from isolated subsystems of the guinea pig brain: Actions of electrical stimulation and of papaverine. *Brain Res.* **53,** 105–116.

Henderson, J. F. (1979) Regulation of adenosine metabolism, in *Physiological and Regulatory Functions of Adenosine and Adenine Nucleotides* (Baer, H. P. and Drummond, G. I., eds.), Raven, New York, pp. 315–322.

Henderson, J. F. (1983) Mechanisms of toxicity of adenosine, 2'-deoxyadenosine, and inhibitors of adenosine deaminase, in *Regulatory Function of Adenosine* (Berne, R. M., Rall, T. W., and Rubio, R., eds.) Martines Nijhoff, The Hague, pp. 223–234.

Henderson, J. F., Scott, F. W., and Lowe, J. K. (1980) Toxicity of naturally occurring purine deoxyribonucleosides. *Pharmacol. Ther.* **8,** 573–604.

Hershfield, M. S. (1979) Apparent suicide inactivation of human lymphoblast S-adenosylhomocysteine hydrolase by 2'-deoxyadenosine and adenine arabinoside. *J. Biol. Chem.* **254,** 22–25.

Hertz, L. (1978) Kinetics of adenosine uptake into astrocytes. *J. Neurochem.* **31,** 55–62.

Hess, D. T. and Hess, A. (1986) 5'-Nucleotidase of cerebellar molecular layer: Reduction in Purkinje cell-deficient mutant mice. *Dev. Brain Res.* **29,** 93–100.

Hirschhorn, R., Papageorgious, P. S., Kesarwala, H. H., and Taft, L. T. (1980) Amelioration of the neurologic abnormalities after enzyme replacement in adenosine deaminase deficiency. *Med. Intelligence* **303,** 377–380.

Holland, M. J. C. (1986) Specificity of 2'-deoxycoformcyin inhibition of adenosine metabolism in intact human skin fibroblasts. *Res. Comm. Chem. Pathol. Pharmacol.* **51,** 311–324.

Holmgren, M., Hedner, J., Mellstrand, T., Nordberg, G., and Hedner, Th. (1986) Characterization of the antinociceptive effects of some adenosine analogues in the rat. *Naunyn Schmiedebergs Arch. Pharmacol.* **334,** 290–293.

Holton, P. (1959) The liberation of adenosine triphosphate on antidromic stimulation of sensory nerves. *J. Physiol.* (Lond.) **145,** 494–504.

Honegger, C. G., Krenger, W., and Langemann, H. (1986) Increased concentrations of uric acid in the spinal cord of rats with chronic relapsing experimental allergic encephalomyelitis. *Neurosci. Lett.* **69,** 109–114.

Hosli, E. and Hosli, L. (1984) Autoradiographic localization of binding sites for [^{3}H]histamine and H_1-and H_2-antagonists on cultured neurones and glial cells. *Neurosci.* **13,** 863–870.

Huang, M. and Daly, J. (1974) Adenosine-elicited accumulation of cyclic AMP in brain slices: Potentiation by agents which inhibit uptake of adenosine. *Life Sci.* **14,** 489–503.

Jahr, C. E. and Jessell, T. M. (1983) ATP excites a subpopulation of rat dorsal horn neurons. *Nature* **304,** 730–733.

Jammohamed, N. S., Young, J. D., and Jarvis, S. M. (1985) Proteolytic cleavage of [^{3}H]nitrobenzylthioinosine-labeled nucleoside transporter in human erythrocytes. *Biochem. J.* **230,** 777–784.

Jarvis, M. F., Jackson, R. H., and Williams, M. (1989) Autoradiographic characterization of high affinity A_2 receptors in the rat brain. *Brain Res.* **484,** 111–118.

Jarvis, S. M. and Ng, A. S. (1985) Identification of the adenosine uptake sites in guinea pig brain. *J. Neurochem.* **44,** 183–188.

Jarvis, S. M., Hammond, J. R., Paterson, A. R. P., and Clanachan, A. S. (1982) Species differences in nucleoside transport: A study of uridine transport and nitrobenzylthioinosine binding by mammalian erythrocytes. *Biochem. J.* **208,** 2202–2208.

Johnston, M. E. and Geiger, J. D. (1989) Sodium-dependent uptake of nucleosides by associated brain cells from the rat. *J. Neurochem.* **52,** 75–81.

Johnston, J. B., Begleiter, A., Pugh, L., Leith, M. K., Wilkins, J. A., Cavers, D. J., and Israels, L. G. (1986) Biochemical changes induced in hairy-cell leukemia following treatment with the adenosine deaminase inhibitor 2'-deoxycoformycin. *Cancer Res.* **46,** 2179–2184.

Jonzon, B. and Fredholm, B. B. (1985) Release of purines, noradrenaline, and GABA from rat hippocampal slices by field stimulation. *J. Neurochem.* **44,** 217–224.

Jordan, W. K., March, R., Boyd Houshin, O., and Popp, E. (1959) Intracellular partition of purine deaminases in rodent brain. *J. Neurochem.* **4,** 170–174.

Kanemitsu, H., Tamura, A., Sano, K., Iwamoto, T., Yoshuira, M., and Iryama, K. (1986) Changes of uric acid level in rat brain after focal ischemia. *J. Neurochem.* **46,** 851–853.

Kleihues, P., Kobayashi, K., and Hossmann, K.-A. (1974) Purine nucleotide metabolism in the cat brain after one hour of complete ischemia. *J. Neurochem.* **23,** 417–425.

Klotz, K.-N. and Lohse, M. J. (1986) The glycoprotein nature of A_1 adenosine receptors. *Biochem. Biophys. Res. Comm.* **140,** 406–413.

Kocsis, J. D., Eng, D. L., and Bhisitkul, R. B. (1984) Adenosine selectively blocks parallel-fiber-mediated synaptic potentials in rat cerebellar cortex. *Proc. Natl. Acad. Sci. USA* **81,** 6531–6534.

Kohn, M. C. and Garfinkel, D. (1977) Computer simulation of ischemic rat heart purine metabolism. II. Model behavior. *Am. J. Physiol.* **232,** H394–H399.

Kreutzberg, G. W., Barron, K. D., and Schubert, P. (1978) Cytochemical localization of 5'-nucleotidase in glial plasma membranes. *Brain Res.* **158,** 247–257.

Kuhar, M. J. (1973) Neurotransmitter uptake: A tool in identifying neurotransmitter-specific pathways. *Life Sci.* **13,** 1623–1634.

Lee, K. S. and Reddington, M. (1986) Autoradiographic evidence for multiple CNS binding sites for adenosine derivatives. *Neurosci.* **19,** 535–549.

Lee, K. S., Schubert, P., Reddington, M., and Kreutzberg, G. W. (1983) Adenosine receptor density and the depression of evoked neuronal activity in the rat hippocampus *in vitro. Neurosci. Lett.* **37,** 81–85.

Lee, K. S., Schubert, P., Reddington, M., and Kreutzberg, G. W. (1986) The distribution of adenosine A_1, receptors and 5'-nucleotidase in the hippocampal formation of several mammalian species. *J. Comp. Neurol.* **246,** 427–434.

Lewin, E. and Bleck, V. (1981) Electroshock seizure in mice: Effect on brain adenosine and its metabolites. *Epilepsia* **22,** 665–668.

Londos, C., Wolff, J., and Cooper, D. M. F. (1979) Actions of adenosine on adenylate cyclase, in *Physiological and Regulatory Functions of Adenosine and Adenine Nucleotides* (Baer, H. P. and Drummond, G. I., eds.), Raven, New York, pp. 271–281.

Lowenstein, M., Naito, Y., and Collinson, A. R. (1986) Regulatory properties of intracellular and ecto 5'-nucleotidases and their role in the production of adenosine. *Pflugers Archiv.* **407,** S9.

MacDonald, W. F. (1988) Endogenous adenosine release from rat brain synaptosomes. Ph.D. Thesis. Dalhousie University, Halifax, Nova Scotia.

MacDonald, W. F. and White, T. D. (1985) Nature of extrasynaptosomal accumulation of endogenous adenosine evoked by K^+ and veratridine. *J. Neurochem.* **45,** 791–797.

Magistretti, P. J., Hof, P. R., and Martin, J.-L. (1986) Adenosine stimulates glycogenolysis in mouse cerebral cortex: A possible coupling mechanism between neuronal activity and energy metabolism. *J. Neurosci.* **6,** 2558–2562.

Marangos, P. J. (1984) Differentiating adenosine receptors and adenosine uptake sites in brain. *J. Recept. Res.* **4,** 231–244.

Marangos, P. J. and Deckert, J. (1987) [³H]Dipyridamole binding to guinea pig brain membranes, possible heterogeneity of central adenosine uptake sites. *J. Neurochem.* **48,** 1231–1237.

Marangos, P. J. Houston, M., and Montgomery, P. (1985) [³H]Dipyridamole: A new ligand probe for brain adenosine uptake sites. *Eur. J. Pharmacol.* **117,** 393, 394.

Marangos, P. J., Clark-Rosenberg, R., and Patel, J. (1982a) [³H]Nitrobenzyl-thioinosine is a photoaffinity probe for adenosine uptake sites in brain. *Eur. J. Pharamcol.* **85**, 359,360.

Marangos, P. J., Patel, J., Clark-Rosenberg, R., and Martina, A. M. (1982b) [³H]Nitrobenzylthioinosine binding as a probe for the study of adenosine uptake sites in brain. *J. Neurochem.* **39**, 183–191.

McCaman, M. W. (1986) Uptake and metabolism of [³H]adenosine by Apylsia ganglia and by individual neurons. *J. Neurochem.* **47**, 1026–1031.

McConnell, W. R., Suling, W. J., Rice, L. S., Shannon, W. M., and Hill, D. L. (1978) Use of microbiologic and enzymatic assays in studies on the disposition of 2'-deoxycoformycin in the mouse. *Cancer Treat. Rep.* **62**, 1153–1159.

Mendelson, W. B., Kuruvilla, A., Watlington, T., Goehl, K., Paul, S. M., and Skolnick, P. (1983) Sedative and electroencephalographic actions of erythro-9-(2-hydroxy-3-nonyl)-adenine (EHNA): Relationship to inhibition of brain adenosine deaminase. *Psychopharmacology* (Berlin) **79**, 126–129.

Miller, D. S. and Horowitz, S. B. (1986) Intracellular compartmentalization of adenosine triphosphate. *J. Biol. Chem.* **261**, 13911–13915.

Mistry, G. and Drummond, G. I. (1986) Adenosine metabolism in microvessels from heart and brain. *J. Mol. Cell. Cardiol.* **18**, 13–22.

Morgan, P. F. and Stone, T. W. (1986) Inhibition by benzodiazepines and beta-carbolines of brief (5 seconds) synaptosomal accumulaton of [³H]adenosine. *Biochem. Pharmacol.* **35**, 1760–1762.

Morii, S., Ngai, C., and Winn, H. R. (1986) Reactivity of rat pial arterioles and venules to adenosine and carbon dioxide: With detailed description of the closed cranial window technique in rats. *J. Cereb. Blood Flow Metab.* **6**, 34–41.

Motley, S. J. and Collins, G. G. S. (1983) Endogenous adenosine inhibits excitatory transmission in the rat olfactory cortex slice. *Neuropharmacol.* **22**, 1081–1086.

Murray, T. F. and Cheney, D. L. (1982) Neuronal localization of N⁶-cyclohexyl ³H-adenosine binding sites in rat and guinea pig brain. *Neuropharmacology* **21**, 575–580.

Mustafa, S. J. and Tewari, C. P. (1970) Latent adenosine deaminase in mouse brain II. Purification and properties of mitochondrial and supernatant adenosine deaminase. *Biochem. Biophys. Acta* **220**, 326–337.

Nagata, H., Mimori, Y., Nakamura, S., and Kameyama, M. (1984) Regional and subcellular distribution in mammalian brain of the enzymes producing adenosine. *J. Neurochem.* **42**, 1001–1007.

Nagy, A. K., Shuster, T. A., and Delgado-Escueta, A. V. (1986a) Ecto-ATPase of mammalian synaptosomes: Identification and enzyme characterization. *J. Neurochem.* **47**, 976–986.

Nagy, J. I. (1982) Capsaicin: A chemical probe for sensory neurone mechanisms, in *Handbook of Psychopharmacology*, vol 15 (Iversen, L. L., Iversen, S. D., and Snyder, S. H., eds.), Plenum, New York, pp. 185–235.

Nagy, J. I. and Daddona, P. E. (1985) Anatomical and cytochemical relationships of adenosine deaminase-containing primary afferent neurons in the rat. *Neuroscience* **15,** 799–813.

Nagy, J. I., Buss, M., LaBella, L. A., and Daddona, P. E. (1984a) Immunohistochemical localization of adenosine deaminase in primary neurons of the rat. *Neurosci.* **48,** 133–138.

Nagy, J. I., Buss, M., and Daddona, P. E. (1986b) On the innervation of trigeminal mesencephalic primary afferent neurons by adenosine deaminase-containing projections from the hypothalamus in the rat. *Neuroscience* **17,** 141–156.

Nagy, J. I., Geiger, J. D., and Daddona, P. E. (1985) Adenosine uptake sites in rat brain: Identification using ^{3}H-nitrobenzylthioinosine and colocalization with adenosine deaminase. *Neurosci. Lett.* **55,** 47–53.

Nagy, J. I., Iversen, L. L., Goedert, M., Chapman, D., and Hunt, S. P. (1983) Dose-dependent effects of capsaicin on primary sensory neurons in the neonatal rat. *J. Neuroscience* **3,** 399–406.

Nagy, J. I., LaBella, L. A., Buss, M., and Daddona, P. E. (1984b) Immunohistochemistry of adenosine deaminase: implication for adenosine neurotransmission. *Science* **224,** 166–168.

Nagy, J. I., LaBella, L. A., and Daddona, P. E. (1984c) Purinergic mechanisms in antidromic vasodilation and neurogenic inflammation, in *Antidromic Vasodilation and Neurogenic Inflammation* (Ghahl, J. A., Szolesanyi, J., and Lembeck, F., eds.), Hungarian Okodemia Kiata, Budapest, Hungary, pp. 193–206.

Nagy, J. I. Yamamoto, T., Dewar, K., Geiger, J. D., and Daddona, P. E. (1988) Adenosine deaminase-'like' immunoreactivity in cerebellar Purkinje cells of rat. *Brain. Res.* **457,** 21–28.

Newby, A. C. (1984) Adenosine and the concept of "retaliatory metabolites." *Trends Biochem. Sci.* **9,** 42–44.

Nimit, Y., Skolnick, P., and Daly, J. (1981) Adenosine and cyclic AMP in rat cerebral cortical slices: Effects of adenosine uptake inhibitor and adenosine deaminase inhibitors. *J. Neurochem.* **36,** 908–912.

Norstrand, I. F. (1985) Histochemical demonstration of adenosine deaminase in the human neuraxis. *Neurochem. Pathol.* **3,** 73–82.

Norstrand, I. F., Siverls, V. C., and Libbin, R. M. (1984) Regional distribution of adenosine deaminase in the human neuraxis. *Enzyme* **32,** 20–25.

Okada, Y. and Saito, M. (1979) Inhibitory action of adenosine, 5-HT (serotonin) and GABA (γ-aminobutyric acid) on the post synaptic potential (PSP) of slices from olfactory cortex and superior colliculus in correlation to the level of cyclic AMP. *Brain Res.* **160,** 368–371.

Paes de Carvalho, R. and de Mello, F. G. (1985) Expression of A₁ adenosine receptors modulating dopamine-dependent cyclic AMP accumulation in the chick embryo retina. *J. Neurochem.* **44,** 845–851.

Parsons, P. G. and Hayward, I. P. (1986) Human melanoma cells sensitive to deoxyadenosine and deoxyinosine. *Biochem. Pharmacol.* **35,** 655–660.

Parsons, P. G., Bowman, E. P. W., and Blakley, R. L. (1986) Selective toxicity of deoxyadenosine analogues in human melanoma cell lines. *Biochem. Pharmacol.* **35**, 4025–4029.

Patel, B. T. and Tudball, N. (1986) Localization of S-adenosylhomocysteine hydrolase and adenosine deaminase immunoreactivities in rat brain. *Brain Res.* **370**, 250–264.

Patel, B. T., Tudball, N., Wada, H., and Watanabe, T. (1986) Adenosine deaminase and histidine decarboxylase coexist in certain neurons of the rat brain. *Neurosci. Lett.* **63**, 185–189.

Paterson, A. R. P., Harley, E. R., and Cass, C. E. (1985) Measurement and inhibition of membrane transport of adenosine, in *Methods in Pharmacology*, vol 6, (Paton, E. M. ed.), Plenum, New York, pp. 165–180.

Paterson, A. R. P., Jakobs, E. S., Ng, C. Y. C., Odegard, R., and Adjei, A. A. (1987) Nucleoside transport inhibition *in vitro* and *in vivo*, in *Topics and Perspectives in Adenosine Research,* (Gerlach, E. and Becker, B. F., eds.), Springer-Verlag, N.Y., pp. 89–101.

Paterson, A. R. P., Kolassa, N., and Cass, C. E. (1981) Transport of nucleoside drugs in animal cells. *Pharmacol. Ther.* **12**, 515–536.

Perez, M. T. R., Ehinger, B. E., Lindstrom, K., and Fredholm, B. B. (1986) Release of endogenous and radioactive purines from the rabbit retina. *Brain Res.* **398**, 106–112.

Phillips, E. and Newsholme, E. A. (1979) Maximum activities, properties and distribution of 5'-nucleotidase, adenosine kinase and adenosine deaminase in rat and human brain. *J. Neurochem.* **33**, 553–558.

Phillis, J. W. and Wu, P. H. (1983) Nitrobenzylthioinosine inhibition of adenosine uptake in the guinea-pig brain. *J. Pharm. Pharmacol.* **35**, 540.

Phillis, J. W., DeLong, R. E., and Towner, J. K. (1985) Adenosine deaminase inhibitors enhance cerebral anoxic hyperemia in the rat. *J. Cereb. Blood Flow Metab.* **5**, 295–299.

Phillis, J. W., Barraco, R. A., DeLong, R. E., and Washington, D. O. (1986) Behavioral characteristics of centrally administered adenosine analogs. *Pharmacol. Biochem. Behav.* **24**, 263–270.

Phillis, J. W., Edstrom, J. P., Kostopoulos, G. K., and Kirkpatrick, J. R. (1979) Effects of adenosine and adenine nucleotides on synaptic transmission in the cerebral cortex. *Can. J. Physiol. Pharmacol.* **57**, 1289–1312.

Phillis, J. W. and Wu, P. H. (1981) The role of adenosine and its nucleotides in central synaptic transmission. *Prog. Neurobiol.* **16**, 187–239.

Plagemann, P. G. W. and Wohlhueter, R. M. (1980) Permeation of nucleosides, nucleic acid bases and nucleotides in animal cells. *Curr. Top. Membr. Transp.* **14**, 225–330.

Pons, F., Bruns, R. F., and Daly, J. W. (1980) Depolarization-evoked accumulation of cyclic AMP in brain slices: The requisite intermediate adenosine is not derived from hydrolysis of released ATP. *J. Neurochem.* **34**, 1319–1323.

Post, C. (1984) Antinociceptive effects in mice after intrathecal injection of 5'-N-ethylcarboxamide adenosine. *Neurosci. Lett.* **51**, 325–330.

Potter, P. and White, T. D. (1980) Release of adenosine 5'-triphosphate from synaptosomes from different regions of rat brain. *Neuroscience* **5,** 1351–1356.

Pull, I. and McIlwain, H. (1974) Rat cerebral-cortex adenosine deaminase activity and its subcellular distribution. *Biochem. J.* **144,** 37–41.

Radulovacki, M., Virus, R. M., Djuricic-Nedelson, M., and Green, R. D. (1983) Hypnotic effects of deoxycoformycin in rats. *Brain Res.* **271,** 392–395.

Ratech, H., Thorbeck, G. J., and Hirschhorn, R. (1981) Metabolic abnormalities of human adenosine deaminase deficiency reproduced in the mouse by 2'-deoxycoformycin, an adenosine deaminase inhibitor. *Clin. Immunol. Immunopathol.* **21,** 119–127.

Ratech, H., Greco, M. A., Gallo, G., Rimoin, D. L., Kamino, H., and Hirschhorn, R. (1985a) Pathologic findings in adenosine-deaminase-deficient-severe combined immunodeficiency. I. Kidney, adrenal, and chondroosseous tissue alterations. *Am. J. Phathol.* **120,** 157–169.

Ratech, H., Hirschhorn, R., and Thorbecke, G. J. (1985b) Effects of deoxycoformycin in mice. III. A murine model reproducing multi-system pathology of human adenosine deaminase deficiency. *Am. J. Physiol.* **119,** 65–72.

Reddington, M. and Pusch, R. (1983) Adenosine metabolism in a rat hippocampal slice preparation: Incorporation into S-adenosylhomocysteine. *J. Neurochem.* **40,** 285–290.

Reddington, M., Erfurth, A., and Lee, K. S. (1986) Heterogeneity of binding sites for N-ethylcarboxamido [³H]adenosine in rat brain. Effects of N-ethylmaleimide. *Brain Res.* **399,** 232–239.

Ribeiro, J. A. and Sebastiao, A. M. (1985) On the type of receptor involved in the inhibitory action of adenosine at the neuromuscular junction. *Br. J. Pharmacol.* **84,** 911–918.

Ribeiro, J. A. and Sebastiao, A. M. (1986) Adenosine receptors and calcium: Basis for proposing a third (A_3) adenosine receptor. *Prog. Neurobiol.* **26,** 179–209.

Robins, E., Smith, D. E., and McCaman, R. E. (1953) Microdetermination of purine nucleoside phosphorylase activity in brain and its distribution within the monkey cerebellum. *J. Biol. Chem.* **204,** 927–937.

Salt, T. E. and Hill, R. G. (1983) Excitation of single sensory neurons in the rat caudal trigeminal nucleus by iontophoretically applied adenosine 5'-triphosphate. *Neurosci. Lett.* **35,** 53–57.

Salter, M. W. and Henry, J. L. (1985) Effects of adenosine 5'-monophosphate and adenosine 5'-triphosphate on functionally identified units in the cat spinal dorsal horn. Evidence for a differential effect of adenosine 5'-triphosphate on nociceptive vs non-nociceptive units. *Neuroscience* **15,** 815–825.

Sawynok, J., Sweeney, M. I., and White, T. D. (1986) Classification of adenosine receptors mediating antinociception in the rat spinal cord. *Br. J. Pharmacol.* **88,** 923–930.

Schaeffer, J. M. and Anderson, S. M. (1981) Nucleoside uptake by rat retina cells. *Life Sci.* **29,** 939–946.

Schrader, W. P. and Bryer, P. J. (1982) Characterization of an insoluble adenosine deaminase complexing protein from human kidney. *Arch. Biochem. Biophys.* **215**, 107–115.

Schrader, W. P., Harder, C. M., and Schrader, D. K. (1983) Adenosine deaminase complexing proteins of the rabbit. *Comp. Biochem. Physiol.* (B)**75**, 119–125.

Schrader, W. P., Harder, C. M., Schrader, D. K., and West, C. A. (1984) Metabolism of different molecular forms of adenosine deaminase intravenously infused into the rabbit. *Arch. Biochem. Biophys.* **230**, 158–167.

Schrader, W. P. and Pollara, B. (1978) Localzation of an adenosine deaminase-binding protein in human kidney. *J. Lab. Clin. Med.* **92**, 656–662.

Schrader, W. P. and Stacey, A. R. (1979) Immunoassay of the adenosine deaminase complexing proteins of human tissues and body fluids. *J. Biol. Chem.* **254**, 11958–11963.

Schrader, W. P. and West, C. A. (1985) Adenosine deaminase complexing proteins are localized in exocrine glands of the rabbit. *J. Histochem. Cytochem.* **33**, 508–514.

Schubert, P., Komp, W., and Kreutzberg, G. W. (1979) Correlation of 5'-nucleotidase activity and selective transneuronal transfer of adenosine in the hippocampus. *Brain Res.* **168**, 419–424.

Schultz, V. and Lowenstein, J. M. (1976) Purine nucleotide cycle. Evidence for the occurrence of the cycle in brain. *J. Biol. Chem.* **251**, 485–492.

Schultz, V. and Lowenstein, J. M. (1978) The purine nucleotide cycle. Studies of ammonia production and interconversions of adenine and hypoxanthine nucleotides and nucleosides by rat brain *in situ*. *J. Biol. Chem.* **253**, 1938–1943.

Scott, T. G. (1967) The distribution of 5'-nucleotidase in the brain of the mouse. *J. Comp. Neurol.* **129**, 97–114.

Senba, E., Daddona, P. E., and Nagy, J. I. (1986a) Immunohistochemical localization of adenosine deaminase in the retina of the rat. *Brain Res. Bull.* **17**, 209–217.

Senba, E., Daddona, P. E., and Nagy, J. I. (1987d) A subpopulation of preganglionic parasympathetic neurons in the rat contain adenosine deaminase. *Neuroscience* **20**, 487–502.

Senba, E., Daddona, P. E., and Nagy, J. I. (1987a) Transient expression of adenosine deaminase in facial and hypoglossal motoneurons of the rat during development. *J. Comp. Neurol.* **255**, 217–230.

Senba, E., Daddona, P. E., and Nagy, J. I. (1987b) Adenosine-containing neurons in olfactory systems of the rat during development. *Brain Res. Bull.* **18**, 635–648.

Senba, E., Daddona, P. E., and Nagy, J. I. (1987c) Development of adenosine deaminase-immunoreactive neurons in the rat brain. *Dev. Brain Res.* **31**, 59–71.

Senba, E., Daddona, P. E., Watanabe, T., Wu, J. Y., and Nagy, J. I. (1985) Coexistence of adenosine deaminase, histidine decarboxylase, and glutamate decarboxylase in hypothalamic neurons of the rat. *J. Neurosci.* **5**, 3393–3402.

Shi, M. M. and Young, J. D. (1986) [³H]Dipyridamole binding to nucleoside transporters from guinea-pig and rat lung. *Biochem. J.* **240**, 879–883.

Shi, M. M., Wu, J. S. R., Lee, C. M., and Young, J. D. (1984) Nucleoside transport. Photoaffinity labeling of high affinity nitrobenzylthioinosine binding sites in rats and guinea pig lung. *Biochem. Biophys. Res. Commun.* **118,** 594–600.

Shimizu, H., Tanaka, S., and Kodama, T. (1972) Adenosine kinase of mammalian brain: partial purification and its role for the uptake of adenosine. *J. Neurochem.* **19,** 687–698.

Silinsky, E. M. (1984) On the mechanism by which adenosine receptor activation inhibits the release of acetylcholine from motor nerve endings. *J. Physiol.* (Lond.) **346,** 243–256.

Simon, L. N., Bauer, R. J., Tolman, R. L., and Robins, R. K. (1970) Calf intestine adenosine deaminase substrate specificity. *Biochem.* **9,** 573–579.

Snyder, S. H. (1985) Adenosine as a neuromodulator. *Annu. Rev. Neurosci.* **8,** 103–124.

Sollevi, A. (1986) Cardiovascular effects of adenosine in man; possible clinical implications. *Prog. Neurobiol.* **27,** 319–349.

Spector, R. (1985) Thymidine transport and metabolism in choroid plexus: Effect of diazepam and thiopental. *J. Pharmacol. Exp. Ther.* **235,** 16–19.

Stafford, A. (1966) Potentiation of adenosine and the adenine nucleotides by dipyridamole. *Br. J. Pharmacol. Chemother.* **28,** 218–227.

Staines, W. A., Yamamoto, T., Daddona, P. E., and Nagy, J. I. (1986a) Adenosine deaminase-containing hypothalamic neurons accumulates 5-HTP: A dual-colour immunofluorescence procedure using a new fluorescence marker. *Neurosci. Lett.* **70,** 1–5.

Staines, W. A., Yamamoto, T., Daddona, P. E., and Nagy, J. I. (1986b) Neuronal colocalization of adenosine deaminase, monoamine oxidase, galanin and 5-hydroxytryptophan uptake in the tuberomammillary nucleus of the rat. *Brain Res. Bull.* **17,** 351–365.

Staines, W. A., Yamamoto, T., Daddona, P. E., and Nagy, J. I. (1987a) The hypothalamus receives major projections from the tuberomammillary nucleus in the rat. *Neurosci. Lett.* **76,** 257–262.

Staines, W. A., Daddona, P. E., and Nagy, J. I. (1987b) The organization and hypothalamic projections of the tuberomammillary nucleus in the rat: An immunohistochemical study of adenosine-positive neurons and fibers. *Neuroscience* **23,** 571–596.

Stefanovich, V. (1983) Uptake of adenosine by isolated bovine cortex microvessels. *Neurochem. Res.* **8,** 1459–1469.

Stone, T. W. (1981) Physiological roles for adenosine and adenosine 5'-triphosphate in the nervous system. *Neuroscience* **6,** 523–555.

Stone, T. W. (1985) Summary of the symposium discussion on purine receptor nomenclature, in *Purines: Pharmacology and Physiological Roles* (Stone, T. W., ed.), Macmillan, London, pp. 1–4.

Subrahmanyam, K., Murthy, B., Prasad, M. S. K., Shrivastaw, K. P., and Sadasivudu, B. (1984) Adenosine deaminase in convulsions along with its regional, cellular and synaptosomal distribution in rat brain. *Neurosci. Lett.* **48,** 327–331.

Sun, M.-C., McIlwain, H., and Pull, I. (1976) The metabolism of adenine derivatives in different parts of the brain of the rat, and their release from hypothalamic preparations on excitation. *J. Neurobiol.* **7,** 109–122.

Sylwestrowicz, T., Piga, A., Murphy, P., Ganeshaguru, K., Russell, N. H., Prentice, H. G., and Hoffbrand, A. V. (1982) The effects of deoxycoformycin and deoxyadenosine on deoxyribonucleotide concentrations in leukaemic cells. *Br. J. Haematol.* **50,** 623–630.

Szentmiklosi, A. J., Nemeth, M., Cseppento, A., Szegi, J., Papp, J. Gy., and Szekeres, L. (1982) Potentiation of the myocardial actions of adenosine in the presence of coformycin, a specific inhibitor of adenosine deaminase. *Arch. Int. Pharmacodyn. Ther.* **256,** 236–252.

Tauc, L. (1982) Nonvesicular release of neurotransmitter. *Physiol. Rev.* **62,** 857–893.

Thampy, K. G. and Barnes, E. M., Jr. (1983a) Adenosine transport by primary cultures of neurons from chick embryo brain. *J. Neurochem.* **40,** 874–879.

Thampy, K. G. and Barnes, E. M., Jr. (1983b) Adenosine transport by cultured glial cells from chick embryo brain. *Arch. Biochem. Biophys.* **220,** 340–346.

Trams, E. G. and Lauter, C. J. (1975) Adenosine deaminase of cultured brain cells. *Biochem. J.* **152,** 681–687.

Tritsch, G. L., Paolini, N. S., and Bielat, K. (1985) Adenosine deaminase activity associated with phagocytic vacuoles. Cytochemical demonstration by electron microscopy. *Histochemistry* **82,** 281–285.

Trotta, P. P., Peterfreund, R. A., Schonberg, R., and Balis, M. E. (1979) Rabbit adenosine deaminase conversion proteins. Purification and characterization. *Biochemistry* **18,** 2953–2959.

Tse, C. M., Belt, J. A., Jarvis, S. M., Paterson, A. R. P., Wu, J. S., and Young, J. D. (1985) Reconstitution studies of the human erythrocyte nucleoside transporter. *J. Biol. Chem.* **260,** 3506–3511.

Ukena, D., Jacobson, K. A., Padgett, W. L., Ayala, C., Shamim, M. T., Kirk, K. L., Olsson, R. O., and Daly, J. W. (1986) Species differences in structure-activity relationships of adenosine agonists and xanthine antagonists at brain A_1 adenosine receptors. *FEBS Lett.* **209,** 122–128.

Van der Weyden, M. B. and Kelley, W. N. (1976) Human adenosine deaminase. Distribution and properties. *J. Biol. Chem.* **251,** 5448–5456.

Van Wylen, D. G. L., Park, T. S., Rubio, R., and Berne, R. M. (1986) Increases in cerebral interstitial fluid adenosine concentration during hypoxia, local potassium infusion and ischemia. *J. Cereb. Blood Flow Metab.* **6,** 522–528.

Verma, A. and Marangos, P. J. (1985) Nitrobenzylthioinosine binding in brain: An interspecies study. *Life Sci.* **36,** 283–290.

Verma, A., Houston, M., and Marangos, P. J. (1985) Solubilization of an adenosine uptake site in brain. *J. Neurochem.* **45,** 596–603.

White, T. D. (1985) The demonstration and measurement of adenosine triphosphate release from nerves, in *Methods Used in Adenosine Research* (Paton, D. M., ed.) (*Methods in Pharmacology*, vol 6) Plenum, New York, pp. 43–63.

White, T. D. and Leslie, R. A. (1982) Depolarization-induced release of adenosine 5'-triphosphate from isolated varicosities derived from the myenteric plexus of the guinea pig small intestine. *J. Neurosci.* **2,** 206–215.

White, T. D., Downie, J. W., and Leslie, R. A. (1985) Characteristics of K⁺- and veratridine-induced release of ATP from synaptosomes prepared from dorsal and ventral spinal cord. *Brain Res.* **334,** 372–374.

Williams, M. (1984) Mammalian central adenosine receptors, in *Receptors in the Nervous System*, (Lajtha, A., ed.) (Handbook of Neurochemistry, vol 6), Plenum, New York, pp. 1–26.

Winn, R. H., Rubio, R., and Berne, R. M. (1980) Brain adenosine production, in the rat during 60 seconds of ischemia. *Circ. Res.* **45,** 486–492.

Wojcik, W. J. and Neff, N. H. (1983a) Location of adenosine release and adenosine A₂ receptors to rat striatal neurons. *Life Sci.* **33,** 755–763.

Wojcik, W. J. and Neff, N. H. (1983b) Adenosine A₁ receptors are associated with cerebellar granule cells. *J. Neurochem.* **41,** 759–763.

Wojcik, W. J. and Neff, N. H. (1983c) Differential location of adenosine A₁ and A₂ receptors in striatum. *Neurosci. Lett.* **41,** 55–60.

Wu, P. H. and Phillis, J. W. (1982) Nucleoside transport in rat cerebral cortical synaptosomal membrane: A high affinity probe study. *Int. J. Biochem.* **14,** 1101–1105.

Wu, P. H. and Phillis, J. W. (1984) Uptake by central nervous tissues as a mechanism for the regulation of extracellular adenosine concentrations. *Neurochem. Int.* **6,** 613–632.

Yamada, Y., Goto, H., and Ogasawara, N. (1980) Purification and properties of adenosine kinase from rat brain. *Biochim. Biophys. Acta* **616,** 199–207.

Yamamoto, T., Geiger, J. D., Daddona, P. E., and Nagy, J. I. (1987) Subcellular, regional and immunohistochemical localization of adenosine deaminase in various species. *Brain Res. Bull.* **19,** 473–484.

Yarbrough, G. G. and McGuffin-Clineschmidt, J. C. (1981) *In vivo* behavioral assessment of central nervous system purinergic receptors. *Eur. J. Pharmacol.* **76,** 137–144.

Yeager, R. E., Nelson, R., and Storm, D. R. (1986) Adenosine inhibition of calmodulin-sensitive adenylate cyclase from bovine cerebral cortex. *J. Neurochem.* **47,** 139–144.

Young, J. D. and Jarvis, S. M. (1983) Nucleoside transport in animal cells. *Biosci. Rep.* **3,** 309–322.

Young, J. D., Jarvis, S. M., Belt, J. A., Gati, W. P., and Paterson, A. R. P. (1984) Identification of the nucleoside transporter in cultured mouse lymphoma cells. *J. Biol. Chem.* **259,** 8363–8365.

Zetterstrom, T., Vernet, L., Ungerstedt, U., Tossman, U., Jonzon, B., and Fredholm, B. B. (1982) Purine levels in the intact rat brain. Studies with an implanted perfused hollow fibre. *Neurosci. Lett.* **29,** 111–115.

Zimmerman, T. P., Gersten, N. B., Foss, A. F., and Miech, R. P. (1971) Adenine as substrate for purine nucleoside phosphorylase. *Can. J. Biochem.* **49,** 1050–1054.

CHAPTER 8

Adenosine and Cardiovascular Function

Kevin M. Mullane and Michael Williams

1. Historical Perspective

The effects of adenosine and its related nucleotides on cardiovascular function were initially reported by Drury and Szent-Györgyi (1929), who noted in studies performed in a variety of mammalian species, that these compounds were potent bradycardic and blood pressure lowering agents with marked vasoconstrictor effects in the kidney.

In the next four years, clinical studies were initiated using adenosine to treat cardiac arrhythmias (Honey et al., 1930; Jezer et al., 1933), but the injection of large boluses induced temporary cardiac arrest, leading to the conclusion that adenosine was not "a useful therapeutic preparation for the treatment of heart diseases." The pervading negative attitude regarding the clinical utility of adenosine limited further clinical development until more recently. However, adenosine is very effective in the treatment of supraventricular arrhythmias, and its potential as the drug of choice for this condition has been discussed because of its short half-life and rela-

Adenosine and Adenosine Receptors Editor: Michael Williams ©1990 The Humana Press Inc.

tively benign side-effect profile (Bellhassen and Pelleg, 1984; Ramkumar et al., 1988), although a site-specific, long acting, A_1-selective agent should be a superior agent.

Adenosine and its various analogs generally have not been regarded as potential therapeutic agents of great significance. This lack of enthusiasm can be attributed, in part, to the ubiquitous distribution of the purine nucleoside and the broad range of activities it elicits, including central nervous system effects, bradycardia, and renal vasoconstriction. Infusions of adenosine in humans produce a high incidence of side effects, such as headaches, flushing, listlessness, an urge to breathe deeply, and angina pectoris-like pain, leading to the suggestion that adenosine produced during myocardial ischemia is the mediator of pain by stimulating chemosensitive C-fibers (Sylven et al., 1986; Robertson et al., 1988). Interestingly, others (Ramkumar et al., 1988) have viewed adenosine as a therapeutically safe agent, its short duration of action and transient side-effect profile making it particularly attractive for clinical evaluation.

Further studies on the ability of adenosine to dilate resistance vessels in heart tissue occurred some 25 years after these initial observations (Green and Stoner, 1950; Winbury et al., 1953; Wolf and Berne, 1956; Berne et al., 1983) and were extended to brain (Berne et al.,1974), skeletal muscle (Bockman et al., 1975), lung (Mentzer et al., 1975), gut (Granger et al., 1978), and fat tissue (Sollevi and Fredholm, 1981). The effects of adenosine on cardiac conductance (negative dromotropic and chronotropic actions) were studied further by Green and Stoner (1950) and others (James, 1965; Chiba and Hashimoto,1972; Bellardinelli et al.,1981, 1983; Jonzon et al., 1986; Pelleg, 1988).

The classic studies of Berne and his colleagues in the 1960s, whereby adenosine formed the link between the metabolic demands of the heart and an increase in the oxygen supply, have progressed little further in the ensuing two decades, despite a multitude of studies that are purported to either support or refute the "Berne hypothesis" (Berne et al., 1987). One potential problem is the failure to define the appropriate source of adenosine involved in coronary vasoregulation. The large potential for adenosine production by myocytes during ATP depletion in ischemia is obviously attractive, but is then dependent on adequate diffusion and/or transport to reach blood vessels without further metabolism. Perhaps a local source, although not so abundant, could play the major role in vaso-regulation, the nucleoside being formed at the site of activity. An isolated strip or ring of arterial tissue suspended in physiological media at 37°C exhibits local regulation in response to changes in pO_2 (Rubanyi and Vanhoutte, 1985). If adenosine contributes to this "metabolic link," it can be derived from the blood vessel itself. The groups of Schrader et al., and Nees, Gerlach, and colleagues have defined the contribution of coronary endothelial cells to cardiac adenosine production in a variety of settings (Deussen et al., 1986; Gerlach et al., 1985; 1987; Nees et al., 1985, 1987), concluding that endothelial cells contribute only about 14% of the total cardiac adenosine that is formed. However, since the endothelium represents only 3% of the total myocardial volume (Anversa et al., 1983), it is clearly an active source at an appropriate site. To our knowledge, there are no reports on the total adenosine contribution by vascular tissue, i.e., endothelium and smooth muscle cells. Vasodilators, such as acetylcholine or prostacyclin, do not influence cardiac metabolism, but stimulate adenosine release, which may contribute to the increase in coronary blood flow (Blass et al., 1980; Edlund et al., 1985; Deussen et al.,1986). These studies indicate that adenosine release in quantities and at a site sufficient to alter vascular tone can be dissociated from ATP breakdown, and that the myocellular ATP may not always be the most appropriate source of this mediator. A second pathway whereby 5-adenosyl-homocysteine is converted to adenosine and homocysteine by 5-adenosylhomocysteine hydrolase has been proposed, but its importance remains undefined (Lloyd and Schrader, 1987).

The excitement over the recognition of a chemical link between cardiac metabolism and coronary flow led to the concept that agents that either mimic or enhance adenosine availability could improve the oxygen supply to the ischemic myocardium and exhibit antianginal properties. However, adenosine preferentially dilates the small arterioles rather than the large vessels (in contrast to nitroglycerin), and agents such as dipyridamole are ineffective in models of angina because they increase flow in the nonischemic region of the heart, leading to the phenomenon of "coronary steal." Indeed, clinically, dipyridamole is used in the thallium scanning of patients with angina to demarcate the zone of defective perfusion.

Examination of the role of adenosine in flow regulation and other cardiovascular settings has relied on the use of drugs and/or agents that lack specificity and selectivity. For example, dipyridamole used to block adenosine uptake has intrinsic vasodilator properties, and these confound appropriate conclusions regarding the contribution of the purine to the vasodilator actions of this compound. The advent of more selective and specific adenosine antagonists and agonists has prompted resurrection of the concepts and considerations proposed in the last 60 years for potential utility as therapeutic agents. Adenosine is already used to treat supraventricular tachyarrhythmia (Di Marco et al., 1985) and is also used to produce a controlled hypotension during surgical treatment of cerebral aneurysms (Sollevi et al., 1984; Owall et al., 1987), whereas adenosine antagonists may supress the bradyarrhythmias provoked by ischemia of the posterior wall of the heart. Thus, the full picture of the clinical potential of adenosine modulators has thus yet to emerge.

2. Biochemical Characterization of Adenosine Receptors in Cardiac Tissues

Like most peripheral tissues, the heart has very low concentrations of adenosine receptors as compared to the central nervous system (Williams, 1987), a fact that has made the application of radioligand binding techniques to the characterization of adenosine receptors in heart and coronary vessels somewhat difficult. Responses to adenosine receptor activation in the heart reflected the initial classification of the two receptor subtypes, namely inhibition (A_1) and activation (A_2) of adenylate cyclase (Londos et al., 1980). However, as with other tissues sensitive to adenosine, this classification is now pharmacological, based on the agonist profile of activity, rather than on the second messenger involved (Hamprecht and Van Calker, 1985).

The effects of adenosine on ion conductance and cardiac contractility are mediated via A_1-type receptors (Evans et al.,1982; Collis, 1983; Leung et al., 1983). Although initial attempts to label these receptors in myocardial membranes using radiolabeled adenosine (Dutta and Mustafa, 1979; Ollinger and Kukovetz, 1983) and 2-chloro-adenosine (Beck et al., 1984; Michaelis et al., 1985) were somewhat disappointing, the development of selective high-affinity ligands, such as [^{125}I]aminobenzyl-adenosine (Linden et al., 1985), CPX (8-cyclopentylxanthine; Bruns et al., 1987), and BW-A 844U (Linden et al.,1988), have permitted characterization of these receptors and their anatomical distribution.

The A_1 receptor from bovine heart has a molecular weight of approximately 35,000 daltons (Stiles et al., 1985; Klotz and Lohse, 1986; Stiles and Jacobson, 1988). CPT has also allowed the tentative labeling of a binding site thought to be the A_2 receptor in human coronary arteries (Ramagopal et al., 1988).

3. Endogenous Adenosine as a Physiological Regulator

Under normal conditions, adenosine levels in the heart are maintained betwen 0.1 and 1 μM (Hanley et al., 1983). During ischemia (Rubio et al., 1974), hypoxia (Rubio and Berne, 1969), and adrenergic stimulation (Dobson and Schrader, 1984), adenosine levels can be elevated, and the purine can thus subserve a physiological role as an autacoid in regulating both coronary blood flow and cardiac workload (Thompson et al., 1980; Bellardinelli et al., 1983; Berne et al., 1983; Berne, 1980, 1985). Conversely, under conditions where there is an excess oxygen supply, such as that which occurs following nitroglycerin administration (Berne et al., 1983), adenosine formation is reduced. Although the protective role of adenosine in the myocardium is intended to prevent damage to the heart, the potential for reduced cardiac output or coronary steal have obvious deleterious effects on tissues other than the heart when oxygen availability is restricted. The cardioprotectant effects of adenosine represent a physiologically relevant homeostatic process. Whether the blood pressure lowering effects of the purine occur under normal conditions in response to a "purinergic tone" is unknown and is a somewhat controversial topic.

Exogenous adenosine causes vasodilation but is not especially efficacious as a blood pressure lowering agent (Schrader et al., 1977), a fact attributed to its uptake by the coronary endothelium

(Nees et al., 1985), which leads to a plasma half-life on the order of 10 s in humans (Klabunde, 1984). In addition, adenosine administration to human subjects is associated with angina (Sylven et al., 1986) and psychic disturbances (Schaumann and Kutscha, 1972; Robertson et al., 1988). More stable adenosine analogs that do not undergo uptake are also effective coronary vasodilators (Hamilton et al., 1987; Oei et al., 1988). Agonists that are active at the A_2 receptor such as CI 936 ([N^6(2,2-diphenylethyl)-adenosine; Hamilton et al., 1987]), a series of N^6-substituted adenosine analogs (Bridges et al., 1988) and CGS 21680 (2-[p-(2-carboxyethyl) phenethylamino-5'-N-ethylcarboxamidoadenosine; Hutchison et al., 1989), are potent hypotensive agents that have no bradycardiac actions, but do tend to cause a transient tachycardia via autonomic mechanisms. Brief increases in systolic and diastolic pressures have been observed in conscious normal volunteers following adenosine administration. These effects were also accompanied by an increase in respiration (Robertson et al., 1988), which was also noted by Drury and Szent-Györgyi (1929) in the rabbit.

From a therapeutic perspective, however, although adenosine represents a novel hypotensive agent via its vasodilatory actions, its ability to reduce cardiac output is a major drawback, especially when compared with agents such as the angiotensin-converting enzyme inhibitors. Current knowledge, obtained almost exclusively in animal preparations, would indicate that A_2-selective adenosine agonists would have all the apparent benefits of adenosine as a vasorelaxant agent while being relatively free of effects on heart rate and cardiac output. This hypothesis is discussed further in the final section of this chapter.

4. Adenosine and Blood Flow

4.1. Coronary

As already noted, the oxygen supply–demand ratio in cardiac tissue can proportionally regulate myocardial adenosine production. The relative importance of the purine as an endogenous modulator is somewhat controversial (Kroll and Feigl, 1985; Sparks and Gorman, 1987); nonetheless, adenosine, acting via A_2 receptors, is a potent coronary vasodilator. Thus when cardiac metabolism is increased, there is a concommitant increase in adenosine production, and blood flow is increased. Biphasic effects on transmural endocardial–epicardial blood-flow ratios have been noted. These effects are independent of global hemodynamic parameters and reflect in the endocardial layer a greater sensitivity to the purine, even though there is a greater capacity to increase flow in the epicardial layer, resulting in a progressive increase in transmural blood flow (Leppo et al., 1984). However, despite exhaustive evaluation, there is still some controversy as to the importance of the purine in the local regulation of blood flow. Much of the discussion revolves around endogenous adenosine levels and experiments designed to limit the potential effects of the purine, either by enzymic degradation or receptor blockade (Berne et al., 1987; Sparks and Gorman, 1987). In many instances, however, the antag-onist used has been theophylline, which is a relatively weak adenosine antagonist (Schrader and Deussen, 1985). Thus the data generated are limited, and continue to be confusing as this xanthine is used rather than more potent and selective entities, such as CGS 15943 (Williams et al., 1987). Recent studies using 8-phenyltheophylline, the xanthine congener XAC (Collis et al., 1983; Fredholm et al., 1987; Oei et al., 1988), and the potent A_1 receptor antagonist PACPX (Oei et al., 1988), have, however, provided data showing that xanthines can act as cardioselective adenosine antagonists.

Infusing labeled adenosine into isolated guinea pig hearts, and then separating the cells, reveals that 80% of the radioactivity is found in endothelial cells, and the remainder is present in the supernatant. These studies led Nees et al. (1987) to conclude that the coronary endothelium functions as a metabolic barrier for adenosine, preventing further penetration, and that adenosine-induced coronary vasodilation must occur as the result of interactions with endothelial cells. Although A_2 receptors have been identified on the endothelial surface, how receptor activation translates into vasodilation is unclear. Evidence argues against the involvement of cyclic AMP. Moreover, high molecular weight derivatives of adenosine, such as polyadenylic acid, an AMP-protein conjugate, or adenosine-oligosaccharide derivatives, elicit a prompt increase in coronary flow, attributable to activation of endothelial receptors.

Since the original description of an "endothelial-derived relaxing factor" (EDRF) by Furchgott and Zawadzki (1980), considerable interest has centered on the role of the endothelium in the control of vascular tone. EDRF is released by a wide variety of vasoactive mediators and has recently been identified as nitric oxide. This agent elicits vasorelaxation by stimulating cGMP production.

Adenosine-induced relaxations of rabbit aorta and both rabbit and canine femoral artery in vitro or in vivo are unaffected by the absence or presence of the endothelial layer. However, relaxation of either the canine coronary artery or the porcine or rat aortae are partially endothelial-dependent (Yen et al., 1988). Thus in some vascular beds, a component of adenosine-induced vasodilation may be endothelial-dependent. Whether EDRF or nitric oxide, another endothelial component is involved, or the response involves the microcirculation remains to be determined.

Vascular smooth muscle cells also have A_2 receptors-activation of which produces vasodilation. If the endothelium is an impermeable barrier to adenosine, there may be compartmentation of the vascular effects of endogenous vs exogenous nucleoside, which could account for some of the discrepancies between the effects observed for antagonists or uptake inhibitors and the effects of the agonist administered into the vascular system.

Adenosine may also have other actions on endothelial cells, independent of modulating vascular tone. The migration of endothelial cells is stimulated by adenosine and hypoxia, responses attenuated by 8-phenyltheophylline (Meininger et al., 1988). Hypoxia-induced release of adenosine may thus act as an angiogenic stimulus facilitating vessel proliferation (Dusseau et al., 1986). Adenosine stimulates cyclic AMP formation in endothelial cells, and agents that also increase cyclic AMP, such as PGI_2, augment cholesterol ester hydroylase activity and decrease cholesterol deposition in the arterial wall (Hajjar, 1985). Consequently an anti-atherosclerotic potential for the nucleoside should be considered, although this avenue of therapy has yet to be evaluated. The ability of adenosine to prevent neutrophil-mediated endothelial cell damage (Cronstein et al., 1987) would offer a synergism with the effects of the purine on cholesterol formation and may be of benefit in postoperative care following angioplasty.

The involvement of adenylate cyclase inhibition in the A_1-receptor mediated effects on cardiac conduction is somewhat controversial, some authors reporting inhibition (Leung et al., 1983; Linden et al., 1985; Martens et al., 1987), some no effect (Bohm et al., 1984; Bruckner et al., 1985) and some stimulation (Anand-Srivastava, 1985). These conflicting data may result from contamination of the sarcolemmal membranes, used to study adenylate cyclase inhibition, with endothelial membranes (Schutz et al., 1986).

Coronary artery vasorelaxation may also involve an additional purine-related mechanism. Several groups (Collis and Brown, 1983; Odawara et al., 1986; Samet and Rutledge, 1985) have invoked an intracellular P-site effect. However, the involvement of adenylate cyclase in this response has been questioned (Bruckner et al., 1985), and inosine has been shown to evoke vasorelaxation via an intracellular mechanism (Collis et al., 1986). The purine may also be useful in maintaining patency of the ductus arteriosus in congenital heart disease in infants (Mentzer et al., 1985).

4.2. Renal

Adenosine and its analogs can either constrict or dilate the renal vasculature (Osswald, 1983, 1988; Churchill and Bidani, 1990). Vasoconstriction is attributable to A_1-receptor activation (Rossi et al., 1988), which also mediates the inhibition of renin secretion (Churchill and Bidani, 1990) via a pertussis toxin-sensitive mechanism that appears to involve cyclic AMP (Arend et al., 1987; Freissmuth et al., 1987; Rossi et al., 1987). Activation of A_2 receptors is associated with renal vasodilation, but no change in glomerular filtration rate. Although studies using kidney slices suggest that A_2-receptor activation may stimulate renin release, elevations in plasma renin activity in vivo following administration of A_2-selective receptor agonists can be attributed to a fall in blood pressure rather than a direct effect on the intrarenal renin–angiotensin system (Miller et al., 1990). Unlike the situation in other vascular beds, where adenosine is a vasodilator and blood flow is dependent on purine production, kidney blood flow apparently regulates metabolism (Osswald, 1988). The effects of the adenosine analogs NECA and 2-CADO are biphasic, an initial transient vasoconstriction being followed by dilation.

Renal ischemia is associated with adenosine release and a post-occlusive vasoconstriction, rather than with the hyperemic response seen in other vascular beds. A_1-receptor mediated vasoconstriction could account for the exacerbation of ischemia, but adenosine also increases the frequency of afferent renal nerve activity (Recordati et al., 1977; Katholi et al., 1983, 1984) to enhance the efferent sympathetic nervous system, leading to a disproportionately greater increase in renal vascular resistance than is found in other vascular beds (Vatner et al., 1971; Katholi et al., 1983). Intrarenal infusions of adenosine produce hypertension as a result of activation of the sympathetic nervous system (Katholi et al., 1983, 1984), leading to the suggestion (Katholi et al., 1988) that the purine, released when renal blood flow is compromised (as in the one-kidney, one-clip rat), produces hypertension. Renal denervation, ganglionic blockade, and intrarenal adenosine deaminase infusion all lower blood pressure in such hypertensive animals, but have no effect in normotensive controls.

Angiotensin II receptor antagonists can attenuate adenosine-induced vasconstrictor effects (Spielman and Osswald, 1979). The pressor agent can act synergistically with adenosine to lower renal blood flow and glomerular filtration rate (GFR; Osswald et al., 1978). Studies using the antagonist 8-phenyltheophylline have indicated that the effects of adenosine on GFR occur only after acute renal failure (Collis, 1988).

Adenosine has also been implicated in tubuloglomerular feedback responses (Osswald et al., 1982). The magnitude of feedback responses can be increased by A_1 agonists, such as CPA, CHA, or R-PIA, at concentrations of 10 μM, whereas adenosine or NECA either reduce the magnitude of the feedback response or reversed its direction (Schonermann, 1988). It has been concluded that if endogenous adenosine mediates tubuloglomerular feedback, the concentration range at which this occurs is very narrow, and the effect must be restricted to activation of preglomerular A_1 receptors.

4.3. Cerebral

Cerebral blood flow is maintained relatively constant over a wide range of arterial blood pressures by autoregulatory mechanisms (Busija and Heistad, 1984; Winn, 1985) that are closely interrelated to cerebral energy utilization (Raichle et al., 1977; McCulloch, 1983). The mechanism mediating autoregulation is unclear, but may involve myogenic, neurogenic or metabolic factors (Winn et al., 1985). Adenosine can enhance cerebral blood flow in primates (Forrester et al., 1979) and in the rabbit (Heistad et al., 1981), and has been proposed as a metabolic regulator of cerebrovascular resistance (Kuschinsky, 1983). These effects of the purine appear to be mediated via A_2 receptors linked to cyclic AMP production (Edvinsson and Jansen, 1985). A sustained (5 min) reduction in blood pressure can cause adenosine production, with a doubling in nucleoside concentration when blood pressure decreases from 135 to 72 mm Hg. The purine causes vasodilation in pial vessels (Berne et al., 1974) and has been proposed to function as a metabolic regulator in response to hypoxia, and to ischemic and hypercapnic situations (Busija and Heisted, 1984; Phillis et al., 1984; Winn et al., 1985). The source of the adenosine and the causal factor(s) in controlling purine availability are unknown (Winn, 1985). Adenosine systemically causes a generalized increase in cerebral blood flow in humans as assessed by PET scanning, with up to a 100% increase in cortical and thalamic areas (Sollevi et al., 1987). Interestingly, the blood-flow increase in response to adenosine can be abolished by hyperventilation, indicating the complexity of the responses to the purine and the homeostatic mechanisms that exist to regulate cardiovascular function in response to changes in blood pressure and altered hemodynamic situations (Katholi et al., 1983; Ohnishi et al., 1986). Adenosine-induced hypotension in dogs to 40 mm Hg can decrease cerebral blood flow, the cerebral cortex and corpus callosum being most affected (Kassell et al., 1983). Xanthines can also modulate cerebral blood flow (Grome and Stefanovich, 1985), increasing glucose utilization in selected brain regions (Nehlig et al., 1984). In humans, caffeine decreases cerebral blood flow (Mathew and Wilson, 1985) and has no effect on cerebral blood flow during acute hypotensive episodes, suggesting that adenosine is not an essential factor under such conditions

(Phillis and De Long, 1986). Theophylline can, however, decrease vascular resistance in the cerebral vessels (Oberdorster et al., 1975). It also has complex effects on feline pial vessel diameter (Edvinsson and Fredholm, 1983).

Whereas it is axiomatic that blood flow is governed by cerebral energy needs, the decrease in cerebral blood flow in response to, theophylline and caffeine is accompanied by an increase in local cerebral glucose utilization (Grome and Stefanovich, 1985). This unusual finding has been interpreted in terms of a "resetting" of the coupling mechanism between cerebral blood flow and glucose utilization. The atypical xanthine propentofylline (HWA 285), which is a cerebral vasodilator, was found to have the opposite effects, increasing cerebral blood flow while decreasing glucose utilization. In relating the uncoupling of blood flow from energy utilization to a disturbance of the normally functional autoregulatory systems, Grome and Stefanovich (1986) have suggested that adenosine antagonism by the xanthines may explain the effects observed.

4.4. Skeletal Muscle

Adenosine acts as an autacoid in eliciting activity hyperemia in skeletal muscle (Bockman et al., 1975), an effect that may be mediated via inhibition of arteriolar sympathetic tone (Fugslang and Crone, 1987). Xanthines can prevent fatigue in skeletal muscle, an effect that appears to involve effects on calcium mobilization rather than adenosine antagonism (Waldeck, 1985).

5. Central Regulation of Cardiovascular Function

Administration of adenosine into brain ventricles can decrease respiration (Eldridge et al., 1984; Eldridge and Millhorn, 1987), with a corresponding decrease in blood pressure and bradycardia (Mueller et al., 1984; Wessberg et al., 1985). The blood pressure lowering activity of CHA has both central and peripheral components (Singer et al., 1986). Dose-dependent decreases in blood pressure occur when adenosine agonists are injected in rat cerebral ventricles (Barraco et al., 1984). These effects can be localized to the NTS and the area postrema of the fourth ventricle (Barraco et al.,1986; Tseng et al., 1988) at doses 100–1000-fold less than those required to decrease blood pressure by femoral administration (Barraco et al., 1987). As already noted, these hypotensive and bradycardiac effects of adenosine following central administration are markedly attenuated in spontaneously hypotensive rats (Ohnishi et al., 1986; Tung et al., 1987, Robertson et al., 1988).

It is of interest, given the potent effects of adenosine in inhibiting the release of glutamate in the CNS (Dolphin and Archer, 1983), that L-glutamate, acting via N-methyl-D-aspartate (NMDA) type receptors in the zona incerta can decrease blood pressure and heart rate (Miller and Felder, 1988; Spenser et al.,1988). In the NTS, NMDA can modulate autonomic discharge thus affecting vagal outflow. Of additional interest in this context is the suggestion (Ramkumar et al., 1988) that increases in cyclic AMP as a result of A_2-receptor acti-vation can facilitate the release of EDRF. Very recently, L-glutamate has been reported to evoke the release of an EDRF-like factor in mammalian brain (Garthwaite et al., 1988).

The fact that adenosine can decrease blood pressure by direct effects in the periphery (cardiac, renal, pulmonary) as well as when administered centrally raises an important issue regarding the selectivity of the actions of adenosine. There has long been controversy (Snyder et al., 1981; Barraco et al., 1984; Barraco, 1988; Jarvis and Williams, 1990) as to whether the CNS-depressant effects of adenosine occur secondarily to an effect on blood pressure reduction or result from direct actions of the purine on CNS receptor-mediated processes. Given that adenosine can affect blood pressure via both direct and indirect peripheral mechanisms as well as through central actions, the resolution of this controversy appears unclear at this time.

6. Adenosine and Cardiac Contractility

The direct actions of adenosine on cardiac contractility, unlike the indirect actions discussed below, are independent of effects on catecholamine-mediated responses. The former involve negative chronotropic, inotropic, and dromotropic actions. These reflect, respectively, slowing of the sinoatrial node, decreased AV node conduction, and a decrease in atrial contractility (Belardinelli et al., 1983; Endoh et al., 1983; Pelleg, 1988) in addition to elimination of intrinsic pacemaker capacity.

In the myocardium, the effects of adenosine are very similar to those seen with acetylcholine (De Gubareff and Sleator, 1965; Rockoff and Dobson, 1980; Belardinelli and Isenberg, 1983), both compounds having a negative inotropic action that reduces action potential duration and induces hyperpolarization. These effects appear to to be caused by K^+ channel activation (Kurachi et al.,1986) via a G-protein-linked mechanism (Cooper and Caldwell, 1990). The involvement of cyclic AMP or a calcium/inositol triphosphate-linked system in these effects is uncertain (Pelleg, 1988). Furthermore, whereas adenosine can elicit the release of prostaglandins from heart tissue (Zehl et al., 1976), there is no evidence to suggest that these agents mediate the effects of the purine, although adenosine has been described as a "prostacyclin promotor" (Parratt et al., 1988). Since the atrial muscle Gi-proteins linked to potassium channels appear to be multifunctional (Yatani et al., 1988), it may be anticipated that additional levels of complexity in neuroeffector-mediated responses involving adenosine will be found.

The nucleoside reduces ischemia-related ventricular arrhythmias (Fagbemi and Parratt, 1984), an effect additional to the anti-arrhythmic actions in AVN-related paroxysmal supraventicular tachycardia (Pelleg et al., 1987). There thus appear to be distinct regional differences in the antiarrythmic effects of adenosine, the His Purkinje system being more sensitive to the purine than either the SN or AV nodes (Pelleg, 1988). Regional differences in the availability of adenosine may also exist (Bardenheuer et al., 1987).

The effects of adenosine on cardiac contractility are A_1 receptor-mediated (Evans et al., 1982; Bellardinelli et al., 1983; Collis, 1983; Haleen and Evans,1985; Clemo and Bellardinelli, 1986; Hamilton et al., 1987; Oei et al., 1988) and have been reported to occur at doses lower than those required to reduce blood pressure (Watt and Routledge, 1986). The sinus bradycardia and sinus arrest occuring as a result of myocardial infarction are thought to occur via locally produced adenosine (James, 1965), the purine again mediating ischemic AVN blockade (Bellardinelli et al., 1981; Wesley et al., 1985). Adenosine can be used as a diagnostic tool and as an acute therapy in supraven-tricular tachyarrhythmia (Di Marco et al., 1985).

Xanthines can decrease peripheral vascular resistance, have cardiotonic activity, and can produce arrhythmias (Ogilvie et al., 1977). These effects can be ascribed to both the adenosine-antagonist properties of these compounds and their phosphodiesterase-inhibiting activity. However, the use of more potent xanthines such as XAC and PACPX (Fredholm et al., 1987; Oei et al., 1988; Webb et al., 1989), which are devoid of phosphodiesterase inhibiting activity at concentrations where they antagonize adenosine receptors, has shown a clear interaction of the xanthines at adenosine receptors. From a therapeutic viewpoint, an A_1-selective adenosine antagonist (of which there are many such entities: *see* Williams, 1989) would be an effective cardiotonic acting to reduce the negative dromotropic and chronotropic actions of adenosine. The cardiotonics sulmazole and amrinone have been described as A_1 receptor antagonists with micromolar activity (Parsons et al., 1988).

The thioxanthine S-caffeine, in contrast to caffeine, has negative inotropic and chronotropic activity (Fassina et al., 1985) and is thus "adenosine-like" in terms of its actions on cardiac contractility. The structural requirements that change an antagonist to an agonist are not known in the case of S-caffeine. However, like caffeine, the thioxanthine is an effective phosphodiesterase inhibitor, indicating that this activity is unlikely to account for the reported effects of xanthines on heart function. It has also been reported, based on comparative structure–activity profiles for a series of alkylxanthes, that the positive inotropic actions of these agents are not mediated by blockade of adenosine receptors (Collis et al., 1984).

7. Indirect Actions of Adenosine

Adenosine also has indirect effects on cardiovascular function. The most direct of these indirect actions is the antiadrenergic effect of the purine (Dobson et al., 1984; Rardon and Bailey, 1984) where the purine can antagonize the cardiostimulant actions resulting from β-adrenoceptor stimulation through presynaptic effects on transmitter release (Fredholm and Dunwiddie, 1988). However the purine has more oblique indirect actions, affecting renin production from the kidney via effects on kidney adenosine receptors (Osswald, 1983; Churchill and Bidani, 1990) and by its effects on the neural pathways related to chemoceptor and baroceptor reflex responses (Ribeiro and McQueen,1983; Ohnishi et al., 1986; Ribeiro et al., 1988; Eldridge and Millhorn, 1987; Tung et al., 1987). It may be noted however that the effects of adenosine on renin release appear to be due to alterations in blood pressure (Miller et al., 1989).

7.1. Antiadrenergic

Adenosine attenuates β-adrenergic receptor mediated contractile and glycogenolytic responses in the ventricular myocardium, an effect mediated via increased cyclic AMP formation (Dobson, 1983; Bohm et al., 1985, 1986). These actions are mediated via a classical A_1 receptor that modulates the G protein transduction mechanisms associated with β-receptor activation. Recent studies (Romano et al., 1988) have shown that the agonist R-PIA can cause a threefold decrease in the binding affinity of isoproterenol for β-receptors labeled with iodocyanopindolol, an effect associated with an apparent loss of high-affinity β-receptor binding. Thus adenosine appears to effectively "uncouple" the β-receptor from adenylate cyclase.

The physiological relevance of the adenosine antagonism of endogenous catecholamines is poorly understood despite the finding that, in humans, exogenous adenosine can antagonize catecholamine-induced hypertension (Sollevi, 1986). Dobson has proposed (Dobson, 1983; Dobson et al., 1987) that the purine acts as an important negative feedback modulator of β-adrenergic receptor mediated responses to protect the heart from excessive activation. Meanwhile, Raberger et al. (1987), using a dog treadmill model, found no evidence for a physiological role of adenosine as an antiadrenergic agent in conscious animals. These conflicting data have been ascribed to differing levels of endogenous adenosine in the various models studied (Gerlach and Becker, 1987). In conscious humans, however, adenosine infusion (10–140 μg/kg/min iv) activated rather than attenuated the sympathoadrenal system, an effect potentiated by the adenosine-uptake blocker, dipyridamole (Biaggioni et al., 1986). Thus while diastolic pressure was lowered, heart rate, systolic blood pressure, and plasma norepinephrine levels were raised. This study was in marked contrast to most preclinical studies and to those performed in humans under anesthesia, in which a hypotensive effect of adenosine and a lack of reflex sympathetic activation was the norm. The authors (Biaggioni et al. 1986), concluded "that in conscious man, the hypotensive effects of adenosine are obscured by an activation of the sympathoadrenal system," reinforcing the need for caution in extrapolating effects from preclinical animal models to humans.

In comparing the effects of adenosine on heart rate and blood pressure in conscious Wistar-Kyoto (WKY) and spontaneously hypotensive (SHR) rats, Ohnishi et al. (1986) found that although the purine was an effective blood pressure lowering agent in both strains of rat, there was a greater hypotensive response to adenosine infusion in the SHR than in the WKY. These differences were attributed to a greater contribution of adenosine-autonomic interactions to the net depressor response in the SHR, since ganglionic blockade attenuated the effects of adenosine in lowering blood pressure in the SHR, but not the WKY rat. Based on differences in the hypotensive potency of adenosine when the purine is administered iv or intraaortically, it was also speculated that pulmonary extraction of the nucleoside could be as great as 50% in the SHR, with essentially zero lung deposition in the WKY. Strain differences in the effect of adenosine on heart rate were also observed. In SHR, intraaortic arch and suprarenal aortic adenosine infusion had no effect on heart rate, whereas iv infusion caused a marked decrease in heart rate. In WKY, intraaortic arch and suprarenal aortic infusion caused slight increases in heart rate. Ganglionic blockade attenuated the adenosine-mediated effects on heart rate in the WKY, but had no effect in the SHR.

The importance of this interstrain study relates to the complexity of the parameters measured and the different routes of administration used, in addition to the use of infusion as opposed to bolus injections of adenosine. The ability to compare the relative importance of the sympathetic, direct, and postulated pulmonary contributions to the overall actions of adenosine in the SHR, and to contrast this to the WKY, also emphasizes the need to consider the cardiovascular actions of adenosine as part of a dynamic spectrum in which individual direct and indirect actions of the purine may subsume relatively different degrees of importance that are dependent on the intrinsic conditions of the tissue under study. As Ohnishi et al. (1986) point out "both the state of anesthesia and the presence or absence of arterial hypertension might be expected to modify cardiovascular responses to adenosine."

7.2. Antihistaminic

Histamine is present in high concentrations in mammalin heart, being released both in the course of systemic acute allergic reactions and in ischemia (Levi, 1988). The effects of histamine on cardiac contractility are also susceptible to modulation by adenosine. Activation of H_2 receptors produces a positive inotropic effect; H_1-receptor activation has negative inotropic actions (Hattori and Levi, 1984). Adenosine is able to block the H_2-mediated effects of histamine via an A_1-type receptor (Genovese et al., 1988), probably at the level of second messenger generation (Hollingsworth et al., 1986), thus producing a histamine-induced negative inotropic effect. The physiological role of the "yin–yang" effects of histamine on cardiovascular function and the role of H_2-receptor activation are unclear at this time, however, it has been suggested (Genovese et al., 1988) that the attenuation of the H_2-mediated responses may decrease myocardial oxygen demand. However, the "uncovering" of the H_1-mediated responses (Hattori and Levi, 1984) may result in an imbalance that results in coronary spasm, arrhythmia, and contractile failure (Genovese et al., 1988). A report describing severe contractile failure in anaphylatic patients without preexisting heart disease (Raper and Fisher, 1988) would support this hypothesis.

7.3. Reflex Responses

Systemic baroreflex control of heart rate plays an important role in maintaining arterial pressure in response to changes in cardiac output and peripheral arterial resistance. The reflex tachycardia associated with a fall in blood pressure in the conscious dog is blunted by iv administration of adenosine (Hintze et al., 1985), the purine thus acting to decrease baroreceptor sensitivity.

Chemoreceptor and stretch receptor reflexes are also important in controlling blood pressure in response to altered cardiac and resistance vessel function. Carotid chemoreceptor activity can be increased by adenosine (Watt and Routledge, 1985; Dixon et al., 1986; Monteiro and Ribeiro, 1987), leading to increased respiration. Intracarotid injection of adenosine and its stable analogs caused dose-dependent increases in tidal volume, respiratory frequency, and minute volume that were dependent on an intact carotid sinus nerve and appeared to be A_2-receptor mediated (Monterio and Ribeiro, 1987), but the administration of the same analogs by the iv route caused an inhibition of respiration (Wessberg et al., 1985; Monteiro and Ribeiro, 1987) that was not affected by carotid sinus-nerve section. In addition, in contrast to the rapid onset and short duration of the adenosine-mediated excitatory response, the inhibitory response was more long-lasting (minutes vs seconds). This latter effect appears to be centrally mediated (Wessberg et al., 1985; Eldridge and Millhorn, 1987), although the former is more direct. Chemoreceptors in the renal pelvis may also be stimulated by adenosine, leading to an increase in sympathetic tone (Katholi et al., 1983). The involvement of adenosine as a mediator of stretch-receptor responses is uncertain (Witzleb, 1983). Reflexes involving these receptors have, however, been invoked to described the global responses to altered hemodynamic parameters in response to adenosine administration (Ohnishi et al., 1986).

Many of the physiological studies of adenosine effects on heart function have tended to ignore the reflex actions that changes in blood pressure can evoke. Thus, decreases in cardiac output and increases in coronary flow tend to alter renal function as well as signal central regulatory mechanisms via baroreceptor and chemoreceptor systems to override the compound-related effects. Many studies related to the understanding of the role of adenosine in regulating hemodynamic parameters use vagally sectioned or anesthized animals in which normal physiological responses may be compromised (Vatner and Braunwald, 1975; Biaggoni et al., 1987).

8. Adenosine and Cardiovascular Pathophysiology

8.1. Hypertension

Although the renal actions of adenosine on the sympathetic nervous system would appear to indicate that the purine is a hypertensive mediator, the purine is well-documented as a hypotensive agent. Adenosine analogs, whether A_1- or A_2-selective in nature, lower blood pressure and systemic vascular resistance. This is attributable to the inhibition of renin release and presynaptic modulation of sympathetic nerve activity via A_1-receptor activation. The contribution of adenosine-related systems to the pathophysiology of hypertension remains relatively unexplored, although altered cardiovascular responses to adenosine (Ohnishi et al., 1986) and adenosine-sensitive adenylate cyclase responses have been noted in cardiac tissue from the SHR (Anand-Srivastava, 1988). The hypotensive and bradycardiac effects occurring following injection of adenosine into the nucleus tractus solitarius (NTS) and area postrema are attenuated in the SHR as compared to WKY controls (Robertson et al., 1988). The ability of adenosine to attenuate sympathetic activity is decreased in SHRs (Kamikawa et al., 1983; Jackson, 1987; Kuan and Jackson, 1988).

Endogenous adenosine may not represent a physiological regulator of sympathetic function, because nerve stimulation does not increase adenosine release from the *in situ* blood-perfused mesenteric vascular preparation of SHR, and xanthine blockade of adenosine receptors does not alter neurotransmission (Jackson, 1987). In contrast, inhibition of neuroeffector function may represent an important component of the antihypertensive actions of A_1-selective agonists. However, since such agents provoke bradycardia and renal vasoconstriction, their usefulness as antihypertensive agents appears remote. Katholi et al. (1988) have proposed that endogenous intrarenally released adenosine controls renal blood flow and blood pressure in renal hypertensive rats, where a blockade of adenosine effects with caffeine or theophylline can increase plasma renin activity up to sevenfold while increasing blood pressure. These observations are difficult to reconcile with previous observations (Katholi et al., 1983) of endogenous intrarenal adenosine as a contributing factor to the development of hypertension in the one-kidney, one-clip rat model.

SHRs are more sensitive to the hypotensive effects of adenosine than WKY controls (Ohnishi et al., 1986). Adenosine or NECA administration iv lowers blood pressure and heart rate, whereas the hypotension induced by more A_2-selective agonists, such as carbocyclic adenosine, CV 1808 or CGS 21680 (Kawazoe et al., 1980; Dunham and Vince, 1986; Hutchison et al., 1989; Webb et al., 1989) is associated with a transient increase in heart rate that is probably baroreflexly-mediated.

8.2. Ischemia/Reperfusion Injury

Acute myocardial ischemia resulting from a reduction in coronary blood flow produces a marked increase in coronary vessel adenosine levels as a result of ATP breakdown (Schrader et al., 1982). The purine can thus act as a vasodilator to increase oxygen availability to cardiac tissue as well as reducing the adrenergic tone. Thus adenosine functions as an antiarrhythmic agent (Fagbemi and Parratt, 1984).

Reperfusion of the ischemic myocardium produces morphological and functional derangements collectively termed "reperfusion injury." Tissue damage occurs via oxygen-derived free radical generation and formation of various modulators of the inflammatory response (Lucchesi and Mullane, 1986; Halliwell, 1987; Simpson et al., 1987; Mullane, 1989). Granulocyte accumulation during the ischemic episode can lead to the obstruction of the microvasculature, and during reperfusion, these blood elements remain responsive to inflammatory mediators that in turn can prevent tissue salvage and lead to reperfusion injury (Mullane, 1989). Antioxidant therapy involving the use of iron chelators such as desferrioxamine, free radical scavengers such as the enzyme superoxide dismutase (SOD), dimethylurea, iloprost, prostacyclin, or various lipoxygenase inhibitors are possible therapeutic approaches to the treatment of this condition.

More recently, however, attention has focused on the potential use of adenosine in the treatment of reperfusion injury. Local intracoronary infusions of adenosine at the time of reperfusion can reduce myocardial necrosis, endothelial damage, and granulocyte accumulation (Olafsson et al., 1987) and improve post-ischemic regional myocardial dysfunction (Wyatt et al., 1987). Whether amelioration of reperfusion injury represents a metabolic effect of adenosine to replenish ATP or can be attributed to receptor-mediated actions requires studies with adenosine analogs that are not incorporated into the nucleotide. At the molecular level, the purine can inhibit granulocyte activation, adhesion, and free radical-mediated endothelial damage (Cronstein et al., 1983; Cronstein et al., 1987; Cronstein and Hirschhorn, 1990) via an A_2 receptor-mediated process (Cronstein et al., 1990). The amount of adenosine locally available appears to be critical, since an amount sufficient to increase flow can actually potentiate reactive hyperemia (Saito et al., 1985).

Although discrete infusions of adenosine or its analogs locally to the ischemic myocardium are not feasible from a therapeutic standpoint, agents that potentiate the effects of adenosine may have potential use. In addition to the classical uptake blockers (O'Regan et al., 1988), a provocative alternative to circumvent the problems associated with systemic administration of a potent coronary vasodilator is the use of the purine biosynthetic intermediate, AICA-riboside (AICAR; 5-amino-4-imidazole carboxamide-ribose; Engler, 1987). This agent is taken up by myocardial cells where it undergoes phosphorylation and enters the *de novo* synthetic pathway to accelerate ATP synthesis. Because ATP depletion may be an important determinant of myocardial injury, its repletion could aid the recovery of the postischemic heart.

The effects of AICAR appear, however, to be controversial. Some investigators (Mitsos et al., 1985; Swain et al., 1982) have reported that the compound enhances ATP repletion and improves cardiac function; others (Hoffmeister et al., 1988; Mentzer et al., 1988) found no improvement in ATP levels while function deteriorated. The accumulation of AICAR phosphates inhibits the enzyme adenylosuccinate lyase (Sabina et al., 1982) that converts IMP to the AMP precursor adenylosuccinate, thereby effectively preventing ATP synthesis. Alternatively, according to the "adenosine potentiator" role (Engler, 1987), AICAR increases the release of adenosine and IMP, thus increasing purine levels in the microenvironment of the ischemic, neutrophil-dependent insult. This novel concept appears validated by the finding that when adenine nucleotide metabolism is normal, AICAR has no effect. Based on data from dogs subjected to a 1-h ischemic episode in which AICAR treatment resulted in an increase in endogenous adenosine levels, increased coronary blood flow, and decreased the incidence of arrhythmias, Engler (1987) has proposed that adenosine can thus act as an endogenous antiinflammatory autacoid by limiting the deleterious actions of neutrophils and thus preventing cardiac cell death.

The AICAR approach is an attractive and novel one from the point of view that the "adenosine potentiator" can function as a site- and event-specific prophylatic agent; however, further work is required to substantiate the usefulness of such a compound, especially when its mechanism of action, efficacy, and side-effect profile have yet to be determined.

The decrease in cerebral blood flow occurring as the result of stroke can lead to cerebral ischemia and attendant neuronal cell loss and brain function. Although a major area of clinical concern, the mechanisms underlying the eventual cell death have only recently become reasonably well-defined and consequently potential therapeutic targets. Decreased blood flow to the brain can lead to a decrease in energy-rich metabolites and a large increase in adenosine levels, a sequence of events similar to that observed in cardiac tissue. Accompanying these changes is an increase in the release of the excitatory amino acid, L-glutamate. Acting at NMDA-type receptors, L-glutamate can cause extensive depolarization. This in turn causes a massive influx of calcium, the deposition of which is the trigger for cell death (Meldrum,1985). Considerable effort has been focused on the use of certain calcium entry blockers and NMDA receptor complex modulators and antagonists as anti-ischemic agents (Onodera and Kogure, 1985; Jarvis et al., 1988). Several adenosine analogs have also been found to prevent the neuronal cell loss normally observed following ischemia in pre-clinical stroke models (Von Lubitz et al., 1988). The potential use of adenosine in this condition is discussed in more detail elsewhere (Jarvis and Williams, 1990).

8.3. Platelet Function

Adenosine is a potent inhibitor of ADP and thrombin induced platelet aggregation (Born, 1964; Born and Cross, 1963; Kien et al., 1971). This effect is cyclic AMP mediated (Haslam et al., 1978) via an A_2-type receptor (Cusack and Hourani, 1982; Huttemann et al., 1984). Conversely, ADP, via a distinct receptor, inhibits cyclic AMP and the actions of adenosine (Cooper and Rodbell, 1979; Mills et al., 1983) while inducing aggregation. In addition to hemostasis, platelets function as "all-purpose inflammatory cells" (Weksler, 1988). This role can be compared to that of the neutrophils and as such, attenuation of platelet actions by adenosine agonists or adenosine "potentiators" may be beneficial under certain circumstances in preventing excessive tissue damage resulting from the production of inflammatory mediators. Platelet aggregation may also play a role in the etiology of atherosclerosis (Spaet et al., 1974) either via thrombogenesis or in response to endothelial cell damage and consequent plaque formation (Ross and Glomset, 1973). Adenosine can prevent endothelial cell damage evoked by neurotrophil-derived toxic oxygen metabolites (Cronstein et al., 1987), suggesting a synergistic role in attenuating the platelet aggregation occurring as a result of endothelial cell damage as well as in reducing the number of oxygen radicals derived from such processes. Adenosine also promotes neutrophil chemotaxis (Rose et al., 1988).

Endothelium is a rich source of adenosine (Gerlach et al., 1987) as a result of both the bidirectional flux of the nucleoside and its formation from ATP via endothelial ectonucleotidases. In intact endothelium, there is thus an intrinsic mechanism by which to prevent platelet aggregation and the associated generation of an inflammatory response. This is theoretically compromised in damaged endothelium, where both adenosine transport and ectonucleotidase activity would be suboptimal.

9. Adenosine in Humans

Adenosine administered intravenously to normal subjects has little effect on mean blood pressure, but produces an increase in heart rate and some marked side effects (Watt et al., 1986). This positive chronotropic response is very sensitive to adenosine, whether administered as a bolus (Watt et al., 1987), by infusion (Biaggioni et al., 1986, 1987; Conradson et al., 1987; Clarke et al., 1988), or by injection (Coupe et al., 1988), and is sometimes preceeded by a transient bradycardia (Di Marco et al., 1985). The increase in heart rate is attributable, at least in part, to inhibition of cardiac vagal tone, because it can be blocked by propranolol and atropine, but not by propranolol alone. Adenosine infusions increase plasma norepinephrine and epinephrine levels, which could contribute to the chronotropic activity. In contrast to the tachycardia, stimulation of respiration, and change in arterial pressure observed in conscious subjects, anesthetized patients respond with hypotension and no tachycardia or respiratory changes. (Biaggioni et al., 1987). Stimulation of reflex autonomic responses via activation of carotid body chemoreceptors is thought to account for the responses in conscious subjects, which are absent in patients with severe autonomic failure or under anesthesia (Biaggioni et al., 1987). Adenosine-mediated increases in cerebral blood flow can be attenuated by hyperventilation in anesthetized humans (Sollevi et al., 1987). The uptake inhibitor dipyridamole can enhance the cardiovascular effects of adenosine (Biaggioni et al., 1986; Conradson et al., 1987; German et al., 1989).

10. Adenosine Agonists as Novel Cardiovascular Drug Entities

The potential role of adenosine in cardiovascular function has, through the 60 years since the effects reported of the purine were reported by Drury and Szent-Györgyi (1929), provided a rich arena for experimentation that, as noted by the repetitive use of the adjective "controversial" throughout this brief review, has resulted in little concensus. This in turn has fostered a somewhat ambivalent attitude as to the therapeutic potential of purine agonists and antagonists, especially in the cardiovascular system. From the viewpoint of an outsider who is familiar with the voluminous literature and many monographs on the subject of adenosine in the past five years, it would be reasonable to assume that the purine nucleoside is an interesting research tool that has been evaluated as a hyoptensive agent and found to be seriously wanting. Yet in reality, whereas the study of adenosine as a physiological modulator in discrete organ systems has resulted in some of the more elegant work in the area of physiology in the past two decades, a cohesive understanding of the pharmacological properties of the purine in a pathophysiological context has yet to materalize.

As this overview has attempted to show, adenosine has potent effects on cardiovascular function, both directly via receptor activation of A_1 and A_2 receptors in the heart and by receptor interactions in the peripheral vasculature, kidney, and CNS, and indirectly via its antiamine effects (norepinephrine and histamine) and its effects on renin production (Fig. 1). The complexity of these responses and their interrelationships within a hierarchical context have, however, yet to be determined. This complexity is underlined by studies with the A_2-selective agonist CGS 21680 (Hutchison et al., 1989). This compound originated from a medicinal chemistry program using in vitro radioligand binding to find more potent and selective A_2-receptor agonists. CGS 21680 had high affinity (IC_{50} value = 22 nM) and selectivity (140-fold) for the rat A_2 receptor, and this activity translated into a similar delineation in A_1- and A_2-receptor mediated effects on heart rate and blood flow in the isolated perfused working preparation. However, when this selective A_2 agonist was evaluated in the conscious SHR, its effects on blood pressure in terms of degree and duration of response were not significantly different from that observed with CV 1808, an adenosine agonist that was some fivefold less active at A_2 receptors than CGS 21680 and had only fivefold selectivity. In this particular example, the synthesis of an adenosine agonist with a theoretically superior profile based on its in vitro activity resulted in an entity that could not be delineated from a less selective and less active reference compound in vivo. The reason for this observation is unclear, although it would appear that compounds with weak A_1 activity are unable to directly affect cardiac contractility, as would be observed in the case of compounds such as CHA. The complexity of the various adenosine mediated responses in will no doubt explain the experimental data obtained with CGS 21680 via homeostatic nuances that have yet to be understood. In addition, the acute in vivo

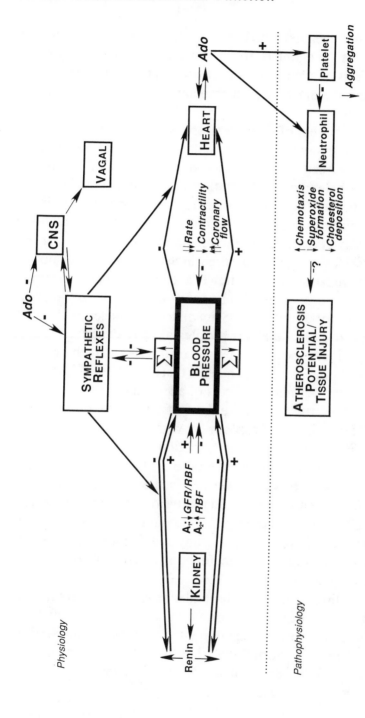

Fig. 1. Schematic diagram of potential effects of adenosine on hemodynamic parameters. For further clarification, *see* accompanying text. The effects of adenosine on blood pressure are complex, occurring via both direct and indirect mechanisms. Ado = Adenosine. Vertical arrows indicate increases or decreases in the parameters indicated.

model has not been extended to an evaluation of the ability of the newer adenosine agonists to induce angina. Furthermore, the chronic effects of adenosine administration that would more accurately reflect the anticipated therapeutic situation have yet to be studied.

With the advent of receptor-selective adenosine agonists and antagonists, it is now possible to enter a new experimental era in studying the effects of adenosine on cardiovascular function, in which the global effects of the nucleoside are studied under various conditions permitting a better clarification of the relative importance of the various organ systems that are susceptible to adenosine modulation and contribute to the regulation of blood pressure and blood flow. However, this basic research component need not preclude the testing of some of the newer adenosine agonists in humans once they have been shown to be safe. In addition to the angina issue, the contribution of CNS-mediated responses to both the cardiovascular and side effect profiles already described will require evaluation, especially in light of the fact that there are no generally accepted preclinical models that enable delineation of the central from the peripheral effects of adenosine.

In addition, the potential effects of adenosine on neutrophil and platelet function need assessment within the context of the nucleoside's blood pressure lowering actions. If adenosine hypotensive agents have no additional effects on cardiovascular parameters than that of blood pressure lowering, such entities will have great difficulty in being accepted as alternative or replacement therapies for the angiotensin-converting enzyme inhibitors. Two attractive areas that have received minimal study in a therapeutic context are adenosine modulation of platelet function and the role of the purine in the etiology of atherosclerosis.

There is thus a considerable challenge for researchers studying the effects of adenosine in cardiovascular function. Developing knowledge of the role of purine-related systems in the SHR and failure to take advantage of the availability of the newer pharmacological tools will ensure that adenosine remains an interesting artifact for another 60 years.

References

Anand-Srivastava, M. B. (1985) Regulation of adenylate cyclase by adenosine and other agonists in rat myocardial sarcolemma. *Arch. Biochem. Biophys.* **243**, 439–446.

Anand-Srivastava, M. B. (1988) Altered responsiveness of adenylate cyclase to adenosine and other agents in the myocardial sarcolemma and aorta of spontaneously-hypertensive rats. *Biochem. Pharmacol.* **37**, 3017–3022.

Anaversa, P., Leuwicky, T., Beghi, C., McDonald, S. L. and Kikkassa, Y. (1983) Morphometry of excercise–induced right ventricular hypertrophy in rat. *Circ. Res.* **52**, 57–64.

Arend, L. J., Sonnenburg, W . K., Smith, W. L. and Spielmann, W. S. (1987) A_1 and A_2 adenosine receptors in rabbit cortical collecting tubule cells. *J. Clin. Invest.* **79**, 710–714.

Barraco, R. A., Marcantonio, D. R., Phillis, J. W. and Campbell, W. R. (1987) The effects of parenteral injection of adenosine and its analogues on blood pressure and heart rate. *Gen. Pharmacol.* **18**, 405–416.

Barraco, R. A. (1988) Adenosinergic modulation of brainstem mechanisms involved in cardiovsacular control, in *Adenosine and Adenine Nucleotides. Physiology and Pharmacology* (Paton, D. M. ed.), Taylor and Francis, London, pp. 233–240.

Barraco, R. A., Aggarwal, A. K., Phillis, J. W., Moran, M. A., and Wu, P. H. (1984) Dissociation of the locomotor and hypertensive effects of adenosine analogs in rat. *Neurosci. Lett.* **48**, 139–143.

Barraco, R. A. Phillis, J. W., Campbell, W. R., Marcantonio, D. R., and Salah, R. S. (1986) The effects of central injection of adenosine analogues on blood pressure and heart rate in rat. *Neuropharmacology* **25**, 675–680.

Bardenheuer, H. and Schrader, J. (1983) Relationship between myocardial oxygen consumption, coronary flow, and adenosine release in an improved isolated working heart preparation of guinea pigs. *Circ. Res.* **51**, 263–271.

Bardenheuer, H., Whelton, B. K., and Sparks, H. V. (1987) Cellular compartmentation of adensine in the heart, in *Topics and Perspectives in Adenosine Research* (Gerlach, E. and Becker, B. F. eds.), Springer-Verlag, Berlin, pp. 480–485.

Beck, D.W., Vinters, H. V., Moore, S. A., Hart, M. N., Henn, F. A., and Cancilla, P. A. (1984) Demonstration of adenosine receptors on mouse cerebral smooth muscle cell membranes. *Stroke* **15**, 725–727.

Bellardinelli, L. and Isenberg, G. (1983) Isolated atrial myocytes: Adenosine and acetylcholine increase potassium conductance. *Am. J. Physiol.* **244**, H734–H737.

Bellardinelli, L., Mattos, E. C., and Berne, R. M. (1981) Evidence for adenosine mediation of atrioventricular block in the ischemic canine myocardium. *J. Clin. Invest.* **68**, 195–205.

Bellardinelli, L., West, A., Crampton, R. and Berne, R. M. (1983) Chronotropic and dromotropic actions of adenosine, in *Regulatory Function of Adenosine* (Berne, R. M., Rall, T. W. and Rubio, R. eds.) Nihoff, Boston, pp. 378–398.

Bellhassen, B. and Pelleg, A. (1984) Electrophysiologic effects of adenosine triphosphate and adenosine in the mammalian heart: Clinical and experimental aspects. *J. Amer. Coll. Cardiol.* **4**, 414–420.

Berne, R. M. (1980) The role of adenosine in the regulation of coronary blood flow. *Circ. Res.* **47**, 807–813.

Berne, R. M. (1985) Criteria for the involvement of adenosine in the regulation of blood flow, in *Methods in Pharmacology vol. 6 Methods Used in Adenosine Research* (Paton, D. M. ed.), Plenum, New York, pp. 331–336.

Berne, R. M., Rubio, R., and Curnish, R. R. (1974) Release of adenosine from ischemic brain: Effect on cerebral vascular resistance and incorporation into cerebral adenine nucleotides. *Circ. Res.* **25**, 262–271.

Berne, R. M., Winn, H. R., Knabb, R. M., Ely, S. W., and Rubio, R. (1983) Blood flow regulation in heart brain and skeletal muscle, in *Regulatory Function of Adenosine* (Berne, R. M., Rall, T. W., and Rubio, R. eds.), Nihoff, Boston, pp. 293–317.

Berne, R. M., Gidday, J. M., Hill, H. E., Curnish, R. R., and Rubio, R. (1987) Adenosine in the local regulation of blood flow; Current controversies, in *Topics and Perspectives in Adenosine Research* (Gerlach, E. and Becker, B. F. eds.), Springer-Verlag, Berlin, pp. 299–308.

Biaggioni, I., Onrot, J., Hollister, A. S., and Robertson, D. (1986) Cardiovascular effects of adenosine infusion in man and their modulation by dipyridamole. *Life Sci.* **39**, 2229–2236.

Biaggioni, I., Olafsson, B., Robertson, R. M., Hollister, A. S., and Robertson, D. (1987) Cardiovascular and respiratory effects of adenosine in conscious man. Evidence for chemoreceptor activation. *Circ. Res.* **61**, 779–786.

Blass, K. E., Fotster, W., and Zehl, U. (1980) Coronary vasodilation: Interactions between prostacyclins and adenosine. *Br. J. Pharmacol.* **69**, 555–559.

Bockman, E. L., Berne, R. M., and Rubio, R. (1975) Release of adenosine and lack of release of ATP from contracting skeletal muscle. *Pflugers Arch.* **335**, 229–241.

Bohm, M., Bruckner, R., Meyer, W., Nose, M., Schmitz, W., Schlotz, H., and Starbatty, J. (1985) Evidence for adenosine receptor-mediated isoprenaline-antagonistic effects of the adenosine analogs PIA and NECA on force of contraction in guinea pig atrial and ventricular cardiac preparations. *Naunyn-Schmidebergs Arch. Pharmacol.* **331**, 131–139.

Bohm, M., Bruckner, R., Neumann, J., Schmitz, W., Schlotz, H., and Starbatty, J. (1986) Role of guanine-nucleotide-binding protein in the regulation by adenosine of cardiac potassium conductance and force of contraction. Evaluation with pertussis toxin. *Naunyn-Schmidebergs Arch. Pharmacol.* **332**, 403–405.

Bohm, M., Bruckner, R., Hackbarth, I., Haubitz, B., Linhart, R., Meyer, W., Schmidt, B., Schmitz, W., and Scholz, H. (1984) Adenosine inhibition of catecholamine-induced increase in force of contraction in guinea pig atrial and ventricular heart preparations. Evidence against a cyclic AMP and cyclic GMP-dependent effect. *J. Pharmacol. Exp. Ther.* **230**, 483–492.

Born, G. V. R. (1964) Strong inhibition by 2-chloroadenosine of the aggregation of blood platelets by adenosine diphosphate. *Nature* **202**, 95, 96.

Born, G. V. R. and Cross, M. J. (1963) The aggregation of blood platelets. *J. Physiol. (Lond.)* **168**, 178–195.

Bridges, A. J., Bruns, R. F., Ortwine, D. F., Priebe, S. R., Szotek, D. L., and Trivedi, B. K. (1988) N[6][2-(3,5-dimethoxyphenyl)-2-(2-methoxypheny) adenosine and its uronomide derivatives. Novel adenosine agonists with both high affinity and high selectivity for the adenosine A$_2$ receptor. *J. Med. Chem.* **31**, 1282– 1285.

Bruckner, R., Fenner, A., Meyer, W., Nobis, T., Schmutz, W., and Scholz, H. (1985) Cardiac effects of adenosine and adenosine analogs in guinea-pig atrial and ventricular preparations: Evidence against a role of cyclic AMP and cyclic GMP. *J. Pharmacol. Exp. Ther.* **234**, 766–774.

Bruns, R. F., Fergus, J. H., Badger, E. W., Bristol, J. A., Santray, L. A., Hartman, J. D., Hays, S. J., and Huang, C. C. (1987) Binding of the A$_1$-selective adenosine antagonist 8-cyclopentyl 1,3, dipropylxanthine to rat brain membranes. *Naunyn-Schmiedebergs Arch. Pharmacol.* **335**, 59–63.

Busija, D. W. and Heistad, D. D. (1984) Factors involved in the physiological regulation of the cerebral circulation. *Rev. Physiol. Biochem. Pharmacol.* **101**, 161–211.

Chiba, S. and Hashimoto, K. (1972) Differences in chronotropic and dromotropic responses of the SA and AV nodes to adenosine and acetycholine. *Jpn. J. Pharmacol.* **22**, 273, 274.

Churchill, P.C. and Bidani, A. (1990) Adenosine and renal function, in *Adenosine and Adenosine Receptors* (Williams, M., ed.), Humana, Clifton, New Jersey, in press.

Clarke, B., Conradson, T-B., Dixon, C. M. S., and Barnes, P. J. (1988) Reproducibility of heart rate changes following adenosine infusion in man. *Eur. J. Clin. Pharmacol.* **5**, 309–311.

Clemo, H. F. and Bellardinelli, L. (1986) Effect of adenosine on atrioventricular conduction. I.: Site and characterization of adenosine action in the guinea pig atrioventricular node. *Circ. Res.* **59**, 427–436.

Collis, M. G. (1983) Evidence for an A$_1$-adenosine receptor in the guinea pig atrium. *Br. J. Pharmacol.* **78**, 207–212.

Collis, M. G. (1988) Cardiac and renal actions of adenosine antagonists in vivo, in *Adenosine and Adenine Nucleotides. Physiology and Pharmacology* (Paton, D. M., ed.), Taylor and Francis, London, pp. 259–268.

Collis, M. J. and Brown, C. M. (1983) Adenosine relaxes the aorta by interacting with an A$_2$ receptor and an intracellular site. *Eur. J. Pharmacol.* **96**, 61–69.

Collis, M. G., Keddie, J. R., and Pettinger, S. J. (1983) 2-Chloroadenosine lowers blood pressure in the conscious dog without reflex tachycardia. *Br. J. Pharmacol.* **79,** 385.

Collis, M. G., Keddie, J. R., and Torr, S. R. (1984) Evidence that positive inotropic effects of the alkylxanthines are not due to adenosine receptor blockade. *Br. J. Pharmacol.* **81,** 401–407.

Collis, M. G., Palmer, D. B., and Baxter, G. S. (1986) Evidence that the intracellular effects of adenosine in the guinea pig aorta are mediated by inosine. *Eur. J. Pharmacol.* **121,** 141–145.

Conradson, T-B. G., Dixon, C. M. S., Clarke, B., and Barnes, P. J. (1987) Cardiovascular effects of infused adenosine in man: Potentiation by dipyridamole. *Acta Physiol. Scand.* **129,** 387–391.

Cooper, D. M. F. and Caldwell, K. K. (1990) Signal transduction mechanisms for adenosine, in *Adenosine and Adenosine Receptors* (Williams, M. ed.), Humana, Clifton, New Jersey, in press.

Cooper, D. M. F. and Rodbell, M. (1979) ADP is a powerful inhibitor of human platelet plasma membrane adenylate cyclase. *Nature* **282,** 517, 518.

Coupe, M.O., Clarke, B., Robson, A., Oldershaw, P. J., and Barnes, P. J. (1988) The cardiovascular effects of nebulized adenosine in man. *Eur. J. Clin. Pharmacol.* **34,** 645–647.

Cronstein, B. N. and Hirschhorn, R. (1990) Adenosine and host defense: modulation through metabolism and receptor-mediated mechanisms, in *Adenosine and Adenosine Receptors* (Williams, M., ed.), Humana, Clifton, New Jersey, in press.

Cronstein, B. N., Duguma, L., Williams, M., and Hutchison, A. J. (1990) The adenosine/neutrophil paradox resolved: adenosine inhibits superoxide anion (O_2) generation and promotes chemotaxis by occupying separate receptors. *J. Clin. Inves.* (in press).

Cronstein, B. N., Kramer, J. B., Weissmann, G., and Hirschhorn, R. (1983) Adenosine: A physiological modulator of superoxide anion generation by human neutrophils. *J. Exp. Med.* **158,** 1160–1177.

Cronstein, B. N., Levin, R. I., Belanoff, J., Weissmann, G., and Hirschhorn, R. (1987) A new function for adenosine: Protection of vascular endothelial cells from neutrophil mediated injury, in *Topics and Perspectives in Adenosine Research* (Gerlach, E. and Becker, B. F. eds.), Springer-Verlag, Berlin, pp. 299–308.

Cusack, N. J. and Hourani, S. M. O. (1982) Adenosine 5'-diphosphate antagonists and human platelets: No evidence that aggregation and inhibition of stimulated adenylate cyclase are mediated by different receptors. *Br. J. Pharmacol.* **76,** 221–227.

De Gubareff, T. and Sleator, W. (1965) Effects of caffeine on mammalian atrial muscle and its interaction with adenosine and calcium. *J. Pharmacol. Exp. Ther.* **148,** 202–214.

Deussen, A., Moser, G., and Schrader, J. (1986) Contribution of coronary endothelial cells to cardiac adenosine production. *Pflugers Arch* **406,** 608–614.

Di Marco, J. P., Sellers, T. D., Lerman, B. B., Greenberg, M. L., Berne, R. M., and Bellardinelli, L. (1985) Diagnostic and therapeutic use of adenosine in patients with supraventricular tachyarrhythmias. *J. Am. Coll. Cardiol.* **6,** 417–415.

Dixon, C. M. S., Fuller, R. W., Hughes, J. M. B., Maxwell, D. L., and Nolop, K. B. (1986) Hypoxic and hypercapnic responses during adenosine infusion in man. *J. Physiol. (Lond.)* **376,** 51.

Dobson, J. G. (1983) Mechanism of adenosine inhibition of catecholamine-induced responses in heart. *Circ. Res.* **52,** 151–160.

Dobson, J. G. and Schrader, J. (1984) Role of extracellular and intracellualr adenosine in the attenuation of catecholamine-evoked responses in guinea pig heart. *J. Mol. Cell Cardiol.* **16,** 813–822.

Dobson, J. G., Fenton, R. A., and Romano, F. D. (1987) The antiadrenergic actions of adenosine in the heart, in *Topics and Perspectives in Adenosine Research* (Gerlach, E. and Becker, B. F. eds.), Springer-Verlag, Berlin, pp. 356–368.

Dolphin, A. C. and Archer, E. A. (1983) An adenosine agonist inhibits and a cyclic AMP analogue enhances the release of glutamate but not GABA from slices of rat dentate gyrus. *Neurosci. Lett.* **43,** 49–54.

Drury, A. N. and Szent-Györgyi, A. (1929) The physiological action of adenine compounds with especial reference to their action on the mammalian heart. *J. Physiol. (Lond.)* **68,** 214–237.

Dunham, E. W. and Vince, R. (1986) Hypotensive and renal vasodilator effects of carbocyclic adenosine (aristeromycin) in anesthetized spontaneously hypertensive rats. *J. Pharmacol. Exp. Ther.* **238,** 954–959.

Dusseau, J. W., Hutchins, P. M., and Malbasa, D. (1986) Stimulation of angiogenesis by adenosine on the chick chorioallantoic membranes. *Circ. Res.* **59,** 163–170.

Dutta, P. and Mustafa, S. J. (1979) Saturable binding of adenosine to the dog heart microsomal fraction: Competitive inhibition by aminophylline. *J. Pharmacol. Exp. Ther.* **211,** 496–501.

Edlund, A., Berglund, B., van Dorne, D. Kaijser, L., Nowak, J. Patroni, C., Sollevi, A., and Wennmalm, A. (1985) Coronary flow regulation in ischemic heart disease: Release of purines and prostacyclin and the effect of inhibitors of prostaglandin formation. *Circulation* **71,** 1113–1121.

Edvinsson, L. and Fredholm, B. B. (1983) Characterization of adenosine receptors in isolated cerebral arteries of cat. *Br. J. Pharmacol.* **80,** 631–637.

Edvinsson, L. and Jansen, I. (1985) Demonstration of adenosine receptors in isolated cerebral arteries, in *Adenosine: Receptors and Modulation of Cell Function* (Stefanovich, V., Rudolphi, K., and Schubert, P., eds.) IRL Press, Oxford, UK, pp. 409–417.

Eldridge, F. L., Millhorn, D. E., and Kiley, J. P. (1984) Respiratory effects of a long acting analog of adenosine. *Brain Res.* **301,** 273–280.

Eldridge, F. L. and Millhorn, D. E. (1987) Role of adenosine in the regulation of breathing, in *Topics and Perspectives in Adenosine Research* (Gerlach, E. and Becker, B. F. eds.), Springer-Verlag, Berlin, pp. 586–596.

Endoh, M., Maruyama, M., and Taira, N. (1983) Adenosine-induced changes in rate of beating and cyclic nucleotide levels in rat atria: modification by islet-activating protein, in *Physiology and Pharmacology of Adenine Derivatives* (Daly, J. W., Kuroda, Y., Phillis, J. W., Shimizu, H., Ui, M., eds.), Raven, New York, pp. 127–141.

Engler, R. (1987) Consequences of activation and adenosine-mediated inhibition of granulocytes during myocardial ischemia. *Fed. Proc.* **46**, 2407–2412.

Evans, D., Schenden, J., and Bristol, J. A. (1982) Adenosine receptors mediating cardiac depression. *Life Sci.* **31**, 2425–2432.

Fagbemi, O. and Parratt, J. R. (1984) Antiarrhythmic actions of adenosine in the early stages of experimental myocardial ischemia. *Eur. J. Pharmacol.* **100**, 243, 244.

Fassina, G., Gaion, R. M., Caparrotta, L., and Carpenedo, F. (1985) A caffeine analog (1,3,7) trimethyl-6-thioxo-2-oxopurine) with negative chronotropic and inotropic activity. *Naunyn-Schmiedebergs Arch. Pharmacol.* **330**, 222–226.

Forrester, T., Harper, A. M., McKenzie, E. T., and Thomsen, E. M. (1979) Effect of adenosine triphosphate and some derivatives on cerebral blood flow and metabolism. *J. Physiol. (Lond.)* **296**, 343–355.

Fredholm, B. and Dunwiddie, T. V. (1988) How does adenosine inhibit neurotransmitter release? *Trends Pharmacol. Sci.* **9**, 130–134.

Fredholm, B. B., Jacobson, K. A., Jonzon, B., Kirk, K. L., Li, Y. O. and Daly, J. W. (1987) Evidence that a novel 8-phenyl-substituted xanthine derivative is a cardioselective adenosine antagonist in vivo. *J. Cardiovasc. Pharmacol.* **9**, 396–400.

Freissmuth, M., Hausleithner, V., Tuisl, E., Nanoff, C., and Schutz, W. (1987) Glomeruli and microvessels of the rabbit kidney contain both A_1 and A_2 adenosine receptors. *Naunyn-Schmiedebergs Arch. Pharmacol.* **335**, 438–444.

Fugslang, A. and Crone, C. (1987) Adenosine-mediated presynaptic inhibition of sympathetic innervation as an explanation of functional hyperemia, in *Topics and Perspectives in Adenosine Research* (Gerlach, E. and Becker, B. F. eds.), Springer-Verlag, Berlin, pp. 533–536.

Furchgott, R. F. and Zawadzki, J. V. (1980) The obligatory role of endothelial cells in the relaxation of arterial smooth muscle by acetylcholine. *Nature* **288**, 373–376.

Garthwaite, J., Charles, S. L. and Chess-Williams, R. (1988) Endothelium-derived relaxing factor release on activation of NMDA receptors suggests role as intracellular messenger in the brain. *Nature* (Lond.) **336**, 385–388.

Genovese, A., Gross, S. S., Sakuma, I., and Levi, R. (1988) Adenosine promotes histamine H_1-mediated negative chronotropic and inotropic effects on human atrial myocardium. *J. Pharmacol. Exp. Ther.* **247**, 844–849.

Gerlach, E. and Becker, B. F. (1987) Joint discussion, in *Topics and Perspectives in Adenosine Research* (Gerlach, E. and Becker, B. F., eds.), Springer-Verlag, Berlin, pp 393, 394.

Gerlach, E., Nees, S., and Becker, B. F. (1985) The vascular endothelium: A survey of some newly evolving biochemical and physiological features. *Basic Res. Cardiol.* **80,** 459–474.

Gerlach, E., Becker, B. F., and Nees, S. (1987) Formation of adenosine by vascular endothelium: A homeostatic and antithrombogenic mechanism? in *Topics and Perspectives in Adenosine Research* (Gerlach, E. and Becker, B. F. eds.), Springer-Verlag, Berlin, pp. 309–320.

German, D. C., Kredich, N. M., and Bjornsson, T. D. (1989) Oral dipyridamole increases plasma adenosine levels in human beings. *Clin. Pharmacol. Ther.* **45,** 80–84.

Green, H. N. and Stoner, H. B. (1950) *Biological Actions of Adenine Nucleotides* (Lewis, London).

Granger, D. N., Valleau, J. D., Parker, R. E., Lane, R. S., and Taylor, A. E. (1978) Effects of adenosine on intestinal hemodynamics, oxygen delivery and capilliary fluid exchange. *Am. J. Physiol.* **235,** H707–H719.

Grome, J. J. and Stefanovich, V. (1985) Differential effects of xanthine derivatives on local cerebral blood flow and glucose utilization in the conscious rat, in *Adenosine: Receptors and Modulation of Cell Function* (Stefanovich, V., Rudolphi, K., and Schubert, P., eds.), IRL Press, Oxford, 453–460.

Grome, J. J. and Stefanovich, V. (1986) Differential effects of methylxanthines on local cerebral blood flow and glucose utilization in the conscious rat. *Naunyn-Schmiedebergs Arch. Pharmacol.* **333,** 172–179.

Hajjar, D. P. (1985) Prostaglandins and cyclic nucleotides: Modulation of arterial cholesterol metabolism. *Biochem. Pharmacol.* **34,** 295–300.

Haleen, S. J. and Evans, D. B. (1985) Selective effects of adenosine receptor agonists upon coronary resistance and heart rate in isolated working rabbit hearts. *Life Sci.* **36,** 127–137.

Halliwell, B. (1987) Oxidants and human diseases: some new concepts. *FASEB J.* **1,** 358–364.

Hamilton, H. W., Taylor, M. D., Steffen, R. P., Haleen S. J., and Bruns, R. F. (1987) Correlation of adenosine receptor affinities and cardiovascular activity. *Life Sci.* **41,** 2295–2302.

Hamprecht, B. and Van Calker, D. (1985) Nomenclature of adenosine receptors. *Trends Pharmacol. Sci.* **6,** 153, 154.

Hanley, F., Messina, L. M., Baer, R. W., Uhlig, P. N., and Hoffman, J. I. E. (1983) Direct measurement of left ventricular interstitial adenosine. *Am. J. Physiol.* **245,** H327–H335.

Haslam, R. J., Davidson, M. M. L., and Desjardins, T. V. (1978) Inhibition of adenylate cyclase by adenosine analogues in preparations of broken and intact human platelets. Evidence for the unidirectional control of platelet function by cyclic AMP. *Biochem. J.* **176,** 83–95.

Hattori, Y. and Levi, R. (1984) Adenosine selectively attenuates H_2- and β-mediated cardiac responses to histamine and norepinephrine: An unmasking of H_1- and α-mediated responses. *J. Pharmacol. Exp. Ther.* **231**, 215–223.

Heistad, D. D., Marcus, M. L., Gourley, D. K., and Busija, D. W. (1981) Effect of adenosine and dipyridamole on cerebral blood flow. *Physiology* **240**, 775–782.

Hintze, T. H., Belloni, F. L., Harrison, J. E., and Shapiro, G. C. (1985) Apparent reduction in baroreflex sensitivity to adenosine in conscious dogs. *Am. J. Physiol.* **249**, H554–H559.

Hoffmeister, H. M., Betz, R., Fiechtner, H., and Siepel, L. (1988) Myocardial and circulatory effects of inosine. *Cardiovasc. Res.* **21**, 65–71.

Hollingsworth, E. B., De La Cruz, R. A., and Daly, J. W. (1986) Accumulations of inositol phosphates and cyclic AMP in brain slices: Synergistic interactions of histamine and 2-chloroadenosine. *Eur. J. Pharmacol.* **122**, 45–50.

Honey, R. M., Ritchie, W. T., and Thomson, W. A. R. (1930) The action of adenosine upon the human heart. *Quart. J. Med.* **23**, 485–490.

Hutchison, A. J., Webb, R. L., Oei, H. H., Ghai, G. R., Zimmerman, M. B., and Williams, M. (1989) CGS 21680C, an A_2 selective adenosine agonist with selective hypotensive activity. *J. Pharmacol. Exp. Ther.* **251**, 47–55.

Hutteman, E., Ukena, D., Lenschow, V., and Schwabe, U. (1984) Ra adenosine receptors in human platelets. Characterization by 5'-Nethylcarboxamido-[^{3}H] adenosine binding in relation to adenylate cyclase activity. *Naunyn-Schmiedebergs Arch. Pharmacol.* **325**, 226–233.

Jackson, E. K. (1987) Role of adenosine in noradrenergic neurotransmission in spontaneously hypertensive rats. *Am. J. Physiol.* **253**, H909–H918.

James, T. N. (1965) The chronotropic action of ATP and related compounds studied by direct perfusion of the sinus node. *J. Pharmacol. Exp. Ther.* **149**, 233– 240.

Jarvis, M. F. and Williams, M. (1989) Adenosine in central nervous system function, in *The Adenosine Receptors* (M. Williams, ed.), Humana, Clifton, New Jersey, in press.

Jarvis, M. F., Murphy, D. E., Williams, M., Gerhardt, S. C., and Boast, C. A. (1988) The novel N-methyl-D-aspartate (NMDA) antagonist, CGS 19755, prevents ischemia-induced reductions of adenosine A-1, NMDA and PCP receptors in gerbil brain. *Synapse* **2**, 577–584.

Jezer, A., Oppenheimer, B. S., and Schwartz, S. P. (1933) The effect of adenosine on cardiac irregularities in man. *Am. Heart J.* **9**, 252–258.

Jonzon, B., Bergquist, A., Li, Y. O., and Fredholm, B. B. (1986) Effects of adenosine and stable adenosine analogues on blood pressure, heart rate and colonic temperature in the rat. *Acta Physiol. Scand.* **126**, 491–498.

Kamikawa, Y., Cline, Jr., W. H., and Su, C. (1983) Possible roles of purinergic modulation in pathogenesis of some diseases: Hypertension and asthma, in *Physiology and Pharmacology of Adenosine Derivatives* (Daly, J. W., Kuroda, Y., Phillis, J. W., Shimizu, H., and Ui, M., eds.), Raven, New York, pp. 189–196.

Kassell, N. F., Boarini, D. J., Olin, J. J., and Spowell, J. A. (1983) Cerebral and systemic circulatory effects of arterial hypotension induced by adenosine. *J. Neurosurg.* **58,** 69–76.

Katholi, R. E., Hageman, G. R., Whitlow, P. L., and Woods, W. T. (1983) Hemodynamic and afferent renal nerve responses to intrarenal adenosine in the dog. *Hypertension* **5,** I-149–I-154.

Katholi, R. E., Whitlow, F. L., Hageman, G. R., and Woods, W. T. (1984) Intrarenal adenosine produces hypertension by activating the sympathetic nervous system via the renal nerves in the dog. *J. Hypertension* **2,** 349–359.

Katholi, R. E., Creek, R. D., and McCann, W. P. (1988) Endogenous intrarenal adenosine preserves renal blood flow in one-kidney, one clip rats. *Hypertension* **11,** 650–656.

Kawazoe, K., Matsumoto, N., Tanabe, M., Fujiwara, S., Yanagimoto, M., Hirata, M., and Kikuchi, K. (1980) Coronary and cardiohemodynamic effects of 2-phenylamino-adenosine (CV 1808) in anesthetized dogs and cats. *Arzneimittelforsch.* **30,** 1083–1087.

Kien, M., Belamarich, F. A., and Shepro, D. (1971) Effect of adenosine and related compounds on thrombocyte and platelet aggregation. *Am. J. Physiol.* **220,** 604–608.

Klabunde, R. E. (1984) Dipyridamole inhibition of adenosine metabolism. *Eur. J. Pharmacol.* **93,** 21–26.

Klotz, K.-N. and Lohse, M. J. (1986) The glycoprotein nature of A_1 adenosine receptors. *Biochem. Biophys. Res. Commun.* **140,** 406–413.

Kroll, K. and Feigl, E. O. (1985) Adenosine is unimportant in controlling coronary blood flow in unstressed dog hearts. *Am. J. Physiol.* **249,** H1176–H1187.

Kuan, C. J. and Jackson, E. K. (1988) Role of adenosine in noradrenergic neurotransmission. *Am. J. Physiol.* **255,** H386–H393.

Kurachi, Y., Nakajima, T., and Sugimoto, T. (1986) On the mechanism of activation of muscarinic K^+ channels by adenosine in isolated atrial cells: Involvement of GTP-binding proteins. *Pflugers Arch.* **407,** 264–274.

Kuschinsky, W. (1983) Coupling between function, metabolism and blood flow in the brain: State of the art. *Microcirc.* **2,** 357–378.

Leppo, J. A., Simons, M., and Hood, Jr., W. B. (1984) Effect of adenosine on transmural flow gradients in normal canine myocardium. *J. Cardiovascul. Pharmacol.* **6,** 1115–1119.

Leung, E., Johnston, C. I., and Woodcock, E. A. (1983) Demonstration of adenylate cyclase coupled adenosine receptors in guinea pig venticular membranes. *Biochem. Biophys. Res. Commun.* **110,** 208–215.

Levi, R. (1988) Cardiac anaphylaxis: Models, medaitors, mechanisms, and clinical considerations, in *Human Inflammatory Disease. Clinical Immunology.* vol 1. (Marone, G., Lichtenstein, L. C., Condorelli, M., and Fauci, A. S., eds.), B. C. Decker, Philadephia, pp. 93–105.

Linden, J., Hollen, C. E., and Patel, A. (1985) The mechanism by which adenosine and cholinergic agents reduce contractility in rat myocardium. Correlation with cyclic adenosine monophosphate and receptor densities. *Circ. Res.* **56**, 728–735.

Linden, J., Patel, A., Earl, C. Q., Craig, R. H., and Daluge, S. M. (1988) ^{125}I-Labeled 8-phenylxanthine derivatives: Anatagonist radioligands for adenosine A₁ receptors. *J. Med. Chem.* **31**, 745–751.

Lloyd, H. G. E. and Schrader, J. (1987) The importance of the transmethylation pathway for adenosine metabolism in the heart, in *Topics and Perspectives in Adenosine Research* (Gerlach, E. and Becker, B. F., eds.), Springer-Verlag, Berlin, pp. 199–210.

Londos, C., Cooper, D. M. F., and Wolfe, J. (1980) Subclasses of external adenosine receptor. *Proc. Natl. Acad. Sci. USA* **77**, 2551–2554.

Lucchesi, B. R. and Mullane, K. M. (1986) Leukocytes and ischemia-induced myocardial injury. *Annu. Rev. Pharmacol. Toxicol.* **26**, 201–224.

Martens, D., Lohse, M. J., Rauch, B., and Schwabe, U. (1987) Pharmacological characterization of A₁ adenosine receptors in isolated rat ventricular myocytes. *Naunyn-Schmiedebergs Arch. Pharmacol.* **336**, 342–348.

Mathew, R. J. and Wilson,W. H. (1985) Caffeine-induced changes in cerebral circulation. *Stroke* **16**, 814–817.

McCulloch, J. (1983) Peptides and the microregulation of blood flow in the brain. *Nature* (Lond.) **304**, 120.

Meininger, C. J., Schelling, M. E., and Granger, H. J. (1988) Adenosine and hypoxia stimulate proliferation and migration of endothelial cells. *Am. J. Physiol.* **255**, H554–H562.

Meldrum, B. (1985) Possible therapeutic applications of antagonists of excitatory amino acid neurotransmitters. *Clin. Sci.* **68**, 113–122.

Mentzer, R. M., Rubio, R., and Berne, R. M. (1975) Release of adenosine by hypoxic canine lung tissue and its possible role in the pulmonary circulation. *Am. J. Physiol.* **229**, 1625–1631.

Mentzer, R. M., Ely, S. W., Lasley, R. D., Mainwaring, R. D., Wright, E. M., and Berne, R. M. (1985) Hormonal role of adenosine in maintaining patency of the ductus arteriosus in fetal lambs. *Ann. Surg.* **202**, 223–230.

Mentzer, R. M., Ely, S. W., Lasley, R. D., and Berne, R. M. (1988) The acute effects of AICAR on purine nucleotide metabolism and postischemic cardiac function. *J. Thorac. Cardiovasc. Surg.* **95**, 286–293.

Michaelis, M. L., Kitos, T. E., and Mooney, T. (1985) Characteristics of adenosine binding sites in atrial sarcolemmal membranes. *Biochim. Biophys. Acta* **816**, 241–250.

Miller, B. D. and Felder, R. D. (1988) Excitatory amino acid receptors intrinsic to synaptic transmission in nucleus tractus solatarius. *Brain Res.* **456**, 333–340.

Miller, M. J. S., Sharif, R., and Field, F. P. (1990) Renal vasoconstrictor responses to adenosine analogues are indirect and dependent on the agent used to raise vascular tone. *Naunyn-Schmidebergs Arch. Pharmacol.*, submitted.

Mills, D. C. B., Macfarlane, D. E., Lemmex, B. W. G., and Haslam, R. J. (1983)

Receptors for nucleosides and nucleotides on blood platelets, in *Regulatory Function of Adenosine* (Berne, R. M., Rall, T. W., and Rubio, R., eds.), Nijhoff, Boston, pp. 277–289.

Mitsos, S. E., Jolly, S. R., and Lucchesi, B. (1985) Protective effects of AICA-riboside in the globally ischemic isolated cat heart. *Pharmacology* **31,** 121–131.

Monterio, E. C. and Riberio, J. A. (1987) Ventilatory effects of adenosine mediated by carotid body chemoreceptors in the rat. *Naunyn-Schmiedebergs Arch. Pharmacol.* **335,** 143–148.

Mueller, R. A., Widerlow, E. W., and Breese, G. R. (1984) Attempted antagonism of adenosine analogue induced depression of respiration. *Pharmacol. Biochem. Behav.* **21,** 289–296.

Mullane, K. M. (1989) Oxygen-derived free radicals and reperfusion injury of the heart, in *Current Perspectives on ACE Inhibitors: From Mechanism to Therapy* (Sonnenblick, E., Laragh, J. H., and Lesch, M., eds.), Excerpta Medica, Boston, pp. 234–270.

Nees, S., Herzog, V., Becker, B. F., Bock, M., Des Rosiers, C., and Gerlach, E. (1985) The coronary endothelium: A highly active metabolic barrier for adenosine. *Basic Res. Cardiol.* **80,** 515–529.

Nees, S., Des Rosiers, J., and Bock, M. (1987) Adenosine receptors in the coronary endothelium: Functional implications, in *Topics and Perspectives in Adenosine Research* (Gerlach, E. and Becker, B. F., eds.), Springer-Verlag, Berlin, pp. 454–469.

Nehlig, A., Lucignani, G., Kadekaro, M., Porrino, L. J., and Sokoloff, L. (1984) Effects of acute administration of caffeine on local cerebral glucose utilization in the rat. *Eur. J. Pharmacol.* **101,** 91–100.

O'Regan, M. H., Phillis, J. W., and Walter, G. A. (1988) Effects of dipyridamole and soluflazine, nucleoside transport inhibitors, on the release of purines from rat cerebral cortex. *Abstr. Soc. Neurosci.* **14,** 994.

Oberdorster, G., Lany, R., and Zimner, R. (1975) Influence of adenosine and lowered cerebral blood flow on the cerebrovascular effects of theophylline. *Eur. J. Pharmacol.* **30,** 197–204.

Odawara, S., Kurahashi, K., Usui, H., Taniguchi, T., and Fujiwara, M. (1986) Relaxations of isolated rabbit coronary artery by purine derivatives: A_2-adenosine receptors. *J. Cardiovasc. Pharmacol.* **8,** 567–573.

Oei, H. H., Ghai, G. R., Zoganas, H. C., Stone, G. A., Zimmerman, M. B., Field, F. P., and Williams, M. (1988) Correlation between binding affinities for brain A_1 and A_2 receptors of adenosine agonists and antagonists and their effect on heart rate and coronary vascular tone. *J. Pharmacol. Exp. Ther.* **247,** 882–888.

Ogilvie, R. I., Fernandez, P. G., and Winsberg, F. (1977) Cardiovascular response to increasing theophylline concentrations. *Eur. J. Clin. Pharmacol.* **12,** 409–415.

Ohnishi, A., Biaggioni, I., Deray, G., Branch, R. A., and Jackson, E. K. (1986) Hemodynamic effects of adenosine in conscious hypertensive and normotensive rats. *Hypertension* **8,** 391–398.

Ohnishi, A., Li, P., Branch, R. A., Biaggioni, I. O., and Jackson, E. K. (1988) Adenosine in renin-dependent renovascular hypertension. *Hypertension* **12,** 152–161.

Olafsson, B., Farman, M. B., Puett, D. W., Pou, A., Cates, C. U., Friesinger, G. C., and Virmani, R. (1987) Reduction of reperfusion injury in the canine preparation by intracoronary adenosine: Importance of the endothelium and the no-reflow phenomenon. *Circulation* **76,** 1135–1145.

Ollinger, P. and Kukovetz, W. R. (1983) [^{3}H]Adenosine binding to bovine coronary arteries and myocardium. *Eur. J. Pharmacol.* **93,** 35–43.

Onodera, H. and Kogure, K. (1985) Autoradiographic visualization of adenosine A$_1$ receptors in the gerbil hippocampus: Changes in the receptor density after transient ischemia. *Brain Res.* **345,** 406–408.

Osswald, H. (1983) Adenosine in renal function, in *Regulatory Function of Adenosine* (Berne, R. M., Rall, T. W., and Rubio, R., eds.), Nihoff, Boston, pp. 399–415.

Osswald, H. (1988) Effects of adenosine analogs on renal hemodynamics and renin release, in *Adenosine and Adenine Nucleotides. Physiology and Pharmacology* (Paton, D. M., ed.), Taylor and Francis, London, pp. 193–202.

Osswald, H., Schmitz, H.-J., and Heidenreich, O. (1978) Adenosine response of the rat kidney after saline loading, sodium restriction and hemorrhagia. *Pfluger's Arch.* **357,** 323–333.

Osswald, H., Hermes, H. H., and Nabokowski, F. (1982) The role of adenosine in signal transmission of tuberoglomerular feedback. *Kidney Int.* **22,** (Supp. 12) S136–S142.

Owall, A., Gordon, E., Lagerkranser, M., Lindquist, C., Rudehill, A., and Sollevi, A. (1987) Clinical experience with adenosine for controlled hypotension during cerebral aneurysm surgery. *Anesth. Analg.* **66,** 229–234.

Parratt, J. R., Boachie-Ansah, G., Kane, K. A., and Wainwright, C. L. (1988) Is adenosine an endogenous antiarrhythmic agent under conditions of myocardial ischemia? in *Adenosine and Adenine Nucleotides. Physiology and Pharmacology* (Paton, D. M., ed.), Taylor and Francis, London, pp. 157–166.

Parsons, W. J., Ramkumar, S., and Stiles, G. L. (1988) The new cardiotonic sumazole is an A$_1$ adenosine receptor antagonist and inotropic agent functionally blocks the inhibitory regulator Gi. *Mol. Pharmacol.* **33,** 441–448.

Pelleg, A. (1988) Cardiac electrophysiologic actions of adenosine and adenosine 5'-triphosphate, in *Adenosine and Adenine Nucleotides. Physiology and Pharmacology* (Paton, D. M., ed.), Taylor and Francis, London, pp. 143–155.

Pelleg, A., Mitsouka, T., and Michelson, E. L. (1987) Adenosine mediates the negative chronotropic actions of adenosine 5'-triphosphate in canine sinus node. *J. Pharmacol. Exp. Ther.* **242,** 791–798.

Phillis, J. W. and De Long, R. E. (1986) The role of adenosine in cerebral vascular regulation during reductions in perfusion pressure. *J. Pharm. Pharmacol.* **38,** 460–462.

Phillis, J. W., Preston, G., and De Long, R. E. (1984) Effects of anoxia on cerebral blood flow in the rat brain: Evidence for a role of adenosine in autoregulation. *J. Cereb. Blood Flow Metab.* 4, 586–592.

Raberger, G., Fischer, G., Krumpl, G., Schneider, W., and Stroissing, H. (1987) Further evidence against adenosine-catecholamine antagonism in vivo: Investigations with treadmill dogs, in *Topics and Perspectives in Adenosine Research.* (Gerlach, E. and Becker, B. F., eds.), Springer-Verlag, Berlin, pp. 383–394.

Raichle, M. E., Grubb, R. L., Gado, M. H., Eichling, J. O., and Ter-Pogossian, M. M. (1977) In vivo correlations between regional cerebral blood flow and oxygen utilization in man. *Acta Neurol. Scand.* 56, 240, 241.

Ramagopal, M. V., Chitwood, R. W., and Mustafa, S. J. (1988) Evidence for A_2 adenosine receptor in human coronary arteries. *Eur. J. Pharmacol.* 151, 483–486.

Ramkumar, V., Pierson, G., and Stiles, G. L. (1988) Adenosine receptors: Clinical implications and biochemical mechanisms. *Prog. Drug. Res.* 32, 195–247.

Raper, R. F. and Fisher, M. (1988) Profound reversible myocardial depression after anaphylaxis. *Lancet i.,* 386–388.

Rardon, D. P. and Bailey, J. C. (1984) Adenosine attenuation of the electrophysiological effects of isoproterenol on canine cardiac Purkinje fibers. *J. Pharmacol. Exp. Ther.* 228, 792–798.

Recordati, G. M., Moss, N. G., and Waselkor, L. E. (1977) Renal chemoceptors. *Circ.* 56, (suppl. III), III-247.

Robertson, D., Biaggioni, I., and Tseng, C. J. (1988) Adenosine and cardiovascular control, in *Adenosine and Adenine Nucleotides. Physiology and Pharmacology.* (Paton, D. M., ed.), Taylor and Francis, London, pp. 241–250.

Riberio, J. A. and McQueen, D. S. (1983) On the neuromuscular depression and carotid chemoreceptor activation caused by adenosine, in *Physiology and Pharmacology of Adenine Derivatives* (Daly, J. W., Kuroda, Y., Phillis, J. W., Shimizu, H., Ui, M., eds.), Raven, New York, pp. 179–188.

Riberio, J. A., Monteiro, E. C., and McQueen, D. S. (1988) The action of adenosine on respiration, in *Adenosine and Adenine Nucleotides. Physiology and Pharmacology* (Paton, D. M., ed.), Taylor and Francis, London, pp. 225–232.

Rockoff, J. B. and Dobson, J. G. (1980) Inhibition by adenosine of actecholamine-induced increase in rat atrial contractility. *Am. J. Physiol.* 239, H365–H370.

Romano, F. D., Fenton, R. A., and Dobson, Jr., J. G. (1988) The adenosine Ri agonist, phenyl isopropyl adenosine reduces high affinity isoproternol binding to the β-adrenergic receptor of rat myocardial membranes. *Second Messengers Phosphoproteins* 12, 29–43.

Rose, F. R., Hirschhorn, R., Weissmann, G., and Cronstein, B. N. (1988) Adenosine promotes neutrophil chemotaxis. *J. Exp. Med.* 167, 1186–1194.

Ross, R. and Glomset, J. A. (1973) Atherosclerosis and the arterial smooth muscle cell. *Science* **180**, 1332–1339.

Rossi, N. F., Churchill, P. C., and Churchill, M. C. (1987) Pertussis toxin reverses adenosine receptor mediated inhibition of renin secretion in rat renal cortical slices. *Life Sci.* **40**, 481–487.

Rossi, N. F., Churchill, P. C., Ellis, V., and Amore, B. (1988) Mechanism of adenosine receptor-induced renal vasoconstriction in rats. *Am. J. Physiol.* **255**, H885–H890.

Rubio, R. and Berne, R. M. (1969) Release of adenosine by the nomal myocardium in dogs and its relationship to the regulation of coronary resistance. *Circ. Res.* **25**, 407–415.

Rubio, R., Widemeier, V. T., and Berne, R. M. (1974) Relationship between coronary flow and adenosine production and release. *J. Mol. Cell. Cardiol.* **6**, 561–566.

Sabina, R. L., Kernstine, K. H., Boyd, R. L., Holmes, E. W., and Swain, J. L. (1982) Metabolism of 5-amino-4-imidazolecarboxamide riboside in cardiac and skeletal muscle. *J. Biol. Chem.* **257**, 10178–10183.

Saito, D., Hyodo, T., Takeda, K., Abe, Y., Tani, H., Yamada, N., Ueeda, M., and Nakatsu, Y. (1985) Intracoronary adenosine enhances myocardial reactive hyperemia after brief coronary occlusion. *Am. J. Physiol.* **248**, H812–H817.

Samet, M. K. and Rutledge, C. O. (1985) Antagonism of the positive chronotropic effect of norepinephrine by purine nucleosides in rat atrai. *J. Pharmacol. Exp. Ther.* **232**, 106–110.

Schaumann, E. and Kutscha, W. (1972) Klinisen pharmakologische untersuchungen mit einem neuen peroral wirksamen adenosinederivat. *Arzneim. Forsch.* **22**, 783–790.

Schnermann, J. (1988) Effect of adenosine analogues on tuboglomerular feedback response. *Am. J. Physiol.* **255**, F33–F42.

Schrader, J. and Deussen, A. (1985) Adenosine-xanthine interactions in the heart, in *Anti-Asthma Xanthines and Adenosine* (Andersson, K.-E. and Persson, C. G. A., eds.), Excerpta Medica, Amsterdam, pp. 273–279.

Schrader, J., Baumann, F., and Gerlach, E. (1977) Adenosine as an inhibitor of the myocardial effects of catecholamines. *Pflugers Arch.* **372**, 29–35.

Schrader, J., Thompson, C. I., Hiendlmayer, G., and Gerlach, E. (1982) Role of purines in acetylcholine-induced coronary vasodilation. *J. Mol. Cell. Cardiol.* **14**, 427–430.

Schutz, W., Freissmuth, M., Hausleithner, V., and Tuisl, E. (1986) Cardiac sarcolemmal purity is essential for the verification of adenylate cyclase inhibition via A_1-adenosine receptors. *Naunyn-Schmiedebergs Arch. Pharmacol.* **333**, 156–162.

Simpson, P. J., Fantone, J. C., and Lucchesi, B. (1987) Myocardial ischemia and reperfusion injury: Oxygen radicals and the role of the neutrophil, in *Oxygen Radicals and Tissue Injury* (Halliwell, B., ed.), FASEB/Upjohn, Bethesda, Maryland, pp. 63–77.

Singer, R. M., Olszewski, B. J., Major, T. C., Soltis, E., Cohen, D. M., Ryan, M. J., and Kaplan, H. R. (1986) Evidence for peripheral mediation of the blood pressure (BP) lowering activity of *N*-cyclohexyl adenosine (CHA). *Pharmacologist* **28**, 154.

Snyder, S. H., Katims, J. J., Annau, Z., Bruns, R. F., and Daly, J. W. (1981) Adenosine and behavioral actions of the methylxanthines. *Proc. Natl. Acad. Sci.* **78**, 3260–3264.

Spenser, S. E., Sawyer, W. B., and Lowey, A. D. (1988) L-glutamate stimulation of the zona incerta in the rat decreases heart rate and blood pressure. *Brain Res.* **458**, 72–81.

Spielman, W. S. and Osswald, H. (1979) Blockade of postocclusive renal vasco-constriction by an angiotensin II antagonist: Evidence for an angiotensin-adenosine interaction. *Am. J. Physiol.* **237**, F463–F467.

Sollevi, A. (1986) Cardiovascular effects of adenosine in man: Possible clinical implications. *Prog. Neurobiol.* **27**, 319–349.

Sollevi, A. and Fredholm, B. B. (1981) Role of adenosine in adipose tissue cir-culation. *Acta Physiol. Scand.* **112**, 293–298.

Sollevi, A., Lagerkranser, T., Irestedt, L., Gordon, E., and Lindquist, C. (1984) Controlled hypotension with adenosine in cerebral aneurysm surgery. *Anesthesiology* **61**, 400–405.

Sollevi, A., Torssell, L., Owall, A., Edlund, A., and Lagerkranser, M. (1987) Levels and cardiovascular effects of adenosine in humans, in *Topics and Perspectives in Adenosine Research* (Gerlach, E. and Becker, B. F., eds.), Springer-Verlag, Berlin, pp. 599–613.

Spaet, T. H., Gaynor, E., and Stermerman, M. B. (1974) Thrombosis, athero-sclerosis and endothelium. *Am. Heart J.* **87**, 661–668.

Sparks, H. V. and Gorman, M. W. (1987) Adenosine in the local regulation of blood flow: Current controversies, in *Topics and Perspectives in Adeno-sine Research* (Gerlach, E. and Becker, B. F., eds.), Springer-Verlag, Ber-lin, pp. 406–415.

Stiles, G. L. and Jacobson, K. A. (1988) A new high affinity iodinated adenosine receptor antagonist as radioligand/photoaffinity crosslinking probe. *Mol. Pharmacol.* **32**, 184–188.

Stiles, G. L., Daly, D. T., and Olsson, R. A. (1985) The A_1 adenosine receptor: Identification of the binding subunit by cross linking. *J. Biol. Chem.* **260**, 10806–10811.

Swain, J. L., Hines, J. J., Sabina, R. L., and Holmes, E. W. (1982) Accelerated repletion of ATP and GTP pools in postischemic canine myocardium using a precusor of purine de novo synthesis. *Circ. Res.* **51**, 102–105.

Sylven, C., Beermann, B., Jonzon, B., and Brandt, R. (1986) Angina-pectoris-like pain provoked by intravenous adenosine in healthy volunteers. *Br. Med. J.* **293**, 227–230.

Thompson, C. I., Rubio, R., and Berne, R. M. (1980) Changes in adenosine and glycogen phosphorylase activity during the cardiac cycle. *Am. J. Physiol.* **238**, H389–H398.

Tseng, C-J., Biaggioni, I., Appalsamy, M., and Robertson, D. (1988) Purinergic receptors in the brainstem mediate hypotension and bradycardia. *Hypertension* **11**, 191–197.

Tung, C.-S., Chu, K.-M., Tseng, C.-J., and Yin, T.-H. (1987) Adenosine in hemorrhagic shock: Possible role in attenuating sympathetic activation. *Life Sci.* **41**, 1375–1382.

Vatner, S. F., Franklin, D., and Braunwald, E. (1971) Effects of anesthesia and sleep on circulatory response to carotid sinus nerve stimulus. *Am. J. Physiol.* **229**, 1249–1255.

Vatner, S. F. and Braunwald, E. (1975) Cardiovascular control mechanisms in the conscious state. *New Eng. J. Med.* **293**, 970–976.

Von Lubitz, Dambrosia, J. E., Kempster, O., and Redmond, D. J. (1988) Cyclohexyladenosine protects against neuronal death following ischemia in the CA-1 region of gerbil hippocampus. *Stroke* **19**, 1133–1138.

Waldeck, B. (1985) A brief review of the effects of xanthines on mammalian smmoth muscle in vitro, in *Anti-Asthma Xanthines and Adenosine* (Andersson, K.-E. and Persson, C. G. A., eds.), Excerpta Medica, Amsterdam, pp. 184,185.

Watt, A. H. and Routledge, P. A. (1985) Adenosine stimulates respiration in man. *Br. J. Clin. Pharmacol.* **20**, 503–506.

Watt, A. H. and Routledge, P. A. (1986) Transient bradycardia and subsequent sinus tachycardia produced by intravenous adenosine in healthy adult subjects. *Br. J. Clin. Pharmacol.* **21**, 533–536.

Watt, A. H., Reid, P. G., Routledge, P., Singh, H., Penny, W. J., and Henderson, A. H. (1986) Angina pectoris-like pains provoked by intravenous adenosine. *Br. Med. J.* **293**, 504, 505.

Watt, A. H., Reid, P. G., Stephens, M. R., and Routledge, P. A. (1987) Adenosine-induced respiratory stimulation in man depends on site of infusion. Evidence for an action on the carotid body. *Br. J. Clin. Pharmacology* **23**, 486–490.

Webb, R. L., McNeal, R. B., Barclay, B. W., and Yasay, G. D. (1989) Hemodynamic effects of adenosine analogs in conscious spontaneously hypertensive rats. *FASEB J.* **3**, A1032.

Weksler, B. B. (1988) Platelets, in *Inflammation: Basic Principles and Clinical Correlates* (Gallin, J. I., Goldstein, I. M., and Snyderman, R., eds.), Raven, New York, pp. 543–557.

Wessburg, P. J., Hedner, J., Hedner, T., Persson, B., and Jonasson, J. (1985) Adenosine mechanisms in the regulation of breathing in the rat. *Eur. J. Pharmacol.* **106**, 59–67.

Wesley, R. C., Boykin, M. T., and Boykin, L. (1985) Role of adenosine as a mediator of bradyarrythymia during hypoxia in isolated guinea pig heart. *Cardiovasc. Res.* **20**, 752–759.

Williams, M. (1987) Purine receptors in mammalian tissues: Pharmacology and functional significance. *Ann. Rev. Pharmacol. Toxicol.* **27**, 315–345.

Williams, M. (1989) Adenosine antagonists. *Med. Res. Rev.* **9**, 219–243.

Williams, M., Francis, J., Ghai, G. R., Braunwalder, A., Psychoyos, S., Stone, G. A., and Cash, W. D. (1987) Biochemical characterization of the triazololquinazoline, CGS 15943, a novel, nonxanthine adenosine antagonist. *J. Pharmacol. Exp. Ther.* **241,** 415–420.

Winn, H. R. (1985) Metabolic regulation of cerebral blood flow by adenosine, in *Purines: Pharmacology and Physiological Roles* (Stone, T. W., ed.), VCH Publications, Deerfield Beach, Florida, pp. 131–141.

Winn, H. R., Morii, S., and Berne, R. M. (1985) The role of adenosine in autoregulation of cerebral blood flow. *Ann. Biomed. Eng.* **13,** 321–328.

Winbury, M. M., Papierski, D. H., Hemmer, M. L., and Hambourger, W. E. (1953) Coronary dilator action of the adenine-ATP series. *J. Pharmacol. Exp. Ther.* **109,** 255–260.

Witzleb, E. (1983) Functions of the vascular system, in *Human Physiology* (Schmidt, R. F. and Thews, G., eds.), Springer-Verlag, Berlin, pp. 355–397.

Wolf, M. M. and Berne, R. M. (1965) Coronary vasodilator properties of purine and pyrimidine derivatives. *Circ. Res.* **4,** 343–348.

Wyatt, D. A., Edmunds, M. C., Lasley, R. D., Berne, R. M., and Mentzer, R. M. (1987) Intracoronary adenosine infusion ameliorates postischemic regional myocardial disfunction. *Circ.* **76** (suppl. IV), IV–21.

Yatani, A., Mattera, R., Codina, J., Graf, R., Okabe, K., Padrell, E., Iyengar, R., Brown, A. M., and Birnbaumer, L. (1988) The G-protein gated K⁺channel is stimulated by three distinct Gia-subunits. *Nature* **336,** 680–682.

Yen, M.-H., Wu, C.-C., and Chiou, W.-F. (1988) Partially endothelium-dependent vasodilator effect of adenosine in rat aorta. *Hypertension* **11,** 514–518.

Zehl, U., Ritter, C., and Forster, W. (1976) Influence of prostaglandins upon adenosine release and of adenosine upon prostaglandin release in isolated rabbit heart. *Acta Bio. Med. Ger.* **35,** K77–K82.

CHAPTER 9

Adenosine
and Renal Function

Paul C. Churchill
and Anil K. Bidani

1. Introduction

Exogenous adenosine has been shown to affect nearly all aspects of renal function: renal blood flow and its distribution within the kidney, glomerular filtration rate, renin secretion, urine flow, sodium excretion, transmitter release from renal efferent nerves, and the activity of renal afferent nerves. Many of these effects are produced by adenosine receptors, since the effects are antagonized by alkylxanthines and mimicked by adenosine analogs that act as adenosine receptor agonists. The orders of potency of agonists in producing some of these effects have been determined, and therefore the subclasses of adenosine receptors that are involved have been established. These observations, taken together with the observation that kidneys produce and release adenosine into extracellular fluids, suggest that variations in the concentration of endogenously released adenosine could play important roles in renal function and/or dysfunction. Indeed, it has been postulated that adenosine is the mediator of several physiological and pathophysiological phenomena: the autoregulation of renal

Adenosine and Adenosine Receptors Editor: Michael Williams ©1990 The Humana Press Inc.

blood flow and glomerular filtration rate, the tubuloglomerular feedback response, the effect of macula densa cells on the adjacent renin-
secreting juxtaglomerular cells, the hemodynamic changes in acute
renal failure, and hypertension in some experimental animal models.

We intend in this review to evaluate critically the present state of
knowledge of these aspects of the "renal adenosine system."

2. Adenosine
Production, Release, and Uptake

2.1. Enzymatic Pathways

The major known biochemical pathways for adenosine production and disposal by the kidney are shown in Fig. 1. Nonspecific
phosphatases and 5'-nucleotidase catalyze adenosine production
from 5'-AMP, and S-adenosylhomocysteine hydrolase catalyzes its
production from S-adenosylhomocysteine. The presence of these enzymes has been demonstrated in the kidney (Finkelstein and Harris,
1973; Eloranta, 1977; Schatz, et al., 1977; Stocker et al., 1977; Miller
et al., 1978). Intracellular adenosine serves as substrate not only for
S-adenosylhomocysteine hydrolase, but also for adenosine kinase
and adenosine deaminase, both of which have been demonstrated in
kidney tissue (Jackson, et al., 1978). No information is available concerning the regulation of any of these pathways in the kidney, or their
relative contributions to determining intracellular adenosine concentration. Moreover, other biochemical pathways for renal adenosine
synthesis (e.g., the breakdown of RNA to 3'-adenylate and its subsequent dephosphorylation) have not been investigated, but could be
important sources of adenosine in some situations.

In addition to these enzymatic pathways that affect intracellular
adenosine concentration, extracellular adenosine can serve either as
a source or as a sink, as indicated in Fig. 1. Renal uptake of extracellular adenosine has been demonstrated in man (Kuttesch and Nelson,
1982), dogs (Thompson et al., 1985), rats (Trimble and Coulson,
1984; Coulson and Trimble, 1986), and mice (Kuttesch and Nelson,
1982; Nelson et al., 1983). At least in renal tubular cells, uptake
occurs via facilitated diffusion (Angielski et al., 1983; Le Hir and
Dubach, 1984; Trimble and Coulson, 1984; Coulson and Trimble,

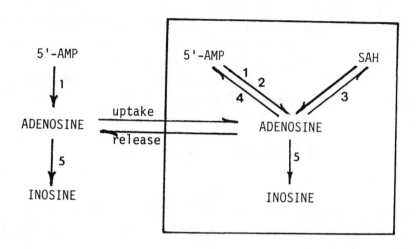

Fig. 1. Pathways of adenosine metabolism in the kidney. Adenosine can be synthesized and degraded in both intra- and-extracellular fluids. The enzymes involved are: (1) 5'-nucleotidase, which has access to intra- and extra-extracellular adenosine; (2) nonspecific phosphatases; (3) SAH (S-adenosylhomocysteine) hydrolase; (4) adenosine kinase; and (5) adenosine deaminase. In addition, both uptake of extracellular adenosine and release of intracellular adenosine have been demonstrated.

1986), and the classical adenosine uptake antagonists—papaverine and dipyridamole—antagonize renal adenosine uptake (Le Hir and Dubach, 1984; Trimble and Coulson, 1984; Arend et al., 1985; Coulson and Trimble, 1986). Conversely, net renal production and release of adenosine into extracellular fluids (plasma and urine) have been demonstrated (Fredholm and Hedqvist, 1978; Miller et al., 1978; Thompson et al., 1985; Ramos-Salazar and Baines, 1986). Renal transmembrane adenosine transport has been characterized particularly well in renal tubular cells, which are capable of both reabsorption of filtered adenosine and adenosine secretion (Kuttesch and Nelson, 1982; Nelson et al., 1983; Trimble and Coulson, 1984; Thompson et al., 1985). As shown in Fig. 1, enzymatic pathways exist for the extracellular synthesis and degradation of adenosine, since both 5'-nucleotidase and cyclic nucleotide phosphodiesterase are partially ectoenzymes, and since plasma normally contains some adenosine deaminase activity.

These observations demonstrate that the kidney has the capacity for adenosine production in intracellular and extracellular fluids, for release of intracellular adenosine, for uptake of extracellular adenosine, and for the degradation and disposal of both intracellular and extracellular adenosine. Theoretically, the adenosine available for occupancy of plasma membrane receptors on a particular target cell could be influenced by all of these factors.

2.2. Adenosine Measurements

Adenosine has been measured in kidney tissue of some mammals. Several recently reported mean values for normal rats are very similar (in nmol/g wet weight): 5.13 ± 0.60 (Osswald et al., 1977); 5.7 ± 1.2 (Osswald et al, 1980); 6.7 ± 0.2 (Miller et al., 1978); 7.05 ± 0.79 (Osswald et al., 1982). Ramos-Salazar and Baines (1986) found a similar value, 5.6 ± 1.4 nmol/g wet weight, in the isolated perfused rat kidney. The mean values reported for cats and dogs are in the same range: 4.1 ± 0.5 and 7.6 ± 3.4 nmol/g wet weight, respectively (Miller et al., 1978). Since approximately 80% of kidney wet weight is water, these figures imply adenosine concentration on the order of $5\text{–}10 \mu M$. However, it is easy to show that uniform distribution between intracellular and extracellular fluids cannot be assumed. Measured values for extracellular fluids are all less than $5\text{–}10 \mu M$: arterial and renal venous concentrations in normal rats and dogs are submicromolar (Arend et al., 1985, 1986; Thompson et al.,1985; Ramos-Salazar and Baines, 1986; Jackson and Ohnishi, 1987), and urinary concentrations are either submicromolar (Arend et al., 1985; Thompson et al., 1985) or approximately $1 \mu M$ (Miller et al., 1978; Katholi et al., 1985; Ramos-Salazar and Baines, 1986). Since extracellular adenosine is $1 \mu M$ or less, and not distributed uniformly, it follows that most tissue adenosine is inside cells.

Since intracellular adenosine is probably unavailable to plasma membrane receptors, one might conclude that such measurements are irrelevant in evaluating the role of extracellular adenosine in renal function. However, changes in tissue adenosine in response to some manipulations seem to be reflected by changes in extracellular adenosine concentrations. For example, renal tissue adenosine increases in response to complete ischemia (Osswald et al., 1977; Miller et al., 1978) and relative hypoxia (Ramos-Salazar and Baines, 1986); both urinary (Miller et al., 1978; Ramos-Salazar and Baines, 1986) and renal venous adenosine (Ramos-Salazar and Baines, 1986) are also elevated in these situations. Similarly, acute iv administration of maleate increases renal tissue adenosine (Osswald et al., 1982) and both urinary and renal venous adenosine (Arend et al., 1986). On the other hand, the relation between sodium balance and either tissue or extracellular adenosine is difficult to assess. Osswald et al. (1980) have shown that acute iv administration of hypertonic saline increases tissue adenosine; however, there appears to be no difference in renal tissue adenosine levels between chronically sodium-deprived rats and rats on a normal sodium diet. In contrast, in the report by Arend et al. (1986), it appears that arterial, renal venous, and urinary adenosine values are higher in chronically sodium-loaded dogs than in sodium-deprived dogs.

2.3. Summary

Kidneys are capable of adenosine production, release, and uptake. Measurements of tissue adenosine and adenosine concentrations in urine and in arterial and renal venous plasma have provided some important and suggestive information, but there have been no measurements of adenosine in the extracellular fluid surrounding the putative target cells of adenosine generated within the kidney.

It follows from the above that only pharmacological tools are available to determine whether or not endogenously released adenosine plays a role in the physiological regulation of renal function. Because drugs are rarely if ever specific, pharmacological tools are imperfect. However, in this review we will assume that the effects of methylxanthines, particularly the in vivo effects, must be mediated by antagonism of plasma membrane adenosine receptors, rather than by inhibition of cyclic nucleotide phosphodiesterase activity. This assumption is supported by the observations that nearly millimolar concentrations of, e.g., theophylline are required to inhibit renal phosphodiesterase activity (Fredholm et al., 1978), yet plasma concentrations that are an order of magnitude lower produce toxic and frequently lethal effects (Nicholson and Chick, 1973). In contrast, μ*M* methyl-xanthines competitively antagonize plasma membrane adenosine receptors (Daly, 1982). Therefore, methylxanthines can be used to determine the role of endogenously released adenosine in renal function. Administration of adenosine deaminase to destroy extracellular adenosine can be used for the same purpose. We assume, further, that adenosine uptake antagonists (e.g., dipyridamole, papaverine) should produce "adenosine-like" effects, consistent with increased extracellular adenosine, and that these effects should be blocked by methylxanthines and/or by adenosine deaminase. Finally, we assume that adenosine analogs (2-chloradenosine [2-ClA]; N^6-cyclohexyladenosine [CHA]; L-N^6-phenylisopropyl adenosine [L-PlA]; 5'-*N*-ethylcarboxamide adenosine [NECA]) are superior to adenosine *per se* in probing the effects mediated by plasma membrane adenosine receptors. Some exhibit selectivity for the A_1 and/or A_2 adenosine receptors; moreover, many of them are substrates for neither adenosine deaminase nor cellular adenosine uptake, and they are inactive at the internal adenosine P-site (Daly, 1982).

3. Renal Hemodynamics

3. 1. Exogenous Adenosine In Vivo

The renal hemodynamic effects of adenosine have been studied in many species, including chickens, rats, rabbits, cats, dogs, and pigs. In these experiments, the mode of administration (bolus injection vs continuous infusion), the site of administration (intravenous; intraaortic, above the origins of the renal arteries; intrarenal-arterial), the dosages were extremely varied. Two additional factors complicate the comparison and interpretation of the results. First, endogenous levels of adenosine were not measured in the experiments, and in some, the animals were subjected to various pretreatments (hemorrhage, sodium deprivation, sodium loading) that might have affected endogenous adenosine levels. Second, adenosine has an extremely short half-life in blood (Sollevi et al., 1984); it is taken up by erythrocytes, and can be degraded to inosine by plasma adenosine deaminase. Therefore, the amounts of administered adenosine that reached putative adenosine receptors and the concentration-dependency of any of the observed effects are unknown. In view of these problems, the results have been surprisingly consistent.

3.2.1. Blood Flow

Drury and Szent-Györgyi (1929) were probably the first to report that although adenosine dilates most vascular beds, it can constrict the renal vasculature. Subsequently it has been shown that intrarenal-arterial bolus injections of adenosine reduce renal blood flow in rats (Osswald et al., 1975, 1978a; Sakai et al., 1981), rabbits (Sakai et al., 1981), cats (Spielman and Osswald, 1978), dogs (Thurau, 1964; Scott et al., 1965; Hashimoto and Kumakura, 1965; Spielman and Osswald, 1979; Sakai et al., 1981; Gerkens et al., 1983a,b), and pigs (Sakai et al., 1981). Although most of the hypothetical roles adenosine are predicated upon adenosine-induced vasoconstriction, the vasoconstrictive response is usually extremely transient. Upon initiation of a continuous infusion of adenosine, blood flow decreases for only 1–3 minutes, whereas at steady-state, blood flow is usually increased above preinfusion levels (Sakai et al., 1968; Hashimoto and Kokubun, 1971; Tagawa and Vander, 1970; Bhanalaph et al., 1973; Osswald et al., 1978b; Spielman et al., 1980; Arend et al., 1984; Beck et al., 1984; Spielman, 1984; Macias-Nunez et al., 1985; Hall et al., 1985; Premen et al., 1985; Hall and Granger, 1986a; Macias-Nunez et al., 1986). In addition to affecting total renal blood flow, adenosine may affect its distribution within the kidney. Osswald et al., (1978b) and Haas and Osswald (1981) found evidence of persistent adenosine-induced vasoconstriction of the outer renal cortex. However, Spielman et al. (1980) found no evidence of decreased blood flow in the outer cortex; rather, blood flow in the inner cortex was increased.

3.1.2. Glomerular Filtration Rate and Filtration Fraction

Glomerular filtration rate and filtration fraction (the ratio between glomerular filtration rate and either renal plasma flow or PAH clearance) are almost invariably reduced during continuous infusions of adenosine (Nechay, 1966; Tagawa and Vander, 1970; Osswald, 1975; Osswald et al., 1975, 1978a, 1978b; Arend et al., 1984; Spielman, 1984; Hall et al., 1985; Premen et al., 1985; Hall and Granger, 1986a). Although there is some evidence of persistent cortical afferent arteriolar constriction (Osswald et al., 1978b; Haas and Osswald, 1981), the evidence is conflicting (Spielman et al., 1980), and in either case, afferent arteriolar constriction alone cannot account for the frequent finding that filtration fraction and glomerular filtration rate are decreased, even if blood flow is increased. The simplest hypothesis that accounts for this set of observations was originally advanced by Tagawa and Vander (1970): adenosine constricts the afferent arteriole and dilates the efferent. Overall renovascular resistance would be affected in opposite directions by these changes. Depending upon the relative magnitude of the resistance changes, blood flow could decrease, remain constant, or increase. However, no matter in which direction blood flow changed, filtration fraction and glomerular filtration rate would decrease.

3.2. Adenosine Analogs In Vivo

Adenosine analogs have been used in a few in vivo experiments, and the results are similar to those obtained with adenosine: variable effects on renal blood or plasma flow, consistent reductions in glomerular filtration rate. Thus, in rats, renal plasma flow and glomerular filtration rate are reduced by continuous iv administration of 2-ClA (Churchill, 1982; Churchill et al., 1984; Churchill and Bidani, 1987), CHA (Cook and Churchill, 1984; Churchill and Bidani, 1987). Bolus iv injections of CHA and NECA increase and decrease, respectively, total renovascular resistance in rats (Dunham and Vince, 1986), whereas continuous iv infusions of NECA (Beck et al., 1984) and a precursor of NECA (Schutz et al., 1983) transiently increase, then decrease, total renovascular resistance in dogs.

Two lines of evidence demonstrate that the renovascular effects of exogenous adenosine are receptor-mediated. First, the effects are mimicked by analogs, and second, the effects of exogenous adenosine (Osswald, 1975; Gerkens et al., 1983a,b; Spielman, 1984; Premen et al., 1985) and of several analogs (Churchill, 1982; Dunham and Vince, 1986; Churchill et al., 1987) are antagonized or blocked by alkylxanthines. However, the order of potency of a series of agonists must be determined in order to pharmacologically characterize adenosine receptors, and this has not been done in in vivo experiments.

One problem with speculations based on available information is that kidneys produce and release adenosine, and the extent to which adenosine receptors are already occupied by endogenously released adenosine is unknown. A second problem is that, although a given analog might be selective (e. g., the affinity constants for CHA at A_1 and A_2 receptors differ by four orders of magnitude in some cases [Daly, 1982]), no agonist is completely specific. Moreover, the disposition and metabolism of analogs in vivo is unknown. To illustrate these problems, virtually identical renovascular effects are produced in rats by infusing NECA at 1 nmol/min/kg body weight and CHA at 10 nmol/min/kg body weight (Churchill and Bidani, 1987). Assuming no cellular uptake, and therefore distribution within the extracellular space (15% of body weight), plasma concentrations of the two agonists would increase at 6.7 and 67 nM/min, respectively, under the best of circumstances (no metabolism, no excretion). A_1 receptors should be half-saturated by CHA within seconds (affinity constant of 3 nM [Daly, 1982]) and by NECA in less than 15 min (affinity constant of 100 nM [Daly, 1982]). On the other hand, to half-saturate the A_2 receptors, NECA would require 1–4 h (affinity constant 0.5–2.0 µM [Daly, 1982]), and CHA would require more than 7 h (affinity constant of 30 µM [Daly, 1982]). Since both agonists produced effects within minutes (the clearance studies were completed in less than 1 h), these calculations suggest that the observed effects were mediated primarily, if not entirely, by A_1 adenosine receptors. However, within 15 min, arterial blood pressure was lowered to an equal extent by both agonists, and the hypotensive effect of adenosine analogs is usually taken to be mediated by A_2 receptors (Jacobson et al., 1985; Fredholm and Sollevi, 1986; Jonzon et al., 1986). Therefore, the in vivo effects of adenosine and its analogs on blood flow and glomerular filtration rate could be mediated by A_1 and/or A_2 receptors.

3.3. Adenosine and Adenosine Analogs In Vitro

The isolated perfused kidney can be used to advantage in studying the hemodynamic effects of substances like adenosine, because reflexes initiated by systemic effects (e. g., hypotension) are eliminated, and the kidney can be perfused either at constant flow (such that

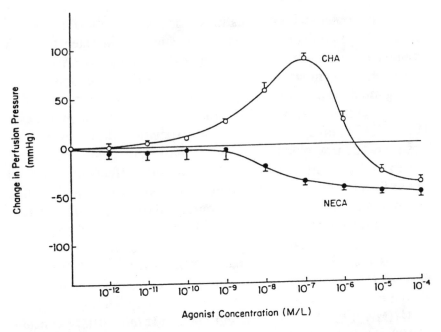

Fig. 2. Effects of CHA and NECA on perfusion pressure of isolated rat kidneys perfused at constant flow. Mean ± SEM; n=4 for each group. Reprinted from Murray and Churchill, 1985.

changes in perfusion pressure reflect changes in resistance) or at constant pressure (such that changes in flow reflect changes in resistance). If a single-pass system is used, and the perfusate is cell-free, two additional problems are avoided: the uptake and metabolism of adenosine by erythrocytes, and the accumulation in the perfusate of endogenously released vasoactive substances, including adenosine. Murray and Churchill (1984, 1985) used the isolated perfused rat kidney to study the hemodynamic effects of adenosine and several analogs. Their results, some of which are shown in Fig. 2, demonstrate that A_1 and A_2 adenosine receptors mediate constriction and dilation, respectively. The fact that vasodilation is observed at high CHA concentrations, despite fully saturated, higher-affinity A_1 receptors mediating vasoconstriction, implies either that there are more A_2 than A_1 receptors or that A_2-induced vasodilation can overcome A_1-induced vasoconstriction. If both A_1 and A_2 receptors are present on the renal vasculature, it is very easy to understand why adenosine and adenosine analogs can decrease, increase, or have no effect on blood flow in vivo. Based on calculated segmental resistance changes in response to adenosine analogs, Murray and Churchill (1984, 1985) suggested that afferent arterioles have both A_1 and A_2 receptors, whereas efferent artioles have only, or mainly, A_2 receptors.

Kenakin and Pike (1987) also used the isolated perfused rat kidney to study the effects of a series of analogs. Like Murray and Churchill (1984, 1985), they found that analogs such as CHA have biphasic effects (vasoconstriction at low concentrations converting to vasodilation at higher concentrations), which is consistent with mediation of vasoconstriction and vasodilation by A_1 and A_2 receptors, respectively. Rossi et al. (1987b) have provided additional evidence that renal vasoconstriction is mediated by A_1 receptors: CHA-induced vasoconstriction of the isolated perfused rat kidney is competitively antagonized by theophylline and by xanthine amine congener (8-[4[[2-aminoethyl]-aminocarbonylmethyloxy]phenyl]-1,3-dipropylxanthine; xac), an A_1-selective antagonist.

3.4. Interactions with Other Vasoactive Substances

3.4.1. Angiotensin II

Since adenosine dilates most vascular beds, some investigators have been reluctant to believe that adenosine *per se* can constrict the renal vasculature. Indeed, it has been hypothesized that the constriction is actually mediated by intrarenally generated angiotensin II (Thurau, 1964; Osswald, 1975; Osswald et al., 1975, 1980; Spielman and Osswald, 1978, 1979; Spielman and Thompson, 1982; Arend et al., 1984; Hall et al., 1985). Several observations have been cited in support; for instance, sodium loading and sodium deprivation suppress and potentiate, respectively, both the activity of the renin–angiotensin system and adenosine-induced renal vasoconstriction in vivo (Thurau, 1964; Osswald, 1975; Osswald et al., 1975; Spielman and Osswald, 1979; Spielman et al., 1980; Arend et al., 1984, 1985). Moreover, exogenous angiotensin II restores adenosine-induced vasoconstriction in sodium-loaded rats (Osswald et al., 1975), and either a competitive angiotensin II antagonist (Spielman and Osswald, 1979) or a converting enzyme inhibitor (Hall et al., 1985) can attenuate adenosine-induced renal vasoconstriction in vivo.

Despite these consistent observations, several more can be cited against the hypothesis that angiotensin II mediates adenosine-induced renal vasoconstriction.

1. The observation that sodium loading attenuates both the activity of the renin–angiotensin system and adenosine-induced renal vasoconstriction cannot be taken as evidence of a causal relationship, since aortic clamping is one of the most potent stimulators of the renin–angiotensin system, yet blocks, rather than enhances, adenosine-induced renal vasoconstriction in vivo (Haas and Osswald, 1981).

2. Adenosine agonists elicit vasoconstriction in isolated kidneys perfused with a nonrecirculating medium devoid of renin substrate (Murray and Churchill, 1984, 1985; Kenakin and Pike, 1987; Rossi et al., 1987b).

3. Adenosine actually inhibits renin secretion, and the inhibitory effect is mediated by A_1 receptors (*see below*). Thus, A_1-induced inhibition of renin secretion and A_1-induced vasoconstriction are simultaneous events in the isolated perfused rat kidney (Murray and Churchill, 1984, 1985), and it is difficult to accept that a decrease in renin secretion, even if substrate were present, could result in an angiotensin-induced vasoconstriction.

4. The renal hemodynamic effects of 2-ClA are completely independent of tissue renin in the two-kidney, one-clip Goldblatt rat preparation (Churchill et al., 1984).

5. Sodium loading suppresses the renin–angiotensin system, but enhances CHA-induced vasoconstriction in the isolated perfused rat kidney (Rossi et al., 1987b).

6. Both CHA and angiotensin II elicit vasoconstriction of the isolated perfused rat kidney, and at equipotent concentrations of these agonists, saralasin (angiotensin II antagonist) completely blocks angiotensin's effect, but fails to attenuate CHA's effect (Rossi et al., 1987b).

These observations support the concepts that A_1 adenosine receptors and angiotensin II receptors are separate, distinct biochemical entities and that occupation of either receptor by its agonist leads to renal vasoconstriction by completely independent mechanisms.

Thus, although changes in dietary sodium have reproducible effects on adenosine-induced renal vasoconstriction in vivo, the mechanism remains to be elucidated.

3.4.2. Prostaglandins

Some vasoconstrictors (e.g., α-adrenergic agonists and angiotensin II) induce the synthesis of prostaglandins, which in turn attenuate the agonist-induced vasoconstriction. Since adenosine can constrict the renal vasculature, it is reasonable to suppose that the constriction might be associated with the production of vasodilator prostaglandins, and therefore that prostaglandin synthesis inhibitors would potentiate the vasoconstriction. Consistently, meclofenamate potentiates renal vasoconstriction in response to bolus injections of adenosine in cats (Spielman and Osswald, 1978), and in both cats (Spielman and Osswald, 1978) and dogs (Spielman and Osswald, 1979), it potentiates posthypoxic renal vasoconstriction, which is mediated by adenosine (*see below*). In dogs, indomethacin potentiates the transient increase in renal resistance in response to continuous iv infusions of adenosine and NECA (Beck et al., 1984), and in rats, indomethancin restores the ability of adenosine to constrict the renal vasculature during ureteral occlusion and aortic clamping (Haas and Osswald, 1981). On the other hand, although ATP and ADP stimulate renal prostaglandin synthesis, adenosine does not, at least not at the concentrations tested (Needleman et al., 1974; Schwartzman et al., 1981). Moreover, even if indomethacin and meclofenamate potentiate adenosine-induced renovascular changes, the mechanism could be unrelated to inhibition of prostaglandin synthesis, since it has been shown that both substances are weak antagonists of cellular adenosine uptake (Phillis and Wu, 1981).

3.5. Second Messengers

Collectively, the above observations indicate that A_1 and A_2 receptors induce constriction of afferent arterioles and dilation of afferent and efferent arterioles, respectively. Increases and decreases in the contractility of vascular smooth muscle are usually taken as evidence of increases and decreases in the intracellular concentration of free calcium. How do A_1 and A_2 receptors increase and decrease intracellular calcium?

A reasonable hypothesis can be based on the observations that A_1 and A_2 receptors inhibit and stimulate, respectively, adenylate cyclase activity in many cells (Daly, 1982), and that cyclic AMP stimulates calcium-activated ATPase activity, the biochemical correlate of active calcium efflux and sequestration (Rasmussen and Barrett, 1984). Thus, A_1 adenosine receptors could increase intracellular calcium and cause contraction by decreasing the rate of cyclic AMP-stimulated removal of calcium from the cytosol, and A_2 adenosine receptors could decrease intracellular calcium and cause relaxation by increasing the rate of this process, as shown in Fig. 3.

Two lines of evidence demonstrate that A_1-induced renal vasoconstriction is not attributable to decreased calcium efflux and sequestration in response to decreased cyclic AMP. First, pertussis toxin blocks receptor-induced inhibition of adenylate cyclase, but does not affect CHA-induced vasoconstriction of the isolated perfused rat kidney (Rossi et al., 1988). Second, methoxyverapamil blocks CHA-induced vasoconstriction of the isolated perfused rat kidney (Rossi et al., 1988), and verapamil blocks the transient adenosine-induced renal vasoconstriction in vivo (Arend et al., 1984; Miklos and Juhasz-Nagy, 1984; Macias-Nunez et al., 1985). These observations indicate that A_1-induced vasoconstriction is mediated by increased calcium influx through voltage-operated calcium channels, rather than by decreased calcium efflux and sequestration. The mechanism by which A_1 receptors activate voltage-operated calcium channels is unknown. Tetrodotoxin, a specific blocker of voltage-operated sodium channels, fails to affect CHA-induced vasoconstriction of the isolated perfused rat kidney (Rossi et al., 1988), which suggests that activation of calcium channels is not dependent upon a receptor-mediated, sodium-dependent membrane depolarization.

Similarly, A_2-induced renal vasodilation may occur by a mechanism that is independent of increased adenylate cyclase activity, increased cyclic AMP, and increased calcium efflux and sequestration. After a lengthy study of the effects of adenosine analogs on several smooth-muscle preparations, Baer and Vriend (1985) concluded that "a classification of smooth muscle adenosine receptors according to criteria established for cyclase-coupled receptors may be inappropriate or misleading, particularly with respect to implications

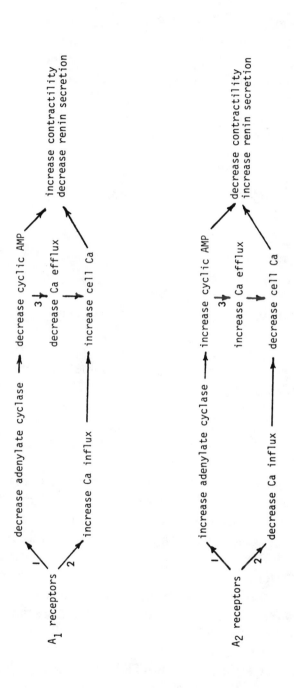

Fig. 3. Second messengers (cyclic AMP and calcium) mediating the effects of adenosine receptors. Pathway 1: adenosine receptors alter cyclase activity, and cyclic AMP acts as second messenger (independently of calcium). Pathway 2: adenosine receptors alter calcium influx, and calcium acts as second messenger (independently of cyclic AMP). Pathway 3: adenosine receptors alter cyclase activity, cyclic AMP alters calcium efflux (cyclic AMP stimulates active calcium efflux and sequestration), and calcium acts as "final" second messenger.

of adenylate cyclase involvement...." Possibly adenosine-induced renal vasodilation is attributable to decreased calcium influx; adenosine is known to act as a (voltage-operated) calcium-channel antagonist in cardiac muscle (Belardinelli et al., 1979), and the effect may be independent of cyclic AMP (Bohm et al., 1984; Bruckner et al., 1985). Alternatively, or in addition, adenosine could desensitize vascular smooth-muscle myofilaments to the existing level of cytoplasmic free calcium (Bradley and Morgan, 1985).

3.6. Regulation of Renal Hemodynamics by Adenosine

3.6.1. Autoregulation

The kidney exhibits autoregulation of both blood flow and glomerular filtration rate; both remain relatively constant despite changes in renal arterial pressure over a range from well below to well above normal (e.g., 80–180 mm Hg). Afferent arterioles are clearly the major, if not the only, site of autoregulatory changes in resistance. They must constrict in response to increased perfusion pressure, thereby simultaneously preventing an increase in blood flow and an increase in glomerular capillary pressure and glomerular filtration rate, and they must dilate in response to decreased perfusion pressure. These autoregulatory changes in afferent arteriolar resistance are intrinsic, since they are independent of the renal nerves and circulating humoral factors, and they are exhibited by isolated kidneys perfused in vitro with physiological saline solutions.

Many theories have been advanced to explain renal autoregulation, and this chapter is not the place to review all of them. However, adenosine is the proposed mediator in one variant of the "metabolic theory" (Thurau, 1964; Schnermann et al., 1977; Osswald et al., 1980, 1982; Spielman and Thompson, 1982; Osswald, 1983, 1984). To explain afferent arteriolar constriction in response to increased renal perfusion pressure according to this theory, it is proposed that the glomerular filtration rate increases slightly, which increases the load of sodium delivered to the macula densa region of the distal tubule. This entails an increase in the rate of active sodium reabsorption, and, as a consequence, there are increases in ATP hydrolysis and adenosine production and release. Adenosine released from macula densa cells acts on the adjacent afferent arteriolar vascular smooth-muscle cells to increase contractility.

Despite a certain plausibility of this theory, and despite all the observations that can be marshalled in its support (Osswald et al., 1980, 1982; Osswald, 1983, 1984), there are several fatal flaws.

1. Metabolic theories of renal autoregulation in general "would be hard pressed to meet some of the time constants (seconds) actually observed" (Visscher, 1964).

2. Equally compelling, isolated perfused kidneys autoregulate in the complete absence of oxygen for more than 10 min (Waugh, 1964).

3. Macula densa cells are as close to efferent as to afferent arterioles, and in any case, they are close to only a small fraction of the total number of arteriolar vascular smooth-muscle cells.

4. In response to decreased renal perfusion pressure, adenosine increases (Scott et al., 1965; Katholi et al., 1985), rather than decreases as the theory would require.

5. Several features of the renovascular response to adenosine are completely incompatible with the theory. In autoregulation, increased perfusion pressure induces a persistent increase in renovascular resistance such that blood flow does not change. In contrast, adenosine-induced changes in renovascular resistance are biphasic with respect to concentration (Fig. 2; vasoconstriction, no effect, vasodilation) and with respect to time (transient decrease in blood flow followed by a prolonged increase). In autoregulation, blood flow and glomerular filtration rate are not dissociated; both remain relatively constant in the face of changes in renal perfusion pressure. In contrast, adenosine usually drives renal blood flow and glomerular filtration rate in opposite directions; flow increases while filtration decreases. Autoregulatory changes in resistance are confined to the afferent arteriole. In contrast, there is strong evidence that adenosine affects both afferent and efferent arterioles.

6. Finally, there are two independent lines of pharmacological evidence against the theory. Adenosine receptor antagonists

do not block renal autoregulation (Premen et al., 1985); to the contrary, theophylline actually restores impaired autoregulation in some circumstances (Osborn et al., 1983). Similarly, adenosine uptake antagonists do not necessarily potentiate renal autoregulatory responses; to the contrary, papaverine has been used classically to "paralyze" renal autoregulation (Thurau, 1964), although the mechanism of this effect has not been shown to be mediated by increased adenosine.

3.6.2. Tubuloglomerular Feedback

Phenomenologically, tubuloglomerular feedback is the sensitivity of the single nephron glomerular filtration rate to changes in the flow of tubular fluid in the macula densa region of the tubule. Micropuncture studies have established that increasing tubular fluid flow rate in this region leads to a decrease in the filtration rate of the same nephron, which is mediated by afferent arteriolar constriction (Wright, 1981). In our perhaps unpopular opinion, it is unlikely that the tubuloglomerular feedback mechanism plays a significant role in the physiological regulation of glomerular filtration rate, for the following reason: As a result of the extensive reabsorption of filtrate in more proximal segments, flow in the macula densa region is normally about one-fifth the glomerular filtration rate. It follows that a gain of more than five (a1-mL/min increase in flow in the macula densa region eliciting more than a 5-mL/min decrease in filtration) would be required to meet the definition of a "feedback" control mechanism. However, the measured gain is approximately one order of magnitude less than five (Wright, 1981). Regardless of this issue, some mediator must transmit information concerning tubular fluid flow and/or composition from the tubular cells to the afferent arteriolar vascular smooth-muscle cells of the same nephron, and Osswald and coworkers have marshalled impressive evidence to support the hypothesis that adenosine is the mediator (Schnermann et al., 1977; Osswald et al., 1980, 1982; Osswald, 1983, 1984). The mechanism is identical to that proposed for adenosine-mediation of renal autoregulation: increased delivery of sodium to macula densa cells elicits increased active sodium reabsorption, increased ATP hydrolysis and adenosine production and release, and increased adenosine-induced constriction of the afferent arteriole. In support, tubuloglomerular feedback is abolished by methylxanthines (Schnermann et al., 1977) and by adenosine deaminase (Osswald e. al., 1980). Furthermore, tubuloglomerular feedback is potentiated by adenosine uptake antagonists (Osswald et al., 1980, 1982). This is virtually definitive evidence that adenosine plays a critical role in the tubuloglomerular feedback response.

Although it is frequently asserted that the tubuloglomerular feedback mechanism accounts at least in part for autoregulation of blood flow and glomerular filtration rate, the compelling evidence that adenosine participates in the tubuloglomerular feedback response, and the even more compelling evidence that adenosine cannot play a role in autoregulation, strongly suggests that tubuloglomerular feedback and autoregulation are separate, and independently mediated, phenomena.

3.6.3. Posthypoxic Renal Vasoconstriction

Endogenously released adenosine is an important local factor that controls blood flow through many vascular beds. In these beds, adenosine dilates the resistance vessels and thereby increases blood flow. Adenosine production and release are stimulated by hypoxia, and many organs exhibit posthypoxic (or postischemic) vasodilation and hyperemia caused by the accumulation of adenosine released during the hypoxic period. In contrast, the kidney frequently exhibits a posthypoxic vasoconstriction, and several observatons are consistent with adenosine mediation.

1. Hypoxia does increase renal adenosine production and release.
2. Exogenous adenosine, particularly if given as a bolus, can increase renovascular resistance.
3. Posthypoxic renal vasoconstriction is potentiated by substances that antagonize adenosine uptake: dipyridamole (Ono et al., 1966; Sakai et al., 1968; Osswald, 1984), indomethacin, and meclofenamate (Spielman and Osswald, 1978, 1979).
4. Posthypoxic renal vasoconstriction is antagonized by theophylline (Osswald et al., 1977; Osswald, 1984).

3.6.4. Acute Renal Failure

Acute renal failure (ARF) is a syndrome of "abrupt decline in renal function sufficient to result in retention of nitrogenous waste" (Anderson and Schrier, 1980). Despite a great deal of investigation, the pathogenesis of the syndrome is poorly understood. The initiating events can be separated into two broad categories of renal injury: "ischemic" and "nephrotoxic" (Stein et al., 1978). The course of the resulting ARF has been divided into initiation, maintenance, and recovery phases, since different pathogenic factors may be involved in the retention of nitrogenous waste during the different phases. The pathogenic mechanisms postulated to mediate the retention of nitrogenous waste (creatinine, urea) can be classified into two broad categories: (a) hemodynamic factors, consisting of a decrease in, and a redistribution of, renal blood flow; a marked fall in filtration fraction and glomerular filtration rate; and a decrease in glomerular capillary ultrafiltration coefficient; and (b) tubular factors, consisting of tubular obstruction and "backleakage" of glomerular filtrate. The relative contributions of these various factors, both in human and in experimental animal models, is a matter of considerable controversy.

Recent investigations have yielded a better understanding of the processes involved in tubular cell injury and death (Humes, 1986). However, hemodynamic changes are invariably associated with tubular cell injury, and the mechanism and significance remain uncertain. These hemodynamic changes are usually observed in the absence of consistent evidence of direct anatomic injury to the renal vasculature, and this has led to the proposal that the altered hemodynamics may represent a functional response to tubular cell injury and dysfunction.

Churchill and Bidani (1982) proposed that adenosine mediates the hemodynamic changes and that the hemodynamic changes are pathogenic in reducing glomerular filtration rate. According to their hypothesis, a state of energy deficit (impaired oxidative phosphorylation resulting from hypoxic, ischemic, or nephrotoxic tubular cell injury) leads to decreased cellular ATP levels and increased adenosine production and release; adenosine then acts on afferent and efferent arterioles to produce the hemodynamic changes. Several previous observations are consistent with this hypothesis. Renal ischemia results in decreased cellular ATP levels and increased adenosine production and release. Nephrotoxic injury of renal tubular cells also decreases cellular ATP levels, but it is not known if this is accompanied by increased adenosine production and release. The hemodynamic changes seen after administration of adenosine and adenosine analogs strikingly mimic the hemodynamic changes of ARF. These include variable changes in renal blood flow and its distribution, but consistent decreases in filtration fraction and glomerular filtration rate. It is of interest that sodium loading, which attenuates the renal hemodynamic effects of adenosine (*see above*), also ameliorates the severity of experimental ARF (Flamenbaum, 1973), and that conversely, sodium deprivation potentiates both the renal hemodynamic response to adenosine (*see above*) and the severity of experimental ARF (Flamenbaum, 1973).

Pharmacological manipulation of the renal adenosine system in experimental models of ischemic ARF has provided additional support for the hypothesis that adenosine mediates the hemodynamic changes, at least in part, and that these changes are of pathogenic significance. Theophylline has protective effects on renal function in an ischemic ARF model, that is produced in rats by unilateral occlusion of a renal artery for 30 or 45 min. Renal plasma flow and glomerular filtration rate are higher in the previously ischemic kidneys of rats treated with theophylline that in rats treated with the vehicle during both the initiation (Lin et al., 1986) and the maintenance phases (Lin et al., 1988). Moreover, dipyridamole enhances the severity of ARF in the initiation phase of this model (Lin et al., 1987). Ischemia is considered to play a significant role in glycerol-induced myoglobinuric ARF in rats, and in this model also, theophylline has protective effects (Bidani and Churchill, 1983) that are dose-dependent and independent of any effects on sodium excretion or tubuloglomerular feedback (Bidani et al., 1987). Bowmer et al. (1986) have shown that 8-phenyltheophylline, a more potent adenosine receptor antagonist, has similar protective effects in the glycerol model. Theophylline has protective effects in a model of endotoxin-induced ARF produced in rats by the administration of *E. coli* lipopolysaccharide (Churchill et al., 1987). On the other hand, adenosine-mediated hemodynamic changes seem to play a role in only some experimental models of nephroxic ARF. Thus, theophylline has protective effects on glomerular filtration rate in amphotericin-induced ARF (Heidemann et al., 1983) and in cisplatinum-induced ARF (Heidemann et al., 1982), but not in cyclosporin-induced ARF (Gerkins and Smith, 1985) or in $HgCl_2$-induced ARF (P. C. Churchill, unpublished). Perhaps adenosine production and release are not elevated in the latter two models.

3.7. Summary

A_1 and A_2 receptors appear to induce constriction and dilation of the renal vasculature, respectively. The mechanisms (altered calcium influx or efflux and sequestration) and second messengers (Cyclic AMP, calcium) remain to be fully elucidated. A_1 receptors appear to be confined to the afferent arterioles, but A_2 receptors appear to be present on both afferent and efferent arterioles. Such a distribution can account for the variable effects that adenosine and its analogs have on blood flow and for the invariant effect they have on glomerular filtration rate. Whether the level of endogenously released adenosine is normally high enough to affect renal hemodynamics cannot be stated with certainty. On the one hand, measured concentrations of adenosine in renal extracellular fluids are on the order of $1\mu M$, and this is more than enough to saturate A_1 receptors (affinity constant, $10 \, nM$ [Daly, 1982]). It follows that theophylline should always produce renal hemodynamic effects. On the other hand, Homer Smith (1951) wrote that the renal hemodynamic effects were varied and unpredictable; indeed, theophylline has been shown in many recent studies to block the renal hemodynamic effects of exogenous adenosine and adenosine analogs, but not to have any hemodynamic effects in their absence (Churchill, 1982; Spielman, 1984; Arend et al., 1985; Premen et al., 1985; Dunham and Vince, 1986). There is evidence that increased endogenous adenosine is involved in the tubuloglomerular feedback response, in posthypoxic renal vasoconstriction, and in some experimental models of ARF. Also, increased adenosine mediates the renal hemodynamic effects of dipyridamole (Arend et al., 1985), of maleic acid (Arend et al., 1986), and, probably, of papaverine. The hemodynamic effects of papaverine are strikingly "adenosine-like": variable changes in blood flow (Hashimoto and Kumakura, 1965; Witty et al., 1971, 1972; Freeman et al., 1974;

Blaine, 1978; Gaal et al., 1978) associated with an invariant reduction in glomerular filtration rate (Witty et al., 1972; Blaine, 1978; Gaal et al., 1978).

4. Renin Secretion

4.1. The Juxtaglomerular Apparatus (JGA)

The renin–angiotensin–aldosterone system plays a central role in electrolyte homeostasis and in the regulation of arterial blood pressure. The level of activity of this system is determined primarily by the rate at which the kidneys secrete renin into the blood. Physiologically, renin secretion is regulated by two intrarenal mechanisms (the baroreceptor and macula densa mechanisms), the renal sympathetic nerves, and several circulating organic and inorganic substances (catecholamines, angiotensin II, K, Mg). Exogenous adenosine and analogs have been shown to affect renin secretion, and since the kidney produces and releases adenosine, it is possible that endogenous adenosine is one of those substances that regulate renin secretion physiologically.

The renin-secreting juxtaglomerular apparatus (JGA) is a specialized region of each nephron, which is composed of three anatomical structures or cell types: granular juxtaglomerular cells (JG cells), macula densa, and mesangial cells. The JG cells, found primarily in the media of the afferent arteriole just adjacent to the glomerulus, are the site of renin synthesis, storage, and release. There is evidence that these cells have receptors for α- and β-adrenergic agonists, for angiotensin II, and for vasopressin (Keeton and Campbell, 1980). Therefore, it is reasonable to suppose that adenosine affects renin secretion by acting directly on JG-cell adenosine receptors. On the other hand, it is well-established that the JG cells respond to some function of afferent arteriolar transmural pressure, such as stretch; increases and decreases in stretch inhibit and stimulate renin secretion, respectively (the baroreceptor mechanism). Since adenosine affects the contractility of afferent arteriolar vascular smooth muscle, it could affect renin secretion not only by acting directly on putative JG-cell adenosine receptors, but also by altering the stretch of the afferent arteriole in the vicinity of the JG cells. Furthermore, the renal sympathetic nerves innervate the JG cells, and renin secretion is stimulated by norepinephrine released from the nerves (the β-adrenergic mechanism). Adenosine has been shown to antagonize transmitter release from the renal sympathetic nerves (*see below*), and therefore, it could affect renin secretion by the β-adrenergic mechanism. The macula densa is the tubular segment that marks the transition between the ascending limb of the loop of Henle and the distal tubule. This region of the tubule lies in close contact with the arterioles of the same nephron, and several observations support the hypothesis that renin secretion is inversely related to sodium reabsorptive flux in the macula densa segment (the macula densa mechanism). Since adenosine affects the tubular sodium reabsorptive rate (*see below*), it could alter renin secretion via the macula densa mechanism by altering the amount of sodium delivered to and/or reabsorbed by the cells in the macula densa segment. To further complicate matters, it has been proposed that adenosine is the chemical signal that the macula densa cells generate to affect the activity of the adjacent renin-secreting JG cells, as will be discussed in greater detail below. The mesangial cells are interstitial cells in contact with both the JG cells and the macula densa cells. There is no evidence that these cells play any role in controlling renin secretion.

4.2. Exogenous Adenosine

That adenosine inhibits renin secretion in a variety of experimental preparations is well-documented. Inhibitory effects have been observed in dogs (Tagawa and Vander, 1970; Arend et al., 1984, 1986; Spielman, 1984; Macias-Nunez et al., 1985, 1986; Premen et al., 1985; Hall and Granger, 1986a; Opgenorth et al., 1986; Deray et al., 1987), in rats (Osswald et al., 1978a), in rat renal cortical slices (Skott and Baumbach, 1985; Barchowsky et al., 1987), in isolated superfused rat glomeruli with attached afferent arterioles (Skott and Baumbach, 1985), and in isolated rabbit afferent arterioles (Itoh et al., 1985). Adenosine-induced inhibition of renin secretion is blocked by theophylline in dogs (Spielman, 1984) and in isolated rabbit afferent arterioles (Itoh et al., 1985), and by 8-phenyltheophylline in rabbit renal cortical slices (Barchowsky et al., 1987).

Although such inhibitory effects are stressed in theories concerning the role of adenosine in controlling renin secretion (Tagawa and Vander, 1970; Osswald et al., 1978a; Osswald, 1983, 1984; Spielman and Thompson, 1982), adenosine also has been observed to have no effect on renin secretion in dogs fed a normal-sodium diet (Winer et al., 1971), in dogs given dibutyryl cyclic AMP to stimulate renin secretion (Deray et al., 1987), in rats fed a high-sodium diet (Osswald et al., 1978a), in superfused rat glomeruli (Skott and Baumbach, 1985), and in isolated rabbit afferent arterioles with attached macula densa cells (Itoh et al., 1985). Furthermore, adenosine actually stimulates renin secretion in isolated perfused rat kidneys (Hackenthal et al., 1983). Thus, adenosine has been reported to inhibit, to have no effect on, and to stimulate renin secretion in a variety of experimental preparations.

4.3. Adenosine Analogs

Similar results have been obtained with adenosine analogs. At submicromolar concentrations, CHA is selective for A_1 receptors (Daly, 1982); it inhibits renin secretion in anesthetized rats (Cook and Churchill, 1984; Churchill and Bidani, 1987), in isolated perfused rat kidneys (Murray and Churchill, 1984, 1985), in rat renal cortical slices (Churchill and Churchill, 1985; Churchill et al., 1987; Rossi et al., 1987a) and in isolated rabbit afferent arterioles (Itoh et al., 1985). At submicromolar concentrations, R-PIA is also A_1-selective, and it inhibits renin secretion in rabbit renal cortical slices (Barchowsky et al., 1987). At μM and higher concentrations, CHA activates A_2 as well as A_1 receptors (Daly, 1982), and high concentrations stimulate renin secretion in isolated perfused rat kidneys (Hackenthal et al., 1983) and in rat renal cortical slices (Churchill and Churchill, 1985; Churchill et al., 1987). At submicromolar concentrations, NECA activates A_2 receptors (Daly, 1982), and it stimulates renin secretion in anesthetized rats (Churchill and Bidani, 1987), in isolated perfused rat kidneys (Hackenthal et al., 1983; Murray and Churchill, 1984, 1985), and in rat renal cortical slices (Churchill and Churchill, 1985). Thus, like adenosine itself, adenosine analogs can either inhibit or stimulate renin secretion in vivo and in various in vitro preparations.

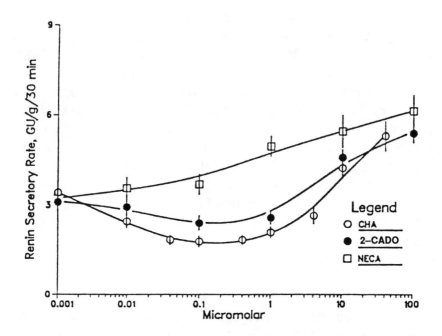

Fig. 4. Renin secretory rate of rat renal cortical slices vs the logarithm of the concentrations of CHA, 2-CADO, and NECA. Means ± SEMs; n=12–30 separate observations for each point. The basal secretory rate, not shown, averaged 3.87 GH/g/30 min. Reprinted from Churchill and Churchill, 1985.

The results that Churchill and Churchill (1985) obtained using rat renal cortical slices are shown in Fig. 4. Over the 10 nM–1µM concentration range, renin secretion was inhibited, and the relative positions of the three curves demonstrate that, with respect to inhibition of renin secretion, the order of potency is CHA > 2-CADO >>> NECA. Both the concentration range (nM–µM) and the potency order are consistent with the hypothesis that activation of A$_1$ receptors inhibits renin secretion. Over the 1–100 µM concentration range, NECA stimulated renin secretion in comparison with the basal rate, and over the same concentration range, the other two agonists began to stimulate secretion in comparison with the inhibited secretory rates at lower concentrations. Above 10 µM, all agonists stimulated in comparison with the basal secretory rate. The relative positions of the curves indicate that the order of potency for stimulation is NECA > 2-

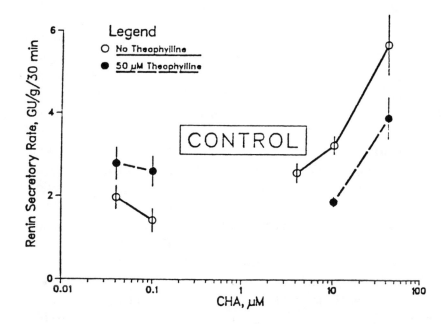

Fig. 5. Antagonistic effects of theophylline on CHA-induced and CHA-stimulated renin secretion of rat renal cortical slices. Means ± SEMs; n=6 observations for each point. The height of the horizontal rectangle represents the mean ± 1 SEM of the basal or control secretory rate. Reprinted with permission from Churchill and Churchill, 1985.

CADO. This order of potency and the concentration range are consistent with the hypothesis that activation of A_2 receptors stimulates renin secretion. Thus, A_1 and A_2 receptors induce inhibition and stimulation of renin secretion, respectively. All of the in vivo observations cited above are consistent; two types of receptors, having opposite effects on renin secretion, can obviously account not only for inhibition and stimulation of renin secretion, but also for no effect at all (A_1-induced inhibition balanced by A_2-induced stimulation). Furthermore, as can be seen in Fig. 5, theophylline blocks both the inhibitory effect of A_1 receptors and the stimulatory effect of A_2 receptors. It would be very difficult to explain the effects of theophylline shown in Fig. 5 on the basis of any single known effect (e.g., phosphodiesterase inhibition) other than antagonism of A_1 and A_2 adenosine receptors.

4.4. Interactions with Other Controlling Mechanisms

Although adenosine could theoretically alter renin secretion indirectly (i. e., by acting on a non-JG cell to produce a change that in turn affected the JG cell), there is no evidence to suggest that this is the case. Adenosine's renin secretory effects in vivo are independent of the baroreceptor mechanism (Arend et al., 1984; Macias-Nunez et al., 1985), the macula densa mechanism (Tagawa and Vander, 1970; Opgenorth et al., 1986), and the renal nerves (Cook and Churchill, 1984). Furthermore, the observations that adenosine and/or adenosine receptor agonists can inhibit and/or stimulate renin secretion in in vitro preparations, such as isolated perfused kidneys (Hackenthal et al., 1983; Murray and Churchill, 1984, 1985), renal cortical slices (Churchill and Churchill, 1985; Skott and Baumbach, 1985), and isolated afferent arterioles (Itoh et al., 1985), virtually exclude explanations based on such mechanisms. Therefore, even if non-JG cells are present in all preparations used to study renin secretion, it seems reasonable to suppose that A_1 and A_2 adenosine receptors exist on the plasma membranes of JG cells and that occupation of these receptors leads directly to inhibition and stimulation of renin secretion, respectively. This does not exclude the possibility that indirect effects might modulate direct effects, however. For example, adenosine-induced inhibition of renin secretion in vivo might have both a direct component (adenosine acting on JG-cell A_1 receptors) and indirect components (adenosine-induced reduction in norepinephrine release from the renal sympathetic nerves).

4.5. Second Messengers

Theoretically, cyclic AMP could mediate the renin secretory effects of adenosine. Although a positive correlation has not been established between JG-cell cyclic AMP and renin secretory activity, pharmacological evidence strongly supports the hypothesis that cyclic AMP is a stimulatory second messenger in the renin secretory process (Churchill, 1985). Since there is considerable evidence that adenylate cyclase activity is inhibited and stimulated, respectively, by A_1 and A_2 adenosine receptor activation (Daly, 1982), inhibitory and stimulatory effects on renin secretion could be attributed to decreases and increases in JG-cyclic AMP.

On the other hand, calcium could be the second messenger. Several lines of evidence support the hypothesis that calcium is an inhibitory second messenger in the renin secretory process (Churchill, 1985). As discussed in section 3.6, occupation of A_1 and A_2 adenosine receptors leads to increased and decreased contractility of the afferent arteriole, respectively, and these effects may be mediated by increased and decreased intracellular calcium, respectively. JG cells, found in the afferent arteriole, are derived from vascular smooth-muscle cells, and it is reasonable to suppose that JG-cell A_1 and A_2 receptors lead to increased and decreased intracellular calcium, respectively, mediating inhibition and stimulation of renin secretion. In this context, it is interesting to compare the contractile responses in Fig. 2 with the renin secretory responses in Fig. 4. CHA inhibits renin secretion and increases vascular smooth-muscle contractility (intracellular calcium increases) over the range of concentrations $0.001–0.1\mu M$; there are inflection points in both dose–response curves at or near $0.1 \mu M$, where renin secretion begins to increase and contractility begins to decrease (intracellular calcium decreases); finally, at $10 \mu M$, renin secretion is above baseline and contractility is below. In contrast, NECA only stimulates renin secretion and decreases contractility (intracellular calcium decreases).

Finally, cyclic AMP and calcium could be sequential second messengers in the renin secretory response to adenosine, since cyclic AMP stimulates active calcium efflux and sequestration (*see* Fig. 3). Thus, A_1 receptor activation could lead to inhibition of renin secretion by the following sequence of events: inhibition of adenylate cyclase, decreased cyclic AMP, decreased active calcium extrusion, and increased intracellular calcium. This hypothetical sequence can account for the observations

1. That pertussis toxin blocks adenosine receptor-induced inhibition of renin secretion (Rossi et al., 1987a), presumably by preventing adenosine-induced inhibition of adenylate cyclase;

2. That calcium channel blockers do not block adenosine receptor-induced inhibition of renin secretion (Arend et al., 1984; Churchill and Churchill, 1985; Macias-Nunez et al., 1985), presumably because adenosine receptors increase intracellu-

lar calcium by decreasing calcium efflux, rather than by increasing calcium influx through a pathway that is affected by channel blockers; and

3. That adenosine receptor-induced inhibition of renin secretion is blocked by chelating extracellular calcium (Churchill and Churchill, 1985), presumably because this blocks the adenosine-induced decrease in calcium efflux.

Conversely, A_2 receptors could stimulate renin secretion by stimulating adenylate cyclase, increasing cyclic AMP, increasing active calcium extrusion, and thereby decreasing intracellular calcium. Such a sequence of events can explain the observations that adenosine receptor-induced stimulation of renin secretion can be blocked by vanadate, by ouabain, and by K-depolarization (Churchill and Churchill, 1985), all of which increase intracellular calcium, but by different mechanisms of action. Similarly, it has been shown that vanadate, ouabain, and K-depolarization also block isoproterenol-stimulated renin secretion (Churchill and Churchill, 1982), which is undoubtedly mediated by increased cyclic AMP.

4.6. Regulation of Renin by Adenosine

4.6.1. The Macula Densa Theory

Since the concentration of endogenously released adenosine is sometimes high enough to affect arteriolar resistance, and since the JG cells are embedded in the afferent arteriolar walls, it seems likely that endogenously released adenosine would affect JG cells. In fact, it has been proposed that adenosine is the chemical signal generated by macula densa cells to affect the renin secretory activity of the adjacent JG cells: increased sodium reabsorptive flux in the macula densa segment entails increased ATP hydrolysis and adenosine production and release, and this leads to increased adenosine-induced inhibition of renin secretion (Tagawa and Vander, 1970; Osswald et al., 1978a; Spielman and Thompson, 1982; Osswald, 1983, 1984). Although this is an attractive proposal, it does not take into account the biphasic effects of adenosine. A_1-induced inhibition of renin secretion would be observed only in the submicromolar range of adenosine concentrations; increasing adenosine concentrations above the point at which the A_1 receptors are fully saturated should lead to a reversal of the inhibitory effect, and finally, to stimulation of renin secretion (Fig. 4). Furthermore, without techniques for measuring the concentration of endogenously released adenosine in the vicinity of the JG cells and for determining that the adjacent macula densa cells are the source of the adenosine, the proposal is impossible to test directly.

Regardless of the cellular source of endogenously released adenosine, there is pharmacological evidence to suggest that it can affect renin secretion. Several investigators have reported that theophylline stimulates renin secretion in vivo (Winer et al., 1969; Allison et al., 1972; Ueda et al., 1978; Spielman 1984). Interestingly, other investigators have reported that theophylline has no effect on renin secretion in vivo (Johns and Singer, 1973). The lack of effect could mean that endogenous adenosine was either very low (below the threshold for activating A_1 receptors) or moderately high (sufficient to saturate A_1 receptors and to activate some A_2 receptors). Theoretically, if endogenous adenosine were sufficiently high to stimulate renin secretion, theophylline would be expected to block the effect and thereby inhibit renin secretion, as in Fig. 5. No inhibitory effects of theophylline have been reported.

As mentioned above, renal hemodynamic effects of papaverine are "adenosine-like," which suggests that they are produced by increased extracellular concentration of endogenously released adenosine. Similarly, papaverine's renin secretory effects are "adenosine-like." It inhibits renin secretion in sodium-deprived dogs (Gotshall et al., 1973), but not in sodium-replete dogs (Witty et al., 1971). Papaverine also antagonizes the renin stimulatory effects of thoracic caval constriction (Witty et al., 1972) and of epinephrine in dogs (Johnson et al., 1971), and it blocks the stimulatory effects of isoproterenol on renin secretion of isolated perfused rat kidneys (Fray, 1978) and rat renal cortical slices (Churchill et al., 1980). The inhibitory effects of papaverine are very difficult to explain on the basis of other known actions (e.g., phosphodiesterase inhibition). It would be of interest to determine if the inhibitory effects, traditionally attributed to blockade of the baroreceptor mechanism, can be antagonized by alkylxanthines.

4.7. Summary

Taken together, the experimental observations are consistent with the following conclusions.

1. JG cells have both A_1 and A_2 receptors, and their activation leads to inhibition and stimulation of renin secretion, respectively.
2. The inhibitory and stimulatory effects seem to be mediated by increased and decreased intracellular calcium, possibly resulting from decreased and increased calcium efflux (cyclic AMP stimulated).
3. Pharmacological evidence indicates that the concentration of endogenously released adenosine in the vicinity of JG cells is normally in the range where it either inhibits or has no effect on renin secretion.

5. Renal Tubular Transport

5.1. Adenosine, Adenosine Analogs, and Methylxanthines

Exogenous adenosine produces very intense antidiuretic and antinatriuretic effects in many species, including chickens (Nechay, 1966), rats (Osswald et al., 1975, 1978a), and dogs (Tagawa and Vander, 1970; Osswald, 1975; Osswald et al., 1978b; Spielman, 1984; Hall et al., 1985). These effects are receptor-mediated, since they are competitively antagonized by theophylline (Osswald, 1975) and mimicked by several adenosine analogs: 2-ClA (Churchill, 1982; Churchill et al., 1984; Churchill and Bidani, 1987), CHA (Cook and Churchill, 1984; Churchill and Bidani, 1987), R L-PIA (Churchill et al., 1987), and NECA (Churchill and Bidani, 1987). Moreover, theophylline blocks the antidiuretic and antinatriuretic effects of 2Cl-A (Churchill, 1982) and of L-PIA (Churchill et al., 1987).

Since exogenous adenosine and adenosine analogs have antidiuretic and antinatriuretic effects, and since methylxanthines are competitive antagonists of adenosine receptors (Daly, 1982), it seems reasonable to assume that the well-known diuretic and natriuretic effects of methylxanthines are produced by antagonism of the effects of endogenously released adenosine (Fredholm, 1980; Persson et al., 1982, 1986; Fredholm, 1984; Johannesson et al., 1985).

5.2. Mechanisms of Action

The diuretic and natriuretic effects of adenosine receptor antagonists, and the opposite effects of adenosine receptor agonists, could be produced by a variety of mechanisms. Explanations of their effects that are based on systemic effects (changes in cardiac output, blood pressure, neural activity, or hormone secretion) seem to be excluded by the observations that isolated perfused kidneys respond predictably to both agonists (Murray and Churchill, 1984, 1985) and antagonists (Verney and Winton, 1930; Viskoper et al., 1977). However, changes in urine flow and sodium excretion could be consequences of changes in renal hemodynamics; adenosine may induce a vasodilation in the juxtamedullary cortex (Spielman et al., 1980), and it is generally believed that juxtamedullary nephrons reabsorb water and sodium more avidly than cortical nephrons. Moreover, adenosine decreases the glomerular filtration rate, and therefore the filtered loads of water and sodium. However, this factor alone cannot easily account for the intensity of the antidiuretic and antinatriuretic effects, since the percentage decreases in urine flow and sodium excretion exceed by far the percentage decrease in glomerular filtration rate (Nechay, 1966; Tagawa and Vander, 1970; Osswald, 1975; Osswald, 1978a,b; Churchill, 1982; Cook and Churchill, 1984; Spielman, 1984; Hall et al., 1985; Churchill and Bidani, 1987). Conversely, methylxanthines have been noted to produce diuresis and natriuresis in the absence of detectable increases in blood flow or glomerular filtration rate (Spielman, 1984; Arend et al., 1985). Therefore, it seems reasonable to propose that adenosine-induced antidiuresis and antinatriuresis (and by inference, methylxanthine-induced diuresis and natriuresis [Davis and Shock, 1949; Brater et al., 1983]) can be mediated by both renal hemodynamic and direct tubular mechanisms. Consistent with direct tubular mechanisms, adenosine analogs stimulate active sodium transport in toad kidney cells (Lang et al., 1985), and adenosine and adenosine analogs increase water permeability, mimicking the effect of vasopressin, in isolated rabbit collecting-ducts (Dillingham and Anderson, 1985). The latter effect, however, requires extremely high concentrations (concentrations less than 100 μM were ineffective), which suggests that the effect cannot account for the antidiuretic effect produced by nM concentrations in vivo.

5.3. *Receptor Subclass and Second Messengers*

Although the renal excretory effects of several quite selective adenosine receptor agonists have been studied, it is impossible to draw any conclusions about the subclass of receptors that mediates antidiuretic and antinatriuretic effects in vivo. All of the difficulties in interpretation of results that were discussed in section 3 are applicable here. In addition, it is very likely that the excretory effects are produced by adenosine action at multiple sites with different subclasses of receptors: e.g., a decrease in glomerular filtration rate, produced by A_1 receptors on afferent arterioles and/or A_2 receptors on efferent arterioles, coupled with stimulated tubular reabsorption of water and sodium (mediated by A_1 and/or A_2 receptors). In any case, the receptors appear to have very high affinity (nM).

Similarly, it is difficult to draw any conclusions concerning cyclic AMP mediation of the transport/excretory effects that are consistent with all the literature. Roy (1984) found that adenosine receptor agonists had biphasic effects on adenylate cyclase activity of homogenates of vasopressin-sensitive pig kidney cells, inhibiting activity at low concentrations (nM–μM) and stimulating it at higher concentrations. In accord with this report, adenosine receptor agonists have biphasic effects on cyclic AMP levels in vasopressin-sensitive rabbit kidney cells (Arend et al., 1987). However, it is hard to reconcile the effects produced by nM concentrations in these studies (inhibited cyclase, decreased cyclic AMP) with the in vivo effects produced by nM concentrations (antidiuresis, antinatriuresis), particularly in view of the fact that cyclic AMP mediates the effects of vasopressin (antidiuretic hormone). One could postulate that the stimulatory effects on cyclase and cyclic AMP levels, by mimicking the effect of vasopressin, would be related to antidiuresis in vivo. However, stimulatory effects, e.g., of NECA on cyclic AMP, were detected at 0.1–1.0 μM (Arend et al., 1987), whereas stimulatory effects of NECA on water permeability required 100 μM. The toad kidney cells present an interesting contrast. Whereas 0.1 μM 2-CADO nearly doubled active sodium transport, a concentration two orders of magnitude higher was required before any change in cyclic AMP could be detected (Lang et al., 1985). The reports that adenosine does not affect cyclic AMP levels in rat cortical and medullary tissue slices (Abboud and Dousa, 1983) and that adenosine analogs have no effects on adenylate cyclase activity in membrane preparations of renal cortex and medulla of both rats (Woodcock et al., 1984) and humans (Woodcock et al., 1986) also do not support cyclic AMP mediation of adenosine-induced changes in tubular transport processes, since both cortex and medulla consist primarily of tubular cells.

5.4. Summary

Adenosine receptor agonists have antidiuretic and antinatriuretic effects. Alkylxanthines block the effects of exogenous agonists, and even in their absence, tend to have diuretic and natriuretic effects. These diuretic and natriuretic effects are probably mediated by antagonism of the effects of endogenously released adenosine. The effects of adenosine receptor agonists and antagonists may, in part, be attributable to renal hemodynamic changes (changes in glomerular filtration rate and the distribution of blood flow within the kidney), but direct effects on tubular reabsorptive processes are likely. The tubular sites of action (proximal, loop of Henle, distal, collecting duct), the receptor subclass which mediates the effects (A_1 and/or A_2), and whether or not cyclic AMP is the second messenger, are all unknown.

6. Renal Nerves

6.1. Efferent Sympathetic Nerves

The renal sympathetic nerves terminate on nearly all types of cells in the kidney, and transmitter release (norepinephrine) influences nearly all aspects of renal function—renal hemodynamics, tubular salt and water transport, and renin secretion. Hedqvist and Fredholm (1976) showed that adenosine antagonizes transmitter release from the renal efferent nerves in rabbits, and this has been confirmed and extended to other species (Hedqvist et al., 1978; Ekas et al., 1981, 1983). The effect is receptor-mediated, since it is antagonized by theophylline (Hedqvist et al., 1978). Such an effect would be predicted to produce renal vasodilation, diuresis, and natriuresis, and to inhibit renin secretion. It remains to be determined to what extent this effect modulates the direct effects of adenosine receptor agonists on renal hemodynamics, renal tubular transport processes, and renin secretion.

6.2. Afferent Nerves

Katholi (1983) has marshalled evidence to suggest that the renal afferent nerves play an important role in the pathogenesis of hypertension, and several experimental observations are consistent with the hypothesis that adenosine, by activating the afferent nerves, is pathogenic (Katholi et al., 1983, 1984, 1985). Others (Jing-Yun et al., 1985) have disputed the importance of the renal afferent nerves in hypertension. Regardless of the issue of hypertension, afferent renal nerve activity is increased by adenosine, whether infused into either the renal artery or the renal pelvis (Katholi et al., 1983, 1984). The difference in onset of the responses suggest the existence of adenosine-sensitive nerve endings in or near the renal pelvis. These adenosine-sensitive "chemoreceptors" should not be confused with "adenosine receptors," however, since theophylline does not antagonize the responses.

7. Conclusions

Adenosine has been shown to affect renal vascular smooth-muscle cells, renin-secreting JG cells, renal tubular cells, and the terminals of both afferent and efferent renal nerves. Many of the renal effects of adenosine are striking, if not completely unique to the kidney: the biphasic character of many effects (vasoconstriction/vasodilation; inhibited/stimulated renin secretion; inhibited/stimulated adenylate cyclase activity; decreased/increased cyclic AMP levels); the constriction of the renal vasculature in contrast to the dilation typical of most vascular beds; the A_1-receptor induced activation of voltage-sensitive calcium channels in the renal vasculature in contrast to the A_1 receptor-induced antagonism of voltage-sensitive calcium channels in the heart. Perhaps this uniqueness of the kidney will provide incentive to complete the work that many have begun: establishing if and when endogenously released adenosine plays an important role in regulating renal function.

Acknowledgments

The research of the authors has been supported by a grant from the National Institutes of Health (HL 24880), by an institutional Biomedical Research Support Grant from Rush Medical College, and by Biomedical Applications of Detroit.

References

Abboud, H. E., and Dousa, T. P. (1983) Action of adenosine on cyclic 3',5'-nucleotides in glomeruli. *Am. J. Physiol.* **244,** F633–F638.

Allison, D. J., Tanagawa, H., and Assaykeen, T. A. (1972) The effects of cyclic nucleotides on plasma renin activity and renal function in dogs, in *Control of Renin Secretion,* (Assaykeen, T. A., ed.), Plenum, New York, pp. 3–47.

Anderson, R. J., and Schrier, R. W. (1980) Clinical spectrum of oliguric and nonoliguric acute renal failure, in *Acute Renal Failure* (Brenner, M. M., and Stein, J. H., eds.), Churchill-Livingstone, New York, pp. 1–16.

Angielski, S., Le Hir, M., and Dubach, U. C. (1983) Transport of adenosine by renal brush border membranes. *Pflugers Arch.* **397,** 75–77.

Arend, L. J., Thompson, C. I., and Spielman, W. S. (1985) Dipyridamole decreases glomerular filtration in the sodium-depleted dog. Evidence for mediation by intrarenal adenosine. *Circ. Res.* **56,** 242–251.

Arend, L. J., Haramati, A., Thompson, C. I., and Spielman, W. S. (1984) Adenosine-induced decrease in renin release: Dissociation from hemodynamic effects. *Am. J. Physiol.* **247,** F447–F452.

Arend, L. J., Sonnenburg, W. K., Smith, W. L., and Spielman, W. S. (1987) A_1 and A_2 adenosine receptors in rabbit cortical collecting tubule cells. Modulation of hormone-stimulated cAMP. *J. Clin. Invest.* **79,** 710–714.

Arend, L. J., Thompson, C. I., Brandt, M. A., and Spielman, W. S. (1986) Elevation of intrarenal adenosine by maleic acid decreases GFR and renin release. *Kidney Int.* **30,** 656–661.

Baer, H. P. and Vriend, R. (1985) Adenosine receptors in smooth muscle: Structure–activity studies and the question of adenylate cyclase involvement in control of relaxation. *Can. J. Physiol. Pharmacol.* **63,** 972–977.

Barchowsky, A., Data, J. L., and Whorton, A. R. (1987) Inhibition of renin release by analogues of adenosine in rabbit renal cortical slices. *Hypertension* **9,** 619–623.

Beck, A., Seitelberger, R., and Raberger, G. (1984) Effects of indomethacin on changes in renal blood flow induced by adenosine and its analogues in conscious dogs. *Naunyn-Schmiedebergs Arch. Pharmacol.* **326,** 75–79.

Belardinelli, L., Rubio, R., and Berne, R. M. (1979) Blockade of Ca^{2+} dependent rat atrial slow action potentials by adenosine and lanthanum. *Pflugers Arch.* **380,** 19–27.

Bhanalaph, T., Mittelman, A., Ambrus, J. L., and Murphy, G. P. (1973) Effect of adenosine and some of its analogs on renal hemodynamics. *J. Med.* **4,** 178–188.

Bidani, A. K. and Churchill, P. C. (1983) Aminophylline ameliorates glycerol-induced acute renal failure in rats. *Can. J. Physiol. Pharmacol.* **61,** 567–571.

Bidani, A. K., Churchill, P. C., and Packer, W. (1987) Theophylline-induced protection in myoglobinuric acute renal failure: Further characterization. *Can. J. Physiol. Pharmacol.* **65,** 42–45.

Blaine, E. H. (1978) Sodium excretion after renal vasodilation by papaverine in conscious and anesthetized sheep. *Proc. Soc. Exp. Biol. Med.* **158,** 250–254.

Bohm, M., Bruckner, R., Hackbarth, I., Haubitz, B., Linhart, R., Meyer, W., Schmidt, B., Schmitz, W., and Scholz, H. (1984) Adenosine inhibition of catecholamine-induced increase in force of contraction in guinea-pig atrial and ventricular heart preparations. Evidence against a cyclic AMP- and cyclic GMP-dependent effect. *J. Pharmacol. Exp. Ther.* **230,** 483–492.

Bowmer, C. J., Collis, M. G., and Yates, M. S. (1986) Effect of the adenosine antagonist 8-phenyltheophylline on glycerol-induced acute renal failure in the rat. *Br. J. Pharmacol.* **88,** 205–212.

Bradley, A. B. and Morgan, K. G. (1985) Cellular Ca^{2+} monitored by aequorin in adenosine-mediated smooth muscle relaxation. *Am. J. Physiol.* **248,** H109–H117.

Brater, D. C., Kaojarern, S., and Chennavasin, P. (1983) Pharmacodynamics of the diuretic effects of aminophylline and acetazolamide alone and combined with furosemide in normal subjects. *J. Pharmacol. Exp. Ther.* **227,** 92–97.

Bruckner, R., Fenner, A., Meyer, W., Nobis, T.-M., Schmitz, W., and Scholz, H. (1985) Cardiac effects of adenosine and adenosine analogs in guinea-pig atrial and ventricular preparations: Evidence against a role of cyclic AMP and cyclic GMP. *J. Pharmacol. Exp. Ther.* **234,** 766–774.

Churchill, P. C. (1982) Renal effects of 2-chloroadenosine and their antagonism by aminophylline in anesthetized rats. *J. Pharmacol. Exp. Ther.* **222,** 319–323.

Churchill, P. C. (1985) Second messengers in renin secretion. *Am. J. Physiol.* **249,** F175–F184.

Churchill, P. C. and Bidani, A. K. (1982) Hypothesis: Adenosine mediates the hemodynamic changes in renal failure. *Med. Hypotheses* **8,** 275–285.

Churchill, P. C. and Bidani, A. K. (1987) Renal effects of selective adenosine receptor agonists. *Am. J. Physiol.* **252,** F299–F303.

Churchill, P. C. and Churchill, M. C. (1982) Isoproterenol-stimulated renin secretion in the rat: Second messenger roles of Ca and cyclic AMP. *Life Sci.* **30,** 1313–1319.

Churchill, P. C. and Churchill, M. C. (1985) A_1 and A_2 adenosine receptor activation inhibits and stimulates renin secretion of rat renal cortical slices. *J. Pharmacol. Exp. Ther.* **232,** 589–594.

Churchill, P. C., Bidani, A. K., and Schwartz, M. M. (1987) Renal effects of endotoxin in the male rat. *Am. J. Physiol.* **253,** F244–F250.

Churchill, P. C., McDonald, F. D., and Churchill, M. C. (1980) Effects of papaverine on basal and on isoproterenol-stimulated renin secretion from rat kidney slices. *Life Sci.* **27,** 1299–1305.

Churchill, P. C., Rossi, N. F., and Churchill, M. C. (1987) Renin secretory effects of N^6-cyclohexyladenosine: Effects of dietary sodium. *Am. J. Physiol.* **252**, F872–F876.

Churchill, P. C. Bidani, A. K., Churchill, M. C., and Prada, J. (1984) Renal effects of 2-chloroadenosine in the two-kidney Goldblatt rat. *J. Pharmacol. Exp. Ther.* **230**, 302–306.

Cook, C. B. and Churchill, P. C. (1984) Effects of renal denervation on the renal responses of anesthetized rats to cyclohexyladenosine. *Can. J. Physiol. Pharmacol.* **62**, 934–938.

Coulson, R. and Trimble, M. E. (1986) Effects of papaverine and theophylline on renal adenosine transport. *J. Pharmacol. Exp. Ther.* **239**, 748–753.

Daly, J. W. (1982) Adenosine receptors: targets for future drugs. *J. Med. Chem.* **25**, 197–207.

Davis, J. O. and Shock, N. W. (1949) The effect of theophylline ethylene diamine on renal function in control subjects and in patients with congestive heart failure. *J. Clin. Invest.* **28**, 1459–1468.

Deray, G., Branch, R. A., Herzer, W. A., Ohnishi, A., and Jackson, E. K. (1987) Adenosine inhibits beta-adrenoceptor but not DBcAMP-induced renin release. *Am. J. Physiol.* **252**, F46–F52.

Dillingham, M. A. and Anderson, R. J. (1985) Purinergic regulation of basal and arginine vasopressin-stimulated hydraulic conductivity in rabbit cortical collecting tubule. *J. Membr. Biol.* **88**, 277–281.

Drury, A. N. and Szent-Györgyi, A. (1929) The physiological activity of adenine compounds with especial reference to their action upon the mammalian heart. *J. Physiol.* (Lond.), **68**, 213–237.

Dunham, E. W. and Vince, R. (1986) Hypotensive and renal vasodilator effects of carbocyclic adenosine (aristeromycin) in anesthetized spontaneously hypertensive rats. *J. Pharmacol. Exp. Ther.* **238**, 954–959.

Ekas, R. D., Jr., Steenberg, M. L., and Lokhandwala, M. F. (1983) Increased norepinephrine release during sympathetic nerve stimulation and its inhibition by adenosine in the isolated perfused kidney of spontaneously hypertensive rats. *Clin. Exp. Hypertens. (A)* **A5 (1)**, 41–48.

Ekas, R. D., Jr., Steenberg, M. L., Eikenburg, D. C., and Lokhandwala, M. R. (1981) Presynatpic inhibition of sympathetic neurotransmission by adenosine in the rat kidney. *Eur. J. Pharmacol.* **76**, 301–307.

Eloranta, T. O. (1977) Tissue distribution of S-adenosylmethionine and S-adenosylhomocysteine in the rat. *Biochem. J.* **166**, 521–529.

Finkelstein, J. D. and Harris, B. (1973) Methionine metabolism in mammals. Synthesis of S-adenosylhomocysteine in rat tissue. *Arch. Biochem. Biophys.* **159**, 160–165.

Flamenbaum, W. (1973) Pathophysiology of acute renal failure. *Arch. Int. Med.* **131**, 911–928.

Fray, J. C. S. (1978) Stretch receptor control of renin release in perfused rat kidney: Effect of high perfusate potassium. *J. Physiol.* **282**, 207–217.

Fredholm, B. B. (1980) Are methylxanthine effects due to antagonism of endogenous adenosine? *Trends Pharmacol. Sci.* **1**, 129–131.

Fredholm, B. B. (1984) Cardiovascular and renal actions of methylxanthines. *Prog. Clin. Biol. Res.* **158**, 303–330.

Fredholm, B. B. and Hedqvist, P. (1978) Release of ^{3}H-purines from [^{3}H]- adenine labelled rabbit kidney following sympathetic nerve stimulation, and its inhibition by α-adrenoceptor blockade. *Br. J. Pharmacol.* **64**, 239–245.

Fredholm, B. B. and Sollevi, A. (1986) Cardiovascular effects of adenosine. *Clin. Physiol.* **6**, 1–21.

Fredholm, B. B., Hedqvist, P., and Vernet, L. (1978) Effect of theophylline and other drugs on rabbit renal cyclic nucleotide phosphodiesterase, 5'-nucleotidase and adenosine deaminase. *Biochem. Pharamacol.* **27**, 2845–2850.

Freeman, R. H., Davis, J. O., Gotshall, R. W., Johnson, J. A., and Spielman, W. S. (1974) The signal perceived by the macula densa during changes in renin release. *Cir. Res.* **35**, 307–315.

Gaal, K., Siklos, J., Mozes, T., and Toth, G. F. (1978) Effect of papaverine on renin release in dogs in vivo and in vitro. *Acta Physiol. Acad. Sci. Hung.* **51**, 305–314.

Gerkens, J. F., Heidemann, H. T., Jackson, E. K., and Branch, R. A. (1983a) Effect of aminophylline on amphotericin B nephrotoxicity in the dog. *J. Pharmacol. Exp. Ther.* **224**, 609–613.

Gerkens, J. F., Heidemann, H. T., Jackson, E. K., and Branch, R. A. (1983b) Aminophylline inhibits renal vasoconstriction produced by intrarenal hypertonic saline. *J. Pharmacol. Exp. Ther.* **225**, 611–615.

Gerkens, J. F. and Smith, A. J. (1985) Effect of captopril and theophylline treatment on cyclosporin-induced nephrotoxicity in rats. *Transplantation* **2**, 213, 214.

Gotshall, R. W., Davis, J. O., Shade, R. E., Spielman, W., Johnson, J. A., and Braverman, B. (1973) Effects of renal denervation on renin release in sodium-depleted dogs. *Am. J. Physiol.* **225**, 344–349.

Haas, J. A. and Osswald, H. (1981) Adenosine induced fall in glomerular capillary pressure. Effect of ureteral obstruction and aortic constriction in the Munich-Wistar rat kidney. *Naunyn-Schmiedebergs Arch. Pharmacol.* **317**, 86–89.

Hackenthal, E., Schwertschlag, U., and Taugner, R. (1983) Cellular mechanisms of renin release. *Clin. Exp. Hypertens. (A)* **A5 (7, 8)**, 975–993.

Hall, J. E. and Granger, J. P. (1986a) Renal hemodynamics and arterial pressure during chronic intrarenal adenosine infusion in conscious dogs. *Am. J. Physiol.* **250**, F32–F39.

Hall, J. E. and Granger, J. P. (1986b) Adenosine alters glomerular filtration control by angiotensin II. *Am. J. Physiol.* **250**, F917–F923.

Hall, J. E., Granger, J. P., and Hester, R. L. (1985) Interactions between adenosine and angiotensin II in controlling glomerular filtration. *Am. J. Physiol.* **248**, F340–F346.

Hashimoto, K. and Kokubun, H. (1971) Adenosine-catecholamine interaction in the renal vascular response. *Proc. Soc. Exp. Biol. Med.* **136**, 1125–1128.

Hashimoto, K. and Kumakura, S. (1965) The pharmacological features of the coronary, renal, mesenteric and femoral arteries. *Jpn. J. Physiol.* **15**, 540–551.

Hedqvist, P. and Fredholm, B. B. (1976) Effects of adenosine on adrenergic neurotransmission. Prejunctional inhibition and postjunctional enhancement. *Naunyn-Schmiedebergs Arch. Pharmacol.* **293**, 217–223.

Hedqvist, P., Fredholm, B. B., and Olundh, S. (1978) Antagonistic effects of theophylline and adenosine on adrenergic neuroeffector transmission in the rabbit kidney. *Circ Res.* **43**, 592–598.

Heidemann, H., Gerkens, J., and Branch, R. (1982) Prevention of cis-platinum (CP) nephrotoxicity by aminophylline, furosemide and high salt diet—Possible role of tubuloglomerular feedback (TGF). *Clin. Res.* **30**, 253A (abstract).

Heidemann, H., Gerkens, J. F., Jackson, E. K., and Branch, R. A. (1983) Effect of aminophylline on renal vasoconstriction produced by amphotericin B in the rat. *Naunyn-Schmiedebergs Arch. Pharmacol.* **324**, 148–152.

Humes, D. H. (1986) Role of calcium in pathogenesis of acute renal failure. *Am. J. Physiol.* **250**, F579–F589.

Itoh, S., Carretero, O. A., and Murray, R. D. (1985) Possible role of adenosine in the macula densa mechanism of renin release in rabbits. *J. Clin. Invest.* **76**, 1412–1417.

Jackson, E. K. and Ohnishi, A. (1987) Development and application of a simple microassay for adenosine in rat plasma. *Hypertension* **10**, 189–197.

Jackson, R. C., Morris, H. P., and Weber, G. (1978) Adenosine deaminase and adenosine kinase in rat hepatomas and kidney tumors. *Br. J. Cancer* **37**, 701–713.

Jacobson, K. A., Kirk, K. L., Daly, J. W., Jonzon, B., Li, Y.-O., and Fredholm, B. B. (1985) A novel 8-phenyl-substituted xanthine derivative is a selective antagonist at adenosine A_1-receptors in vivo. *Acta Physiol. Scand.* **125**, 341, 342.

Jing-Yun, P., Bishop, V. S., Ball, N. A., and Haywood, J. R. (1985) Inability of dorsal spinal rhizotomy to prevent renal wrap hypertension in rats. *Hypertension* **7**, 722–728.

Johannesson, N., Andersson, K.-E., Joelsson, B., and Persson, C. G. A. (1985) Relaxation of lower esophageal sphincter and stimulation of gastric secretion and diuresis by antiasthmatic xanthines. Role of adenosine antagonism. *Am. Rev. Respir. Dis.* **131**, 26–31.

Johns, E. J. and Singer, B. (1973) Effect of propranolol and theophylline on renin release caused by furosemide in the cat. *Eur. J. Pharmacol.* **23**, 67–73.

Johnson, J. A., Davis, J. O., and Witty, R. T. (1971) Effects of catecholamines and renal nerve stimulation on renin release in the nonfiltering kidney. *Circ. Res.* **29**, 646–653.

Jonzon, B., Bergquist, A., Li, Y.-O. and Fredholm, B. B. (1986) Effects of adenosine and two stable adenosine analogues on blood pressure, heart rate and colonic temperature in the rat. *Acta Physiol. Scand.* **126**, 491–498.

Katholi, R. E. (1983) Renal nerves in the pathogenesis of hypertension in experimental animals and humans. *Am. J. Physiol.* **245**, F1–F14.

Katholi, R. E., McCann, W. P., and Woods, W. T. (1985) Intrarenal adenosine produces hypertension via renal nerves in the one-kidney, one clip rat. *Hypertension* **7** (Suppl. 1), 88–93.

Katholi, R. E., Hageman, G. R., Whitlow, P. L., and Woods, W. T. (1983) Hemodynamic and afferent renal nerve responses to intrarenal adenosine in the dog. *Hypertension* **5** (Suppl. 1), 149–154.

Katholi, R. E., Whitlow, P. L., Hageman, G. R., and Woods, W. T. (1984) Intrarenal adenosine produces hypertension by activating the sympathetic nervous system via the renal nerves in the dog. *J. Hypertens.* **2**, 349–359.

Keeton, T. K. and Campbell, W. B. (1980) The pharmacologic alteration of renin release. *Pharmacol. Rev.* **32**, 81–227.

Kenakin, T. P. and Pike, N. B. (1987) An in vitro analysis of purine-mediated renal vasoconstriction in rat isolated kidney. *Br. J. Pharmacol.* **90**, 373–381.

Kuttesch, J. F., Jr. and Nelson, J. A. (1982) Renal handling of 2'-deoxyadenosine and adenosine in humans and mice. *Cancer Chemother. Pharmacol.* **8**, 221–229.

Lang, M. A., Preston, A. S., Handler, J. S., and Forrest, J. N., Jr. (1985) Adenosine stimulates sodium transport in kidney A6 epithelia in culture. *Am. J. Physiol.* **249**, C330–C336.

Le Hir, M. and Dubach, U. C. (1984) Sodium gradient-energized concentrative transport of adenosine in renal brush border vesicles. *Pflugers Arch.* **401**, 58–63.

Lin, J.-J., Churchill, P. C., and Bidani, A. K. (1986) Effect of theophylline on the initiation phase of postischemic acute renal failure in rats. *J. Lab. Clin. Med.* **108**, 150–154.

Lin, J.-J., Churchill, P. C., and Bidani, A. K. (1987) The effect of dipyridamole on the initiation phase of post-ischemic acute renal failure in rats. *Can. J. Physiol. Pharmacol.* **65**, 1491–1495.

Lin, J.-J., Churchill, P. C., and Bidani, A. K. (1988) Theophylline in rats during the maintenance phase of post-ischemic acute renal failure. *Kidney Int.* **33**, 24–28.

Macias-Nunez, J. F., Garcia-Iglesias, C., Santos, J. C., Sanz, E., and Lopez-Novoa, J. M. (1985) Influence of plasma renin content, intrarenal angiotensin II, captopril, and calcium channel blockers on the vasoconstriction and renin release promoted by adenosine in the kidney. *J. Lab. Clin. Med.* **106**, 562–567.

Macias-Nunez, J. R., Revert, M., Fiksen-Olsen, M., Knox, F. G., and Romero, J. C. (1986) Effect of allopurinol on the renovascular responses to adenosine. *J. Lab. Clin. Med.* **108**, 30–36.

Miklos, E. S. and Juhasz-Nagy, A. (1984) Calcium antagonist verapamil inhibits adenosine-induced renal vasoconstriction in the dog. *Acta Physiol. Hung.* **63**, 161–165.

Miller, W. L., Thomas, R. A., Berne, R. M., and Rubio, R. (1978) Adenosine production in the ischemic kidney. *Circ. Res.* **43**, 390–397.

Murray, R. D. and Churchill, P. C. (1984) The effects of adenosine receptor agonists in the isolated-perfused rat kidney. *Am. J. Physiol.* **247**, H343–H348.

Murray, R. D. and Churchill, P. C. (1985) The concentration-dependency of the renal vascular and renin secretory responses to adenosine receptor agonists. *J. Pharmacol. Exp. Ther.* **232**, 189–193.

Nechay, B. R. (1966) Renal effects of exogenous adenosine derivatives in the chicken. *J. Pharmacol. Exp. Ther.* **153**, 329–336.

Needleman, P., Minkes, M. S., and Douglas, J. R., Jr. (1974) Stimulation of prostaglandin biosynthesis by adenine nucleotides. Profile of prostaglandin release by perfused organs. *Circ. Res.* **34**, 455–460.

Nelson, J. A., Kuttesch, J. R., Jr., and Herbert, B. H. (1983) Renal secretion of purine nucleosides and their analogs in mice. *Biochem. Pharmacol.* **32**, 2323–2327.

Nicholson, D. P. and Chick, T. W. (1973) A re-evaluation of parenteral aminophylline. *Am. Rev. Respir. Dis.* **108**, 241–247.

Ono, H., Inagaki, K., and Hashimoto, K. (1966) A pharmacological approach to the nature of the autoregulation of the renal blood flow. *Jpn. J. Physiol.* **16**, 625–634.

Opgenorth, T. J., Burnett, J. C., Jr., Granger, J. P., and Scriven, T. A. (1986) Effects of atrial natriuretic peptide on renin secretion in nonfiltering kidney. *Am. J. Physiol.* **250**, F798–F801.

Osborn, J. E., Hoversten, L. G., and DiBona, G. F. (1983) Impaired blood flow autoregulation in nonfiltering kidneys: Effects of theophylline administration. *Proc. Soc. Exp. Biol. Med.* **174**, 328–335.

Osswald, H. (1975) Renal effects of adenosine and their inhibition by theophylline in dogs. *Naunyn-Schmiedebergs Arch. Pharmacol.* **288**, 79–86.

Osswald, H. (1983) Adenosine and renal function, in *Regulatory Function of Adenosine* (Berne, R. M., Rall, T. W., and Rubio, R. eds.), Martinus Nijhoff, Boston, pp. 399–415.

Osswald, H. (1984) The role of adenosine in the regulation of glomerular filtration rate and renin secretion. *Trends Pharmacol. Sci.* **5**, 94–97.

Osswald, H., Hermes, H. H., and Nabakowski, G. (1982) Role of adenosine in signal transmission of tubuloglomerular feedback. *Kidney Int.* **22** (Suppl.12), 136–142.

Osswald, H., Nabakowski, G., and Hermes, H. (1980) Adenosine as a possible mediator of metabolic control of glomerular filtration rate. *Int. J. Biochem.* **12**, 263–267.

Osswald, H., Schmitz, H.-J., and Heindenreich, O. (1975) Adenosine response of the rat kidney after saline loading, sodium restriction and hemorrhagia. *Pflugers Arch.* **357**, 323–333.

Osswald, H., Schmitz, H.-J., and Kemper, R. (1977) Tissue content of adenosine, inosine and hypoxanthine in the rat kidney after ischemia and postischemic recirculation. *Pflugers Arch.* **371**, 45–49.

Osswald, H., Schmitz, H.-J., and Kemper, R. (1978a) Renal action of adenosine: Effect on renin secretion in the rat. *Naunyn-Schmiedebergs Arch. Pharmacol.* **303**, 95–99.

Osswald, H., Spielman, W. S., and Knox, F. G. (1978b) Mechanism of adenosine-mediated decreases in glomerular filtration rate in dogs. *Circ. Res.* **43**, 465–469.

Persson, C. G. A., Andersson, K.-E., and Kjellin, G. (1986) Minireview. Effects of enprofylline and theophylline may show the role of adenosine. *Life Sci.* **38**, 1057–1072.

Persson, C. G. A., Erjefalt, I., Edholm, L.-E., Karlsson, J.-A., and Lamm, C.-J. (1982) Tracheal relaxant and cardiostimulant actions of xanthines can be differentiated from diuretic and CNS-stimulant effects. Role of adenosine antagonism? *Life Sci.* **31**, 2673–2681.

Phillis, J. W. and Wu, P. H. (1981) Indomethacin, iboprofen and meclofenamate inhibit adenosine uptake by rat brain synaptosomes. *Eur. J. Pharmacol.* **72**, 139, 140.

Premen, A. J., Hall, J. E., Mizelle, H. L., and Cornell, J. E. (1985) Maintenance of renal autoregulation during infusion of aminophylline or adenosine. *Am. J. Physiol.* **248**, F366–F373.

Ramos-Salazar, A. and Baines, A. D. (1986) Role of 5'-nucleotidase in adenosine-mediated renal vasoconstriction during hypoxia. *J. Pharmacol. Exp. Ther.* **236**, 494–499.

Rasmussen, H. amd Barrett, P. Q. (1984) Calcium messenger system: An integrated view. *Physiol. Rev.* **64**, 938–984.

Rossi, N., Churchill, P., Ellis, V., and Amore, B. (1988) Mechanism of adenosine receptor-induced renal vasoconstriction in rats. *Am. J. Physiol. Ther.* **255**, H885–H890.

Rossi, N. F., Churchill, P. C., and Churchill, M. C. (1987a) Pertussis toxin reverses adenosine receptor-mediated inhibition of renin secretion in rat renal cortical slices. *Life Sci.* **40**, 481–487.

Rossi, N. F., Churchill, P. C., Jacobson, K. A., and Leahy, A. E. (1987b) Further characterization on the renovascular effects of N^6-cyclohexyladenosine in the isolated perfused rat kidney. *J. Pharmacol. Exp. Ther.* **240**, 911–915.

Roy, C. (1984) Regulation by adenosine of the vasopressin-sensitive adenylate cyclase in pig-kidney cells (LLC-PF$_{1L}$) grown in defined media. *Eur. J. Biochem.* **143**, 243–250.

Sakai, K., Aono, J., and Haruta, K. (1981) Species differences in renal vascular effects of diypridamole and in the potentiation of adenosine action by dipyridamole. *J. Cardiovasc. Pharmacol.* **3**, 420–430.

Sakai, K., Yasuda, K., and Hashimoto, K. (1968) Role of catecholamine and adenosine in the ischemic response following release of a renal artery occlusion. *Jpn. J. Physiol.* **18**, 673–685.

Schatz, R. A., Vunnam, C. R., and Sellinger, O. Z. (1977) Species and tissue differences in the catabolism of S-adenosyl-L-homocysteine: A quantitative, chromatographic study. *Life Sci.* **20**, 375–383.

Schnermann, J., Osswald, H., and Hermle, M. (1977) Inhibitory effect of methylxanthines on feedback control of glomerular filtration rate in the rat kidney. *Pflugers Arch.* **369**, 39–48.

Schutz, W., Kraupp, O., Bacher, S., and Raberger, G. (1983) The effect of a long-acting adenosine analog on blood flow through various organs in the dog. *Basic Res. Cardiol.* **78**, 679–684.

Schwartzman, M., Pinkas, R., and Raz, A. (1981) Evidence for different purinergic receptors for ATP and ADP in rabbit kidney and heart. *Eur. J. Pharmacol.* **74**, 167–173.

Scott, J. B., Daugherty, R. M. Jr., Dabney, J. M., and Haddy, F. J. (1965) Role of chemical factors in regulation of flow through kidney, hindlimb, and heart. *Am. J. Physiol.* **208**, 813–824.

Skott, O. and Baumbach, L. (1985) Effects of adenosine on renin release from isolated rat glomeruli and kidney slices. *Pflugers Arch.* **404**, 232–237.

Smith, H. (1951) *The Kidney. Structure and Function in Health and Disease.* (Oxford, New York).

Sollevi, A., Lagerkranser, M., Andreen, M., and Irestedt, L. (1984) Relationship between arterial and venous adenosine levels and vasodilatation during ATP- and adenosine-infusion in dogs. *Acta Physiol. Scand.* **120**, 171–176.

Spielman, W. S. (1984) Antagonistic effect of theophylline on the adenosine-induced decrease in renin release. *Am. J. Physiol.* **247**, F246–F251.

Spielman, W. S. and Osswald, H. (1978) Characterizaiton of the postocclusive response of renal blood flow in the cat. *Am. J. Physiol.* **235**, F286–F290.

Spielman, W. S. and Osswald, H. (1979) Blockade of postocclusive renal vasoconstriction by an angiotensin II antagonist: Evidence for an angiotensin-adenosine interaction. *Am. J. Physiol.* **237**, F463–F467.

Spielman, W. S. and Thompson, C. I. (1982) A proposed role for adenosine in the regulation of renal hemodynamics and renin release. *Am. J. Physiol.* **242**, F423–F435.

Spielman, W. S., Britton, S. L., and Fiksen-Olsen, M. J. (1980) Effect of adenosine on the distribution of renal blood flow in dogs. *Circ. Res.* **46**, 449–456.

Stein, J. H., Lifschitz, M. D., and Barnes, L. D. (1978) Current concepts on the pathophysiology of acute renal failure. *Am. J. Physiol.* **234**, F171–F181.

Stocker, W., Roos, G., Lange, H. W., and Hempel, K. (1977) Monitoring the specific radioactivity of S-adenosylmethionine in kidney in vivo. *Eur. J. Biochem.* **73**, 163–169.

Tagawa, H. and Vander, A. J. (1970) Effects of adenosine compounds on renal function and renin secretion in dogs. *Circ. Res.* **26**, 327–338.

Thompson, C. I., Sparks, H. V., and Spielman, W. S. (1985) Renal handling and production of plasma and urinary adenosine. *Am. J. Physiol.* **248**, F545–F551.

Thurau, K. (1964) Renal hemodynamics. *Am. J. Med.* **36**, 698–719.

Trimble, M. E. and Coulson, R. (1984) Adenosine transport in perfused rat kidney and renal cortical membrane vesicles. *Am. J. Physiol.* **246**, F794–F803.

Ueda, J., Nakanishi, H., and Abe, Y. (1978) Effect of glucagon on renin secretion in the dog. *Eur. J. Pharmacol.* **52**, 85–92.

Verney, E. G. and Winton, F. R. (1930) The action of caffeine on the isolated kidney of the dog. *J. Physiol. (Lond.)* **69**, 1153–1170.

Viskoper, R. J., Maxwell, M. H., Lupu, A. N., and Rosenfeld, S. (1977) Renin stimulation by isoproterenol and theophylline in the isolated perfused kidney. *Am. J. Physiol.* **232**, F248–F253.

Visscher, M. B. (1964) Concluding remarks. *Circ. Res.* **14/15** (Suppl. I), 288–291.

Waugh, W. H. (1964) Circulatory autoregulation in the fully isolated kidney and in the humorally supported, isolated kidney. *Circ. Res.* **14/15** (Suppl. I), 156–169.

Winer, N., Chokshi, D. S., and Walkenhorst, W. G. (1971) Effects of cyclic AMP, sympathomimetic amines, and adrenergic receptor antagonists on renin secretion. *Circ. Res.* **29,** 239–248.

Winer, N., Chokshi, D. S., Yoon, M. S., and Freedman, A. D. (1969) Adrenergic receptor mediation of renin secretion. *J. Clin. Endocrinol.* **29,** 1168–1175.

Witty, R. T., Davis, J. O., Johnson, J. A., and Prewitt, R. L. (1971) Effects of papaverine and hemorrhage on renin secretion in the nonfiltering kidney. *Am. J. Physiol.* **221,** 1666–1671.

Witty, R. T., Davis, J. O., Shade, R. E., Johnson, J. A., and Prewitt, R. L. (1972) Mechanisms regulating renin release in dogs with thoracic caval constriction. *Circ. Res.* **31,** 339–347.

Woodcock, E. A. Leung, E., and Johnston, C. I. (1986) Adenosine receptors in papilla of human kidneys. *Clin. Sci.* **70,** 353–357.

Woodcock, E. A., Loxley, R., Leung, E., and Johnson, C. I. (1984) Demonstration of R_A-adenosine receptors in rat renal papillae. *Biochem. Biophys. Res. Commun.* **121,** 434–440.

Wright, F. S. (1981) Characteristics of feedback control of glomerular filtration rate. *Fed. Proc.* **40,** 87–92.

CHAPTER 10

The Role of Adenosine in Respiratory Physiology

Timothy L. Griffiths and Stephen T. Holgate

1. Introduction

Adenosine is a purine nucleoside, derived predominantly from the cleavage of 5'adenosine monophosphate (AMP) by 5'nucleotidase (Arch and Newsholme, 1978). The nucleoside subserves both intra- and extracellular functions, the latter being effected through specific purine receptors. Two types of extracellular purine receptors have been described: A_1 and A_2, which can, respectively, inhibit and stimulate adenylate cyclase, modifying intracellular levels of the second messenger cyclic 3',5'-adenosine monophosphate (Londos and Wolff, 1977; Londos, et al., 1980). More recently, however, these receptors have been defined in terms of their agonist pharmacology (Hamprecht and Van Calker, 1985).

Adenosine has been suggested to play a role in four areas of respiratory physiology:

1. The modulation of immune responses and caliber of the air-ways;

Adenosine and Adenosine Receptors Editor: Michael Williams ©1990 The Humana Press Inc.

2. Control of ventilation by the central nervous system;
3. The functioning of peripheral chemoreceptors d.
4. The regulation of pulmonary vascular resistance.

2. Adenosine and Bronchoconstriction

2.1. Physiological Control of Airway Caliber

Bronchoconstriction is brought about by contraction of the smooth muscle of the airways, reducing airway caliber in the presence of potentially noxious physical or immunological factors. Relaxation may also occur, for instance, under conditions of exercise, when increased adrenergic drive results in bronchodilation. Therefore, airway caliber at any given time depends on the balance of bronchodilating and constricting factors and the sensitivity of the bronchial smooth muscle to these influences.

In an effort to elucidate the mechanisms by which the various bronchoconstrictor agonists and antagonists work, much attention has been focused on the airways in asthma. In this group of disorders, the airways are more sensitive to bronchoconstricting factors than they are in normal lungs. Asthma is defined clinically as variable airflow obstruction and has been identified in a group of patients, some of whom respond with bronchoconstriction to inhaled allergen (extrinsic or allergic asthma) and some whose bronchoconstriction is not linked with identifiable sensitivity to inhaled allergen (intrinsic or nonallergic asthma). In both types of asthma, increased responsiveness of the bronchial smooth muscle to a variety of physical stimuli (e.g., cold, dust, exhaust fumes) and known inflammatory mediators (e.g., histamine) is invariably present (McFadden, 1984). Pathologically, the airways exhibit disruption of the bronchial epithelium, mucosal edema, and infiltration of the mucosa and submucosa with eosinophils, neutrophils, and mononuclear cells (Hogg et al., 1977; Ellwood et al., 1982). These inflammatory features are probably the result of the varied actions of chemical mediators released in the airway mucosa (Lewis and Austen, 1981; Fish et al., 1983). Evidence has accumulated to suggest that bronchial mast cells are the source of some of these mediators and have a key role to play in the pathogenesis of the disordered airway function in asthma. Other factors resulting in increased bronchial smooth-muscle tone are muscarinic effects mediated via terminations of the vagus nerve on the smooth muscle (Nadel, 1973) and the ramifications of the "third" or "peptidenergic" neural plexuses that have been identified in animals (Coburn and Tomita, 1973; Richardson, 1981). A tonic antispasmodic influence is exerted on the bronchial smooth muscle by circulating catecholamines that act as agonists at β_2-adrenergic receptors, as illustrated by the bronchospasm produced by β-adrenergic receptor blockade in asthmatic subjects (McNeil and Ingram, 1966; Langer, 1967; Richardson and Sterling, 1969).

The precise role played by adenosine in the pathogenesis of asthma remains obscure, but the observed effects of adenosine on bronchial caliber have been related to several of the above processes.

2.2. The Effect of Adenosine on Human Airways

Observations of the effects of adenosine and its more stable analogs on guinea pig tracheal smooth muscle first suggested that adenosine might have a modulating influence on airway caliber. The response observed was one of transient contraction followed by a more powerful and sustained relaxation (Fredholm et al., 1979; Karlsson et al., 1982). Whereas the contractile phase may be mediated by arachidonic-acid metabolites, the relative potency of adenosine analogs on the relaxant phase indicates an A_2 receptor mechanism involving an increase in smooth-muscle intracellular cyclic AMP concentrations (Brown and Collis, 1982).

Prompted by these findings, work was undertaken to determine the effect of adenosine on human airways. Using specific airway conductance as a sensitive index of airway caliber, inhalation of adenosine in concentrations up to 6.7 mg/mL was shown to have no effect in normal subjects. In contrast, a dose-related bronchoconstriction was seen in both allergic and nonallergic asthmatic subjects. Surprisingly, the relaxation that had been seen in the guinea pig was not apparent (Cushley et al., 1983). A degree of selectivity for the adenine moiety was suggested by the finding that another purine nucleoside, guanosine, when inhaled over the same concentration range, had no measurable effect on airway caliber in either normal or asthmatic subjects. Subsequently, inhalation of the adenine nucleotides, AMP and adenosine 5'-diphosphate (ADP), in the same molar concentrations as adenosine, produced similar effects on the airways, whereas the adenosine deamination product, inosine, was ineffective. This suggested that the purine receptors involved were selective for the adenosine moiety (Mann et al., 1986a).

Inhaled adenosine produces bronchoconstriction in asthmatics, but intravenous infusion of adenosine up to 50 μg/Kg/min does not affect bronchial tone as determined by airways conductance or bronchial reactivity to inhaled methacholine (Larsson and Sollevi, 1988). The reason for this discrepancy is not clear, but it seems that adenosine in the pulmonary circulation at that level does not influence the effector site that is accessible to inhaled adenosine.

2.2.1. Nature of the Adenosine Receptor Involved in Bronchoconstriction

When administered as an inhaled aerosol to patients with asthma, adenosine produces a rapid bronchoconstriction, which is maximal 2–5 min after inhalation and gradually subsides over the subsequent hour (Cushley et al., 1983). The time course of bronchoconstriction following inhalation of AMP or ADP is almost identical to that following adenosine (Cushley et al., 1983; Mann et al., 1986a). Other studies have shown that systemically administered AMP is rapidly hydrolyzed to adenosine by intra- and extracellular 5'nucleotidases. It is likely, therefore, that many of the pharmacological actions of AMP are mediated through adenosine following hydrolysis of the nucleotide (Arch and Newsholme, 1978; Bellardinelli et al., 1984). In the light of this, and the fact that AMP is more soluble than adenosine in aqueous solvents, hence can be prepared in higher concentrations for aerosol administration by nebulizer, much work on the effect of stimulation of bronchoconstriction by adenosine has utilized AMP rather than the nucleoside.

The rank order of potency of adenosine and its synthetic analogs as bronchial smooth-muscle contractile agonists has been determined on isolated 4-mm preparations of human bronchi (Holgate et al., 1986). The order of N-ethylcarboxamido adenosine (NECA) > adenosine > R-phenylisopropyl adenosine (R-PIA) suggests an interaction with A_2 receptors. This finding supports the concept that adenosine acts on a cell surface receptor, probably of the A_2 type, to bring about bronchoconstriction. Interestingly, ATP has a greater efficacy than adenosine as a contractile agonist in human bronchial strips in vitro (Finney et al., 1985), suggesting that purine nucleotides may have an effect on bronchial caliber that is independent of their initial hydrolysis to adenosine in that preparation. However, Mann et al. (1986a) showed that adenosine and its nucleotides are equipotent bronchoconstrictor agonists on an equimolar basis in asthmatic patients. The apparent disparity between the results of the in vivo and in vitro studies suggests that the mechanism of airway smooth-muscle stimulation may depend on the experimental model used.

More direct evidence for adenosine-provoked bronchoconstriction being mediated at a specific purine receptor was provided by the observation that the methylxanthine, theophylline—a competitive adenosine antagonist at cell surface receptors (Daly, 1982)—when administered either orally (Mann and Holgate, 1985; Crimi et al., 1989) or by inhalation (Cushley et al., 1984), inhibited bronchoconstriction provoked by inhaled adenosine 3–5x more effectively than bronchoconstriction provoked by histamine. The difference is a result of combined functional antagonism (phosphodiesterase inhibition) and specific antagonism (at purine receptors) in the case of adenosine, as opposed to purely functional antagonism in the case of histamine. Activity of adenosine at the level of cell surface receptors would also explain the observation that systemically administered dipyridamole enhances the bronchoconstrictor effect of inhaled adenosine when compared with that of histamine (Cushley et al., 1985).

2.2.2. Evidence for Endogenous Adenosine Release from the Lung

Adenosine is released into the extracellular space in a variety of tissues, e.g., brain (Berne et al., 1974), heart (Berne, 1980), and skeletal muscle (Fuchs et al., 1986), under circumstances of hypoxia or increased metabolic work. Hypoxia has also been shown to stimulate adenosine release in isolated canine lungs (Mentzer et al., 1975). It has also been shown that adenosine release from rat lungs may be stimulated by exposure to antigen or to the mast cell secretagogue, compound 48/80 (Fredholm, 1981). Evidence that at least some of the adenosine released following IgE-dependent challenge might be of mast cell origin accrues from the finding that adenosine is released in

increased amounts following allergen challenge of cultured rat serosal and mouse interleukin 3 (IL-3)-dependent bone-marrow-derived mast cells (Marquardt et al., 1984a). Human leucocytes activated in vitro by the calcium ionophore, A23187, or the chemotactic
peptide, formyl-methionine-leucine-phenylalanine (FMLP), also increase their release of adenosine (Holgate and Mann, 1984; Mann et
al., 1986b). Thus, adverse conditions that are present in the asthmatic
airways, produced by mast cell activation with mediator secretion,
recruitment of inflammatory leucocytes, smooth-muscle contraction,
and hypoxia, have been shown in vitro to be accompanied by adenosine release (Fig. 1). The question of whether adenosine is released
by human lungs in vivo when challenged by inhaled antigen has been
addressed (Mann et al., 1986a). It was found that both allergen provocation (house dust mite extract) and selective muscarinic agonist
(methacholine) challenge were followed by two- to fivefold increases
in the plasma concentrations of adenosine in asthmatic human subjects. However, the profile of plasma adenosine levels after challenge
does not seem to be related to the time course of the bronchospasm
provoked. This is demonstrated by the observation that the onset of
bronchoconstriction following allergen challenge is considerably
slower than that seen following methacholine inhalation, whereas the
rise in adenosine concentration is more rapid. Furthermore, following
both types of provocation, plasma adenosine levels continue to rise
during relaxation of the bronchial smooth muscle. This observation,
together with the reportedly short adenosine whole-blood half-life of
< 10 s (Klabund, 1983), is suggestive of on-going adenosine release
into the circulation following the initial challenge. Interpretation of
these results is complicated by the fact that the venous blood samples
in these experiments did not exclusively measure adenosine produced
within the lungs, but would also have detected any produced by circulating blood cells activated by pulmonary events. One possible
source of adenosine in the circulation is the platelet whose activation
has been demonstrated following allergen provocation in asthma
(Knauer et al., 1981) and that is known to release adenosine.

If the adenosine released during bronchoconstriction does originate in the lungs, it is unclear which process or combination of processes are involved in its production, since stimulation of the immune
system and direct stimulation of smooth muscle both lead to an ele-

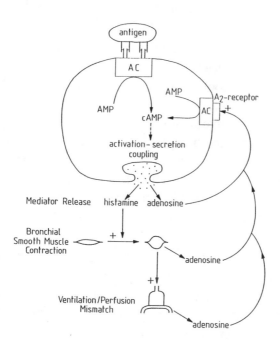

Fig. 1. Hypothetical model of the role of adenosine in mast-cell mediator release and bronchoconstriction. Following activation of the mast cell by antigen binding at IgE on the cell surface, degranulation occurs. Bronchoconstrictor and inflammatory mediators (e.g., histamine) are released in addition to adenosine. Histamine acts on the bronchial smooth muscle stimulating contraction, with consequent further adenosine release. The resultant derangement in local ventilation: perfusion ratios might produce regional hypoxia and further adenosine release. Extracellular adenosine could then exert positive feedback by activating adenylate cyclase (AC) thus enhancing further mast-cell degranulation.

vation of levels. This leaves indeterminate the question of whether adenosine is produced as a result of mast cell activation, smooth-muscle contraction, local hypoxia as a consequence of regional ventilation–perfusion inequality produced by bronchospasm, or an as yet unknown process. The sampling of pulmonary venous blood during provocation would provide a clearer picture of whether adenosine is released in the lung following antigen challenge and also define its time course.

2.2.3. The Mechanism of Adenosine-Induced Bronchoconstriction

By analogy with other pharmacological and physiological sites of action of bronchoconstrictor agents in asthma, inhaled adenosine may act at the level of smooth-muscle contraction, neural reflexes, or mast cell function. The techniques used to investigate which of these are important in the action of adenosine in vivo fall into two categories. In the first, adenosine bronchial challenge is performed in the presence of pharmacological agents that interfere with the functioning of one of the above systems. In the second, the efficacy of adenosine in provoking bronchoconstriction is compared with the ability of other agents with specificity for one of the mechanisms to produce the same response.

2.2.3.1. Smooth Muscle. The bronchoconstrictor response to inhaled methacholine or histamine is regarded as an index of nonspecific bronchial responsiveness, or the excitability of the airways. Whereas correlation between the responses to these agonists is good, the degree of bronchoconstriction provoked by adenosine or AMP has been shown to correlate only weakly with these other indices of airways responsiveness (Mann et al., 1986a; Cushley and Holgate, 1985). Furthermore, adenosine at millimolar concentration has been shown to produce only weak contraction in isolated normal human bronchial smooth-muscle preparations (Davis et al., 1982). These two sets of observations suggest that the bronchoconstrictor action of the purine is not mediated primarily by a direct effect on normal airway smooth muscle. However, a more powerful and consistant response has been seen in airway preparations from two birch-pollen-sensitive asthmatics (Dahlen et al., 1983).

The possibility of adenosine acting on smooth muscle by interfering with the tonic influence of circulating catecholamines acting at β_2-adrenergic receptors has also been investigated (Mann et al., 1985). In a group of asthmatic subjects, β_2-adrenergic receptor responsiveness of the airways was not reduced by preinhalation of adenosine, making it unlikely that adenosine acts as an antagonist at the level of adenylate cyclase.

2.2.3.2. Neural Reflexes. The bronchial mucosa is supplied by vagal afferent receptors that, when stimulated, excite reflex bronchoconstriction, the efferent arm being subserved by the vagus nerve at muscarinic cholinergic receptors on bronchial smooth-muscle cells. Several inhaled bronchoconstrictor stimuli have been shown to be at least partially antagonized by inhaled muscarinic antagonists indicating their action via this reflex. Such stimuli include cold air (Sheppard et al., 1982), exercise (Chan-Yeung, 1977), sulfur dioxide (Sheppard et al., 1980), hypotonic solutions (Sheppard et al., 1983), prostaglandin $F_{2\alpha}$ (Orehek et al., 1977), and histamine (Eiser and Guz, 1982). The possibility of adenosine also acting in this way has been investigated by administering large inhaled doses of the potent muscarinic antagonist, ipratropium bromide, prior to adenosine or methacholine bronchial challenge (Mann et al., 1985). In doses that inhibited methacholine-induced bronchoconstriction, ipratropium bromide was found not to attenuate the bronchoconstrictor response to inhaled adenosine. This strongly suggests that vagal reflex pathways are of little importance in the pathophysiology of adenosine-induced bronchoconstriction, but does not exclude an involvement of the peptidergic system.

2.2.3.3. IMMUNE SYSTEM. As noted previously, inflammation is of prime importance in the pathophysiology of asthma. An important component of the immune system of mucosal surfaces such as the airways is the presence of mast cells (Jeffrey and Corni, 1984), which make up 0.01–0.1% of lung tissue (Kaliner, 1980). Approximately 80% of lung mast cells are situated in the airways. Whereas the majority of these are subepithelial, mast cells do compose 0.12–0.42% of bronchial epithelial cells (Lamb and Lumsden, 1982). The interaction of specific antigen with IgE bound to the surface of mast cells (and possibly other mediator secretory cells, e.g., basophils, eosinophils, and macrophages) results in the release of a variety of inflammatory mediators. Some of these are preformed and stored in granules within the cells, whereas others are newly formed metabolites of arachidonic acid, synthesized after cell activation. Preformed mediators stored in human lung mast cells include histamine, exoglycosidases, proteases, chemotactic factors, superoxide dismutase, peroxidase, and heparin. Newly formed arachidonic acid derived mediators include the cyclooxygenase product PGD_2 (Lewis et al., 1982; Schleimer et al., 1982) and the lipooxygenase products, leukotrienes (LT) B_4, C_4, D_4, and E_4 (Robinson and Holgate, 1985). These mediators subsequently bring about smooth-muscle contraction, edema, vasodilatation, and recruitment and activation of other immunologically active cells. Platelet activating factor (PAF), an ether-linked phospholipid generated with inflammatory cell activation, is also a highly potent vaso- and bronchoconstrictive mediator. Interest has focused on adenosine in relation to these processes because of its ability to modulate the calcium- and IgE-dependent secretion of granule-derived inflammatory mediators from human basophil leucocytes (Church et al., 1985) and lung mast cells (Holgate et al., 1980a; Marquardt et al., 1978).

2.2.4. Modulation of Mast-Cell Mediator Release by Adenosine

Observations of the effect of mast-cell-stabilizing drugs on the response to inhaled adenosine has led to the suggestion that adenosine exerts its action on the airways by augmenting mast cell degranulation. A study comparing the protective effect of inhaled sodium cromoglycate against bronchoconstriction induced by nebulized adenosine and histamine in asthmatic subjects showed a striking inhibition of the response to adenosine, whereas that to histamine was unaffected (Cushley and Holgate, 1985). This suggests that mast cell degranulation itself is important in the mechanism of adenosine's action, rather than a modulation of the effect of released mast-cell-derived mediators. These initial observations have been supported by work carried out using the related mast-cell-stabilizing drug, nedocromil sodium. In this study, both sodium cromoglycate and nedocromil sodium were shown to antagonize the bronchoconstrictor effect of inhaled AMP (Altounyan et al., 1986). Adenosine may influence mast-cell mediator release by acting either at cell surface receptors (A_1 or A_2) or intracellularly.

The effect of inhaled adenosine is unlikely to involve action on calcium metabolism, since Crimi et al. (1988) have shown that, although sodium cromoglycate attenuates induced bronchoconstriction, the calcium channel blocker nifedipine had no effect.

Mast cell and basophil mediator release is the final event in a complex series of biochemical reactions that follow immunological stimulation. Crosslinkage of IgE-Fc receptors with antigen or anti-

IgE leads to activation of adenylate cyclase, which in turn gives rise to an early increase in intracellular cyclic AMP levels and activation of cyclicAMP-dependent protein kinases, both of which have been implicated in activation–secretion coupling in rodent mast cells (Holgate et al., 1980b). Activation of cell surface adenosine receptors of either the A_1 or A_2 subtypes may modulate intracellular levels of cyclic AMP. This may, therefore, be a mode of action of extracellular adenosine augmenting stimulated mast cell degranulation, A_2 receptor activation leading to a rise in the pool of intracellular cyclic AMP levels linked to activation–secretion coupling (Fig. 1).

The effect of adenosine on mediator release from bronchial mast cells has been investigated in vitro. Human lung mast cells obtained by mechanical dispersal were incubated with anti-IgE, and histamine release was measured spectrophotometrically (Hughes et al., 1984). It was found that the effect of adding adenosine to the mast-cell preparation depended on the timing of the addition in relation to the anti-IgE challenge. Addition before anti-IgE was found to inhibit IgE-dependent histamine release, whereas if adenosine addition was delayed until after anti-IgE, histamine release was enhanced. Maximum potentiation was obtained if adenosine was added 5 min after challenge, at a time coinciding with the end of the rapid phase of histamine release (Holgate et al., 1980b). The order of potency of synthetic adenosine analogs, inhibition of the effects of adenosine by theophylline, and a lack of inhibition by dipyridamole all suggest that adenosine modulation of histamine release in this preparation was mediated through an A_2 purine receptor. Similar results have been obtained in studies of the effect of adenosine on histamine release following anti-IgE challenge of human peripheral blood basophils (Church et al., 1983). In both cell types, the effect of adenosine is more pronounced at low levels of anti-IgE challenge (Church et al., 1983; Hughes et al., 1984).

The mechanism by which the timing of adenosine stimulation results in different effects is not understood. However, it is noteworthy that the maximal effect is seen if adenosine stimulation occurs when the rapid phase of histamine release is nearing completion. This may suggest that an A_2 receptor-mediated rise in intracellular cyclic AMP, causing activation of protein kinases at a critical time in the mediator release process, may facilitate greater histamine release than would have been possible subsequent to allergen stimulation alone.

This augmentation would not be seen if premature activation of cyclic AMP-dependent protein kinases occured at other times during the IgE-dependent secretory sequence (Ishizaka and Ishizaka, 1984).

The inhibitory effect on histamine release of preincubation of mast cells with adenosine may be related to an indirect effect of elevated intracellular levels of adenosine-inhibiting membrane phospholipid reactions, linked to activation–secretion coupling (Ishizaka and Ishizaka, 1984; Benyon et al., 1984).

2.2.5. Mediators Released by Mast Cell Adenosine Receptor Stimulation

In a study of mouse bone-marrow-derived mast cells, incubation with adenosine was associated with release of the granule-derived mediator β-hexosaminidase, whereas the release of arachidonic acid metabolites was unaffected (Marquardt et al., 1984b). Further in vitro studies using guinea pig and human bronchial strips have shown that the histamine H_1-receptor antagonist, diphenhydramine, inhibits the early bronchoconstrictor response to antigen, whereas FPL 55712, a sulfidopeptide LT D_4 antagonist, selectively inhibited the prolonged phase of the response (Adams and Lichtenstein, 1979). The early phase of the response of airways to antigen challenge is considered to relate to histamine release, whereas a major component of the later response is mediated by leukotrienes.

Terfenadine is a potent, selective short-acting histamine H_1-receptor antagonist that, when given orally in a dose of 180 mg, completely protects against bronchoconstriction caused by a dose of inhaled histamine, which would produce a 30% fall in forced expiratory volume in 1 s (FEV_1) after oral placebo (Holgate et al., 1987). The same dose of terfenadine has been shown to inhibit allergen-provoked fall in FEV_1 by about 50%, the major effect being seen in the first 10 min after allergen inhalation, a time when mast-cell histamine release is at its maximum (McFadden, 1984). The bronchoconstrictor response to challenge of these subjects with AMP, inhaled in a dose producing a 30% fall in FEV_1, was more than 86% inhibited by terfenadine, suggesting that a major component of the response to this nucleotide (and by implication adenosine) was mediated by histamine released endogenously in the airways. Similar results have been found using the chemically unrelated H_1 antagonist astemizole.

Bronchoconstriction provoked by AMP in nonallergic asthmatics is also inhibited by terfenadine (180 mg), suggesting that in this variant of asthma, adenosine may activate mast cells as described in vitro (Holgate et al., 1987). Further evidence of adenosine's ability to stimulate stored-mediator release is the phenomenon of refractoriness of mast cells upon repeated challenge of asthmatic airways with AMP (Dazsun et al., 1989). In this study, asthmatic subjects were exposed to bronchial AMP challenge. Subsequent challenges shortly after the first challenge revealed either refractoriness to the effect of the challenge or a greatly diminished response—all subjects were refractory at the third AMP challenge, but at that stage responded normally to histamine challenge. Responsiveness to AMP was only regained 4–6 h after the last challenge and in this respect is similar to refractoriness to exercise as a provoking stimulus for asthma. These results would also support the contention that AMP acts by stimulating the release of preformed, stored mast cell mediators.

3. Adenosine and Ventilatory Control

The central generator of the respiratory rhythm is situated in the medulla oblongata, probably in the nucleus of the solitary tract. It has connections to the *nucleus retroambigualis* and to the *nucleus ambiguus*. The output from these centers descends to phrenic, intercostal, and accessory muscle motor neurons, producing the movements associated with ventilation (Wyman, 1977). The medullary centers receive afferents from the pontine *nucleus parabrachialis medialis* (pneumotaxic center), pulmonary receptors, the carbondioxide-sensitive central chemoreceptor on the ventral medulla, and the peripheral chemoreceptors that are sensitive to hypoxia and increase in hydrogen ion (H^+) concentration and hypercapnia (Bruce and Cherniack, 1987).

Adenosine has been described to act within the respiratory control system in two ways. In the central nervous system (CNS), adenosine has a depressant effect (Moss et al., 1986), but in the peripheral chemoreceptors, its action is excitatory (McQueen, 1983). This dichotomy of effects has complicated the study of the role of adenosine in respiratory control and necessitated the adoption of different approaches in investigating its two functions. To elucidate the central effects of adenosine, the experimental models used have been: (i) a glomectomized, vagotomized, anesthetized animal preparation; (ii) preterm or neonatal mammals in which central and peripheral components of the ventilatory response to hypoxia may be distinguished. Investigation of adenosine's action on the peripheral chemoreceptors has involved recording from the carotid sinus nerve and monitoring ventilation during injections of adenosine agonists or antagonists into the carotid circulation of humans and laboratory animals.

3.1. Adenosine as a Central Respiratory Neuromodulator

Adenosine has been shown to depress various central neurological and behavioral functions (Phillis et al., 1979; Snyder et al., 1981; Jarvis and Williams, 1990), many of these effects being antagonized by alkylxanthines (Snyder et al., 1981). The medullary respiratory control center is one such adenosine-sensitive system (Hedner et al., 1984; Wessburg et al., 1985; Eldridge et al., 1985). Since adenosine has been shown to be produced in brain tissue under conditions of hypoxia (Berne et al., 1974; Rubio et al., 1975; Winn et al., 1981; Zetterstrom et al., 1982), it has been postulated that it could act as an autacoid by providing inhibitory neuromodulation (Moss et al., 1986) in these circumstances. Indeed, the adenosine antagonists, caffeine and theophylline, have long been familiar to clinicians for their respiratory stimulant properties (Richmond, 1949). Central respiratory depression in response to hypoxia was demonstrated in early experiments in dogs, which showed that in the absence of peripheral chemoreceptor afferent activity, hypoxia depressed central respiratory mechanisms (Watt et al., 1943). These findings have subsequently been confirmed in crossed circulation experiments in which the carotid bodies of an hypoxic dog were perfused with blood from a well-oxygenated donor. In this case, respiration was found to be depressed at moderate levels of cerebral hypoxia (arterial partial pressure of oxygen—Pa_{O_2} 6.0–7.5 kPa; Lee and Milhorn, 1975). Similar effects have been described in the goat (Tenney and Brooks, 1966) and the cat (Millhorn et al., 1984). In the latter study, the response to a 10-min period of hypoxia in glomectomized cats was characterized as one of respiratory depression, lasting for over 60 min after the restoration of hyperoxia. This occurred in spite of a reduction in medullary extracellular fluid pH, which would, classically, have stimulated respiration. Pretreatment with theophylline ameliorated the respiratory depression seen during hypoxia and abolished the long-lasting depression that had followed after the return to full oxygenation. This indicated that at least part of the hypoxic central respiratory depression might be mediated by adenosine.

3.1.1. Action of Methylxanthines on the Respiratory Center

The conclusion that methylxanthines stimulate respiration by acting at cell surface adenosine receptors has been arrived at primarily by a process of exclusion. In a study using the paralyzed, glomectomized, vagotomized anesthetized cat preparation, Eldridge and colleagues (1983) excluded several previously postulated mechanisms for the respiratory stimulant action of methylxanthines: muscular or mechanical factors in the periphery and changes in whole-body metabolic rate, since their animals were paralyzed; carotid body or cardiopulmonary reflexes, by the nature of the preparation; spinally mediated mechanisms, by cord transection at C7 level; changes in arterial P_{CO_2}, by ventilating the animals to maintain a constant arterial P_{CO_2}; adrenal catecholamine release, since the animals were adrenalectomized, and changes in medullary extracellular fluid pH. The latter observation is particularly important, since theophylline-induced cerebral arterial constriction has been postulated to cause an increase in CSF H^+ concentration, which might then stimulate the medullary chemoreceptors (Wechsler et al., 1950). However, when monitored throughout a period of theophylline infusion, the pH of ventral medullary ECF did not change significantly, which makes this mechanism of action highly unlikely (Eldridge et al., 1983).

The effective site of action of theophylline therefore appears to be on the neurons of the brain itself. Two mechanisms have been suggested. First, it has been postulated that neural stimulation might be related to cyclic nucleotide phosphodiesterase inhibition, with consequent increases in intracellular cyclic AMP concentration (Butcher and Sutherland, 1962). However, the concentrations of theophylline required to produce CNS stimulation are well below those required to inhibit cellular phosphodiesterase activity (Snyder et al., 1981). The second mechanism, for which evidence is accumulating, is antagonism of the neurodepressant effects of adenosine by methylxanthines (Hedner et al., 1984; Wessberg et al., 1985; Eldridge et al., 1985). In mice, the potency of the central actions of different methylxanthines on behavior has also been found to correlate with their affinities for neuronal adenosine receptors (Snyder et al., 1981), again suggesting that methylxanthines act on adenosine receptors present on the surface of CNS neurons.

Following from studies of this type, indirect evidence that endogenous adenosine might be of importance in respiratory control has been deduced from observations of the respiratory effects of alkylxanthines. Theophylline given intravenously or into the cerebral ventricles of glomectomized, vagotomized, anesthetized cats not only reversed the respiratory depressant effect of the intraventricularly administered adenosine agonist R-PIA, but also induced a level of respiratory activity above the previous control level (Eldridge et al., 1985). This finding may be interpreted as a reversal of a pre-existant tonic adenosine "down regulation" of the respiratory center. In this preparation, intravenous theophylline alone induced significant increases in respiratory activity above baseline levels. A similar effect of theophylline on respiration has been reported in the rat (Winn et al., 1981) and is a well-recognized phenomenon in intact humans (Richmond, 1949; Dowell et al., 1965; Lakshminarayan et al., 1978), in whom intravenous infusion of aminophylline augments the respiratory response to hypoxia. These observations suggest a tonic inhibitory influence of adenosine on the central respiratory control system.

3.1.2. Action of Adenosine on the Respiratory Center

In a more direct investigation of the central effects of adenosine and its analogs undertaken using a glomectomized, vagotomized, anesthetized cat preparation, Eldridge et al. showed that the adenosine analog R-PIA, administered either intravenously or into the third cerebral ventricle, depressed phrenic nerve output (Eldridge et al., 1985). They also found that theophylline, given either intravenously or intraventricularly, could prevent this depression if given before R-PIA, or reverse it if given after the R-PIA. Similar results have been found with the same synthetic adenosine analog in the preterm rabbit (Hedner et al., 1984). These observations were extended by Wessberg et al. (1985), who studied adenosine-related mechanisms in the regulation of breathing in the rat. They found a rank order of potency for respiratory depression by adenosine analogs acting at cell surface receptors of NECA > N^6-cyclohexyladenosine (CHA)> 2-chloroadenosine (2-CADO)=R-PIA. When compared with the relative potencies of these agonists at the two types of extracellular adenosine receptors set out by Fredholm (1982), this order suggests a mixture of A_1 and A_2 effects antagonized by theophylline. Burr and Sinclair (1988) have shown respiratory depression after intraperitoneal injection of PIA in conscious intact rats. Hypoxic ventilatory stimulation was preserved, but at a lower level than in controls. This observation suggests a relative lack of peripheral chemoreceptor stimulation and a central depression by PIA in the rat. Hypercapnic responsiveness was increased in these animals, which might be owing to the suppression of central inhibitory mechanisms acting on the central CO_2 sensing mechanisms. The observation that enprofylline (3-propylxanthine), a xanthine derivative with antibronchoconstrictor properties, but negligible effect on extracellular adenosine receptors, did not antagonize R-PIA-induced respiratory depression further suggests that this CNS function of purine nucleosides is mediated at specific purine receptors (Wessberg et al., 1985).

Respiratory depression by infused adenosine has been observed in a patient following bilateral carotid endarterectomy in whom normal carotid body function was absent. Thus, similar central mechanisms may also be present in humans (Griffiths et al., 1989).

3.1.3. Adenosine in Neonatal Respiratory Control

Another approach to the study of adenosine's role in central respiratory control utilizes a peculiarity of the respiratory control system in neonates. Whereas the peripheral chemoreceptors in several species are active soon after birth (Bureau and Begin, 1982; Darnall, 1982; Fagenholz et al., 1976), and the immediate response to hypoxia in all species studied is one of hyperpnea, in certain species, namely pig (Darnall, 1982), cat (Fagenholz et al., 1976), monkey (Woodrum et al., 1981), rabbit (Grunstein et al., 1981), and humans (Rigatto et al., 1975), this is followed after 30–90 s by a depression of ventilation. In humans this biphasic response persists until the third week of life, after which time the characteristic sustained stimulation of breathing by hypoxia is seen (Rigatto et al., 1975).

Clinically, hypoxia is well-recognized to cause apnea in preterm infants (Rigatto et al., 1975; Cross and Oppe, 1952). For many years, methylxanthines have been known to be effective in the treatment of the periodic breathing and apnea of the newborn (Davi et al., 1978; Aranda and Turmen, 1979) and have been shown to antagonize the late hypoxic fall in ventilation in neonatal piglets (Darnall, 1982; 1983). It now appears that these drugs are active because of their capacity to act as adenosine antagonists. Such an action is supported by the finding that, in neonatal rabbits, theophylline is an antagonist of the respiratory-depressant effects of R-PIA (Winn et al., 1981; Lagercrantz et al., 1984). However, adenosine may not be the only neuromodular mediating the central respiratory depression of hypoxia. Endorphines have been implicated, since they also are produced in the brain under hypoxic conditions, and their respiratory-depressant effect under these circumstances can be antagonized by naloxone (Grunstein et al., 1981; DeBoeck et al., 1983). The relative importance of endorphines and adenosine as central respiratory neuromodulators is unknown.

3.2. Action of Adenosine
on the Peripheral Chemoreceptors

In contrast to its central depressant action, adenosine is reported to exert a stimulant effect on ventilation when infused intravenously into intact animals or humans. The site of stimulation is thought to be the carotid bodies. The effect was first described as an incidental finding in a study of the cardiac effects of adenosine in animals by Drury and Szent-Györgyi (1929), repeated in man by Honey et al. (1930). The present decade has seen much interest in the mechanism of the respiratory-stimulant action of exogenous adenosine, the main focus of attention being its action at the peripheral chemoreceptors and, more particularly, the carotid bodies.

3.2.1. Organization of the Carotid Bodies

The carotid bodies are bilateral structures, found at the bifurcation of the common carotid arteries. Each derives its arterial blood supply from a small glomic artery that most commonly arises from the angle of the carotid bifurcation (Smith et al., 1982). The glomic artery is furnished with large intimal cushions located at its origin and surrounded by smooth-muscle cells. By the action of this sphincter-like device, the blood flow in the glomic artery may be controlled independently of systemic blood flow (McDonald and Haskell, 1983). The parenchyma of the carotid bodies is composed of two cell types: glomus and sustentacular cells. The most prominent structure is the glomus, or type I cell. In addition to their large nucleus and abundant mitochondria, these cells contain plentiful dense core granules and exhibit formaldehyde-induced fluorescence, indicating the presence of stored catecholamines (Hansen, 1985). Like other neuroendocrine cells, glomus cells are probably derived embryologically from the neural crest (Kondo et al., 1982). The sustentacular, or type II cells, invest the glomus cells and nerve endings with their profuse cytoplasmic projections. The morphology and histochemical characteristics of these cells suggest that they are analogous to glial (connective tissue) cells in the CNS (Kondo et al., 1982; Hess, 1968).

The mammalian carotid body is innervated by the glossopharyngeal and vagus nerves with a sympathetic supply derived from the superior cervical ganglion. Whereas some cranial nerve fibers are efferent, the majority of fibers of the glossopharyngeal nerve carried in the carotid sinus nerve are sensory and abut the glomus cells (Hess and Zapata, 1972). Sympathetic innervation is directed predominantly to the vasculature of the carotid body (Konche and Kienecker, 1977; McDonald and Mitchell, 1981).

The site of chemoreception and the mechanisms of transduction of sensory information to neural impulses are still poorly understood. However, given the relationship of the glomus cells to the nerve endings, their possession of dense core granules, and the necessity of their presence for transduction to take place (Zapata et al., 1976), the glomus cells probably represent primary sensory cells of chemoreception.

Various hypotheses have been formulated to explain the mechanisms of chemoreception in the carotid bodies, but they fall into three major categories (Eyzaguirre and Zapata, 1984).

1. The "acidic hypothesis" relates H^+ concentration in the sensory cleft to neuronal discharge.
2. The "protein receptor hypothesis" suggests a low oxygen affinity hemoglobin-like chromophore containing sulfhydryl

groups located in cell membranes, which could excite neuronal activity when in a reduced state (Lahiri, 1977).

3. The "metabolic hypothesis" (Anichkov and Belen'kii, 1963) links neuronal discharge to the adenosine nucleotide content of the glomus cell.

Implicit in the last hypothesis, is that changes in ambient P_{O_2} alter the ratio of ATP to the lower phosphates of adenosine within the cell, brought about by the presence of a low oxygen affinity cytochrome a_3 in the carotid body's cytochrome oxidase system (Jobsis, 1977).

Afferent traffic in the carotid sinus nerve increases approximately hyperbolically with fall in Pa_{O_2} (Biscoe et al., 1970). The relationship between neural activity and arterial hemoglobin saturation is linear over the range of saturation that coincides with the steep part of the oxygen/hemoglobin dissociation curve (Rebuck and Campbell, 1974). However, above 90% saturation (where large changes in Pa_{O_2} produce a small change in saturation), neuronal discharge is progressively attenuated, with little further increase in oxygen saturation. These findings indicate that, in hypoxia, the adequate stimulus to carotid chemoreception is provided by P_{O_2} rather than hemaglobin oxygen saturation or oxygen content of the blood (Lahiri et al., 1983).

Chemoreception of hypercapnia or acidaemia is mediated by a separate process. This has been shown by rendering the carotid bodies insensitive to hypoxia by treatment with the metabolic poison, oligomycin, and finding that the responsiveness to P_{CO_2} is preserved (Mulligan and Lahiri, 1982).

3.2.2. Adenosine Stimulation of Peripheral Chemoreceptors

In the early 1950s, Jarisch et al. (1952) described an increase in chemoreceptor discharge following injection of ATP into the carotid circulation. However, it was not until almost 30 years later that McQueen and Ribeiro (1981a) reported stimulation of carotid chemoreceptor activity in the cat, following intracarotid administration of adenosine itself. By recording from the carotid sinus nerve in pentobarbitone-anesthetized cats, they found a prompt dose-dependent increase in afferent nerve discharge upon intracarotid bolus injection of adenosine. Injections of adenosine into the carotid circulation in the rat (Monteiro and Ribeiro, 1987) and in humans (Biaggioni et al., 1987; Watt et al., 1987) have been shown to stimulate ventilation. Intravenous injection of adenosine in the rabbit (Buss et al., 1986) and in humans (Watt and Routledge, 1985; Maxwell, 1986; Biaggioni et al., 1987; Fuller et al., 1987) is also effective in stimulating breathing. The seeming contradiction between these observations and those described earlier involving central respiratory depression may be the result of peripheral chemoreceptor afferents overriding the central depressant effect of adenosine or poor penetration by adenosine across the blood–brain barrier, as has been previously suggested (Berne et al., 1974).

3.2.2.1. SPECIFICITY OF THE CAROTID BODY ADENOSINE RECEPTOR. The specificity of the receptor involved in carotid body stimulation by adenosine has been addressed by the groups in Edinburgh and Oeiras. They have found that in anesthetized cats, intracarotid injection of 0.1–100 µg adenosine elicits a rapid increase in chemoreceptor discharge as recorded from filaments of the carotid sinus nerve (McQueen and Ribeiro, 1981a). However, when 1–100 µg ATP was injected using a similar technique, a latency of 2–8 s was observed before increased carotid sinus nerve discharge was evident (McQueen and Ribeiro, 1983). The same investigators also showed that the stable ATP analogue α-β-methylene ATP (which cannot be metabolized to adenosine), when given over a dose range of 10–100 µg by the same route, produced a dose-related reduction in spontaneous afferent carotid sinus nerve discharge, whereas β-γ-methylene ATP, which may be hydrolyzed to AMP and adenosine, produced a chemoexcitation quantitively similar to that observed with native ATP. These results suggest that the role of ATP as a chemostimulant may be contingent upon it first being metabolized to adenosine. These observations were extended by a comparison of the effects of different adenine nucleotides, adenosine and coenzyme A, in the anesthetized ventilated cat (Ribeiro and McQueen, 1984). In this study, it was shown that the chemoreceptor stimulant effects of ATP, ADP, AMP, and adenosine were quantitatively similar when compared on an equimolar basis.

Further evidence that ATP acts in the same way as adenosine was shown by the fact that concomitant infusion of adenosine reduces the effect of intracarotid ATP administration (Ribeiro and McQueen, 1984). Specificity of the mechanism of chemoexcitation for adenosine has been shown by the absence of effect of its deamination product, inosine (McQueen and Ribeiro, 1983; Reid et al., 1987), and lack of effect of other nucleosides: guanosine, cytosine, and the purine bases, uridine and adenine (McQueen and Ribeiro, 1983).

3.2.2.2. CHARACTERIZATION OF THE ADENOSINE RECEPTOR. Evidence that a receptor on the cell surface rather than within the cell is responsible for the observed stimulatory effects of adenosine on the carotid body is of two types. First, McQueen and Ribeiro (1983) gave intracarotid injections of various adenosine agonists and monitored their effects on spontaneous afferent neural discharge. The cell surface receptor agonists N^6-methyladenosine and 2-CADO gave effects similar to those seen with adenosine itself, whereas 2'-deoxyadenosine, active at the intracellular P site on the catalytic subunit of adenylate cyclase, had little or no effect. 3'-deoxyadenosine, which has both A- and P-receptor agonist activity, but a higher affinity for A receptors, caused a stimulation of afferent discharge. The second line of evidence derives from the observation that intravenous or intracarotid administration of dipyridamole elevates the spontaneous discharge rate in the carotid sinus nerve of anesthetized cats. Moreover, in the presence of dipyridamole, chemoreceptor responses to injections of adenosine are greatly potentiated (McQueen and Ribeiro, 1983; Ribeiro and McQueen, 1984). As dipyridamole blocks the facilitated uptake of adenosine by cells, this evidence supports the view that adenosine acts on an extracellular receptor in the carotid body. Dipyridamole when administered orally to humans also shifts the ventilatory dose–response curve of intravenous adenosine significantly to the left (Watt and Routledge, 1987).

Studies of the effect of methylxanthines on adenosine-induced chemoexcitation have yielded less clear-cut results. In the cat, the comparatively high bolus dose of 1 mg of theophylline given into the carotid artery by McQueen and Ribeiro (1981a) initially produced a short-lived decrease in spontaneous carotid sinus nerve discharge; however, the excitatory action of adenosine given subsequently was enhanced. This sensitizing effect of theophylline has not been observed using smaller doses of theophylline given concurrently with adenosine or its analogs, or given as an infusion during challenge with adenosine. Thus a reason for enhancement of the response to adenosine by theophylline in this study may have been the central actions of theophylline described in the previous section, together with dilution of the theophylline still present in the carotid body. Intracarotid infusion of theophylline has been shown to reduce the chemoexcitation produced by concurrent injection of the adenosine analog, NECA, in the cat (McQueen and Ribeiro, 1986). In rats, theophylline has also been shown to produce a shift to the right in the dose–response curve of 2-CADO to ventilation, indicating activity at a xanthine-sensitive receptor (Monteiro and Ribeiro, 1987). Infusion of the potent adenosine receptor antagonist, 8-phenyltheophylline (8-PT), which is free from phosphodiesterase-inhibitory activity, into the anesthetized cat resulted in a reduction of spontaneous carotid sinus nerve firing and antagonized the stimulatory effect of adenosine (McQueen and Ribeiro, 1986).

In the intact human, the effect of methylxanthines on adenosine-induced stimulation of ventilation is unclear. This may be because of the combination of central and peripheral actions of theophylline. Biaggioni et al. found no effect of infused aminophylline on the ventilatory response to either intravenous boluses (Biaggioni et al., 1986a) or continuous intravenous infusion (Biaggioni et al., 1986b) of adenosine. Routledge and Watt (1986) have reported a possible attenuation of the ventilatory response to intravenous adenosine boluses, but their results did not reach statistical significance. More recently, Maxwell et al. (1987a,b) and Smits et al. (1987) have shown significant attenuation of respiratory stimulation induced by intravenous adenosine infusion by pretreatment with theophylline and caffeine, respectively. Enprofylline, a xanthine without adenosine antagonist activity, did not show this effect. The latter result, together with the enhancing effect of dipyridamole (Watt and Routledge, 1987), would be compatible with adenosine acting at A_1 or A_2 receptors to bring about respiratory stimulation in humans.

The rank order of potency of adenosine analogs in stimulating carotid chemosensory discharge in the cat (McQueen and Ribeiro, 1986) has been found to be NECA > adenosine > R-PIA > S-PIA and that in stimulating ventilation in the anesthetized rat to be NECA > 2-CADO > S-PIA = R-PIA, with no marked stereoselectivity displayed for the two isomers of PIA (Monteiro and Ribeiro, 1987). Both of these sets of results strongly indicate that the chemosensory stimulation brought about by adenosine is mediated at A_2-type cell surface receptors.

These purine receptors are probably located in the carotid bodies, since the increased ventilatory response to adenosine infusion is ablated by section of the carotid sinus nerve in the rat (Monteiro and Ribeiro, 1987) and the rabbit (Buss et al., 1986), and has been shown not to be present when adenosine is infused intravenously into patients deprived of their hypoxic chemosensitivity following bilateral carotid endarterectomy (Griffiths et al., 1989). These observations, together with recordings made from the carotid sinus nerve in cats, suggest that the ventilatory stimulant effect of infused adenosine is mediated at an A_2 cell surface receptor located in the peripheral chemoreceptors.

3.2.3. Adenosine Infusion and Ventilation in Humans

Stimulation of ventilation by intravenous bolus doses of adenosine was first described in humans by Watt and Routledge (1985). They showed that boluses of 60 µg/kg or greater produced a dose-related but short-lived increase in ventilation, primarily as a result of increased tidal volume, with no significant change occurring in the respiratory frequency. Fuller et al. (1987) have extended these observations to show that the mean inspiratory flowrate V_t/T_I is increased during adenosine infusion at 50 µg/kg/min, with a significant increase in V_t (where V_t represents the tidal volume, and T_I is the duration of inspiration). At higher infusion rates (100 µg/kg/min), the ratio of inspiratory duration to the duration of the complete respiratory cycle (T_I/T_{tot}) was significantly reduced as a result of the shortening of T_I in most subjects and of the lengthening of expiratory duration in half of the subjects. This change in respiratory timing may represent a central respiratory depressant activity of the infused adenosine, similar to that seen in the glomectomized cat (Eldridge et al., 1984) and in the rat (Wessberg et al., 1985), or could be the result of hypocapnia secondary to the hyperventilation induced (Milic-Emili and Grunstein, 1976). In contrast, the pure peripheral chemoreceptor stimulus of hypoxia augments ventilation by an effect on respiratory drive, as evidenced by an increased V_t/T_p but does not alter T_I/T_{tot} (Mannix et al., 1984).

Another observation regarding the effect of intravenous infusion of adenosine has been that of differential effects on the ventilatory responses to hypoxia and hypercapnia. Using standard rebreathing techniques, Maxwell et al. (1986) found that the slope of the line representing the relationship between ventilation and arterial hemoglobin oxygen saturation was increased during adenosine infusion at a rate of 70–80 µg/kg/min, whereas the slope of the line representing the relationship of ventilation to P_{CO_2} in hyperoxia was unaltered, although the whole response curve was shifted to the left. These results would be compatible with action of adenosine at the carotid bodies. This conclusion follows because the hypoxic ventilatory response in humans is dependent upon functioning carotid bodies (Lugliani et al., 1971), and peripheral chemoreceptors are unresponsive to hypercapnia in hyperoxic conditions. The pattern of response to adenosine infusion suggests an action on either the peripheral chemoreceptors or their central connections. However, since adenosine is a central respiratory depressant, the effect is almost certainly a peripheral one.

3.2.4. Physiological Significance of Adenosine as a Peripheral Chemoreceptor Stimulant

With the weight of evidence suggesting that adenosine acts as a respiratory stimulant at the carotid body level, it is tempting to tie in this fact with the metabolic theory of chemoreception alluded to earlier (Fig. 2). A fall in P_{CO_2} would be "sensed" by the low-affinity cytochrome a_3, with a resulting fall in the ATP content and a rise in that of the lower phosphates and adenosine within the cell, which could then enter the extracellular space by facilitated diffusion down its concentration gradient (Arch and Newsholme, 1978) and stimulate A_2 receptors. However, further research is needed in this area, since the evidence available on ATP levels after stimulation of the carotid body by hypoxia is conflicting; Acker and Starlinger (1984) found no change, and Obeso et al. (1987) reported a decrease. Once in the interstitial space, (*see* Fig. 2) adenosine might:

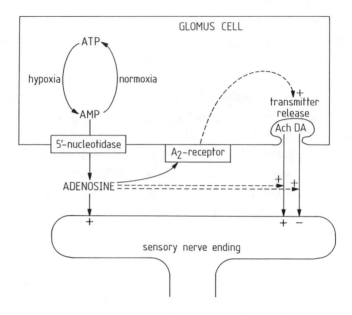

Fig. 2. Hypothetical model of adenosine production and action in the carotid body. Hypoxic conditions favor the production of AMP, with a consequent rise in extracellular adenosine concentration. Extracellular adenosine might stimulate nerve endings directly, enhance glomus cell transmitter release via A_2 receptors, or enhance the action of released acetylcholine (ACh) or dopamine (DA).

1. Stimulate release of transmitter autocoids from glomus cells;
2. Stimulate nerve endings directly; or
3. Modify the action of other transmitters at the nerve endings.

   Information currently available does not allow precise localization of the site of action of adenosine within the carotid body; effects on glomus cells, sustentacular cells, and nerve endings are all possible. Indeed, action on the blood vessels supplying the glomoid tissue is also possible, although the rapidity of the onset of adenosine's effect in chemoexcitation and the comparatively slower, smaller effect of vasoactive drugs on carotid sinus nerve discharge (McQueen, 1983) make this site of action unlikely. Further studies using ligand binding and iontophoretic techniques may eventually determine the site of adenosine's action in the carotid body. In spite of the observation that adenosine sensitizes the carotid chemoreceptors to both the excitatory effects of acetylecholine and the depressant effects of dopamine (McQueen and Ribeiro, 1981b), direct evidence of adenosine acting on the "postsynaptic" nerve ending modulating these responses must await appropriate biophysical experiments.

Stimulation of A_2 receptors might be expected to increase the cyclic AMP content of the cells affected through the known interaction of these receptors with adenylate cyclase. However, intraperitoneal administration of 2-CADO in the rat was not shown to produce a change in carotid body cyclic AMP content (Mir et al., 1983). It may be that in this experiment the peak effect of the adenosine analog was missed or that activation of A_2 receptors in the carotid body is not necessarily linked to the activation of adenylate cyclase (Stone, 1984).

A further intriguing aspect of the respiratory stimulant action of adenosine is the possibility that its production at a distant site and liberation into the blood stream might elicit a chemosensory response at the carotid body. Such a mechanism could operate during myocardial ischemia and during muscular exercise. Both these conditions are recognized to be associated with hyperpnea. Adenosine has been shown to be released into the effluent blood from exercising skeletal muscle (Dobson et al., 1971; Bockman et al., 1975; Proctor and Duling, 1982); the same is also true in the ischemic heart (Rubio et al., 1973). Concentrations of adenosine achieved in the systemic arterial circulation are however unknown. Such a hormonal effect might partially explain the increase in carotid-body sensitivity seen in exercise (Cotes, 1979) and the dyspnea of cardiac ischemia.

4. Adenosine and the Pulmonary Circulation

The final area of pulmonary physiology in which adenosine or adenine nucleotides may have a role is the control of pulmonary vascular resistance.

4.1. Pulmonary Vascular Response to Hypoxia

The normal physiological response of the pulmonary vascular bed to hypoxia in humans is one of vasoconstriction, which if sustained (e.g., secondary to diseases of the lung parenchyma or airways), may lead to pulmonary hypertension and pulmonary vascular disease involving abnormal muscularization of the terminal portions of the pulmonary vascular tree (Heath and Kay, 1976). In several experimental animals, isolated lung preparations exposed to hypoxia exhibit a transient vasoconstriction followed by a more prolonged vasodilatation (Harabin et al., 1981; Peake et al., 1981; Wetzel and Sylvester, 1983). The knowledge that adenosine may be a mediator of metabolic vasodilatation in other vascular beds, such as skeletal and cardiac muscle (Berne et al., 1971), has led to the suggestion that adenosine may serve as a regulatory autacoid of pulmonary vascular resistance.

4.2. Pulmonary Vascular Response to Adenosine and Adenine Nucleotides

It has been shown that concentrations of adenosine and AMP rise in canine lungs ventilated with anoxic gas, while there is a parallel fall in ATP levels (Mentzer et al., 1975). These workers also found that infusion of adenosine into the pulmonary artery *in situ* produced vasodilatation in individual pulmonary lobes, and that such an infusion or treatment with dipyridamole inhibited the rise in vascular resistance that was otherwise produced by ventilating the lobe concerned with anoxic gas. Interestingly, when infused into the pulmonary artery, the higher phosphates ATP and ADP produced an opposite effect, i.e., vasoconstriction. In this model, the action of adenosine might be expected to counteract the stronger, opposite effects on pulmonary arterial caliber that are operative during hypoxia. A similar vasoconstricting effect of ATP has been reported in isolated perfused lungs in the rabbit (Lande et al., 1968) and rat (Hauge, 1968). In a study involving intravenous bolus injections of ATP, ADP, or AMP in the goat, Reeves et al. (1967) found an increase in pulmonary artery pressure with ATP and ADP, but no effect with AMP. Interestingly, histological changes of the intima and media in the pulmonary arteries seen on repeated daily ADP injection over 1–4 wk resemble changes seen in pulmonary embolism—another important cause of pulmonary hypertension. In spite of this evidence, it is unlikely that the vasoconstriction of hypoxemia seen in humans is mediated by ATP. First, there are no studies reporting increased ATP production in the lung under conditions of hypoxia; if anything, the opposite occurs (Mentzer et al., 1975). In isolated rat lung, the vasoconstriction caused by ATP can be inhibited with 2,4-xylenol, an isomer of dimethylphenol that can inhibit vasoconstriction in isolated, blood-perfused rabbit lungs. However, the pressor response to hypoxia in that preparation was unaffected by treatment with 2,4-xylenol at a level associated with potent inhibition of the response to ATP (Hauge, 1968). A pressor effect of ATP in humans undergoing the hypoxia of chronic obstructive airways disease is unlikely, since in these subjects, intravenous infusion of ATP at a low rate (1–2 μmol in 20 min) led to a fall, rather than a rise, in vascular resistance (Gaba et al., 1986). This change was associated with a deterioration in the patients' Pa_{O_2} and Pa_{CO_2}, presumably as a result of increased perfusion to inadequately ventilated regions of the lung.

Human pulmonary vessels have been studied in vitro (McCormack et al., 1989). Adenosine caused a dose related vasodilatation of 5-Hydroxytryptamine-constricted small pulmonary arteries that was attenuated by the adenosine antagonist 8-phenyl theophylline. The rank order of potency of NECA > Adenosine > PIA producing vasodilatation in this preparation suggests action at an A_2 purine receptor. Conversely, Wiklund et al. (1987) found that adenosine, NECA, and PIA enhanced the force of contractions in guinea pig pulmonary arterial strips induced by transmural electrical stimulation and noradrenaline. The order of potency of the agonists in this preparation suggested action at an A_1 purine receptor. Therefore, it seems possible that adenosine may act within the pulmonary vasculature on different receptors to modulate the effects of neurotransmitters active on the vascular smooth muscle.

However, there is as yet no evidence to support a physiological role for adenosine in hypoxic pulmonary vasoconstriction. In isolated ferret lungs, the response to severe hypoxia was shown to be one of initial vasoconstriction followed by vasodilatation. That neither response was affected by perfusion with adenosine deaminase was taken to exclude an important role of adenosine in the primary pulmonary vascular responses to hypoxia (Gottlieb et al., 1984). These findings are consistent with the lack of observed response in chronically hypoxic men in whom adenosine infusion had no effect on pulmonary vascular resistance (Gaba et al., 1986).

5. Summary and Conclusions

Evidence for the involvement of adenosine in various mechanisms of normal and abnormal functioning of the respiratory system has been presented. It has been satisfactorily demonstrated that adenosine is capable of inducing immediate bronchoconstriction in asthmatic airways. Several experimental observations suggest that this action is the result of augmented mast cell release of preformed mediators, initiated by IgE or other mechanisms. Whereas adenosine has been shown to be released into the circulation during provoked bronchoconstriction, the cellular origin of the nucleoside has not been defined.

It follows that a role for mast-cell-derived adenosine in augmenting mast cell degranulation might amplify the response, providing a positive feedback to enhance ongoing mediator release (Fig. 1). This proposition, however, cannot be validated until the site of adenosine production during asthmatic bronchoconstriction can be identified. Until that time, the physiological significance of adenosine as a bronchoconstrictor agonist in the asthmatic airway will remain unknown.

The role of adenosine in the control of breathing depends upon its site of action. The present state of knowledge regarding the central and peripheral effects of adenosine on ventilatory control is summarized in Fig. 3. Within the CNS, adenosine exerts a tonic inhibitory influence on the respiratory center and may be acting as an autoacoid produced during hypoxia and mediating the central respiratory depression that results from cerebral hypoxia. Peripherally, adenosine stimulates ventilation by an action on the carotid bodies. This action is probably mediated at an extracellular A_2 purine receptor, although the cellular location of these receptors has yet to be defined. These findings are compatible with the metabolic theory of peripheral chemoreception, but the importance of carotid body adenosine levels in human health and disease remains to be elucidated.

Finally, adenosine and the adenine nucleotides have been shown to be vasoactive in the pulmonary vasculature. The physiological significance of this is unknown, but the apparently opposite effects of ATP and adenosine form an intriguing aspect that merits further investigation.

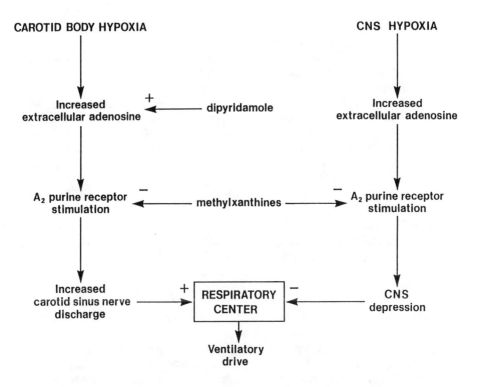

Fig. 3. The contrasting peripheral and central actions of adenosine in the control of breathing, and the observed actions of dipyridamole and methylxanthines at each site.

References

Acker, H. and Starlinger, H. (1984) Adenosine triphosphate content in the cat carotid body under different arterial O_2 and CO_2 conditions. *Neurosci. Lett.* **50,** 175–179.

Adams, G. K., III and Lichtenstein, L. (1979) *In vitro* studies of antigen-induced bronchospasm: Effect of anti-histamine and SRB-A antagonist on response of sensitized guinea pig and human airways to antigen. *J. Immunol.* **122,** 555–562.

Altounyan, R. E. C., Lee, T. B., Rocchiccioli, K. M. S., and Shaw, C. L. (1986) A comparison of the inhibitory effects of nedocromil sodium and sodium cromoglycate on adenosine monophosphate-induced bronchoconstriction in atopic subjects. *Eur. J. Respir. Dis. (Suppl. 147),* **69,** 277–279.

Anichkov, S. V. and Belen'kii, M. L. (1963) Pharmacology of the carotid body chemoreceptors. (Pergamon, Oxford).

412   Griffiths and Holgate

Aranda, J. V. and Turmen, T. (1979) Methylxanthines in apnoea of prematurity.
    Clin. Perinatol. 6, 87–108.
Arch, J. R. S. and Newsholme, E. A. (1978) The control of the metabolism and the
    hormonal role of adenosine. Essays Biochem. 14, 82–123.
Bellardinelli, L., Shryock, J., West, G. A., Clemo, H. F., DiMarco, J. P., and Berne,
    R. M. (1984) Effects of adenosine and adenine nucleotides on the atrio-
    ventricular node of isolated guinea pig hearts. Circulation 70, 1083–1091.
Benyon, R. C., Church, M. K., and Holgate, S. T. (1984) The effect of methyl-trans-
    ferase inhibitors on histamine release from human dispersed lung mast cells
    activated with anti-human IgE and calcium ionophore A23187. Biochem.
    Pharmacol. 33, 2881–2886.
Berne, R. M. (1980) The role of adenosine in the regulation of coronary blood flow.
    Circ. Res. 47, 807–813.
Berne, R. M., Rubio, R., and Curnish, R. R. (1974) Release of adenosine from the
    ischaemic brain: Effect on cerebral vascular resistance and incorporation in-
    to cerebral adenine nucleotides. Circ. Res. 35, 262–271.
Berne, R. M., Rubio, R., Dobson, J. G., Jr., and Curnish, R. R. (1971) Adenosine and
    adenine nucleotides as possible mediators of cardiac and skeletal muscle
    blood flow regulation. Circ. Res. 28, Suppl. 1, 115–119.
Biaggioni, I., Olafsson, B., Robertson, R., Hollister, A. S., and Robertson, D.
    (1986a) Failure of aminophylline to antagonize the haemodynamic and res-
    piratory effects of adenosine boluses. Pflugers Arch. 407, Suppl. 1, S54.
Biaggioni, I., Olafsson, B., Robertson, R., Hollister, A. S., and Robertson, D.
    (1986b) Characteristics of adenosine-induced stimulation of respiration in
    man. Pflugers Arch. 407, Suppl. 1, S55.
Biaggioni, I., Olafsson, B., Robertson, R. M., Hollister, A. S., and Robertson, D.
    (1987) Cardiovascular and respiratory effects of adenosine in conscious man.
    Evidence for chemoreceptor activation. Circ. Res. 61, 779–786.
Biscoe, T. J., Bradley, G. W., and Purves, M. J. (1970) The relation between carotid
    body chemoreceptor discharge, carotid sinus pressure and carotid body ven-
    ous flow. J. Physiol. 208, 99–120.
Bockman, E. L., Berne, R. M., and Rubio, R. (1975) Release of adenosine and lack
    of release of ATP from contracting skeletal muscle. Pflugers Arch. 355,
    229–241.
Brown, C. M. and Collis, M. G. (1982) Evidence for an A_2/R_a adenosine receptor
    in guinea-pig trachea. Br. J. Pharmacol. 76, 381–387.
Bruce, E. N. and Cherniack, N. S. (1987) Central chemoreceptors. J. Appl. Physiol.
    62, 389–402.
Bureau, M. A. and Begin, R. (1982) Postnatal maturation of the respiratory re-
    sponse to O_2 in awake newborn lambs. J. Appl. Physiol. 52, 428–433.
Burr, D. and Sinclair, J. D. (1988) The effect of adenosine on respiratory chemosen-
    sitivity in the awake rat. Respir. Physiol. 72, 47–57.
Buss, D. C., Routledge, P. A., and Watt, A. H. (1986) Respiratory effects of intra-
    venous adenosine in anaesthetised rabbits before and after carotid nerve sec-
    tion. Br. J. Pharmacol. 88, 413p.

Butcher, R. W. and Sutherland, E. W. (1962) Adenosine 3',5'-phosphate in biological materials. *J. Biol. Chem.* **237**, 1244–1250.

Chan-Yeung, M. (1977) The effect of SCH 1000 and disodium cromoglycate on exercise-induced asthma. *Chest* **71**, 320–323.

Church, M. K., Holgate, S. T., and Hughes, P. J. (1983) Adenosine inhibits and potentiates IgE-dependent histamine release from human basophils from an A_2-receptor mediated mechanism. *Br. J. Pharmacol.* **80**, 719–726.

Church, M. K., Benyon, R. C., Hughes, P. J., Cushley, M. J., Mann, J. S., and Holgate, S. T. (1985) Adenosine as a putative mediator in asthma: Its role in bronchoconstriction and modulation of histamine release from human lung mast cells and basophil leukocytes, in *Purines: Pharmacology and Physiological Roles* (Stone, T. W., ed.), Macmillan, London, pp. 175–184.

Coburn, R. F. and Tomita, T. (1973) Evidence for non-adrenergic inhibitory nerves in guinea-pig trachealis muscle. *Am. J. Physiol.* **224**, 1072–1080.

Cotes, J. E. (1979) Control of respiration, in *Lung Function Assessment and Application in Medicine*, 4th Ed. (Cotes, J. E., ed.) Blackwell, Oxford, pp. 251–264.

Crimi, N., Palermo, F., Ciccarello, C., Oliveri, R., Vancheri, C., Palermo, B., and Mistretta, A. (1989) Effect of theophylline on adenosine-induced bronchoconstriction. *Ann. Allergy.* **62**, 123–127.

Crimi, N., Palermo, F., Vancheri, C., Oliveri, R., Distefano, S. M., Polosa, R., and Mistretta, A. (1988) Effect of sodium cromoglycate and nifedipine on adenosine-induced bronchoconstriction. *Respiration* **53**, 74–80.

Cross, K. W. and Oppe, T. E. (1952) Effect of inhalation of high and low concentration of oxygen on the respiration of the premature infant. *J. Physiol.* **117**, 38–55.

Cushley, M. J. and Holgate, S. T. (1985) Adenosine-induced bronchoconstriction in asthma: Role of mast cell-mediator release. *J. Allergy Clin. Immunol.* **75**, 272–278.

Cushley, M. J., Tallant, N., and Holgate, S. T. (1985) The effect of dipyridamole on histamine- and adenosine-induced bronchoconstriction in normal and asthmatic subjects. *Eur. J. Respir. Dis.* **67**, 185–192.

Cushley, M. J., Tattersfield, A. E., and Holgate, S. T. (1983) Inhaled adenosine and guanosine on airway resistance in normal and asthmatic subjects. *Br. J. Clin. Pharmacol.* **15**, 161–165.

Cushley, M. J., Tattersfield, A. E., and Holgate, S. T. (1984) Adenosine-induced bronchoconstriction in asthma: Antagonism by inhaled theophylline. *Am. Rev. Respir. Dis.* **129**, 380–384.

Dahlen, S.-E., Hansson, G., Heqvist, P., Bjorck, T., and Ganstrom, E. (1983) Allergen challenge of lung tissue from asthmatics elicits bronchial contraction that correlates with the release of leukotrienes C_4, D_4, and E_4. *Proc. Natl. Acad. Sci. USA* **80**, 1712–1716.

Daly, J. W. (1982) Adenosine receptors: Targets for future drugs. *J. Med. Chem.* **25**, 197–207.

Darnall, R. A. (1982) Theophylline reduces ventilatory depression in hypoxic newborn piglets. *Pediatr. Res.* **16**, 347A.

Darnall, R. A. (1983) The effect of opioid and adenosine antagonists on hypoxic ventilatory depression in the newborn piglet. *Pediatr. Res.* **17**, 374A.

Davi, M. J., Sankaran, K., Simons, K. J., Simons, F. E. R., Seshia, M. M., and Rigatto, H. (1978) Physiologic changes induced by theophylline in the treatment of apnoea in preterm infants. *J. Pediatr.* **92**, 91–95.

Davis, C., Kannan, M. S., Jones, T. R., and Daniel, E. E. (1982) Control of human airway smooth muscle: In vitro studies. *J. Appl. Physiol.* **53**, 1080–1087.

Dazsun, Z., Rafferty, P., Richards, R., Summerell, S., and Holgate, S. T. (1989) Airways refractoriness to adenosine 5'-monophosphate after repeated inhalation. *J. Allergy Clin. Immunol.* **83**, 152–158.

DeBoeck, C., Van Reempts, P., Rigatto, H., and Chernick, V. (1983) Endorphins and the ventilatory depression during hypoxia in newborn infants. *Pediatr. Res.* **17**, 374A.

Dobson, G., Rubio, R., and Berne, R. M. (1971) Role of adenine nucleotides, adenosine and inorganic phosphate in the regulation of skeletal muscle blood flow. *Circ. Res.* **29**, 375–384.

Dowell, A. R., Heyman, A., Sieker, H. O., and Tripathy, K. (1965) Effect of aminophylline on respiratory-center sensitivity in Cheyne-Stokes respiratory and in pulmonary emphysema. *New Eng. J. Med.* **273**, 1447–1453.

Drury, A. N. and Szent-Györgyi, A. (1929) The physiological action of adenosine compounds with special reference to their action upon mammalian heart. *J. Physiol.* **68**, 213–237.

Eiser, N. M. and Guz, A. (1982) Effect of atropine on experimentally-induced airway obstruction in man. *Bull. Eur. Physiopathol. Respir.* **18**, 449–460.

Eldridge, F. L., Millhorn, D. E., and Kiley, J. P. (1984) Respiratory effects of a long-acting analog of adenosine. *Brain Res.* **301**, 273–280.

Eldridge, F. L., Millhorn, D. E., and Kiley, J. P. (1985) Antagonism by theophylline respiratory inhibition induced by adenosine. *J. Appl. Physiol.* **59**, 1428–1433.

Eldridge, F. L., Millhorn, D. E., Waldrop, T. G., and Kiley, J. P. (1983) Mechanism of respiratory effects of methylxanthines. *Respir. Physiol.* **53**, 239–261.

Ellwood, R. K., Belzberg, A., Hogg, J. C., and Pare, P. D. (1982) Bronchial mucosal permeability in asthma. *Am. Rev. Respir. Dis.* **125**, Suppl. 63.

Eyzaguirre, C. and Zapata, P. (1984) Perspectives in carotid body research. *J. Appl. Physiol.* **57**, 931–957.

Fagenholz, S. A., O'Connell, K., and Shannon, D. C. (1976) Chemoreceptor function and sleep state in apnoea. *Pediatrics* **58**, 31–36.

Finney, M. J. B., Karlsson, J.-A., and Persson, C. G. A. (1985) Effects of bronchoconstrictors and bronchodilators on a novel human small airway preparation. *Br. J. Pharmacol.* **85**, 29–36.

Fish, J. E., Lenfant, C., and Newball, H. H. (1983) In vivo immunopharmacology of the lung, in *Immunopharmacology of the Lung,* vol 19 of *Lung Biology in Health and Disease* (Newball, H. H., ed.), Marcel Dekker, New York, pp. 273–346.

Fredholm, B. B. (1981) Release of adenosine from rat lung by antigen and compound 48/80. *Acta. Physiol. Scand.* **111**, 507,508.

Fredholm, B. B. (1982) Adenosine receptors. *Med. Biol.* **60**, 289–293.

Fredholm, B. B., Brodn, K., and Strandberg, K. (1979) On the mechanism of relaxation of tracheal muscle by theophylline and other cyclic nucleotide phosphodiesterase inhibitors. *Acta. Pharmacol. Toxicol.* **45**, 336–344.

Fuchs, B. D., Gorman, M. W., and Sparks, H. V. (1986) Adenosine release into venous plasma during free flow exercise. *Proc. Soc. Exp. Biol. Med.* **181**, 364–370.

Fuller, R. W., Maxwell, D. L., Conradson, T.-B. G., Dixon, C. M. S., and Barnes, P. J. (1987) Circulatory and respiratory effects of infused adenosine in conscious man. *Br. J. Clin. Pharmacol.* **24**, 309–317.

Gaba, S., Trigui, F., Dujols, P., Godard, P., Michel, F. B., and Prefaut, C. (1986) Compared effects of ATP vs adenosine on pulmonary circulation of COPD. *Eur. J. Respir. Dis. (Suppl. 146),* **69**, 515–522.

Gottlieb, J. E., Peake, M. D., and Sylvester, J. T. (1984) Adenosine and hypoxic pulmonary vasodilatation. *Am. J. Physiol.* **247**, H541–H547.

Griffiths, T. L., Warren, S. J., Chant, A. D. B., and Holgate, S. T. (1989) Ventilatory effects of hypoxia and adenosine infusion in patients after bilateral carotid endarterectomy. *Clin. Sci.,* in press.

Grunstein, M. M., Hazinski, T. A., and Schlueter, M. A. (1981) Respiratory control during hypoxia in newborn rabbits: Implied action of endorphins. *J. Appl. Physiol.* **51**, 122–130.

Hamprecht, B. and Van Calker, D. (1985) Nomenclature of adenosine receptors. *Trends Pharmacol. Sci.* **6**, 153,154.

Hansen, J. T. (1985) Ultrastructure of the primate carotid body: A morphometric study of the glomus cells and nerve endings in the monkey (Macaca fascicularis). *J. Neurocytol.* **14**, 13–32.

Harabin, A. L., Peake, M. D., and Sylvester, J. T. (1981) Effect of severe hypoxia on the pulmonary vascular response to vasoconstrictor agents. *J. Appl. Physiol.* **50**, 561–565.

Hauge, A. (1968) Role of histamine in hypoxic pulmonary hypertension in the rat. I. Blockade or potentiation of endogenous amines, kinins and ATP. *Circ. Res.* **22**, 371–383.

Hauge, A., Lunde, P. K. M., and Waaler, B. A. (1966) Vasoconstriction in isolated blood-perfused rabbit lungs and its inhibition by cresols. *Acta. Physiol. Scand.* **66**, 226–240.

Heath, D. and Kay, J. M. (1976) Respiratory system, in *Muir's Textbook of Pathology,* 10th Ed. (Anderson, J. R., ed.) Edward Arnold, London, pp. 378–449.

Hedner, T., Hedner, J., Jonason, J., and Wessberg, P. (1984) Effects of theophylline on adenosine-induced respiratory depression in the preterm rabbit. *Eur. J. Respir. Dis.* **65**, 153–156.

Hess, A. (1968) Electron microscopic observations of normal and experimental cat carotid bodies, in *Arterial Chemoreceptors* (Torrance, R. W., ed.) Blackwell, Oxford and Edinburgh, pp. 51–56.

Hess, A. and Zapata, P. (1972) Innervation of the cat carotid body: Normal and experimental studies. *Fed. Proc.* **31**, 1365–1382.

Hogg, J. C., Pare, P. D., Boucher, R. C., Michoud, M. C., Guerzon, G., and Moroz, L. (1977) Pathologic abnormalities in asthma, in *Asthma, Physiology, Immunopharmacology and Treatment* (Lichtenstein, L. M., Austen, K. F., and Simon, A. S., eds.) Academic, New York, pp. 1–19.

Holgate, S. T. and Mann, J. S. (1984) Release of adenosine and its metabolites from human leukocytes activated with the calcium ionophore A23187. *Br. J. Pharmacol.* **82**, 262p.

Holgate, S. T., Lewis, R. A., and Austen, K. F. (1980a) Role of adenylate cyclase in immunologic release of mediators from rat mast cells: Agonist and antagonist effects of purine- and ribose-modified adenosine analogs. *Proc. Natl. Acad. Sci. USA* **77**, 6800–6804.

Holgate, S. T., Lewis, R. A., and Austen, K. F. (1980b) 3',5'-cyclic adenosine monophosphate-dependent protein kinase of the rat serosal mast cell and its immunologic activation. *J. Immunol.* **124**, 2093–2099.

Holgate, S. T., Cushley, M. J., Mann, J. S., Hughes, P., and Church, M. K. (1986) The action of purines on human airways. *Arch. Int. Pharmacodyn.* **280**, Suppl., 240–252.

Holgate, S. T., Cushley, M. J., Rafferty, P., Beasley, R., Phillips, G., and Church, M. K. (1987) The bronchoconstrictor activity of adenosine in asthma, in *Topics and Perspectives in Adenosine Research* (Gerlach, S. and Becker, B. F., ed.), Springer-Verlag, Berlin and Heidelberg, pp. 614–624.

Honey, R. M., Ritchie, W. T., and Thomson, W. A. R. (1930) The action of adenosine upon human heart. *PE Q. J. Med.* **23**, 485–489.

Hughes, P. J., Holgate, S. T., and Church, M. K. (1984) Adenosine inhibits and potentiates IgE-dependent histamine release from human lung mast cells by an A_2-purinoceptor mediated mechanism. *Biochem. Pharmacol.* **33**, 3847–3852.

Ishizaka, T. and Ishizaka, K. (1984) Activation of mast cells for mediator release through IgE receptors. *Prog. Allergy* **34**, 188–235.

Jarisch, A., Landgren, A., Neil, E., and Zotterman, Y. (1952) Impulse activity in the caritod sinus nerve following intra-carotid injection of potassium chloride, veratrine, sodium citrate, adenosine-triphosphate and α-dinitrophenol. *Acta. Physiol. Scand.* **25**, 195–211.

Jarvis, M. F. and Williams, M. (1990) Adenonine in central nervous system function, in *Adenonine and Adenonine Receptor* (William, M., ed.), Humana, Clifton, NJ, in press.

Jeffrey, P. and Corni, B. (1984) Structural analysis of the respiratory tract, in *Immunology of the Lung and Upper Respiratory Tract* (Bienenstock, J., ed.), McGraw Hill, New York and Toronto, pp. 1–27.

Jobsis, F. F. (1977) What is a molecular oxygen sensor? What is a transduction process? *Adv. Exp. Med. Biol.* **78**, 3–18.

Kaliner, M. A. (1980) Mast-cell derived mediators and bronchial asthma, in *Airway Reactivity, Mechanisms and Clinical Relevance* (Hargreave, F. E., ed.), Astra, Mississauga, Ontario, pp. 175–187.

Karlsson, J.-A., Kjellin, G., and Persson, C. G. A. (1982) Effect on tracheal smooth muscle of adenosine and methylxanthines and their interaction. *J. Pharm. Pharmacol.* **34,** 788–793.

Klabund, R. E. (1983) Dipyridamole inhibition of adenosine metabolism in human blood. *Eur. J. Pharmacol.* **93,** 21–26.

Knauer, K. A., Lichtenstein, L. M., Adkinson, N. F., Jr., and Fish, J. E. (1981) Platelet activation during antigen-induced airway reactions in asthmatic subjects. *New Engl. J. Med.* **304,** 1404–1407.

Konche, H. and Kienecker, E.-W. (1977) Sympathetic innervation of the carotid bifurication in the rabbit and cat: Blood vessels, carotid body and carotid sinus. A fluorescence and electron microscopic study. *Cell Tissue Res.* **184,** 103–112.

Kondo, H., Iwanaga, T., and Nakajima, T. (1982) Immunocytochemical study on the localization of neuron-specific enolase and S-100 protein in the carotid body of rats. *Cell Tissue Res.* **227,** 291–295.

Lagercrantz, H., Yamamoto, Y., Fredholm, B. B., Prabhakar, N. R., and von Euler, C. (1984) Adenosine analogues depress ventilation in rabbit neonates. Theophylline stimulation of respiration via adenosine receptors? *Pediatr. Res.* **18,** 387–390.

Lahiri, S. (1977) Introductory remarks: Oxygen linked response of carotid chemoreceptors. *Adv. Exp. Med. Biol.* **78,** 185–202.

Lahiri, S., Smatresk, N. J., and Mulligan, E. (1983) Responses of peripheral chemoreceptors to natural stimuli, in *Physiology of the Peripheral Arterial Chemoreceptors* (Acker, H. and O'Regan, R. G., eds.), Elsevier BV, Amsterdam, pp. 221–256.

Lakshminarayan, S., Sahn, S. A., and Weil, J. V. (1978) Effect of aminophylline on ventilatory responses in normal man. *Am. Rev. Respir. Dis.* **117,** 33–38.

Lamb, D. and Lumsden, A. (1982) Intra-epithelial mast cells in human airway epithelium: Evidence for smoking-induced changes in their frequency. *Thorax* **37,** 334–342.

Lande, P. K. M., Waaler, B. A., and Walloe, L. (1968) The inhibitory effect of various phenols upon ATP-induced vasoconstriction in isolated perfused rabbit lungs. *Acta Physiol. Scand.* **72,** 331–337.

Langer, I. (1967) The bronchoconstrictor action of propranolol aerosol in asthmatic subjects. *J. Physiol. (Lond.)* **190,** 41 pp.

Larsson, K. and Sollevi, A. (1988) Influence of infused adenosine on bronchial tone and bronchial reactivity in asthma. *Chest* **93,** 280–284.

Lee, L.-Y. and Milhorn, H. T., Jr. (1975) Central ventilatory responses to O_2 and CO_2 at three levels of carotid chemoreceptor stimulation. *Respir. Physiol.* **25,** 319–333.

Lewis, R. A. and Austen, K. F. (1981) Mediation of local homeostasis and inflammation by leukotrienes and other mast cell-dependent compounds. *Nature* **293,** 103–108.

Lewis, R. A., Soter, N. A., Diamond, P. T., Austen, K. F., Oates, J. A., and Roberts J. L., II (1982) Prostaglandin D_2 generation after activation of rat and human mast cells with anti-IgE. *J. Immunol.* **129,** 1627–1631.

Londos, C. and Wolff, J. (1977) Two distinct adenosine-sensitive sites on adenylate cyclase. *Proc. Natl. Acad. Sci. USA* **74**, 5482–5486.

Londos, C., Cooper, D. M. F., and Wolff, J. (1980) Subclasses of external adenosine receptors. *Proc. Natl. Acad. Sci. USA* **77**, 2551–2554.

Lugliani, R., Whipp, B. J., Seard, C., and Wasserman, K. (1971) Effect of bilateral carotid-body resection on ventilatory control at rest and during exercise in man. *New Engl. J. Med* **285**, 1105–1111.

McCormack, D. G., Clarke, B., and Barnes, P. J. (1989) Characterization of the adenosine receptors in human pulmonary arteries. *Am. J. Physiol.* H41–H46.

McDonald, D. M. and Haskell, A. (1983) Morphology of connections between arterioles and capillaries in the rat carotid body analysed by reconstructing serial sections, in *The Peripheral Arterial Chemoreceptors* (Pallot, D. J., ed.), Croom Helm, London and Canberra, pp. 195–206.

McDonald, D. M. and Mitchell, R. A. (1981) The neural pathway involved in "efferent inhibition" of chemoreceptors in the cat carotid body. *J. Comp. Neurol.* **201**, 457–476.

McFadden, E. R. (1984) Pathogenesis of Asthma. *J. Allergy Clin. Immunol.* **73**, 413–424.

McNeil, R. S. and Ingram, C. G. (1966) Effect of propranolol on ventilatory function. *Am. J. Cardiol.* **18**, 473–475.

McQueen (1983) Pharmacological aspects of putative transmitters in the carotid body, in *Physiology of the Peripheral Arterial Chemoreceptors* (Acker, H. and O'Regan, R. G., eds.), Elsevier, Amsterdam, pp. 149–195.

McQueen, D. S. and Ribeiro, J. A. (1981a) Effect of adenosine on carotid chemoreceptor activity in the cat. *Br. J. Pharmacol.* **74**, 129–136.

McQueen, D. S. and Ribeiro, J. A. (1981b) Excitatory action of adenosine on cat carotid chemoreceptors. *J. Physiol.* **315**, 38 pp.

McQueen, D. S. and Ribeiro, J. A. (1983) On the specificity and type of receptor involved in carotid body chemoreceptor activation by adenosine in the cat. *Br. J. Pharmacol.* **80**, 347–354.

McQueen, D. S. and Ribeiro, J. A. (1986) Pharmacological characterization of the receptor involved in chemoexcitation induced by adenosine. *Br. J. Pharmacol.* **88**, 615–620.

Mann, J. S. and Holgate, S. T. (1985) Specific antagonism of adenosine-induced bronchoconstriction in asthma by oral theophylline. *Br. J. Clin. Pharmacol.* **19**, 685–692.

Mann, J. S., Cushley, M. J., and Holgate, S. T. (1985) Adenosine induced bronchoconstriction in asthma: Role of parasympathetic stimulation and adrenergic inhibition. *Am. Rev. Respir. Dis.* **132**, 1–6.

Mann, J. S., Holgate, S. T., Renwick, A. G., and Cushley, M. J. (1986a) Airway effects of purine nucleosides and nucleotides and release with bronchial provocation in asthma. *J. Appl. Physiol.* **61**, 1667–1676.

Mann, J. S., Renwick, A. G., and Holgate, S. T. (1986b) Release of adenosine and its metabolites from activated human leukocytes. *Clin. Sci.* **70**, 460–468.

Mannix, S. E., Bye, P., Hughes, J. M. B., Cover, D., and Davies, E. E. (1984) Effect of posture on ventilatory response to steady-state hypoxia and hypercapnia. *Respir. Physiol.* **58**, 87–99.

Marquardt, D. L., Gruber, H. E., and Wasserman, S. I. (1984a) Adenosine release from stimulated mast cells. *Proc. Natl. Acad. Sci. USA* **81**, 6192–6196.

Marquardt, D. L., Parker, C. W., and Sullivan, T. J. (1978) Potentiation of mast cell mediator release by adenosine. *J. Immunol.* **120**, 871–878.

Marquardt, D. L., Walker, L. L., and Wasserman, S. I. (1984b) Adenosine receptors on mouse bone marrow-derived mast cells: Functional significance and regulation by animophylline. *J. Immunol.* **133**, 932–937.

Maxwell, D. L. (1986) Effect of adenosine infusion on resting ventilation in man. *J. Physiol.* **374**, 23 pp.

Maxwell, D. L., Fuller, R. W., Nolop, K. B., Dixon, C. M. S., and Hughes, J. M. B. (1986) Effects of adenosine on ventilatory responses to hypoxia and hypercapnia in humans. *J. Appl. Physiol.* **61**, 1762–1766.

Maxwell, D. L., Fuller, R. W., Conradson, T.-B., Dixon, C. M. S., Aber, V., Hughes, J. M. B., and Barnes, P. J. (1987a) Contrasting effects of two xanthines, theophylline and enprofylline, on the cardio-respiratory stimulation of infused adenosine in man. *Acta. Physiol. Scand.* **131**, 459–465.

Maxwell, D. L., Fuller, R. W., Conradson, T.-B., Dixon, C. M. S., Hughes, J. M. B., and Barnes, P. J. (1987b) Oxygen and theophylline reduce the cardio-respiratory effects of adenosine infusion in man. *Clin. Sci.* **72**, 13 pp.

Mentzer, R. M., Jr., Rubio, R., and Berne, R. M. (1975) Release of adenosine by hypoxic canine lung tissue and its possible role in pulmonary circulation. *Am. J. Physiol.* **229**, 1625–1631.

Milic-Emili, J. and Grunstein, M. M. (1976) Drive and timing components of ventilation. *Chest* **70**, Suppl. 131–133.

Millhorn, D. E., Eldridge, F. L., Kiley, J. P., and Waldrop, T. G. (1984) Prolonged inhibition of respiration following acute hypoxia in glomectomized cats. *Respir. Physiol.* **57**, 331–340.

Mir, A. K., Pallot, D. J., and Nahorski, S. R. (1983) Biogenic amine-stimulated cyclic adenosine 3',5'-monophosphate formation in the rat carotid body. *J. Neurochem.* **41**, 663–669.

Monteiro, E. C. and Ribeiro, J. A. (1987) Ventilatory effects of adenosine mediated by carotid body chemoreceptors in the rat. *Naunyn-Schmiedebergs Arch. Pharmacol.* **335**, 143–148.

Moss, I. R., Denavit-Saubie, M., Eldridge, F. L., Gillis, R. A., Herkenham, M., and Lahiri, S. (1986) Neuromodulators and transmitters in respiratory control. *Fed. Proc.* **45**, 2133–2147.

Mulligan, E. and Lahiri, S. (1982) Separation of carotid body chemoreceptor responses to O_2 and CO_2 by oligomycin and by antimycin A. *Am. J. Physiol.* **242**, C200–C206.

Nadel, J. A. (1973) Neurophysiologic aspects of asthma, in *Asthma, Physiology, Immunopharmacology, and Treatment* (Austen, K. F. and Lichtenstein, L. M., eds.), Academic, New York, pp. 29–38.

Obeso, A., Almaraz, L., and Gonzalez, C. (1987) ATP content in the cat carotid body under different experimental conditions. Support for the metabolic hypothesis, in *Chemoreceptors in Respiratory Control* (Ribeiro, J. A. and Pallot, D. J., eds.), Croom Helm, London and Sydney, pp. 78–90.

Orehek, J., Gayrard, P., Grimaud, C., and Charpin, J. (1977) Bronchial response to inhaled prostaglandin $F_{2\alpha}$ in patients with common or aspirin-sensitive asthma. *J. Allergy Clin. Immunol.* **59**, 414–419.

Peake, M. D., Harabin, A. L., Brennan, N. J., and Sylvester, J. T. (1981) Steady-state vascular responses to graded hypoxia in isolated lungs of five species. *J. Appl. Physiol.* **51**, 1214–1219.

Phillis, J. W., Edstrom, J. P., Kostopoulos, G. K., and Kirkpatrick, J. R. (1979) Effects of adenosine and adenosine nucleotides on synaptic transmission in the cerebral cortex. *Can. J. Physiol. Pharmacol.* **57**, 1289–1312.

Proctor, K. G. and Duling, B. R. (1982) Adenosine and free-flow functional hyperaemia in striated muscle. *Am. J. Physiol.* **242**, H688–H697.

Rebuck, A. S. and Campbell, E. J. M. (1974) A clinical method for assessing the ventilatory response to hypoxia. *Am. Rev. Respir. Dis.* **109**, 345–350.

Reeves, J. T., Jokl, P., Merida, J., and Leathers, J. E. (1967) Pulmonary vascular obstruction following administration of high-energy nucleotides. *J. Appl. Physiol.* **22**, 475–479.

Reid, P. G., Watt, A. H., Routledge, P. A., and Smith, A. P. (1987) Intravenous infusion of adenosine but not inosine stimulates respiration in man. *Br. J. Clin. Pharmacol.* **23**, 331–338.

Ribeiro, J. A. and McQueen, D. S. (1984) Effects of purines on carotid chemoreceptors, in *Peripheral Arterial Chemoreceptors* (Pallot, D. J., ed.), Oxford University Press, Oxford, pp. 383–390.

Richardson, J. B. (1981) Nonadrenergic inhibitory innervation of the lung. *Lung* **159**, 315–322.

Richardson, P. S., and Sterling, G. M. (1969) Effects of β-adrenergic receptor blockade on airway conductance and lung volume in normal and asthmatic subjects. *Br. Med. J.* **3**, 143–145.

Richmond, G. H. (1949) Action of caffeine and aminophylline as respiratory stimulants in man. *J. Appl. Physiol.* **2**, 16–23.

Rigatto, H., Brady, J. P., and de la Torre Verduzco, R. (1975) Chemoreceptor reflexes in preterm infants: 1. The effect of gestational and postnatal age on the ventilatory response to inhalation of 100% and 15% oxygen. *Pediatrics* **55**, 604–613.

Robinson, C. and Holgate, S. T. (1985) Mast cell-dependent inflammatory mediators and their putative role in bronchial asthma. *Clin. Sci.* **68**, 103–112.

Routledge, P. A. and Watt, A. H. (1986) Effect of aminophylline on respiratory stimulation and heart rate changes produced by intravenous adenosine boluses in man. *Br. J. Pharmacol.* **89**, 711p.

Rubio, R., Berne, R. M., and Dobson, J. G., Jr. (1973) Sites of adenosine production in cardiac and skeletal muscle. *Am. J. Physiol.* **225**, 938–953.

Rubio, R., Berne, R. M., Bockman, E. L., and Curnish, R. R. (1975) Relationship between adenosine concentration and oxygen supply in rat brain. *Am. J. Physiol.* **228**, 1896–1902.

Schleimer, R. P., MacGlashan, D. W., Jr., Schulman, E. S., Peters, S. P., Adkinson, N. F., Jr., Newball, H. H., Adams, G. K., III, and Lichtenstein, L. M. (1982) Effects of glucocorticoids on mediator release from human basophils and mast cells. *Fed. Proc.* **41**, 487.

Sheppard, D., Rizk, N. W., Boushey, H. A., and Bethel, R. A. (1983) Mechanism of cough and bronchoconstriction induced by distilled water aerosol. *Am. Rev. Respir. Dis.* **127**, 691–694.

Sheppard, D., Epstein, J., Holtzman, M. J., Nadel, J. A., and Boushey, H. A. (1982) Dose-dependent inhibition of cold air-induced bronchoconstriction by atropine. *J. Appl. Physiol.* **53**, 169–174.

Sheppard, D., Wong, W. S., Uehara, C. F., Nadel, J. A., and Boushey, H. A. (1980) Lower threshold and greater bronchomotor responsiveness of asthmatic subjects to sulfur dioxide. *Am. Rev. Respir. Dis.* **122**, 873–878.

Smith, P., Jago, R., and Heath, D. (1982) Anatomical variation and quantative histology of the normal and enlarged carotid body. *J. Pathol.* **137**, 287–304.

Smits, P., Schonten, J., and Thien, T. H. (1987) Respiratory stimulant effects of adenosine in man after caffeine and enprofylline. *Br. J. Clin. Pharmacol.* **24**, 816–819.

Snyder, S. H., Katims, J. J., Annau, Z., Bruns, R. F., and Daly, J. W. (1981) Adenosine receptors and behavioral actions of methylxanthines. *Proc. Natl. Acad. Sci. USA* **78**, 3260–3264.

Stone, T. W. (1984) Purine receptors classification: A point for discussion. *Trends. Pharmacol. Sci.* **5**, 492,493.

Tenney, S. M. and Brooks, J. G., III. (1966) Carotid bodies, stimulus interaction, and ventilatory control in unanesthetized goats. *Respir. Physiol.* **I**, 211–224.

Watt, A. H. and Routledge, P. A. (1985) Adenosine stimulates respiration in man. *Br. J. Clin. Pharmacol.* **20**, 503–506.

Watt, A. H. and Routledge, P. A. (1987) Dipyridamol modulation of heart rate and ventilatory changes produced by intravenous adenosine boluses in man. *Br. J. Clin. Pharmacol.* **23**, 632p,633p.

Watt, A. H., Reid, P. G., Stephens, M. R., and Routledge, P. A. (1987) Adenosine-induced respiratory stimulation in man depends on site of infusion. Evidence for an action on the carotid body? *Br. J. Clin. Pharmacol.* **23**, 486–490.

Watt, J. G., Dumke, P. R., and Comroe, J. H., Jr. (1943) Effects of inhalation of 100 percent and 14 percent oxygen upon respiration of unanaesthetized dogs before and after chemoreceptor denervation. *Am. J. Physiol.* **138**, 610–617.

Wechsler, R. L., Kleiss, L. M., and Kety, S. S. (1950) The effects of intravenously administered aminophylline on cerebral circulation and metabolism in man. *J. Clin. Invest.* **29**, 28–30.

Wessberg, P., Hedner, J., Hedner, T., Person, B., and Jonasen, J. (1985) Adenosine mechanisms in the regulation of breathing in the rat. *Eur. J. Pharmacol.* **106**, 59–67.

Wetzel, R. C. and Sylvester, J. T. (1983) Gender differences in hypoxic vascular response of isolated sheep lungs. *J. Appl. Physiol.* **55**, 100–104.

Wiklund, N. P., Cederqvist, B., Matsuda, H., and Gustafsson, L. E. (1987) Adenosine can stimulate pulmonary artery. *Acta Physiol. Scand.* **131**, 477,478.

Winn, H. R., Rubio, R., and Berne, R. M. (1981) Brain adenosine concentration during hypoxia in rats. *Am. J. Physiol.* **241,** H235–H242.

Woodrum, D. E., Standaert, T. A., Maryock, D. E., and Guthrie, R. D. (1981) Hypoxic ventilatory response in the newborn monkey. *Pediatr. Res.* **15,** 367–370.

Wyman, R. J. (1977) Neural generation of the breathing rhythm. *Annu. Rev. Physiol.* **39,** 417–448.

Zapata, P., Stenaas, L. J., and Eyzaguirre, C. (1976) Axon regeneration following a lesion of the carotid nerve: Electrophysiological and ultrastructural observations. *Brain Res.* **113,** 235–253.

Zetterstrom, T., Vernet, L., Ungerstedt, U., Tossman, U., Jonzon, B., and Fredholm, B. B. (1982) Purine levels in the intact rat brain. Studies with an implanted perfused hollow fibre. *Neurosci. Lett.* **29,** 111–115.

CHAPTER 11

Adenosine in Central Nervous System Function

Michael F. Jarvis and Michael Williams

1. Introduction

Brain adenosine receptors, like those in other tissues (Williams, 1989), can be delineated into two major subclasses, termed A_1 and A_2 (Hamprecht and Van Calker, 1985), and differentiated by pharmacological and functional activity as well as differences in regional distribution (Snyder, 1985; Williams, 1987; Jarvis, 1988). Adenosine receptors in brain tissue, like other receptors, occur in the greatest density in this organ as compared to their distribution in peripheral tissues. The precise physiological contribution of these receptors to central nervous system function remains unclear; however, the many documented inhibitory effects of adenosine on neurotransmitter release in mammalian tissue coupled with the psychomotor-stimulant effects of adenosine receptor antagonists (alkylxanthines) have led to the suggestion that adenosine mediates an "inhibitory tone" in the CNS (Harms et al., 1978).

Adenosine and Adenosine Receptors Editor: Michael Williams ©1990 The Humana Press Inc.

Adenosine can alter neurotransmitter release (Stone, 1981; Fredholm and Hedqvist, 1980) and transmembrane ion fluxes (Dunwiddie, 1985; Phillis and Wu, 1981), and can modulate several kinds of second-messenger systems (Londos and Wolff, 1977; Daly, 1982; Fredholm and Dunwiddie, 1988). Receptor activation produces effects on sleep, seizure activity, anxiety, analgesia, and psychomotor activation (*see* reviews by Snyder, 1985; Dunwiddie, 1985; Williams, 1987). Such findings have contributed greatly to the idea that adenosine functions as a specific modulator of central neurotransmission despite its abundant and ubiquitous availability in brain tissue (Williams, 1990). Specificity in these physiological and behavioral effects has been indicated by the identification of discrete adenosine receptor subtypes and their heterogeneous distribution in mammalian brain (Goodman and Snyder, 1982; Jarvis, 1988). Furthermore, immunohistochemical studies on the distribution of adenosine-metabolizing enzymes such as adenosine deaminase (ADA) have suggested evidence for the existence of discrete "adenosinergic" pathways in the CNS (Geiger and Nagy, 1984; Braas et al., 1986; Geiger and Nagy, 1990). Correlation of the functional effects of purine agonists and antagonists, with their pharmacological activity at central adenosine receptors (Snyder et al., 1981; Katims et al., 1983; Choi et al., 1988) has provided further support for a discrete contribution of adenosine in CNS function.

The remainder of this chapter will be devoted to a review of the evidence for the purinergic modulation of central neurotransmission and its relationship to brain function, especially in regard to mental disease. Particular emphasis will be given to the autoradiographic localization of adenosine receptors in mammalian brain and how these receptor distributions may contribute to the behavioral pharmacology of the purines and their antagonists.

2. Brain Adenosine Receptors

Receptor autoradiography has offered an efficient and reliable means for the evaluation of receptor distribution in various tissues (Unnerstall et al., 1982; Kuhar et al., 1986). The application of this technique to the study of brain cell surface receptors has been particularly interesting. Early autoradiographic studies of brain neurotransmitter receptors offered essentially qualitative descriptions of receptor densities; however, increased sophistication in the generation and analysis of autoradiograms has allowed for quantitative comparisons of binding receptor densities and pharmacology (Kuhar et al., 1986; Jarvis, 1988).

2.1. Adenosine A_1 Receptors

Early autoradiographic studies of brain adenosine receptors involved the localization of A_1 receptors using [^{3}H]cyclohexyladenosine ([^{3}H]CHA) (Lewis et al., 1981; Goodman and Snyder, 1982). In these studies, the highest densities of [^{3}H]CHA recognition sites were found in the molecular layer of the cerebellum and in the CA-1 and CA-3 regions of the hippocampus. Moderate binding levels were observed in the thalamus, caudate-putamen, septum, and cerebral cortex; however, little or no specific ligand binding was found in the hypothalamus and brain stem (Fig. 1). Analysis of the kinetic parameters of [^{3}H]CHA binding to rat whole brain sections revealed high affinity ($K_d = .77$ nM) and limited capacity ($B_{max} = 423$ fmol/mg protein) binding with an approximately 100-fold separation in the activity of the stereoisomers of N^6-phenylisopropyladenosine (PIA) (Goodman and Snyder, 1982).

More recent autoradiographic studies have confirmed the regional distribution and kinetic parameters of [^{3}H]CHA binding to brain adenosine A_1 receptors (Lee and Reddington, 1986; Snowhill and Williams, 1986; Fastbom et al., 1987a). In these studies, the affinity (K_d) of [^{3}H]CHA for A_1 receptors did not appear to vary significantly across brain regions (Snowhill and Williams, 1986). Similarly, characterization of A_1 receptors in several mammalian species, using both autoradiographic and homogenate binding techniques, has not shown major species differences in brain A_1 receptor affinity or density (Ferkany et al., 1986; Fastbom et al., 1987a,b). However, species differences in A_1 receptor pharmacology have been noted (Ferkany et al., 1986). Adenosine antagonists were found to be 10–400-fold more potent in inhibiting [^{3}H]CHA binding to bovine cortex than to human and guinea pig cortical tissue (Ferkany et al., 1986). Interestingly, selective differences in brain A_1 receptor density have also been reported

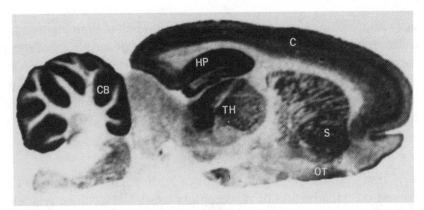

Fig. 1. Representative autoradiographic image of specific [³H]CHA (1 n*M*) binding to rat brain sagittal sections (Jarvis et al., 1989). Specific binding was revealed by digital subtraction autoradiography, in which the image of specific binding was obtained through subtraction of the linearized nonspecific binding image from the total binding image. Abbreviations are: C, cortex; S, striatum; OT, olfactory tubercle; TH, thalamus; HP, hippocampus; CB, cerebellum.

in inbred strains of mice that show differential behavioral responses to alkylxanthines (Jarvis and Williams, 1988). Although the exact physiological significance of these differences in brain adenosine A_1 receptors remains unclear, there is some evidence to indicate that these differences are mediated through complex genetic determinants (Seale et al., 1985) and are reflected in overt behavioral performance (Seale et al., 1986).

[³H]CHA has been the radioligand of choice for the autoradiographic study of brain A_1 receptors; however, efforts to further characterize A_1 receptors have led to the synthesis and evaluation of new radioligands with higher specific activities and greater signal-to-noise ratios. [¹²⁵I]H-phenylisopropyl adenosine ([¹²⁵I]HPIA has recently been reported to specifically label A_1 receptors in rat brain (Weber et al., 1988). This ligand offers several methodological advantages in autoradiographic studies, including short exposure times and a reduction in the differential quenching between gray and white matter found with tritiated radioligands (Kuhar et al., 1986). The high specific activity of this iodinated radioligand (2000 Ci/mmol) also has permitted the visualization of peripheral A_1 receptors (Weber et al., 1988; Leid et al., 1988).

Several antagonist radioligands recently have been developed that selectively label A_1 receptors with high affinity. These compounds include the functionalized congeners of 1,3-dipropylxanthine, [³H]XAC and [³H]XCC (Jacobson et al., 1985, 1986), [³H]CPX (8-cyclopentyl-1,3-dipropylxanthine [Lohse et al., 1987a]—also known as DPCPX or PD 116,948 (Bruns et al., 1987), [³H] PA-PAXAC (Ramkumar et al., 1988), and the A_1-selective radioiodinated antagonist ligand [¹²⁵I]BW-A844U (Patel et al., 1988). All of these ligands appear to offer distinct advantages over the only previously available antagonist radioligand, [³H]1,3-diethyl-8-phenylxanthine (DPX) (Bruns et al., 1980). These ligands have been shown to label A_1 receptors under different experimental conditions; however, the ability of these compounds to interact at A_2 receptors has not been fully characterized. For instance, [³H]XAC has been shown to label rat and bovine brain A_1 receptors (Jacobson et al., 1986); however, it also can be used to label A_2 receptors in human platelets (Ukena et al., 1986). In contrast, autoradiographic studies using [³H]XCC have demonstrated that this ligand specifically binds to rat brain sections with high affinity (Jarvis et al., 1987) and with a pharmacological profile consistent with the specific labeling of A_1 receptors (Jarvis, 1988).

[³H]CHA binding in both brain sections and membrane homogenate preparations has typically been best described by a one-site binding component model (i.e., linear Scatchard plots and Hill coefficients that do not differ from unity) (Ferkany et al., 1986; Lee and Reddington, 1986). However, there is currently some evidence of a more complex agonist ligand interaction at brain A_1 receptors. Curvilinear Scatchard plots recently have been reported for [³H]CHA binding in rat hypothalamic and bovine cortical membranes (Stiles, 1986; Anderson et al., 1987,1988). Additionally, agonist inhibition curves obtained with several novel radioligands, including [¹²⁵I]BW-A-844U, [¹²⁵I]HPIA, and the slightly A_2-selective antagonist, [³H]CGS 15943A, have revealed interactions at high and low affinity binding components with pharmacological profile, consistent with the labeling of A_1 receptors (Patel et al., 1988; Stiles, 1988; Jarvis et al., 1988b; Leid et al., 1988). Further analysis of the binding of these and other selective A_1 antagonist ligands may prove useful in characterizing the low affinity state of the A_1 receptor.

2.2. Adenosine A_2 Receptors

Characterization of adenosine A_2 receptors has not been widespread because of the lack of available radioligands that selectively label this receptor subtype. Currently, the radioligand of choice for the study of A_2 receptors is [³H]5'-N-ethylcarboxamidoadenosine ([³H]NECA), a radioligand that has equivalent nanomolar affinity for both the A_1 and A_2 receptor subtypes (Bruns et al., 1986; Stone et al., 1988). A_2 receptors can be selectively labeled with this ligand only when some method is employed to block [³H]NECA binding to A_1 receptors. Brain A_2 receptors also have been labeled with [³H]NECA following the inactivation of A_1 receptors with N-ethylmalemide (NEM), which uncouples the G_i protein receptor subunit (Yeung and Green, 1984). There is some evidence, however, that this method may actually overestimate [³H]NECA binding to A_2 receptors through a NEM-induced stimulatory action on the G_s binding protein of the A_2 receptor (Bruns et al., 1986). Furthermore, A_2 receptors have been studied in a cell line that contains only the A_2 receptor subtype. Membrane preparations from the human pheochromocytoma (PC-12) cell line have been shown to specifically bind [³H]NECA, whereas no specific binding was observed with [³H]CHA (Williams et al., 1987a). In addition, the pharmacological activity of adenosine agonists to inhibit [³H]NECA in PC-12 membranes is consistent with the specific labeling of A_2 receptors (Williams et al., 1987a).

Mammalian brain adenosine A_2 receptors can also be specifically labeled with [³H]NECA following the saturation of A_1 receptors with a high-affinity A_1-selective compound. [³H]NECA, in the presence of 50 nM cyclopentyladenosine (CPA) using both membrane and brain section preparations, has been shown to specifically label the A_2 receptor subtype in rat striatum and olfactory tubercle (Bruns et al., 1986; Stone et al., 1988; Jarvis et al., 1989) (*see* Fig. 2). This concentration of CPA has been demonstrated to selectively block [³H]NECA binding to the striatal A_1 receptor subtype in a number of mammalian species (Stone et al., 1988). Using this method, two components of [³H]NECA binding have been reported in striatal membranes, a high affinity state ($K_d = 4$ nM) and a lower affinity component ($K_d = 120$ nM ; Bruns et al ., 1986; Stone et al., 1988). Some controversy exists as to whether these binding components represent

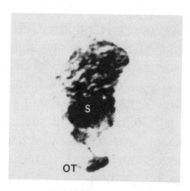

Fig. 2. Representative autoradioagraphic image of specific [³H]NECA (1 n*M*) binding to adenosine A$_2$ receptors in rat brain sagittal sections (Jarvis et al., 1989). [³H]NECA binding to A$_1$ receptors was blocked by the inclusion of 50 n*M* CPA in the binding assay. Specific [³H]NECA binding was revealed, as described in Fig. 1. Abbreviations are: S, striatum; OT, olfactory tubercle.

different affinity states of the A$_2$ receptor or different receptor proteins (Bruns et al., 1986; Stone et al., 1988). Whereas 5'-guanylyimido-phosphate (GppNHp), a nonhydrolizable analog of GTP, has been demonstrated to decrease [³H]NECA binding in rat striatal membranes and to reduce the ability of adenosine agonists to inhibit binding (Stone et al., 1988), a recent study has indicated that [³H]NECA can bind an additional protein that is not the A$_2$ receptor (Lohse et al., 1988). A novel radioligand, the 2-substituted NECA analog CGS 21680 (Hutchison et al., 1989) can directly label brain A$_2$ receptors (Jarvis et al., 1989). This ligand has recently been used to confirm the regional distribution of A$_2$ receptors in rat brain using autoradiographic techniques (Jarvis and Williams, 1989).

Recent autoradiographic studies using [³H]NECA (+ 50 n*M* CPA) have confirmed the selective localization of high-affinity A$_2$ receptors in striatum and olfactory tubercle (Jarvis et al., 1989; Jarvis, 1988). Competition studies using autoradiographic techniques have demonstrated that [³H]NECA binding in these brain structures represents the specific labeling of the A$_2$ receptor subtype (Jarvis et al., 1989). However, other autoradiographic studies have reported the localization of"non-A$_1$" receptors using [³H]NECA in the presence of 1 μ*M* R-PIA (Lee and Reddington, 1986) and, in a separate study, 50

nM 1,3-dipropyl-8-cyclopentylxanthine (CPX) (Lee and Reddington, 1987). The inclusion of these compounds to block [³H]NECA
binding to A_1 receptors produced regional distributions of binding
that were distinctly different from that obtained with [³H]NECA alone
or in combination with 50 nM CPA, as well as that obtained for A_1
receptors labeled with [³H]CHA. In both studies, "non-A_1" receptors
were most highly concentrated in the striatum, thalamus, and cerebral
cortex (Lee and Reddington, 1986; 1987). Since pharmacological
and/or functional characterization of these "non A_1" [³H]NECA recognition sites remains unknown, it is difficult to determine what portion of [³H]NECA binding actually represents binding to brain A_2
receptors and what portion represents binding to the "non-adenosine
receptor " similar to that described by Lohse et al. (1988).

As was seen with A_1 receptors, there do not appear to be any
major species differences in the affinity of [³H]NECA for the A_2 receptor, and only slight differences in A_2 receptor density have been
reported for several mammalian species (Stone et al., 1988). Interestingly, the xanthine-sensitive CBA/J mouse has been shown to have
a significantly greater density of striatal A_2 receptors than the xan-
thine-insensitive SWR/J mouse (Jarvis and Williams, 1988). These
observations suggest that a genetically-mediated variation in xanthine responsiveness is (a) manifest, at least in part, through brain
adenosine receptor densities and (b) greater within a given species
than between species. The interspecies sensitivity to the alkylxanthines is a complex issue and, therefore, may not be entirely accounted
for in terms of agonist radioligand/receptor kinetics. For example,
the nonselective xanthine 1,3-dipropyl-8-(2-amino-4-chloro)phenyl-
xanthine (PACPX) has been found to be a potent, noncompetitive antagonist at both A_1 and A_2 receptors in the rat brain, with essentially
equal affinity for the A_1 receptor in mouse, rat, guinea pig, rabbit, calf,
and human cortical membranes (Williams et al., 1987b; Ferkany et al.,
1986). However, PACPX has significantly greater affinity for A_2 receptors in the rabbit, human and calf as than for those in the rat,
mouse, and guinea pig (Stone et al., 1988). Clearly, a significant
advance in the ability to delineate phylogenetic differences in purinergic sensitivity will have to await the use of more selective A_2 receptor agonists, such as CGS 21680, and the discovery of A_2-selective
antagonists.

2.3. Synaptic Localization
of Adenosine Receptors in the CNS

Based on the above autoradiographic and membrane binding data, the mammalian striatum appears to contain equal concentrations of A_1 and A_2 receptors. There are brain regions, however, that appear to contain only one adenosine receptor subtype. The CA-1 and CA-3 regions of hippocampus contain very high concentrations of A_1 receptors and, essentially, no high-affinity A_2 receptors (Goodman and Snyder, 1982; Jarvis and Williams, 1989). The olfactory tubercle contains a high density of A_2 receptors and minimal quantities of A_1 receptors (Jarvis, 1988; Jarvis and Williams, 1988). Given the apparent lack of regionally specific mechanisms for the regulation of adenosine availability, these discrete distributions of adenosine receptors provide support for a specific contribution of adenosine to central neurotransmission. The question as to how this neuromodulatory activity may be mediated at the synaptic level has received recent attention.

Several studies have used neurotoxins to selectively eliminate synaptic afferent and/or efferent processes in order to examine the synaptic localization of adenosine receptors (Lloyd and Stone, 1985; Geiger, 1986; Jarvis and Williams, 1986). The striatum has been studied most extensively because of the presence of large concentrations of both adenosine receptor subtypes. Striatal adenosine receptors and adenosine uptake sites are not significantly altered following the intrastriatal administration of 6-hydroxydopamine (6-OHDA), which selectively destroys presynaptic dopaminergic terminals, decreases dopamine concentrations, and increases dopamine receptor density (Wojcik and Neff, 1983; Lloyd and Stone, 1985; Geiger, 1986; Jarvis and Williams, 1986) In contrast, the selective destruction of striatal postsynaptic terminals with an excitotoxin, such as kainic acid or quinolinic acid, produces a marked reduction in the number of striatal adenosine receptors, adenosine uptake sites, and 2-deoxyglucose uptake (Geiger, 1986; Lloyd and Stone; 1985; Jarvis and Williams, 1986). These results indicate that striatal adenosine receptors are localized on intrinsic striatal interneurons and/or on corticostriatal terminals. These results are consistent with other observations indicating that adenosine receptors are not localized on striatal dopa-minergic terminals (Murray and Cheney, 1982; Wojcik and Neff, 1983).

It now appears that adenosine receptors are localized on, or adjacent to, excitatory neurons in the mammalian brain (Goodman et al., 1983). Data from autoradiographic studies of rodent brain indicate that adenosine A_1 receptors are highly concentrated in forebrain regions (e.g., hippocampus) that also contain high densities of excitatory amino acid (EAA) receptors (Jarvis et al., 1988a; Deckert and Jorgensen, 1988; Goodman et al., 1983). Moderate concentrations of A_1 receptors are found in the granule layer of the cerebellum, an area in which N-methyl-D-aspartate (NMDA) receptors are also concentrated (Snowhill and Williams, 1986; Jarvis et al., 1987). Direct administration of kainic acid or transient forebrain ischemia have been shown to markedly reduce adenosine A_1 and NMDA receptors in the hippocampal CA-1 region (Deckert and Jorgensen, 1988, Jarvis et al., 1988a; Onodera and Kogure, 1988: Onodera et al., 1987). These manipulations can result in essentially complete destruction of the cell bodies in the pyramidal cell layer of the CA-1 region. However, a significant proportion of A_1 receptors remain unaffected in this brain region, indicating that A_1 receptors are also localized postsynaptically to these excitatory neurons (Jarvis et al., 1988a; Deckert and Jorgensen, 1988; Onodera and Kogure, 1988). These data, coupled with the observations that large amounts of adenosine and EAA are released during an ischemic episode (von Lubitz et al., 1988; Hagberg et al., 1988; Goldberg et al., 1988; Onodera et al., 1986) and that adenosine is a potent inhibitor of the release of the EAA agonist, L-glutamate (Dolphin and Prestwich, 1985), indicate that adenosine may be an important endogenous anticonvulsant that also can modulate the potential neurotoxicity of glutamate.

3. Behavioral Pharmacology of Adenosine

The identification of discrete adenosine receptor populations in brain has mitigated much of the early skepticism regarding a specific purinergic contribution to neurotransmission processes (Williams, 1987). However, the physiological significance of these receptor subtypes can be fully understood only when the functional correlates of receptor activation or blockade have been described. In this regard, the experimental analysis of adenosine's effects on the behaving organism is of fundamental importance to an understanding of purine involvement in neuronal communication. Since adenosine is found in the brain in very high concentrations (Daly, 1982; Williams, 1987), metabolically stable adenosine analogs as well as adenosine antagonists have become useful tools in the characterization of the behavioral effects of adenosine. Several different experimental paradigms have been used to characterize the behavioral pharmacology of adenosine. Of these, the ability of adenosine agonists and antagonists to alter schedule-controlled behavior, function as interoceptive cues, and alter locomotor activity have received the most attention. In addition, the reinforcing effects, physical dependence liability, and the effects of chronic exposure to adenosine antagonists have also received recent experimental evaluation.

3.1. Schedule-Controlled Behavior Studies

The ability of purinergic compounds to alter scheduled-controlled responding has been evaluated in a variety of mammalian species, including New and Old World monkeys, rabbits, rats, and mice. In general, adenosine agonists have been found to decrease operant response rates (Glowa and Spealman, 1984; Coffin and Spealman, 1987). The alkylxanthines, in contrast, produce both response-rate- and dose-dependent effects, increasing low rates of responding and decreasing high rates of responding (Glowa and Spealman, 1984; Carney et al., 1985c; Coffin and Spealman, 1987). There appears to be a competitive interaction between the behavioral effects of the purines and methylxanthines, since the behavioral depressant effects of adenosine agonists can be attenuated by the subsequent administration of methylxanthines (Glowa and Spealman, 1984; Spealman, 1988). That these behavioral effects are mediated by adenosine receptors is indicated by observations that the potency of methylxanthines to increase response rates is positively correlated with their ability to attenuate the rate suppressant-effects of the adenosine agonist, NECA (Spealman, 1988). Fixed-ratio operant responding in monkeys has been disrupted by adenosine agonists with the following order of potency: NECA > 2-CADO = R-PIA > CHA > S-PIA, thus indicating an involvement of adenosine A_2 receptors (Coffin and Spealman, 1987). The same relative potency order was obtained for these compounds in their ability to reduce blood pressure in these animals. Interestingly, the relative potency of these compounds in reducing heart rate was indicative of activity at A_1 receptors, with CPA > 2-CADO > R-PIA > CHA > S-PIA (Coffin and Spealman, 1987), which is consistent with data obtained in vitro for these responses (Evans et al., 1982; Hamilton et al.,1987; Oei et al., 1988). Further evidence that the operant behavior-disrupting effects of purinergic compounds may be mediated via the A_2 receptor can be found in the observation that 8-cyclopentlytheophylline (CPT), which is a highly selective A_1 receptor antagonist in vitro (Bruns et al., 1988), was only slightly more potent than caffeine (a nonselective antagonist) in increasing operant responding.

In contrast to these findings, however, there are several reports that the behavioral actions of the methylxanthines cannot be fully explained through an interaction at adenosine receptors (Glowa et al., 1985; Goldberg et al., 1985). Caffeine has been shown to effectively restore operant responding that has been reduced by R-PIA or NECA in monkeys (Spealman, 1988), rats (Logan and Carney, 1984), and mice (Glowa et al., 1985). However, the topography of operant responding obtained when both of these compounds are present in the animal has been reported to be different than that obtained with either compound alone (Goldberg et al., 1985). This discrepancy may be accounted for by the fact that methylxanthines can exert behavioral effects that are more indicative of phosphodiesterase inhibition than of adenosine receptor antagonism. The rate-increasing effects of caffeine may reflect its ability to block central adenosine receptors, whereas the rate-decreasing effects, which require higher doses, may be mediated through an action at peripheral adenosine receptors and/ or through phosphodiesterase inhibition (Glowa and Spealman, 1984). In this regard, it should be noted that not all alkylxanthines produced rate-dependent effects on operant responding. Enprofylline (Spealman, 1988) and isobutylmethylxanthine (IBMX; Kleven and Sparber, 1987) have been shown to produce only decreases in operant responding, which may be attributed to their potency as phosphodiesterase inhibitors.

3.2. Drug Discrimination Studies

It is now well established that relatively low doses of caffeine (10–100 mg/kg) can readily engender in laboratory animals a stimulus cue that is distinguishable from vehicle injections (Carney and Christensen, 1980; Modrow et al., 1981; Holtzman, 1986, 1987). Similar discriminative stimuli have also been obtained with other methylxanthines; however, these compounds have typically produced stimuli that are either less potent and/or less generalizable to a caffeine cue (Carney et al., 1985a,c). Evaluation of the specificity of the caffeine-induced stimulus cue has been the major focus of these studies. In general, there appears to be a dose-dependent function in the ability of other methylxanthines and psychomotor stimulants to generalize to a caffeine cue (Carney et al., 1985a,c; Holtzman, 1986, 1987). Theophylline, paraxanthine, and 3-methylxanthine, but not theobromine, produced dose-dependent generalization in rats trained to discriminate caffeine (32 mg/kg) from saline (Carney et al., 1985a). In rats trained with a high dose of theophylline (56 mg/kg), stimulus generalization engendered by these compounds was much less dramatic (Carney et al., 1985a). These data indicate that a strict structure–activity relationship best describes the discriminative stimuli engendered by the methylxanthines, which may reflect their activity as adenosine receptor antagonists.

Adenosine agonists also have been shown to produce discriminative stimuli (Holloway et al., 1985b) and to antagonize methylxanthine discriminations (Holtzman, 1986). The stimulus properties of adenosine agonists appear to be centrally mediated, since it has been reported that the peripherally acting adenosine antagonist, 8-sulfophenyl-theophylline, was ineffective in blocking an R-PIA discrimination (Holloway et al., 1985b). Caffeine has been shown to effectively block a stimulus cue produced by R-PIA (0.03–0.08 mg/kg), but R-PIA was relatively ineffective in antagonizing the stimulus cues engendered by high doses of caffeine and theophylline (Holloway et al., 1985a). This result indicates that at least some component of the caffeine discriminative cue may not be caused by an antagonist interaction at A_1 receptors (Holloway et al., 1985a). Interestingly, the cyclic nucleotide phosphodiesterase inhibitors, IBMX and papaverine, have been shown to produce stimulus generalization in caffeine-trained animals (Holtzman, 1986; Holloway et al., 1985a). However, this issue is further complicated by the fact that R-PIA also produced dose-related decreases in drug–lever responding (Holloway et al., 1985a) that may be attributable to a nonspecific motor deficit or, alternately, to the amphetamine-like moiety at the N^6 position of this molecule.

The ability of other psychomotor stimulants to produce generalizable stimuli in caffeine-trained animals remains controversial. In early studies in which rats were trained to discriminate caffeine (32 mg/kg) from saline, no stimulus generalization was obtained with amphetamine, methylphenidate, nicotine, or thyrotropin releasing hormone (TRH); however, theophylline and other methylxanthines were generalizable (Modrow et al., 1981; Carney et al., 1985c). More recent studies using a different discriminative methodology, however, have reported varying degrees of stimulus generalization to both methylxanthines and several monoaminergic indirect agonists in caffeine-trained rats (Holtzman, 1986,1987). Using a two-choice discrete-trial avoidance paradigm, rats trained to discriminate caffeine from saline displayed partial stimulus generalization to D-amphetamine and ephedrine and complete generalization to caffeine, cocaine, and methylphenidate (Holtzman, 1986). In these animals, only partial stimulus generalization was obtained with theophylline, theobromine, and IBMX. This caffeine cue was antagonized by the adenosine agonists, R-PIA and 2-CADO, as well as the alpha-adrenergic receptor blockers, phentolamine, prazosin, and yohimbine (Holtzman, 1986). Additional data indicative of a functional link between monoaminergic and purinergic systems comes from the observation that caffeine in combination with phenylpropylnolamine and/or ephedrine afforded complete stimulus generalization to D-amphetamine in D-amphetamine-trained animals, whereas any of these compounds given alone produced only partial stimulus generalization (Holloway et al., 1985a). Caffeine has also been shown to potentiate the effects of a low dose of D-amphetamine (0.5 mg/kg) in producing stimulus generalization in animals trained to discriminate a higher dose of D-amphetamine (0.8 mg/kg) from saline (Schechter, 1977). In animals trained to discriminate apomorphine from saline, caffeine (7.5–30 mg/kg) was not found to produce drug–lever responding (Schechter, 1980). However, caffeine was observed to potentiate the discriminability of low doses of apomorphine (Schechter, 1980).

Purinergic involvement has also been implicated in the stimulus properties of the noncompetitive NMDA antagonist, phencyclidine (PCP). PIA has been shown to block stereospecifically the ability to discriminate PCP from saline (Browne and Welch, 1982). The antagonist actions of PIA against a PCP stimulus also were shown to be blocked by the subsequent administration of theophylline (Browne and Welch, 1982). Further examination of this phenomenon showed, however, that PIA altered the pharmacokinetics of PCP (Browne et al., 1983).

The discriminability of the methylxanthines may also involve some interaction with central benzodiazepine receptors. Chlordiazepoxide (CDP) has been reported to decrease drug–lever responding in animals trained to discriminate caffeine from saline (Holloway et al., 1985b). Specificity in this interaction is difficult to determine, since caffeine (0.1–56 mg/kg) was found to have no significant effect on the discriminability of CDP (Holloway et al., 1985b).

From the above discussion, it appears that the stimulus properties of caffeine, as well as other methylxanthines, probably are mediated through a variety of mechanisms. The present data indicate a fundamental interaction with central purinergic receptors, since both adenosine agonists and antagonists can be discriminated and can antagonize their respective stimulus cues. However, additional neurochemical mechanisms almost certainly contribute to the stimulus properties of the methylxanthines. Inhibition of cyclic nucleotide phosphodiesterase activity appears to be sufficient to engender a specific and discriminable stimulus cue. The present data also indicate that interactions with other neurotransmitter systems also may contribute to the unique stimulus properties of caffeine, suggesting that some neuromodulatory interaction between purinergic and other neurochemical systems is capable of engendering a discrete stimulus event, not a surprising conclusion given the effects of adenosine on neurotransmitter release.

3.3. Locomotor Activity Studies

Of the many behavioral effects of adenosine, the ability of purinergic compounds to alter spontaneous motor activity has been the most widely characterized. Adenosine agonists administered either centrally (Barraco et al., 1984; Phillis et al., 1986) or systemically (Snyder et al., 1981; Carney, 1982; Katims et al., 1983; Buckholtz and Middaugh, 1987) have been found to cause dose-dependent reductions in spontaneous motor activity. This action appears to be centrally mediated, since the locomotor-depressing effects of purinergic agonists can be dissociated from their cardiovascular effects (Barraco et al., 1984). However, there is still some controversy related to these effects (Mullane and Williams, 1990). Direct injections of NECA or R-PIA into the lateral ventricle can reduce blood pressure at doses that are 10–100-fold greater than are required for significant reductions in spontaneous motor activity (Barraco et al., 1986). Similarly, systemic administration of R-PIA decreases locomotion at doses that do not alter cardiovascular function and that are sufficient to occupy more than 50% of brain adenosine receptors (Katims et al., 1983). When administered intraperitoneally, the peripherally acting adenosine antagonist, 8-PST, can effectively block the hypothermic effects of adenosine agonists without significantly altering purinergic-induced behavioral depression (Seale et al., 1988). The relative activity of purinergic agonists in decreasing locomotion appears to be consistent with an interaction at A_2 receptors, since NECA is more potent than either R-PIA or adenosine (Barraco et al., 1984; Coffin et al., 1984; Phillis et al., 1986).

The prototypical adenosine antagonist, caffeine, causes biphasic and dose-dependent effects on locomotor behavior, with low doses increasing and high doses decreasing motor activity (Thithapandha et al., 1972; Waldeck, 1975; Snyder et al., 1981; Fredholm et al., 1983; Choi et al., 1988). The motor-increasing effects of caffeine appear to result from a competitive interaction at adenosine receptors, since caffeine can dose-dependently block the behavioral depression induced by adenosine agonists (Snyder et al., 1981; Coffin et al., 1984; Phillis et al., 1986; Seale et al., 1986). In addition, compounds that are relatively selective phosphodiesterase inhibitors (i.e., IBMX and papaverine) have been found to produce only decreases in motor activity (Coffin et al., 1984; Phillis et al., 1986; Choi et al., 1988). Support for a central locus in the locomotor-stimulating effects of caffeine can be found in the observation that the relative activity of various methylxanthine analogs in producing increased motor activity correlates well with their ability to inhibit [^{3}H]CHA binding in brain (Snyder et al., 1981). Although most methylxanthines have nonselective antagonist actions at both adenosine receptor subtypes, the novel A$_2$ selective methylxanthine, 3,7-dimethyl-1-propargylxanthine (DMPX), has been shown to have a 10-fold greater potency in blocking decreases in locomotion induced by NECA as compared to CHA and is more potent than caffeine in increasing locomotor activity (Seale et al., 1988). In contrast, the relatively A$_1$-selective receptor antagonist, theobromine, has been shown to completely block a CHA-induced decrease in locomotor activity at doses that were ineffective against a NECA-induced suppression of motor activity (Carney et al., 1986).

As mentioned above, the variation in the behavioral sensitivity to methylxanthines across species is generally small and may be restricted to particular xanthine molecules that exhibit complex interactions at adenosine receptors (Murphy and Snyder, 1982; Ferkany et al., 1986; Stone et al., 1988). In contrast, the variation in behavioral sensitivity to caffeine appears to vary greatly within species. This intraspecies variation in caffeine sensitivity is particularly evident in humans: some individuals can consume 10–15 cups of coffee per day with no apparent ill effects; in others a single cup produces mild tremors (Grant et al., 1978). Possible reasons for such behavioral variation have been explored experimentally in inbred strains of mice that exhibit differential sensitivities to the behavioral and lethal effects of methylxanthines (Carney et al., 1985c; Logan et al., 1986; Seale et al., 1985, 1986; Buckholtz and Middaugh, 1987). CBA/J mice have been found to be dramatically more sensitive to the locomotor-stimulating, hyperthermic, and lethal effects of caffeine than are SWR/J mice (Carney et al., 1985c; Seale et al., 1984, 1985, 1986). These differences cannot be explained on the basis of differential metabolism or pharmacokinetics (Carney et al., 1985b, c); however, behavioral sensitivity to caffeine in these mice has been demonstrated to be genetically determined, with different genetic components controlling sensitivity to different methylxanthines (Seale et al., 1986). Mechanistically, the genetically determined sensitivity to caffeine apears to be manifest, at least in part, through the expression of brain adenosine receptors (Jarvis and Williams, 1988). Saturation binding experiments have revealed that CBA/J mice display significantly greater densities of striatal A_2 receptors and hippocampal and cerebellar A_1 receptors than SWR/J mice (Jarvis and Williams, 1988). These results indicate that the differential behavioral sensitivity to methylxanthines between these inbred mouse strains may be mediated through genetically determined differences in the regional distributions of brain adenosine receptors (Jarvis and Williams, 1988).

Caffeine can also effectively potentiate the locomotor-stimulating effects of other psychomotor stimulants. In a number of studies, caffeine has been shown to potentiate D-amphetamine-and apomorphine-induced hyperactivity and stereotypy (Klawans et al., 1974; Waldeck, 1975; White and Keller, 1984; Fredholm et al., 1983). Caffeine can also potentiate the motor-activity-increasing effects of the dopamine precursor, L-dopa (Waldeck, 1975). Interestingly, treatment with the dopamine neurotoxin, 6-hydroxydopamine (6-OHDA), has been found to attenuate (Erinoff and Snodgrass, 1986) and potentiate (Criswell et al., 1988) the locomotor-stimulating effects of caffeine, opposing effects that may depend upon the degree of 6-OHDA-induced receptor supersensitivity. Although caffeine is ineffective alone, it has been shown to potentiate the reversal of reserpine-induced suppression of locomotor activity by the dopaminergic precursor, L-dopa (Waldeck, 1975). Caffeine-induced locomotor-stimulation can be differentially blocked by the neuroleptic, pimozide, whereas the adrenergic receptor blockers, propanolol and phenoxybenzamine, are ineffective (Waldeck, 1975). From these data it is evident that the locomotor-stimulating effects of caffeine are dependent on an intact dopaminergic system. Although the exact mechanism for the interaction between purinergic and dopaminergic systems in mediating locomotion remains unclear, the evidence described above suggests a specific antagonist action at central adenosine receptors.

3.4. Chronic Effects of Adenosine Antagonists

Whereas caffeine is the most widely consumed psychoactive compound in the world (Gilbert, 1981), its reinforcing effects in humans and laboratory animals, as well as its abuse liability, have not been widely characterized. Caffeine has been shown to be self-administered by laboratory animals (*see* Griffiths and Woodson, 1988a for review); however, such behavior is not as robust as that obtained with other psychomotor stimulants (e.g., amphetamine and cocaine) (Griffiths and Woodson, 1988b). This behavioral profile may be related to some of the subjective effects of caffeine in humans, which can include feelings of "nervousness and anxiety" (Griffiths and Woodson, 1988b). In a recent experimental analysis of caffeine self-administration in humans, caffeine was found to dose-dependently engender positive reinforcement at low doses and caffeine avoidance at higher doses, as measured in a choice behavior paradigm and from subjects' self-reports (Griffiths and Woodson, 1988b).

Tolerance to the acute behavioral effects of various adenosine antagonists develops upon chronic exposure (Carney, 1982; Ahiljanian and Takemori, 1986: Finn and Holtzman, 1987, 1988). Such tolerance has been demonstrated in a variety of behavioral paradigms, including schedule-controlled operant responding (Carney, 1982), locomotor activity (Finn and Holtzman, 1987, 1988; File et al., 1988), and drug discrimination procedures (Holtzman, 1987). In all of these procedures, drug tolerance is typically defined as a rightward shift in drug dose–response curves (surmountable tolerance) or a downward shift in drug–dose response curves (insurmountable tolerance) (Finn and Holtzman, 1987).

Using locomotor activity as the behavioral measure, chronic caffeine exposure has been shown to result in complete tolerance to the biphasic effects of caffeine and in symmetrical cross-tolerance to the stimulant effects of theophylline and 7-(2-chloroethyl)theophylline (Finn and Holtzman, 1988). This caffeine tolerance also appears to be pharmacologically specific, since similar cross-tolerance was not observed with other psychomotor stimulants, including D-amphetamine, methylphenidate, and cocaine (Finn and Holtzman, 1987). Interestingly, in rats that displayed complete tolerance to caffeine, the adenosine agonists, NECA and R-PIA, were found to be only 10-fold less active in producing decreases in locomotor activity (Finn and Holtzman, 1987, 1988). This last observation stands in contrast to a previously reported leftward shift in adenosine agonist dose–response curves in caffeine-tolerant mice (Ahiljanian and Takemori, 1986) and to several demonstrations of brain adenosine receptor up-regulation following chronic exposure to methylxanthines (Boulenger et al., 1983; Green and Stiles, 1986; Sanders and Murray, 1988).

Using drug discrimination procedures, chronic caffeine treatment has been found to produce rightward shifts in caffeine stimulus generalization curves (Holtzman, 1987). Additionally, chronic treatment with either caffeine or methylphenidate resulted in symmetrical cross-tolerance between both drugs. This demonstration of symmetrical cross-tolerance indicates that caffeine and methylphenidate produce salient interoceptive stimuli by some common mechanism (Holtzman, 1987). The mechanistic similarity between these psychomotor stimulants appears to be limited to their stimulus properties, since similar cross-tolerance in their locomotor-stimulant effects does not occur (Finn and Holtzman, 1987, 1988). One possible explanation for these observations is that the dosing regimen required to produce behavioral tolerance in locomotor activity is much greater than that required to obtain tolerance to caffeine-induced interoceptive cues (Holtzman, 1987; Finn and Holtzman, 1987). This fact may result in chronic caffeine-induced inhibition in nucleotide phosphodiesterase activity as well as an upregulation of central adenosine receptors as contributory mediators of tolerance to the locomotor-stimulant effects of caffeine. The observation that prenatal exposure to caffeine can result in long-lasting alterations in central cyclic AMP activity is consistent with this hypothesis (Concannon et al., 1983).

It has been well documented that chronic exposure to adenosine receptor antagonists can result in increased numbers of central adenosine A_1 receptors (Murray, 1982; Boulenger et al., 1983; Chou et al., 1985; Green and Stiles, 1986; Szot et al., 1987). In the majority of these studies, chronic methylxanthine treatment results in an approximately 10–20% increase in brain A_1 receptors without a significant alteration in receptor affinity. These caffeine-induced increases in A_1 receptors appear to be greatest in the cerebral cortex and cerebellum, with more moderate increases occurring in the hippocampus (Szot et al., 1987). It should also be noted that there is at least one report that chronic caffeine treatment can also alter the proportion of high and low affinity states of the cortical A_1 receptor (Green and Stiles, 1986).

Whereas the acute administration of adenosine receptor antagonists can produce marked effects on central monoaminergic function (Fredholm and Hedqvist, 1980; Stone, 1981), the effects of chronic exposure to caffeine on brain monoamines remain controversial. Chronic caffeine treatment has been shown to produce a transitory increase in benzodiazepine receptors (Boulenger et al., 1983) as well as a longer-lasting downregulation of central beta-adrenergic receptors (Goldberg et al., 1982). However, in a more recent study, prolonged caffeine exposure in rats was not observed to significantly alter steady state levels of dopamine or norepinepherine and their respective metabolites (Zielke and Zielke, 1986). Pre- and postnatal exposure to caffeine has also been shown to significantly inhibit cyclic AMP activity in cerebellum; such treatment did not alter whole brain levels of dopamine or norepinepherine (Concannon et al., 1983). These results indicate that chronic caffeine treatment may not produce long-lasting alterations in brain monoaminergic function in the adult or developing organism. However, such treatment may result in long-term alterations in central purin function. This conclusion is further supported by the observation that pre- and postnatal caffeine treatment can produce age-dependent alterations in the locomotor activity of developing rats (Concannon et al., 1983; Hughes and Beveridge, 1986).

From the present discussion, it appears that chronic exposure to adenosine receptor antagonists clearly results in a decreased sensitivity to the acute effects of adenosine antagonists and agonists. Chronic caffeine treatment can produce complete tolerance to the locomotor-stimulant effects of the compound, as well as symmetrical cross-tolerance to other methylxanthines, yet adenosine agonists still retain full efficacy in their ability to reduce locomotor activity (Finn and Holtzman, 1987). From these observations, it has been argued that tolerance to the locomotor-stimulating effects of caffeine cannot be explained solely on the basis of a caffeine-induced upregulation of central adenosine receptors (Finn and Holtzman, 1987). Given that the rate-decreasing effects of the methylxanthines on operant responding and locomotor activity may involve an inhibition of nucleotide phosphodiesterase activity (Choi et al., 1988), as well as adenosine receptor antagonism, it is quite conceivable that the observed tolerance to caffeine in some behavioral paradigms may also involve a reduced sensitivity to caffeine-induced phosphodiesterase inhibition. In behavioral paradigms that may be more sensitive to drug tolerance effects (i.e., drug discrimination procedures [Holtzman, 1987]), an upregulation of central adenosine receptors may be sufficient for tolerance phenomena to occur. Interestingly, tolerance to caffeine-induced interoceptive stimuli also appears to be less pharmacologically strict than that for locomotor activity, since caffeine-induced cross-tolerance to other classes of psychomotor stimulants can be obtained in drug discrimination procedures.

4. Significance of Adenosine in CNS Function

4.1. Anticonvulsant Activity of Purines

That neurochemical and behavioral stimulation are obtained upon adenosine receptor antagonism has provided support for the notion that adenosine mediates a homeostatic "inhibitory tone" in mammalian physiology (Harms et al., 1978). This hypothesis has been further supported by observations that adenosine availability is increased during trauma in both peripheral tissue (Berne, 1963; Berne et al., 1983) and the brain (Berne et al., 1974; Winn et al., 1981; Onodera et al., 1986). With respect to central nervous system function, the possibility that adenosine functions primarily as an endogenous anticonvulsant and/or anxiolytic, with a spectrum of activity similar to the benzodiazepines, has received much attention (Phillis and O'Regan, 1988b; Dunwiddie and Worth, 1982; Dragunow et al., 1985; Marangos and Boulenger, 1985).

Adenosine effectively reduces epileptic seizure activity produced by a variety of chemical and electrical stimuli (Dragunow et al., 1985; Dunwiddie and Worth, 1982; Albertson et al., 1983; Barraco et al., 1984; Murray et al., 1985; Murray and Szot, 1986). Consistent with this observation, methylxanthines have been found to increase seizure activity or to act as proconvulsant agents (Dragunow et al., 1985; Albertson et al., 1983; Albertson, 1986), whereas chronic caffeine exposure has been shown to result in a decreased sensitivity to chemoconvulsants (Szot et al., 1987).

The use of electrically kindled seizures has proven particularly valuable in the assessment of the anticonvulsant effects of adenosine. This procedure allows for the efficient measurement of both the electrical and behavioral components of seizure development, the ictal event, and the subsequent postictal depression (Jarvis and Freeman,

1983). The anticonvulsant effects of adenosine agonists are primarily characterized by their ability to reduce seizure severity and duration without significantly altering seizure threshold (Dragunow et al., 1985). Both centrally (Barraco et al., 1984) and perpherially (Dragunow et al., 1985) administered adenosine agonists (NECA, R-PIA, and 2-CADO) have been found to dose-dependently reduce kindled seizure activity. This anticonvulsant activity of adenosine analogs can be blocked by doses of methylxanthines that, when given alone, have no observable effect on seizure activity (Barraco et al., 1984).

The proconvulsant effects of adenosine antagonists are characterized by a facilitation of kindled electrical afterdischarges, motor seizure durations, and potentiation of partially kindled seizures (Dragunow et al., 1985). Treatment with caffeine has been shown to extend epileptic activity in both developing and fully kindled seizures without significantly affecting electrical seizure threshold or the number of electrical stimulations required for seizure development (Albertson et al., 1983). High doses of methylxanthines can reduce postictal depression, thus facilitating the occurrence of secondary seizure activity (Albertson, 1986). In contrast, administration of R-PIA can dramatically prolong postictal depression and reduce postictal spiking in amygdaloid kindled rats (Rosen and Berman, 1985).

Although not extensively examined, there appear to be regional differences in the anticonvulsant potencies of adenosine agonists. When injected into the lateral ventricle, NECA was found to be slightly more potent than R-PIA in reducing the severity of amygdaloid-kindled seizures (Barraco et al., 1984). However, when injected at the kindled-seizure locus, R-PIA effectively reduced the severity of seizures kindled in the amygdala, hippocampus, and caudate nucleus, but NECA was found to be effective only when drug administration and kindled seizures occured in the caudate nucleus (Rosen and Berman, 1985). These results indicate that purinergic attenuation of kindled seizure activity probably involves both adenosine receptor subtypes, and that the differential efficacy of these compounds may be related to the regional distributions of brain adenosine receptor subtypes.

Purinergic agonists have also been shown to have anticonvulsant activity against seizures produced by a variety of chemoconvulsants, including pentylenetetrazol (Dunwiddie and Worth, 1982; Murray et al., 1985; Murray and Szot, 1986), picrotoxin, strychnine (Dunwiddie and Worth, 1982), penicillin (Niglio et al., 1988), and caffeine (Seale et al., 1986; Popoli et al., 1988). The primary anticonvulsant effect of adenosine agonists is to produce an increase in the dose of the chemoconvulsants that are required to elicit seizure activity (Dunwiddie and Worth, 1982; Murray et al., 1985). Both the pharmacological profile and stereoselectivity of adenosine agonists indicate that their anticonvulsant activity may be mediated at adenosine A_1 receptors. Adenosine agonists exhibit the following rank order of potency against pentylenetetrazol-induced seizures; R-PIA > CHA > 2-CADO > NECA > S-PIA (Murray and Szot, 1986). Additionally, R-PIA has been found to be approximately 80-fold more active than its stereoisomer, S-PIA (Murray and Szot, 1986).

These data clearly indicate that adenosine analogs can effectively attenuate chemically-induced seizure activity; however, whether these anticonvulsant actions are mediated entirely through adenosine A_1 receptors remains controversial. Methylxanthines have been shown to competitively block the anticonvulsant effects of adenosine agonists (Murray et al., 1985); however, their potency in this regard is greatly reduced when compared to their activity in blocking purine-induced sedation and hypothermia (Dunwiddie and Worth, 1982). The anticonvulsant potency of adenosine agonists also appear to vary with different chemoconvulsants (Dunwiddie and Worth, 1982). CHA and 2-CADO have been found to be markedly less potent than R-PIA in attenuating pentylenetetrazol-induced seizures, whereas they were twice as potent as R-PIA in reducing picrotoxin-induced seizure activity (Dunwiddie and Worth, 1982). Taken together, these results suggest that the anticonvulsant efficacy of purinergic agonists is probably not mediated solely by agonist interactions at adenosine A_1 receptors. Involvement of adenosine A_2 receptors, as well as the location of the primary epileptic focus in brain, may also be important factors in purinergic anticonvulsant efficacy.

4.2. Purinergic Involvement
in the Anticonvulsant Activity of Carbamazepine

Carbamazepine is an anticonvulsant with anxiolytic and antidepressant activity (Skerritt et al., 1982, 1983a,b). The mechanism of action of this clinically efficacious agent is unknown, but it has been found to interact competitively with central adenosine receptors (Skerritt et al., 1982; Marangos and Boulenger, 1985; Gasser et al., 1988), leading to the suggestion that this compound may exert its anticonvulsant effects through an interaction with purinergic receptors (Skerritt et al., 1982, 1983a,b).

Carbamazepine is a weak inhibitor of the binding of several A_1 selective agonist radioligands and the nonselective adenosine antagonist ligand, [^{3}H]DPX (Skerritt et al., 1983a,b). However, there appears to be no correlation between the ability of a series of carbamazepine analogs to inhibit adenosine A_1 receptor ligand binding and their anticonvulsant efficacy (Marangos et al., 1983). Similarly, carbamazepine has little activity at central A_2 receptors (Jarvis and Stone, unpublished). Additionally, the anticonvulsant profile of carbamazepine appears to be markedly different from that of adenosine analogs. The anticonvulsant effects of purinergic agonists appear to be mediated through their ability to attenuate the duration of the ictal event (Dragunow et al., 1985), possibly through a potentiation of compensatory inhibitory processes (Albertson et al., 1983; Rosen and Berman, 1985). In contrast, the anticonvulsant activity of carbamazepine is characterized by its ability to increase the threshold for epileptic afterdischarges (Dragunow et al., 1985). With respect to chemically-induced seizures, carbamazepine has been shown to be significantly more efficacious against pentylenetetrazol-induced seizures than the most potent adenosine agonist, R-PIA (Popoli et al., 1988).

From the above data, purinergic involvement in the anticonvulsant actions of carbamazepine appear to be somewhat limited. Yet there are data indicating that carbamazepine and adenosine agonists can interact synergistically in the attenuation of both caffeine- and pentylenetetrazol-induced seizure activity (Popoli et al., 1988). High doses of theophylline also have been shown to block the anticonvulsant actions of carbamazepine (Bernard et al., 1983). The mechanistic confusion concerning purinergic involvement in the anticonvulsant actions of carbamazepine may rest in the fact that this compound can have a mixed agonist/antagonist interaction at adenosine receptors (Marangos and Boulenger, 1985). This action has been indicated by demonstrations that chronic carbamazepine treatment results in a significant, though slight, upregulation of central adenosine receptors (Marangos and Boulenger, 1985). This result is also consistent with the observation that high doses of carbamazepine antagonize adenosine A_2 receptor mediated responses (Weir et al., 1984). Since the anticonvulsant profile of carbamazepine is markedly different from that of the purines, and this compound interacts only weakly with central adenosine receptors, it would appear that purinergic interactions do not contribute to the primary anticonvulsant actions of carbamazepine.

Alternatively, as discussed in the following section, the involvement of adenosine in the molecular events underlying convulsive activity, although unclear, has resulted in a substantial body of literature that circumstantially links the compound to the actions of a number of pharmacological agents with activity in the convulsive axis.

4.3. Adenosine/Benzodiazepine Interactions

A mechanistic involvement of adenosine in anticonvulsant activity is implied by findings that benzodiazepines can potentiate adenosine-induced neuronal inhibition (an effect blocked by adenosine antagonists [Phillis, 1979; Phillis and Wu, 1982]) and by observations that the purine metabolites, inosine, hypoxanthine, and guanosine, have significant pharmacological activity at central benzodiazepine receptors (Snyder, 1985; Marangos and Boulenger, 1985; Marangos et al., 1987). Although these compounds have IC_{50} values in the micromolar–millimolar range (Marangos et al., 1979), they share a behavioral profile similar to that of the benzodiazepine anticonvulsants. Additional support for an interaction between adenosine and the benzodiazepines in neuronal function came from observations that not all of the central effects of the benzodiazepines can be explained by their interaction at the central benzodiazepine/GABA-chloride channel receptor complex (Phillis and O'Regan, 1988 a,b; Williams and Olsen, 1988) and by the fact that benzodiazepines can prolong extracellular adenosine availability (Phillis, 1979). Thus, the possibility exists that a functional interaction between adenosine and benzodiazepine receptor activation underlies some of the anticonvulsant and/or anxiolytic actions of both classes of compounds.

However, this functional interaction does not appear to be receptor mediated since benzodiazepine analogs have been shown to have little pharmacological activity at central adenosine receptors (Williams et al., 1981). One mechanism by which benzodiazepines and adenosine functionally interact may be the ability of benzodiazepines to block adenosine reuptake (Hammond et al., 1981; Phillis and Wu, 1981; Patel et al., 1982). A reasonable correlation has been demonstrated between the ability of benzodiazepines to block the uptake of adenosine and their anxiolytic potency (Phillis and Wu, 1982). As noted above, adenosine agonists and benzodiazepines share the acute behavioral effects of muscle relaxation, sedation, and hypothermia, and also can interact synergistically to produce these effects (Phillis and O'Regan, 1988b).

The adenosine antagonists, theophylline and caffeine, block the sedative effects of the benzodiazepines (Phillis and O'Regan, 1988b) and have anxiogenic properties (Loke et al., 1985; Charney et al., 1985; Griffiths and Woodson, 1988a). Chronic methylxanthine treatment has been shown to increase the number of benzodiazepine receptors (Boulenger et al., 1983) and has recently been reported to uncouple the functional state of the benzodiazepine receptor complex (Roca et al., 1988). Interestingly, the prototypic benzodiazepine antagonist, Ro 15-1788, can block caffeine-induced seizures (Albertson et al., 1983). The endogenous anxiogenic β-carbolines also can functionally block the inhibitory actions of adenosine (Phillis and O'Regan, 1988b). Although the sedative properties of the barbiturates have been ascribed to their actions at the chloride channel of the benziodiazepine/GABA receptor complex (Williams and Olsen, 1988), the excitatory effects of these compounds may be related to their ability to function as adenosine receptor antagonists (Lohse et al., 1985, 1987b).

The benzodiazepine receptor antagonist, Ro 15-1788, alters central purinergic function (Phillis and O'Regan, 1988b) and blocks adenosine uptake, although it is ineffective in blocking benzodiazepine effects on adenosine turnover (Phillis and Stair, 1987; Phillis and O'Regan, 1988b). The peripheral benzodiazepine ligand, Ro 5-4864, also has been shown to block adenosine uptake (Phillis and O'Regan, 1988b). Both compounds can functionally block NECA-induced depression of neuronal activity (Phillis and O'Regan, 1988b).

A considerable body of data supports a relationship between adenosine and the benzodiazepines. However, there is still no tangible information to indicate the molecular mechanism of this circumstantial symbiosis.

4.4. Purinergic Involvement in Stress and Depression

There is some recent evidence that central purinergic systems are altered by chronic exposure to environmental stressors. Sustained exposure to footshock, restraint, and sleep deprivation all cause an upregulation in the number of adenosine A_1 receptors in the rat hypothalamus (Anderson et al., 1987,1988). These effects are relatively small, however, and they are highly localized in the brain; adenosine receptors in other brain regions are not affected by the stress situation (Anderson et al., 1988). Administration of R-PIA reduces the incidence in rats of stomach ulcers produced by chronic restraint stress (Geiger and Glavin, 1985; Westerberg and Geiger, 1987). This effect appears to be centrally mediated, since the central administration of 8-PST or the peripheral administration of 8-PT can block the protective effects of R-PIA (Westerberg and Geiger, 1987).

In a forced swim model of depression (i.e., behavioral immobility), adenosine and the adenosine uptake blocker, dipyridamole, potentiated the immobilization period in mice (Kulkarni and Mehta, 1985). These effects can be blocked by either methylxanthines or the tricyclic antidepressants, imipramine and desipramine (Kulkarni and Mehta, 1985). Mechanistically, these effects may be mediated by the ability of both methylxanthines (Fredholm and Dunwiddie, 1988) and tricyclic antidepressants (Porsolt et al., 1977) to release brain catecholamines. However, this conclusion remains speculative, since acutely administered antidepressants also have been found to increase the release of adenosine (Stone, 1983), whereas chronic antidepressant treatment does not result in long-term changes in brain A_1 receptors (Williams et al., 1983). Interestingly, another therapeutic intervention for depression, chronic electroconvulsive therapy (ECT), can produce an increase in A_1 receptors in the rat brain (Newman et al., 1984).

The data discussed above indicate that central purinergic systems are affected by or possibly mediate some aspects of the behavioral consequences of exposure to chronic stress or depression. Given the known pharmacological actions of the typical antidepressant compounds, it is conceivable that adjunct therapy with methylxanthines may have beneficial effects (Williams and Jarvis, 1988). However, possible contributions of adenosine to the etiology and development of behavioral depression remain unclear at this time.

4.5. Cerebral Blood Flow and Ischemia and Cognition

Whereas the psychomotor-stimulant effects of the methylxanthines primarily have been attributed to their actions as central adenosine receptor antagonists, xanthines have also been found to increase cerebral blood flow (Winn et al., 1981; Sollevi, 1986) and the availability of oxygen and glucose to the brain (Grome and Stefanovich, 1986). These actions appear to be mediated by purinergic-induced vasodilation via receptor-mediated processes (Hardebo et al., 1987). Whether such effects occur directly as a result of adenosine receptor activiation, or are the result of an indirect modulatory action, is unclear (Phillis and DeLong, 1986). However, purinergic effects on the cerebral microvasculature have been suggested in the etiology of migraine (Burnstock, 1985).

Adenosine release following transient hypoxic (Berne et al., 1974; Onodera et al., 1986; Evans et al., 1987; Goldberg et al., 1988) and convulsive (Winn et al., 1981) episodes provides important evidence for a homeostatic function of endogenous adenosine. The pathological consequences of ischemia and/or seizure activity may result from the release of neurotoxic quantities of EAA, since competitive and noncompetitive EAA-receptor antagonists can effectively block the neuronal death associated with these events (Duncan et al., 1982; Simon et al., 1984; Foster et al., 1987; Jarvis et al., 1988a). In this regard, the excitatory amino acid neurotransmitter, L-glutamate, appears to be colocalized with adenosine (Deckert and Jorgensen, 1988; Jarvis et al., 1988a) and its neuronal release can be modulated by adenosine (Dolphin and Prestwich, 1985). Thus, the anticonvulsive and antiischemic actions of endogenous adenosine may be mediated through an inhibitory action on glutamate release. It is also interesting to consider that these protective effects of adenosine in brain may be related to a potential antiinflammatory action of adenosine, which has been demonstrated in peripheral tissues (Cronstein et al., 1983).

The ability of methylxanthines to enhance behavioral performance has been well documented for both locomotor activity and operant behavior paradigms. However, purinergic contributions in
cognitive processing have not been extensively studied, and currently
available data yield contradictory results. As discussed earlier, caffeine is the most widely consumed psychoactive agent, with its primary reinforcing quality being self-reports of arousal (Griffiths and
Woodson, 1988b). In a recent study, however, it was found to reverse
the inhibitory effects of R-PIA on the aqusition of a conditioned
response (Winsky and Harvey, 1987). This effect appears to be mediated by central adenosine receptors, since the selective phosphodiesterase inhibitor, rolipam, did not block an inhibitory effect of R-
PIA on associative learning (Winsky and Harvey, 1987).

The ability of methylxanthines to increase arousal may be mediated by their actions at central adenosine receptors or may be related
to their ability to increase cerebral blood flow as well as glucose
utilization (Grome and Stefanovich, 1986; Nehlig et al., 1984). Be-
haviorally, caffeine has been shown to block the sedative effects of
purine analogs and to shorten thiopental-induced sleep (Louie et al.,
1986). These effects may be indicative of a global increase in functional activity. The xanthine, HWA 285, has been found to have "cognitive enhancing" properties in humans (Hindmarch and Subhan,
1985); however, this compound has only weak activity at adenosine
receptors and may have agonist actions in vivo (Grome and
Stefanovich, 1986). It should be noted that some compounds, which
have either mixed agonist/antagonist actions or inverse agonist actions at central benzodiazepine receptors and which have relatively
weak activity at adenosine receptors, have been reported to enhance
vigilance in some behavioral paradigms (Bennett and Petrack, 1984;
Venault et al., 1986).

4.6. Adenosine and Analgesia

Adenosine may also be of physiological importance in the modulation of nociception and opiate-induced analgesia. Adenosine analogs increase hot plate and tail-flick latencies in mice, effects that can
be competitively and selectively blocked by methylxanthines
(Vapaatalo et al., 1975; Yarbrough and McGuffin-Clineschmidt,

1981; Holmgren et al., 1986; DeLander and Hopkins, 1987; DeLander and Whal, 1988). The pharmacological profile of these antinociceptive effects is consistent with activity at A_2 receptors, with NECA being approximately 10–20 times more active than A_1 selective agonists (Holmgren et al., 1986; DeLander and Hopkins, 1987).

The recent localization of both adenosine receptor subtypes in the spinal cord (Choca et al., 1987) adds some degree of mechanistic specificity to the above observations. The involvement of adenosine is further reinforced by the findings that pertussis toxin pretreatment can decrease the antinociceptive effects of the purine and that phosphodiesterase inhibiton also reduces the antinociceptive effects of CHA, but not those of NECA (Sawynok and Reid, 1988). Adenosine availability in the dorsal spinal cord may be regulated by serotonergic neuronal activity, since serotonin has been shown to competitively modulate the release of the nucleoside in this neural structure (Sweeney et al., 1988). Interestingly, brain adenosine A_1 receptors have been reported to be increased in morphine-dependent mice (Ahiljanian and Takemori, 1986).

The modulation of pain mechanisms by adenosine remains controversial (Marangos and Boulenger, 1985; Williams and Jarvis, 1988). Yet there are many reports that adenosine agonists and antagonists can both potentiate and/or antagonize nociceptive responses (Gourley and Beckner, 1973; Ho et al., 1973; Vapaataalo et al., 1975; Yarbrourgh and McGuffin-Clineschmidt, 1981; Ahiljanian and Takmori, 1986). These variable data may be related to differences in the experimental paradigms used to measure nociception, as well as possible differences in the physiological effects on purinergic acti-vation. In this regard, caffeine has been used clinically as an analgesic adjuvant (Laska et al., 1984), and the prototypic phosphodiesterase inhibitor, IBMX, has been found to produce a "quasi-morphine withdrawal syndrome" (Collier et al., 1981).

4.7. Purinergic Involvement in Basal Ganglia Function

Early support for a functional interaction between purinergic and dopaminergic systems in striatum came from demonstrations that methylxanthines could induce and/or potentiate rotational behavior in rats with unilateral striatal lesions (Fuxe and Ungerstedt, 1974;

Fredholm et al., 1976). It has also been demonstrated that purine nucleosides decrease dopamine synthesis and release in both in vivo and in vitro preparations (Michaelis et al., 1979; Myers and Pugsley, 1986). The methylxanthines, in contrast, increase dopamine release and can inhibit monoamine oxidase activity (Berkowitz et al., 1970; Michaelis et al., 1979). These neurochemical changes may mediate the observed adenosine-induced behavioral inhibition and the stimulant effect of the methylxanthines.

Some methylxanthine-induced behaviors appear to be dependent upon intact striatal dopaminergic functioning. 6-OHDA treatment has been shown to both increase (Criswell et al., 1988) and decrease (Erinoff and Snodgrass, 1986) caffeine-stimulated motor activity. These differential effects may be dependent on test parameters, age of the animals, and/or extent of the dopaminergic receptor supersensitivity. Consistent with a decreased responsivity to methylxanthines following 6-OHDA treatment, α-methyltyrosine administration has been found to attenuate the effects of caffeine on motor activity (White et al., 1978; Criswell et al., 1988). Methylxanthines have also been found to potentiate the psychomotor-stimulant effects of amphetamine, cocaine, and methylphenidate (Waldeck, 1975). Additionally, the dopaminergic receptor antagonists, pimozide and haloperidol, can effectively block the locomotor-stimulant effects of caffeine (Waldeck, 1975).

Although it is clear that methylxanthines can potentiate the effects of dopaminergic agonists, it is also apparent that methylxanthines can mimic the effects of direct-acting dopamine agonists. Theophylline administration alone has been shown to induce rotation in rats with unilateral striatal lesions (Fredholm et al., 1976; Watanabe et al., 1981). Interestingly, contralateral rotation to apomorphine has been demonstrated in rats following the unilateral intrastriatal administration of NECA (Green et al., 1982). In these animals, the ip administration of theophylline blocked this apomorphine-induced rotation (Green et al., 1982). These results indicate that dopamine agonist-induced rotation can be obtained following a purine-induced alteration in striatal function. This observation is further supported by demonstrations that NECA can increase serum prolactin, an effect that is competetively blocked by methylxanthines (Stewart and Pugsley, 1985).

Methylxanthines, when administered to food-deprived rats (Ferrer et al., 1982) or in very large doses (Boyd et al., 1965), can induce self-mutilatory behavior. Similar behavior can also be produced by other psychomotor-stimulants, including L-dopa, pemoline, amphetamine, apomorphine, and SKF-38393 (Muller and Nyhan, 1982; Goldstein et al., 1986; Criswell et al., 1988). This self-mutilatory behavior can be reduced by the intrastriatal administration of adenosine analogs (NECA being more potent than 2-CADO or CPA), which indicates that self-mutilatory behavior may be modulated by agonist interactions at striatal A_2 receptors (Criswell et al., 1988).

Self-mutilatory behavior, along with motor dysfunction and mental retardation, is the hallmark symptoms of the Lesch-Nyhan syndrome (Nyhan, 1973). The Lesch-Nyhan syndrome is an X-linked recessive disorder of purine metabolism in which the lack of hypoxanthine-guanine phosphoribosyl transferase (HGPRT) activity results in decreased levels of adenosine in brain as well as an overproduction of uric acid (Nyhan, 1973). This disorder is also characterized by marked decreases in striatal dopamine concentrations (Lloyd et al., 1981; Kopin, 1981). These observations have led to several suggestions that perturbations in HGPRT activity may contribute to decreased dopaminergic functioning in the basal ganglia (Green et al., 1982). It remains unclear, however, to what extent there are reciprocal pathological interactions between adenosine and dopamine in this disease (Jarvis and Williams, 1987). Furthermore, there appears to be no relation between HGPRT activity and dopa-mine metabolism in the PC-12 cell line (Bitler and Howard, 1986) and that caffeine, at doses that elicit self-destructive behavior, can produce an increase in HGPRT activity in rats (Minana et al., 1984).

Self mutilatory behavior appears to be mediated by supersensitive dopamine receptors (Goldstein et al., 1986; Criswell et al., 1988), rather than by a direct result of purine dysfunction (Bitler and Howard, 1986). Interestingly, the dopaminergic dysfunction in the Lesch-Nyhan syndrome, characterized by a selective destruction of dopamine terminals, is remarkably similar to that found in methamphetamine-induced neurotoxicity (Lloyd et al., 1981; Ricaurte et al., 1982). It is conceivable, therefore, that a purinergic contribution to the self-destructive behavior of the Lesch-Nyhan Syndrome could occur through reduced availability of adenosine in the brain, resulting in a disinhibitory effect on remaining striatal dopaminergic terminals and thereby facilitating dopamine release upon supersensitive receptors (Criswell et al., 1988).

The evidence discussed above indicates that purinergic manipulations of basal ganglia function can result in profound behavioral effects that appear to be mediated through adenosine-induced alterations in dopamine metabolism. Since neither purines nor xanthines directly interact with striatal dopamine receptors (Watanabe and Uramoto, 1986), the actions of endogenous adenosine on dopamine functions appear to be modulatory in nature. As mentioned earlier, selective lesions of pre- and postsynaptic striatal neurons (Geiger, 1986; Jarvis and Williams, 1986) have revealed that adenosine receptors are probably present on interneurons, glia, or corticostriatal terminals, rather than on dopaminergic neurons orginating in the substantia nigra (Wojcik and Neff, 1983; Geiger, 1986; Jarvis and Williams, 1986). These observations suggest that a transsynaptic negative feedback loop within the extrapyramidal system might explain the purinergic contribution in striatal dopamine function. Alternatively, adenosine has also been shown to have pronounced effects on transmembrane Ca^{2+} flux in monoamergic neurons (Dunwiddie, 1985; Fredholm and Dunwiddie, 1988). Modulation of dopaminergic activity through purinergic effects on Ca^{2+} may provide mechanistic insights on methylxanthine-induced dopamine release. For instance, 6-OHDA, α-methyltyrosine, and reserpine have shown to attenuate the locomotor-stimulant effects of caffeine (Waldeck, 1975; Erinoff and Snodgrass, 1986). Amphetamine-induced hyperactivity appears to be mediated via a Ca^{2+}-independent release of newly uptaken or synthesized dopamine that is attenuated by 6-OHDA and by α-methyltyrosine, but not by reserpine treatment (*see* Wagner et al., 1983).

The clinical significance of a purinergic modulation of dopamine function is highlighted in in the Lesch-Nyhan syndrome, in which a distruption of purine metabolism may contribute to an overstimulation of stiratal dopamine receptors. However, as with many of the putative actions of adenosine in central nervous system function, much of the available evidence is circumstantial in nature. Nevertheless, the behavioral pharmacology of adenosine, as well as the selective localization of adenosine A_2 receptors in the striatum, indicates a specific modulatory action of adenosine on dopamine function. This relationship has received further support from the recent observations that A_2-selective adenosine agonists can affect locomotor activity and apomorphine-induced climbing behavior in a manner similar to that of classical neuroleptic compounds such as haloperidol (Heffner et al., 1985,1989).

5. Conclusion

As has been illustrated in the present chapter, adenosine appears to be intimately involved in a wide variety of CNS processes. Although purinergic contributions to mammalian physiology have been extensively documented over the last 60 years, the experimental analysis of adenosine's effects on neuronal function has developed only during the last 10–20 years (Williams, 1987). The apparent skepticism regarding specific purinergic contributions in neurotransmission stems, in large part, from this previous body of data indicating a role for adenosine in many different cellular functions. This situation is not unlike that which existed for the excitatory amino acid L-glutamate before the discovery of specific EAA receptor subtypes and specific antagonist compounds that could effectively block the physiological consequences of receptor activation (Watkins, 1981). Both adenosine and L-glutamate exist in the brain in relatively large quantities and subserve a variety of physiological processes. However, the study of adenosine involvement in neural communication is still hampered by the lack of specific antagonists that have high affinity and selectivity for the known adenosine receptor subtypes. In this regard, elucidation of the therapeutic significance of adenosine has centered on the discovery of compounds that are selective for one receptor subtype.

At present, pharmacological activity at the adenosine A_1 receptor has been best characterized with selective agonists and, to a lesser extent, antagonist compounds. Similar characterization of the adenosine A_2 receptor has not been achieved because of the lack of compounds that selectively interact with this receptor subtype. In both in vitro and in vivo studies, pharmacological activity at the A_2 receptor has been traditionally inferred from the effects of the nonselective

458

Jarvis and Williams

agonist, NECA, by establishing pharmacological profiles for adenosine agonists or, in the less desirable instance, through factoring out
the effects of A_1 selective compounds. Clearly, compounds such as
CGS 21680 (Hutchison et al., 1989), which have high affinity and
selectivity for the A_2-adenosine receptor subtype, will be important
tools for the further evaluation of the role of adenosine in CNS function.

The documented neurochemical and behavioral effects of purines in CNS function highlight the potential physiological significance of adenosine receptor modulators. Mechanistic specificity in
the ability of adenosine to affect different neurochemical pathways
may be determined by the discrete populations of adenosine receptor
subtypes in the mammalian brain. From the data discussed in this
chapter, adenosine A_2 receptors, which are selectively localized in the
striatum, appear to be closely associated with neuronal mechanisms
that mediate locomotion (Spealman and Coffin, 1988) and possibly
with the hypermotility associated with seizure activity (Rosen and
Berman, 1985). In contrast, adenosine A_1 receptors, which are heterogeneously distributed in the brain, may be specifically involved in
limiting seizure activity at the neuronal level, since these receptors
have been localized on the terminals of excitatory neurons (Goodman
et al., 1983) and have been shown to reduce L-glutamate availability
in nervous tissue (Dolphin and Prestwich, 1985).

Adenosine has been considered to function as a neuromodulator
in mammalian brain, rather than as a classical neurotransmitter, primarily because of the lack of identifiable mechanisms that regulate
adenosine availability during neuronal communication (Snyder,
1985; Marangos and Boulenger, 1985; Williams, 1987, 1990a). This
neuromodulatory role for adenosine in neurotransmission is also indicated from the many documented synergistic, rather than additive,
interactions with other neurochemical systems. Such synergistic effects are illustrated in the puringeric modification of seizure activity,
nociception, stress, and various states of psychomotor activation. The
general lack of demonstrated additive interactions between purinergic agonists or antagonists and other neurotransmitters in the many
central effects of adenosine indicates that the nuceloside probably
functions as a modifier rather than as a mediator of synaptic commu-

nication. The facts that adenosine exerts inhibitory effects on a variety of different neurochemical systems and is increased in the brain following excitatory trauma lend further support to the idea that adenosine provides an "inhibitory tone" in central nervous system function and serves as the prototypical neuromodulator (Williams, 1990a).

References

Ahiljanian, M. and Takemori, A. E. (1986) Changes in adenosine receptor sensitivity in morphine-tolerant and-dependent mice. *J. Pharmacol. Exp. Ther.* **236,** 615–620.

Albertson, T. E. (1986) Effects of aminophylline on amygdaloid-kindled postictal depression. *Pharmacol. Biochem. Behav.* **24,** 1599–1603.

Albertson, T. E., Joy, R. M., and Stark, L. G. (1983) Caffeine modification of kindled amygdaloid seizures. *Pharmacol. Biochem. Behav.* **19,** 339–343.

Anderson, S. M., Leu, J. R., and Kant, G. J. (1987) Effects of stress on [³H]cyclohexyladenosine binding to rat brain membranes. *Pharmacol. Biochem. Behav.* **26,** 829–833.

Anderson, S. M., Leu, J. R., and Kant, G. J. (1988) Chronic stress increases the binding of the A_1 adenosine receptor agonist, [³H]cyclohexyladenosine, to rat hypothalamus. *Pharmacol. Biochem. Behav.* **30,** 169–175.

Barraco, R. A., Aggarawai, A. K., Phillis, J. W., Moran, M. A., and Wu, P. H. (1984) Dissociation of locomotor and hypertensive effects of adenosine analogs in rat. *Neurosci. Lett.* **48,** 139–144.

Barraco, R. A., Swanson, T. H., Phillis, J. W., and Berman, R. F. (1986) Anticonvulsant effects of adenosine analogs on amygdaloid-kindled seizures in the rat. *Neurosci. Lett.* **46,** 317–322.

Bennett, D. A. and Petrack, B. (1984) CGS 9896: A nonbenzodiazepine, non-sedating potential anxiolytic. *Drug Dev. Res.* **4,** 75–82.

Berkowitz, B. A., Tarver, J. H., and Spector, S. (1970) Release of norepinepherine in the central nervous system by theophylline and caffeine. *Eur. J. Pharmacol.* **10,** 64–71.

Bernard, P. D., Wilson, D., Pastor, G., Brown, W., and Glenn, T. W. (1983) Possible involvement of adenosine receptors in the electroshock anticonvulsant effects of carbamazepine, diphenylhydantoin, phenobarbital and diazepam. *Pharmacologist* **25,** 164.

Berne, R. M. (1963) Cardiac nucleosides in hypoxia: Possible role in regulation of coronary flow. *J. Physiol. (Lond.)* **204,** 317–322.

Berne, R. M., Rubio, R., and Cornish, R. R. (1974) Release of adenosine from ischemic brain. *Circ. Res.* **32,** 262–271.

Berne, R. M., Knabb, R. M., Ely, S. W., and Rubio, R. (1983) Adenosine in the local regulation of blood flow: A brief review. *Fed. Proc.* **42,** 3136–3142.

Bitler, C. M. and Howard, B. D. (1986) Dopamine metabolism in hypoxanthine-guanine phosphoribosyltransferase-deficient variants of PC12 cells. *J. Neurochem.* **47**, 107–112.

Boulenger, J. P., Patel, J., Post, R. M., Parma, A. M., and Marangos, P. J. (1983) Chronic caffeine consumption increases the number of brain adenoisne receptors. *Life Sci.* **32**, 1135–1142.

Boyd, E. M., Dolman, M., Knight, L. M., and Sheppard, E. P. (1965) The chronic oral toxicity of caffeine. *Can. J. Pharmacol.* **43**, 995–1007.

Braas, K. M., Newby, A. C., Wilson, V. S., and Snyder, S. H. (1986) Adenosine containing neurons in the brain localized by immunocytochemistry. *J. Neurosci.* **6**, 1952–1961.

Browne, R. G. and Welch, W. M. (1982) Stereoselective antagonism of phencyclidine's discriminative properties by adenosine receptor agonists. *Science* **217**, 1157,1158.

Browne, R. G., Welch, W. M., Kozlowski, M. R., and Duthu, G. (1983) Antagonism of PCP discrimination by adenosine analogs, in *Phencyclidine and Related Arylcyclohexylamines: Present and Future Applications* (Kamenka, J. M., Domino, E. G., and Eneste, G., eds.), Brooks, Ann Arbor, pp. 639–666.

Bruns, R. F., Daly, J. W., and Snyder, S. H. (1980) Adenosine receptors in brain membranes: Binding of N^6-cyclohexyl[^{3}H]adenosine and 1,3-diethyl-8-[^{3}H] phenylxanthine. *Proc. Natl. Acad. Sci. USA* **77**, 5547–5551.

Bruns, R. F., Lu, G. H., and Pugsley, T. A. (1986) Characterization of the A_2 adenosine receptor labeled by [^{3}H]NECA in rat striatal membranes. *Mol. Pharmacol.* **29**, 331–346.

Bruns, R. F., Davis, R. E., Nineteman, F. W., Poschel, B. P., Wiley, J. N., and Heffner, T. G. (1988) Adenosine antagonists as pharmacological tools, in *Adenosine and Adenine Nucleotides, Physiology and Pharmacology* (Paton, D.M., ed.) Taylor and Francis, London, pp. 39–49.

Bruns, R. F., Fergus, J. H., Badger, E. W., Bristol, J. A., Santay, L. A. Hartman, J. D., Hays, S. J., and Huang, C. C. (1987) Binding of the A_1-selective adenosine antagonist 8-cyclopentyl-1,3-dipropylxanthine to rat brain membranes. *Naunyn-Schmiedebergs Arch. Pharmacol.* **335**, 59–63.

Buckholtz, N. S. and Middaugh, L. D. (1987) Effects of caffeine and L-phenylisopropyladenosine on locomotor activity of mice. *Pharmacol. Biochem. Behav.* **28**, 179–185.

Burnstock, G. (1985) Neurochemical control of blood vessels: Some future directions. *J. Cardiovascul. Pharmacol.* **7**, S137–S146.

Carney, J. M. (1982) Effects of caffeine, theophylline and theobromide on schedule controlled responding in rats. *Br. J. Pharmacol.* **75**, 451–454.

Carney, J. M. and Christensen, H. D. (1980) Discriminative stimulus properties of caffeine: Studies using pure and natural products. *Pharmacol. Biochem. Behav.* **13**, 313–318.

Carney, J. M., Holloway, F. A., and Modrow, H. E. (1985a) Discriminative stimulus properties of methylxanthines and their metabolites in rats. *Life Sci.* **36**, 913–920.

Carney, J. M., Seale, T. W., Logan, L., and McMaster, S. B. (1985b) Sensitivity of inbred mice to methylxanthines is not determined by plasma xanthine concentration. *Neurosci. Lett.* **56**, 27–31.

Carney, J. M., Holloway, F. A., Williams, H. L., and Seale, T. W. (1985c) Behavioral pharmacology of caffeine in experimental subjects, in *Behavioral Pharmacology: The Current Status* (Balster, R. and Seiden, L., eds.), A. R. Liss, New York, pp. 281–293.

Carney, J. M., Cao, W., Logan, L., Rennert, O. M., and Seale, T. W. (1986) Differential antagonism of the behavioral depressant and hypothermic effects of 5'(*N*-ethylcarboxamide) adenosine by theobromine. *Pharmacol. Biochem. Behav.* **25**, 769–773.

Charney, D. S., Heninger, G. R., and Jatlow, P. I. (1985) Increased anxiogenic effects of caffeine in panic disorders. *Arch. Gen. Psychiat.* **42**, 233–243.

Choca, J. L., Proudfit, H. K., and Green, R. D. (1987) Identification of A_1 and A_2 adenosine receptors in the rat spinal cord. *J. Pharmacol. Exp. Ther.* **242**, 905–910.

Choi, O. H., Shamim, M. T., Padgett, W. L., and Daly, J. W. (1988) Caffeine and theophylline analogues: Correlation of behavioral effects with activity as adenosine receptor antagonists and as phosphodiesterase inhibitors. *Life Sci.* **43**, 387–398.

Chou, D. T., Kan, S., Forde, J., and Hirsh, K. R. (1985) Caffeine tolerance: Behavioral electrophysiological and neurochemical evidence. *Life Sci.* **36**, 2347–2358.

Coffin, V. L. and Spealman, R. D. (1987) Behavioral and cardiovascular effects of analogs of adenosine in cynomolgus monkeys. *J. Pharmacol. Exp. Ther.* **241**, 76–83.

Coffin, V. L., Taylor, J. A., Phillis, J. W., Altman, H. J., and Barraco, R. A. (1984) Behavioral interaction of adenosine and methylxanthines on central purinergic systems. *Neurosci. Lett.* **47**, 91–98.

Collier, H. O. J., Cuthbert, N. J., and Francis, D. L. (1981) Character and meaning of quasimorphine withdrawal phenomena elicited by methylxanthines. *Fed. Proc.* **40**, 1513–1518.

Concannon, J. T., Braughler, M., and Schechter, M. D. (1983) Pre- and postnatal effects of caffeine on brain biogenic amines, cyclic nucleotides and behavior in developing rats. *J. Pharmacol. Exp. Ther.* **226**, 673–679.

Criswell, H., Mueller, R. A., and Breese, G. R. (1988) Assessment of purine-dopamine interactions in 6-hydroxydopamine-lesioned rats: Evidence for pre- and postsynaptic influences by adenosine. *J. Pharmacol. Exp. Ther.* **244**, 493–500.

Cronstein, B. N., Kramer, S. B., Rosenstein, E. D., Weissmann, G., and Hirschhorn, R. (1983) Adenosine: A physiological modulator of superoxide anion generation by human neutrophils. *J. Exp. Med.* **158**, 1160–1177.

Daly J. W. (1982) Adenosine receptors: Target sites for drugs. *J. Med. Chem.* **25**, 197–207.

Deckert, J. and Jorgensen, M. B. (1988) Evidence for pre- and postsynaptic localization of adenosine A_1 receptors in the CA1 region of rat hippocampus: A quantitative autoradiography study. *Brain Res.* **446**, 161–164.

DeLander, G. E. and Hopkins, C. J. (1987) Involvement of A_2 adenosine receptors in spinal mechanisms of nocioception. *FASEB J.* **2**, A1132.

DeLander, G. E. and Wahl, J. J. (1988) Behavior induced by putative nociceptive neurotransmitters is inhibited by adenosine or adenosine analogs coadministered intrathecally. *J. Pharmacol. Exp. Ther.* **246**, 565–570.

Dolphin, A. C. and Prestwich, S. A. (1985) Pertussis toxin reverses adenosine inhibition of neuronal glutamate release. *Nature* **316**, 148–150.

Dragunow, M., Goddard, G. V., and Laverty, R. (1985) Is adenosine an endogenous anticonvulsant? *Epilepsia* **26**, 480–487.

Duncan, P. M., Jarvis, M. F., and Freeman, F. G. (1982) Phencyclidine raises kindled seizure thresholds. *Pharmacol. Biochem. Behav.* **16**, 1009–1011.

Dunwiddie, T. V. (1985) The physiological role of adenosine in the central nervous system. *Int. Rev. Neurobiol.* **27**, 63–139.

Dunwiddie, T. V. and Worth, T. (1982) Sedative and anticonvulsant effects of adenosine in mouse and rat. *J. Pharmacol. Ther.* **220**, 70–76.

Erinoff, L. and Snodgrass, S. R. (1986) Effects of adult or neonatal treatment with 6-hydroxydopamine or 5,7-dihydroxytryptamine on locomotor activity, monoamine levels, and response to caffeine. *Pharmacol. Biochem. Behav.* **24**, 1039–1045.

Evans, D. B., Schenden, J. A., and Bristol, J. A. (1982) Adenosine receptors mediating cardiac depression. *Life Sci.* **31**, 2425–2432.

Evans, M. C., Swan, J. H., and Meldrum, B. S. (1987) An adenosine analog, 2-chloroadenosine protects against long term development of ischemic cell loss in the rat hippocampus. *Neurosci. Lett.* **83**, 287–292.

Fastbom, J., Pazos, A., and Palacios, J. M. (1987a) The distribution of adenosine A_1 receptors and 5'-nucleotidase in the brain of some commonly used experimental animals. *Neuroscience* **27**, 813–826.

Fastbom, J., Pazos, A., Probst, A., and Palacios, J. M. (1987b) Adenosine A_1 receptors in human brain: A quantitative autoradiography study. *Neuroscience* **22**, 827–839.

Ferkany, J. W., Valentine, H. L., Stone, G. A., and Williams, M. (1986) Adenosine A_1 receptors in mammalian brain: Species differences in their interactions with agonists and antagonists. *Drug Dev. Res.* **9**, 85–93.

Ferrer, I., Costell, M., and Grisolia, S. (1982) Lesch-Nyhan syndrome-like behavior in rats from caffeine ingestion: Changes in HGPRTase activity, urea and some nitrogen metabolism enzymes. *FEBS Lett.* **141**, 275–278.

File, S. A., Baldwin, H. A., Johnston, A. L., and Wilks, L. J. (1988) Behavioral effects of acute and chronic administration of caffeine in the rat. *Pharmacol. Biochem. Behav.* **30**, 809–815.

Finn, I. B. and Holtzman, S. G. (1987) Pharmacologic specificity of tolerance to caffeine-induced stimulation of locomotor activity. *Psychopharmacology* **93**, 428–434.

Finn, I. B. and Holtzman, S. G. (1988) Tolerance and cross-tolerance to theophylline-induced stimulation of locomotor activity in rats. *Life Sci.* **42,** 2475–2482.

Foster, A. C., Gill, R., Iversen, L. L, and Woodruff, G. N. (1987) Systemic administration of MK-801 protects against ischemia-induced hippocampal neurodegeneration in the gerbil. *Br. J. Pharmacol.* **90,** 90.

Fredholm, B. and Dunwiddie, T. V. (1988) How does adenosine inhibit transmitter release? *Trends in Pharmacol. Sci.* **9,** 130–134.

Fredholm, B. B. and Hedqvist, P. (1980) Modulation of neurotransmission by purine nucleotides and nucleosides. *Biochem. Pharmacol.* **29,** 1635–1643.

Fredholm, B. B., Fuxe, K., and Agnati, L. (1976) Effects of some phosphodiesterase inhibitors on central dopamine mechanisms. *Eur. J. Pharmacol.* **38,** 31–38.

Fredholm, B. B., Herrara-Marschits, A., Jonzon, B., Lindstrom, K., and Ungerstedt, U. (1983) On the mechanism by which methylxanthines enhance apomorphine induced rotational behavior in the rat. *Pharmacol. Biochem. Behav.* **19,** 535–541.

Fuxe, K. and Ungerstedt, U. (1974) Action of caffeine and theophylline on supersensitive dopamine receptors: Considerable enhancement of receptor responses to treatment with dopa and dopamine. *Med. Biol.* **52,** 48–54.

Gasser, T., Reddington, M., and Schubert, P. (1988) Effect of carbamazepine on stimulus-evoked Ca^{2+} fluxes in rat hippocampal slices and its interaction with A_1-adenosine receptors. *Neurosci. Lett.* **91,** 189–193.

Geiger, J. D. (1986) Localization of [^{3}H]cyclohexyladenosine and [^{3}H]nitrobenzylthioinosine binding sites in rat striatum and superior colliculus. *Brain Res.* **363,** 404–408.

Geiger, J. D. and Glavin, G. B. (1985) Adenosine receptor activiation in brain reduces stress-induced ulcer formation. *Eur. J. Pharmacol.* **115,** 185–190.

Geiger, J. D. and Nagy, J. I. (1984) Heretogenous distribution of adenosine transport sites labelled by [^{3}H]nitrobenzylthioinosine in rat brain: An autoradiographic and membrane binding study. *Brain Res.* **13,** 657–666.

Geiger, J. D. and Nagy, J. I. (1990) Adenosine deaminase and [^{3}H] nitrobenzylthioinosine as markers of adenosine metabolism and transport in central purinergic systems, in *Adenonine and Adenonine Receptors* (Williams, M., ed.) Humana, Clifton, New Jersey, in press.

Gilbert, R. M. (1981) Caffeine: Overview and anthology, in *Nutrition and Behavior,* (Miller, S. A. ed.), Franklin Inst., Philadelphia, pp. 145–166.

Glowa, J. R. and Spealman, R. D. (1984) Behavioral effects of caffeine, N^6-(L-phenylisopropyl)adenosine and their combination in the squirrel monkey. *J. Pharmacol. Exp. Ther.* **231,** 665–670.

Glowa, J. R., Sobel, E., Malaspina, S., and Dews, P. B. (1985) Behavioral effects of caffeine, (-)N-((R)-1-methyl-2-phenylethyl)-adenosine (PIA) and their combination in the mouse. *Psychopharmacology* **87,** 421–424.

Goldberg, M. P., Monyer, H., Weiss, J. H., and Choi, D. W. (1988) Adenosine reduces cortical neuronal injury induced by oxygen or glucose deprivation. *Neurosci. Lett.* **89,** 323–327.

Goldberg, M. R., Curatolo, P. W., Tung, C. S., and Robertson, D. (1982) Caffeine down-regulates beta adrenoceptors in rat forebrain. *Neurosci. Lett.* **31,** 47–52.

Goldberg, S. R., Prada, J. A., and Katz, J. L. (1985) Stereoselective behavioral effects of N^6-phenylisopropyl-adenosine and antagonism by caffeine. *Psychopharmacology* **87,** 272–277.

Goldstein, M., Kuga, S., Kusano, N., Meller, E., Dancis, J., and Schwarcz, R. (1986) Dopamine agonist induced self-mutilative biting behavior in monkeys with unilateral ventromedial tegmental lesions of the brainstem: Possible pharmacological model for Lesch-Nyhan syndrome. *Brain Res.* **367,** 114–120.

Goodman, R. R. and Snyder, S. H. (1982) Autoradiographic localization of adenosine receptors in rat brain using [^{3}H]cyclohexyladenosine. *J. Neurosci.* **2,** 1230–1241.

Goodman, R. R., Kuhar, M. J., Hester, L., and Snyder, S. H. (1983) Adenosine receptors: Autoradiographic evidence for their location on axon terminals of excitatory neurons. *Science* **220,** 967–969.

Gourley, D. R. H. and Beckner, S. K. (1973) Antagonism of morphine analgesia by adenine, adenosine and adenine nucleotides. *Proc. Soc. Exp. Biol. Med.* **144,** 774–780.

Grant, D. M., Tang, B. K., and Kalow, W. (1978) Variability of caffeine metabolism. *Clin. Pharmacol. Ther.* **33,** 591–602.

Green, R. M. and Stiles, G. L. (1986) Chronic caffeine ingestion sensitizes the A_1 adenosine receptor-adenylate cyclase system in rat cerebral cortex. *J. Clin. Invest.* **77,** 222–227.

Green, R. M., Proudfit, H. K., and Yeung, S. H. (1982) Modulation of striatal dopaminergic function by local injection of 5'-N-ethylcarboxamide adenosine. *Science* **218,** 58–61.

Griffiths, R. R. and Woodson, P. P. (1988a) Reinforcing effects of caffeine in humans. *J. Pharmacol. Exp. Ther.* **246,** 21–29.

Griffiths, R. R. and Woodson, P. P. (1988b) Caffeine physical dependence: A review of human and laboratory animal studies. *Psychopharmacology* **94,** 437–451.

Grome, J. J. and Stefanovich, V. (1986) Differential effects of methylxanthines on local cerebral blood flow and glucose utilization in the conscious rat. *Naunyn-Schmiedebergs Arch. Pharmacol.* **333,** 172–179.

Hagberg, H., Andersson, P., Lacarewicz, J., Jacobson, I., Butcher, S., and Sandberg, M. (1987) Extracellular adenosine, inosine, hypoxanthine and xanthine in relation to tissue nucleotides and purines in rat striatum during transient ischemia. *J. Neurochem.* **44,** 227–231.

Hamilton, H. W., Taylor, M. D., Steffen, R. P., Haleen, S. J., and Bruns, R. F. (1987) Correlation of adenosine receptor affinities and cardiovascular activity. *Life Sci.* **41,** 2295–2302.

Hammond, J. R., Paterson, A. R. P., and Clanachan, A. S. (1981) Benzodiazepine inhibition of site-specific binding of nitrobenzylthioinosine, an inhibitor of adenosine transport. *Life Sci.* **29,** 2207–2214.

Hamprecht, B. and Van Calker, D. (1985) Nomenclature of adenosine receptors. *Trends in Pharmacol. Sci.* **6,** 153,154.

Hardebo, J. E., Kahrstrom, J., and Owman, C. (1987) P_1- and P_2 purine receptors in brain circulation. *Eur. J. Pharmacol.* **144**, 343–352.

Harms, H. H., Wardeh, G., and Mulder, A. H. (1978) Adenosine modulates depolarization-induced release of ^{3}H-noradrenaline from slices of rat brain neocortex. *Eur. J. Pharmacol.* **49**, 305–309.

Heffner, T. G., Downsa, D. A., Bristol, J. A., Bruns, R. F., Harrigan, S. E., Moos, W. H., Sledge, K. L., and Wiley, J. N. (1985) Antipsychotic-like effects of adenosine receptor agonists. *The Pharmacologist* **21**, 293.

Heffner, T. G., Wiley, J. N., Williams, A. E., Bruns, R. F., Coughenour, L. L., and Downs, D. A. (1989) Comparison of the behavioral effects of adenosine agonists and dopamine antagonists in mice. *Psychopharmacology* **98**, 31–37.

Hindmarch, I. and Subhan, Z. (1985) A preliminary investigation of "Albert 285" HWA 285 on psychomotor performance, mood, and memory. *Drug Dev. Res.* **5**, 379–386.

Ho, I. K., Lo, H. H., and Way, E. L. (1973) Cyclic adenosine monophosphate antagonism of morphine analgesia. *J. Pharmacol. Exp. Ther.* **185**, 334–346.

Holloway, F. A., Michaelis, R. C., and Huerta, P. L. (1985a) Caffeine-phenylethylamine combinations mimic the amphetamine discriminative cue. *Life Sci.* **36**, 723–730.

Holloway, F. A., Modrow, H. E., and Michaelis, R. C. (1985b) Methylxanthine discrimination in the rat: Possible benzodiazepine and adenosine mechanisms. *Pharmacol. Biochem. Behav.* **22**, 815–824.

Holmgren, M., Hedner, J., Mellstrand, T., Nordberg, G., and Hedner, T. (1986) Characterization of the antinociceptive effects of some adenosine analogues in the rat. *Naunyn-Schmiedebergs Arch. Pharmacol.* **334**, 290–293.

Holtzman, S. G. (1986) Discriminative stimulus properties of caffeine in the rat: Noradrenergic mediation. *J. Pharmacol. Exp. Ther.* **239**, 706–714.

Holtzman, S. G. (1987) Discriminative stimulus effects of caffeine: Tolerance and cross-tolerance with methylphenidate. *Life Sci.* **40**, 381–389.

Hughes, R. N. and Beveridge, I. J. (1986) Behavioral effects of prenatal exposure to caffeine in rats. *Life Sci.* **38**, 861–868.

Hutchison, A. J., Webb, R. L., Oei, H. H., Ghai, G. R., Zimmerman, M. B., and Williams, M. (1989) CGS 21680, an A_2 selective adenosine receptor agonist with preferential hypotensive activity. *J.Pharmacol. Exp. Ther.* **251**, 47–55.

Jacobson, K. A., Kirk, L., Padgett, W. L., and Daly, J. W. (1985) Functionalized congeners of 1,3-dialkylxanthines: Preparation of analogs with high affinity for adenosine receptors. *J. Med. Chem.* **28**, 1334–1350.

Jacobson, K. A., Ukena, D., Kirk, K. L., and Daly, J. W. (1986) [^{3}H]Xanthine amine congener of 1,3-dipropyl-8-phenylxanthine: An antagonsit radioligand for adenosine receptors. *Proc. Natl. Acad. Sci. USA* **83**, 4089–4092.

Jarvis, M. F. (1988) Autoradiographic localization and characterization of brain adenosine receptor subtypes, in *Receptor Localization: Ligand Autoradiography,* (Leslie, F. and Altar, C. A., eds.), Alan R. Liss, New York, pp. 95–113.

Jarvis, M. F. and Freeman, F. G. (1983) The effects of naloxone and interstimulation interval on postictal depression in kindled seizures. *Brain Res.* **288**, 235–241.

Jarvis, M. F. and Williams, M. (1986) Intrastriatal administration of quinolinic acid but not 6-hydroxydopamine (6-OHDA) reduces adenosine receptor density. *Neurosci. Abstr.* **12**, 223.6.

Jarvis, M. F. and Williams, M. (1987) Adenosine and dopamine function in the CNS. *Trends Pharmacol. Sci.* **8**, 330–332.

Jarvis, M. F. and Williams, M. (1988) Differences in adenosine A_1 and A_2 receptor density revealed by autoradiography in methylxanthine-sensitive and insensitive mice. *Pharmacol. Biochem. Behav.* **30**, 707–714.

Jarvis, M. F. and Williams, M. (1989) Direct autoradiographic localization of adenosine A_2 receptor in the rat brain usine the A_2-selective agonist [3H]CGS 21680. *Eur. J. Pharmacol.* **168**, 243–246.

Jarvis, M. F., and Williams, M. (1989) Direct autoradiographic localization of Adenosine A_2 receptors in the rat brain using the A_2 = selective agonist, CGS 21680 (1989) *Eur. J> Pharmacol.* **168**, 243–246.

Jarvis, M. F., Jackson, R. H., and Williams, M. (1989) Autoradiographic characterization of high affinity adenosine A_2 receptors in the rat brain. *Brain Res.* **484**, 111–118.

Jarvis, M. F., Schulz, R., Auchison, A. J., Do, U. H., Sills, M. A., and Williams, M. (1989b) [^{3}H] CGS 21680, a selective A_2 adenosine receptor agonist directly labels A_2 receptors in rat brain. *J. Pharmacol. Exp. Ther.* **251**, in press.

Jarvis, M. F., Jacobson, K. A., and Williams, M. (1987) Autoradiographic localization of adenosine A_1 receptors in the rat brain using [^{3}H]XCC, a functionalized congener of 1,3-dipropylxanthine. *Neurosci. Lett.* **81**, 69–74.

Jarvis, M. F., Murphy, D. E., Williams, M., Gerhardt, S. C., and Boast, C. A. (1988a) The novel *N*-methyl-D-aspartate (NMDA) antagonist CGS 19755 prevents ischemia-induced reductions of adenosine A_1, NMDA, and PCP receptors in gerbil brain. *Synapse* **2**, 577–584.

Jarvis, M. F., Stone, G. A., Williams, M., and Sills, M. A. (1988b) Characterization of [^{3}H]CGS 15943A binding to adenosine A_1 receptors in the rat cortex. *The Pharmacologist* A210.

Katims, J. J., Annau, Z., and Snyder, S. H. (1983) Interactions in the behavioral effects of methylxanthines and adenosine dervatives. *J. Pharmacol. Exp. Ther.* **227**, 167–173.

Klawans, H. L., Moses, H., and Beaulieu, D. M. (1974) The influence of caffeine on D-amphetamine and apomorphine-induced stereotyped behavior. *Life Sci.* **14**, 1493–1500.

Kleven, M. S. and Sparber, S. B. (1987) Attenuation of isobutylmethylxanthine-induced supression of operant behavior by pretreatment of rats with clonidine. *Pharmacol. Biochem. Behav.* **28**, 235–241.

Kopin, I. J. (1981) Neurotransmitters in Lesch-Nyhan syndrome. *New Eng. J. Med.* **305**, 1148–1150.

Kuhar, M. J., De Souza, E. B., and Unnerstall, J. R. (1986) Neurotransmitter receptor mapping by autoradiography and other methods. *Annu. Rev. Neurosci.* **9**, 27–59.

Kulkarni, S. K. and Mehta, A. K. (1985) Purine nucleoside-mediated immobility in mice: Reversal by antidepressants. *Psychopharmacology* **85**, 460–463.

Laska, E. M., Sunshine, A., Mueller, F., Elvers, W. B., Siegel, C., and Rubin, A. (1984) Caffeine as an analgesic adjuvant. *J. Am. Med. Assoc.* **251,** 1711–1718.

Lee, K. S. and Reddington, M. (1986a) Autoradiographic evidence for multiple CNS binding sites for adenosine derivatives. *Neuroscience* **19,** 535–549.

Lee, K. S. and Reddington, M. (1986b) 1,3-Dipropyl-8-cyclopentylxanthine (DPCPX) inhibition of [³H]*N*-ethylcarboxamidoadenosine (NECA) binding allows the visualization of putative non-A₁ adenosine receptors. *Brain Res.* **368,** 394–398.

Leid, M., Schimerlik, M. I., and Murray, T. F. (1988) Characterization of agonist radioligand interactions with porcine atrial A₁ adenosine receptors. *Mol. Pharmacol.* **34,** 334–339.

Lewis, M. E., Patel, J., Edley, S. M., and Marangos, P. J. (1981) Autoradiographic visualization of rat brain adenosine receptors using *N*⁶-cyclohexyl-[³H]adenosine. *Eur. J. Pharmacol.* **73,** 109,110.

Lloyd, H. G. E. and Stone, T. W. (1985) Cyclohexyladenosine binding in rat stiratum. *Brain Res.* **334,** 385–388.

Lloyd, K. G., Hornykiewicz, O., Davidson, L., Shannak, K., Farley, I., Coldstein, M., Shibuya, M., Kelley, W. N., and Fox, I. H. (1981) Biochemical evidence of dysfunction of brain neurotransmitters in Lesch-Nyhan syndrome. *N. Eng. J. Med.* **305,** 1106–1111.

Logan, L. and Carney, J. M. (1984) Antagonism of the behavioral effects of L-phenylisopropyladenosine (L-PIA) by caffeine and its metabolites. *Pharmacol. Biochem. Behav.* **21,** 375–379.

Logan, L., Seale, T. W., and Carney, J. M. (1986) Inherent differences in sensitivity to methylxanthines among inbred mice. *Pharmacol. Biochem. Behav.* **24,** 1281–1286.

Lohse, M. J., Klotz, K. N., Jakobs, K. H., and Schwabe, U. (1985) Barbiturates are selective antagonists at A₁ adenosine receptors. *J. Neurochem.* **45,** 1761–1770.

Lohse, M. J., Klotz, K., Lindenborn-Fotinos, J., Reddington, M., Schwabe, U., and Olson, R. (1987a) 8-Cyclo-1,3-dipropylxanthine (DPCPX)—A selective high affinity antagonist radioligand for A₁ receptors. *Naunym-Schmiedebergs Arch. Pharmacol.* **336,** 204–210.

Lohse, M. J., Boser, S., Klotz, K. N., and Schwabe, U. (1987b) Affinities of barbiturates for the GABA-receptor complex and A₁ adenosine receptors: A possible explanation of their effects. *Naunyn-Schmiedebergs Arch. Pharmacol.* **336,** 211–217.

Loke, W. H., Hinrichs, J. V., and Ghoneim, M. M. (1985) Caffeine and diazepam: Separate and combined effects on mood, memory, and psychomotor performance. *Psychopharmacology* **87,** 344–350.

Londos, C. and Wolff, J. (1977) Two distinct adenosine-sensitive sites on adenylate cyclase. *Proc. Natl. Acad. Sci. USA* **74,** 5482–5486.

Louie, G. L., Prokocimer, P. G., Nicholls, E. A., and Maze, M. (1986) Aminophylline shortens thiopental sleep time and enhances noradrenergic neurotransmission in rats. *Brain Res.* **383,** 377–381.

Marangos, P. J. and Boulenger, J. P. (1985) Basic and clinical aspects of adrenergic neuromodulation. *Neurosci. Biobehav. Rev.* **9**, 421–430.

Marangos, P. J., Deckert, J., and Bisserbe, J. C. (1987) Central sites of adenosine action and their interaction with various drugs, in *Topics and Perspectives in Adenosine Research*, (Gerlach, E. and Becker, B. F., eds.), Springer-Verlag, Berlin, pp. 74–89.

Marangos, P. J., Paul, S. M., Parma, A. M., and Skolnick, P. (1979) Purinergic inhibition of diazepam binding to rat brain in vitro. *Life Sci.* **72**, 269–273.

Marangos, P. J., Post, R. M., Patel, J. Zander, K., Parma, A., and Weiss, S. (1983) Specific and potent interactions of carbamazepine with brain adenosine receptors. *Eur. J. Pharmacol.* **93**, 175–182.

Michaelis, M. L., Michaelis, E. K., and Myers, S. L. (1979) Adenosine modulation of synaptosmal dopamine release. *Life Sci.* **24**, 2083–2092.

Minana, M. D., Portoles, M., Jorda, G., and Grisolia, S. (1984) Lesch-Nyhan syndrome. Caffeine model: Increase of purine and pyrimidine enzymes in rat brain. *J. Neurochem.* **51**, 642–647.

Modrow, H. E., Holloway, F. A., and Carney, J. M. (1981) Caffeine discrimination in the rat. *Pharmacol. Biochem. Behav.* **14**, 683–688.

Mullane, K. and Williams, M. (1990) Adenosine and cardiovascular function, in *Adenosine and Adenosine Receptors* (Williams, M., ed.), Humana, Clifton, New Jersey, this volume.

Muller, K. and Nyhan, W. (1982) Pharmacologic control of pemoline self injurious behavior in rats. *Pharmacol. Biochem. Behav.* **16**, 957–963.

Murphy, K. M. and Snyder, S. H. (1982) Heterogenity of adenosine A_1 receptor binding in brain tissue. *Mol. Pharmacol.* **22**, 260–267.

Murray, T. F. (1982) Upregulation of rat cortical adenosine receptors following chronic administration of theophylline. *Eur. J. Pharmacol.* **82**, 113,114.

Murray, T. F. and Cheney, D. L. (1982) Neuronal location of N^6-cyclohexyl [^{3}H]adenosine binding sites in rat and guinea-pig brain. *Neuropharmacol.* **21**, 575–580.

Murray, T. F. and Szot, P. (1986) A_1 adenosine receptor mediated modulation of seizure susceptibility, in *Neurotransmitters, Seizures, and Epilepsy III* (Nistico, G., ed.), Raven, New York, pp. 341–353.

Murray, T. F., Blaker, W. D., Cheney, D. L., and Costa, E. (1982) Inhibition of acetylcholine turnover rate in rat hippocampus and cortex by intraventricular injection of adenosine analogs. *J. Pharmacol. Exp. Ther.* **222**, 550–554.

Murray, T. F., Sylvester, D., Schultz, C. S., and Szot, P. (1985) Purinergic modulation of the seizure threshold for pentylenetetrazol in the rat. *Neuropharmacology* **24**, 761–766.

Myers, S. and Pugsley, T. A. (1986) Decrease in rat striatal dopamine synthesis and metabolism in vivo by metabolically stable adenosine receptor agonists. *Brain Res.* **375**, 193–197.

Nehlig, A., Luicignani, G., Kadekaro, M., Porrino, L. J., and Sokoloff, L. (1984) Effects of acute administration of caffeine on local cerebral glucose utilization in the rat. *Eur. J. Pharmacol.* **82**, 113,114.

Newman, M. E., Zohar, J., Kalian, M., and Belmaker, R. F. (1984) The effects of chronic lithium and ECT on A_1 and A_2 adenosine receptor systems in the rat brain. *Brain Res.* **291,** 188–192.

Niglio, T., Popoli, P., Caporali, M. G., and de Carolis, S. (1988) Antiepileptic effects of N^6-L-phenylisopropyladenosine (L-PIA) on penicillin-induced epileptogenic focus in rabbits. *Pharmacol. Res. Commun.* **20,** 561–572.

Nyhan, W. L. (1973) The Lesch-Nyhan syndrome. *Annu. Rev. Med.* **24,** 41–61.

Oei, H. H., Ghai, G. R., Zoganas, H. C., Stone, G. A., Field, F. P., and Williams, M. (1989) Correlation between binding affinities for brain A_1 and A_2 receptors of adenosine agonists and antagonists and their effects on heart rate and coronary vascular tone. *J. Pharmacol. Exp. Ther.* **247,** 882–888.

Onodera, H. and Kogure, K. (1988) Differential localization of adenosine A_1 receptors in the rat hippocampus: Quantitative autoradiography study. *Brain Res.* **458,** 212–217.

Onodera, H., Sato, K., and Kogure, K. (1986) Lesions of Schaeffer's collaterals prevent ischemic death of CA1 pyramidal cells. *Neurosci. Lett.* **24,** 169–174.

Onodera, H., Sato, G., and Kogure, K. (1987) Quantitative autoradiographic analysis of muscrinic cholinergic and adenosine A_1 binding sites after transient forebrain ischemia in the gerbil. *Brain Res.* **415,** 309–322.

Patel, A., Craig, R. H., Daluge, S. M., and Linden, J. (1988) ^{125}I-BW-A844U, an antagonist radioligand with high affinity and selectivity for adenosine A_1 receptors, and ^{125}I-azido-BW-A844U, a photoaffinity label. *Mol. Pharmacol.* **33,** 585–591.

Patel, J., Marangos, P. J., Skolnick, P., Paul, S. M., and Martino, A. M. (1982) Benzodiazepines are weak inhibitors of [³H]nitrobenzylthioinosine binding to adenosine uptake sites in brain. *Neurosci. Lett.* **29,** 79–82.

Phillis, J. W. (1979) Diazepam potentiation of purinergic depression of central neurons. *Can. J. Physiol. Pharmacol.* **57,** 432–435.

Phillis, J. W. and DeLong, R. (1986) The role of adenosine in cerebral vascular regulation during reduction in perfusion pressure. *J. Pharm. Pharmacol.* **38,** 460–462.

Phillis, J. W. and O'Regan, M. H. (1988a) Benzodiazepine interaction with adenosine systems explains some anomalies in GABA hypothesis. *Trends Pharmacol. Sci.* **9,** 153,154.

Phillis, J. W. and O'Regan, M. H. (1988b) The role of adenosine in the central actions of the benzodiazepines. *Prog. Neuropsychopharmacol. Biol. Psychiat.* **12,** 389–404.

Phillis, J. W. and Stair, R. E. (1987) Ro 15-1788 both antagonizes and potentiates adenosine-evoked depression of cerebral cortical neurons. *Eur. J. Pharmacol.* **136,** 151–156.

Phillis, J. W. and Wu, P. H. (1981) The role of adenosine and its nucleotides in central synaptic transmission. *Prog. Neurobiol.* **16,** 187–193.

Phillis, J. W. and Wu, P. H. (1982) Adenosine in benzodiazepine action, in *The Pharmacology of Benzodiazepines* (Usdin, E., Skolnick, P., Tallman, J., Greenblatt, D., and Paul, S., eds.), Macmillian, London, pp. 497–507.

Phillis, J. W., Barraco, R. A., DeLong, R. E., and Washington, D. O. (1986) Behavioral characteriztics of centrally administered adenosine analogs. *Pharmacol. Biochem. Behav.* **24,** 263–270.

Popoli, P., Benedetti, M., and de Carolis, A. (1988) Anticonvulsant activity of carbamazepine and N^6-L-phenylisopropyladenosine in rabbits. Relationship to adenosine receptors in central nervous system. *Pharmacol. Biochem. Behav.* **29,** 533–539.

Porsolt, R. D., Le Pichon, M., and Jalfre, M. (1977) Depression: A new animal model sensitive to antidepressant treatments. *Nature* **226,** 730–732.

Ramkumar, V., Pierson, G., and Stiles, G. L. (1988) Adenosine receptors: Clinical implications and biochemical mechanisms. *Prog. Drug Res.* **32,** 195–247.

Ricaurte, G. A., Guillery, R. W., Seiden, L. S., Schuster, C. R., and Moore, R. Y. (1982) Dopamine nerve terminal degeneration produced by high doses of methylamphetamine in the rat brain. *Brain Res.* **235,** 93–103.

Roca, D. J., Schiller, G. D., and Farb, D. H. (1988) Chronic caffeine or theophylline exposure reduces GABA/benzodiazepine receptor interactions. *Mol. Pharmacol.* **30,** 481–485.

Rosen, J. B. and Berman, R. F. (1985) Prolonged postictal depression in amygdala-kindled rats by the adenosine analog, L-phenylisopropyladenosine. *Exp. Neurol.* **90,** 549–557.

Sanders, R. C. and Murray, T. F. (1988) Chronic theophylline exposure increases agonist and antagonist binding to A_1 adenosine receptors in rat brain. *Neuropharmacology* **27,** 757–760.

Sawynok, J. and Reid, A. (1988) Role of G-proteins and adenylate cyclase in antinociception produced by intrathecal purines. *Eur. J. Pharmacol.* **156,** 25–34.

Schechter, M. D. (1977) Caffeine potentiation of amphetamine: Implications for hyperkinetic therapy. *Pharmacol. Biochem. Behav.* **6,** 359–361.

Schechter, M. D. (1980) Caffeine potentiation of apomorphine discrimination. *Pharmacol. Biochem. Behav.* **13,** 307–309.

Seale, T. W., Johnson, P., Carney, J. M., and Rennert, O. M. (1984) Interstrain variation in acute toxic responses to caffeine among inbred mice. *Pharmacol. Biochem. Behav.* **20,** 567–573.

Seale, T. W., Johnson, P., Roderick, T. H., and Carney, J. M. (1985) A single gene difference determines relative susceptibility to caffeine-induced lethality in SWR and CBA inbred mice. *Pharmacol. Biochem. Behav.* **23,** 275–278.

Seale, T. W., Abla, K. A., Shamin, M. T., Carney, J. M. and Daly, J. W. (1988) 3,7-dimethyl-1-propargylxanthine: A potent and selective in vivo antagonist of adenosine analogs. *Life Sci.* **43,** 1671–1684.

Seale, T. W., Roderick, T. H., Johnson, P., Logan, L., Rennert, O. M., and Carney, J. M. (1986) Complex genetic determinants of susceptibitily to methylxanthine-induced locomotor activity changes. *Pharmacol. Biochem. Behav.* **24,** 1333–1341.

Skerritt, J. H., Davies, L. P., and Johnson, G. A. R. (1982) A purinergic component in the anticonvulsant action of carbamazepine. *Eur. J. Pharmacol.* **82,** 195–197.

Skerritt, J. H., Davies, L. P., and Johnson, G. A. R. (1983a) Interactions of the anticonvulsant carbamazepine with adenosine receptors. *Epilepsia* 24, 634–642.

Skerritt, J. H., Johnson, G. A. R., and Chow, S. C. (1983b) Interactions of the anticonvulsant carbamazepine with adenosine receptors 2. *Epilepsia* 24, 643–652.

Simon, R. P., Swan, J. H., Griffiths, T., and Meldrum, B. S. (1984) Blockade of NMDA receptors may protect against ischemic brain damage in the brain. *Science* 226, 850–852.

Snowhill, E. W. and Williams, M. (1986) [³H]Cyclohexyladenosine binding in rat brain: A pharmacological analysis using quantitative autoradiography. *Neurosci. Lett.* 68, 41–46.

Snyder, S. H. (1985) Adenosine as a neuromodulator. *Annu. Rev. Neurosci.* 8, 103–124.

Snyder, S. H., Katims, J. J., Annau, Z., Bruns, R. F., and Daly, J. W. (1981) Adenosine receptors and behavioral actions of methylxanthines. *Proc. Natl. Acad. Sci. USA* 78, 3260–3264.

Sollevi, A. (1986) Cardiovascular effects of adenosine in man, possible clinical implications. *Prog. Neurobiol.* 227, 319–349.

Spealman, R. D. (1988) Psychomotor stimulant effects of methylxanthines in squirrel monkeys: Relation to adenosine antagonism. *Psychopharmacology* 95, 19–24.

Spealman, R. D. and Coffin, V. L. (1988) Discriminative-stimulus effects of adenosine analogs: Mediation by adenosine A_2 receptors. *J. Pharmacol. Exp. Ther.* 246, 610–617.

Stewart, S. F. and Pugsley, T. A. (1985) Increase of rat serum prolactin by adenosine analogs and their blockade by the methylxanthine aminophylline. *Naunyn-Schmiedebergs Arch. Pharmacol.* 331, 140–145.

Stiles, G. L. (1986) A_1 adenosine receptor-G protein coupling in bovine brain membranes: Effects of guanine nucleotides, salt, and solubilization. *J. Neurochem.* 51, 1592–1598.

Stone, G. A., Jarvis, M. F., Sills, M. A., Weeks, B., Snowhill, E. W., and Williams, M. (1988) Species differences in high-affinity adenosine A_2 binding sites in striatal membranes from mammalian brain. *Drug Dev. Res.* 15, 31–46.

Stone, T. W. (1981) Physiological roles for adenosine and adenosine 5'-triphosphate in the nervous system. *Neuroscience* 6, 523–545.

Stone, T. W. (1983) Interactions of adenosine with other agents, in *Regulatory Function of Adenosine*, (Berne, R. M., Rall, T. W., and Rubio, R., eds.), Nijhoff, Boston, pp. 467–477.

Sweeney, M., White, T., and Sawynok, J. (1988) 5-Hydroxytryptamine releases adenosine from primary afferent nerve terminals in the spinal cord. *Brain Res.* 462, 346–349.

Szot, P., Sanders, R. C., and Murray, T. F. (1987) Theophylline-induced upregulation of A_1-adenosine receptors associated with reduced sensitivity to convulsants. *Neuropharmacology* 26, 1173–1180.

Thithapandha, A., Maling, H. M., and Gillette, G. R. (1972) Effects of caffeine and theophylline on activity of rats in relation to brain xanthine concentrations. *Proc. Soc. Exp. Biol. Med.* **139,** 582–586.

Ukena, D., Olsson, R. A., and Daly, J. W. (1987) Definition of subclasses of adenosine receptors associated with adenylate cyclases: Interaction of adenosine with inhibitory A_1 receptors and stimulatory A_2 receptors. *Can. J. Physiol.* **65,** 365–377.

Ukena, D., Jacobson, K. A., Kirk, K. L., and Daly, J. W. (1986) A [^{3}H]amine congener of 1,3-dipropyl-8-phenyl-xanthine: A new radioligand for A_2 receptors in human platelets. *FEBS Lett.* **199,** 269–274.

Unnerstall, J. R., Niehoff, D., Kuhar, M. J., and Palacios, J. M. (1982) Quantitative receptor autoradiography using [^{3}H]Ultrofilm: Application to multiple benzodiazepine receptors. *J. Neurosci. Methods* **6,** 59–73.

Vapaatalo, H., Onken, D., Neuvonen, P. J., and Westermann, E. (1975) Stereospecificity in some central and circulatory effects of phenylisopropyladenosine (PIA). *Arzeneimittelforsch* **25,** 407–410.

Venault, P., Chapouthier, G., Prado de Carvalho, L., Siminad, J., Morre, R., Dodd, H., and Rossier, J. (1986) Benzodiazepines impair and beta-carboline enhances performance in learning and memory tasks. *Nature* **321,** 864–866.

von Lubitz, D. K. J. E., Dambrosia, J. M., Kempski, O., and Redmond, D. J. (1988) Cyclohexyladenosine protects against neuronal death following ischemia in the CA-1 region of gerbil hippocampus. *Stroke* **19,** 1133–1139.

Wagner, G. C., Lucot, J. B., Schuster, C. R., and Seiden, L. S. (1983) Alpha-methyltryosine attenuates and reserpine increases methamphetamine-induced neuronal changes. *Brain Res.* **270,** 285–288.

Waldeck, B. (1975) Effect of caffeine on locomotor activity and central catecholamine mechanisms: A study with special reference to drug interaction. *Acta Pharmacol. Toxicol.* **36,** Supp 4, 1–23.

Watanabe, H. and Uramoto, H. (1986) Caffeine mimics dopamine receptor agonists without stimulation of dopamine receptors. *Neuropharmacology* **25,** 577–581.

Watanabe, H., Ikeda, M., and Watanabe, K. (1981) Properties of rotational behavior produced by methylxanthine derivatives in mice with unilateral striatal 6-hydroxydopamine-induced lesions. *J. Pharm. Dyn.* **4,** 301–307.

Watkins, J. C. (1981) Pharmacology of excitatory amino acid receptors, in *Glutamate: Transmitter in the Central Nervous System*, (Roberts, P. J., Strom-Mathisen, J., and Johnston, G. A. R., eds.), Wiley, New York, pp. 1–24.

Weber, R. G., Jones, C. R., Palacios, J. M., and Lohse, M. J. (1988) Autoradiographic visualization of A_1-adenosine receptors in brain and peripheral tissues of rat and guinea-pig using ^{125}I-HPIA. *Neurosci. Lett.* **87,** 215–220.

Weir, R. L., Padgett, W., Daly, J. W., and Anderson, S. M. (1984) Interaction of anticonvulsant drugs with adenosine receptors in the central nervous system. *Epilepsia* **25,** 492–498.

Westerberg, V. S. and Geiger, J. D. (1987) Central effects of adenosine analogs on stress-induced gastric ulcer formation. *Life Sci.* **41,** 2201–2205.

White, B. C. and Keller, G. E. (1984) Caffeine pretreatment: Enhancement and attenuation of D-amphetamine-induced activity. *Pharmacol. Biochem. Behav.* **20**, 383–386.

White, B. C., Simpson, C. C., Adams, J. E., and Harkins, D. (1978) Monoamine synthesis and caffeine-induced locomotor activity. *Neuropharmacology* **17**, 511–513.

White, B. C., Haswell, K. L., Kassab, C. D., Harkins, D., and Crumbie, P. M. (1984) Caffeine reduces amphetamine-induced activity in asymmetrical interaction. *Pharmacol. Biochem. Behav.* **20**, 387–389.

Williams, M. (1984) Mammalian central adenosine receptors, in *Handbook of Neurochemistry*, (Lajtha, A. ed.), Plenum, New York, vol 6, pp. 1–26.

Williams, M. (1987) Purine receptors in mammalian tissues: Pharmacology and physiological significance. *Annu. Rev. Pharmacol. Toxicol.* **27**, 315–345.

Williams, M. (1989) Adenosine: The prototypic neuromodulator. *Neurochem. Int.* **14**, 249–264.

Williams, M. (1990) Adenosine: A historical overview, in *Adenosine and Adenosine Receptors*, (Williams, M., ed.), Humana, Clifton, New Jersey, this volume.

Williams, M. and Jarvis, M. F. (1988) Adenosine antagonists as potential therapeutic agents. *Pharmacol. Biochem. Behav.* **29**, 433–441.

Williams, M. and Olsen, R. A. (1988) Benzodiazepine receptors and tissue function, in *Receptor Pharmacology and Function* (Williams, M., Glennon, R.A., and Timmermans, P. B. M. W. M., eds.), Marcel Dekker, New York, pp. 385–413.

Williams, M., Risley, E. A., and Huff, J. R. (1981) Interaction of putative anxiolytic agents with central adenosine receptors. *Can. J. Physiol. Pharmacol.* **59**, 897–900.

Williams, M., Risley, E. A., and Robinson, J. R. (1983) Chronic in vivo treatment with desmethylimipramine and mianserin does not alter adenosine A_1 radioligand binding in rat cortex. *Neurosci. Lett.* **35**, 47–51.

Williams, M., Abreu, M. E., Jarvis, M. F., and Noronha-Blob, C. (1987a) Characterization of adenosine receptors in the PC12 pheochromocytoma cell line using radioligand binding-evidence for A_2 selectivity. *J. Neurochem.* **48**, 498–502.

Williams, M., Jarvis, M. F., Sills, M. A., Ferkany, J. W., and Braunwalder, A. F. (1987b) Biochemical characterization of the antagonist actions of the xanthines, PACPX (1,3-dipropyl-8(2-amino-4-chloro)phenylxanthine) and 8-PT (8-phenyltheophylline) at adenosine A_1 and A_2 receptors in rat brain tissue. *Biochem. Pharmacol.* **36**, 4024–4027.

Winn, R. H., Rubio, G. R., and Berne, R. M. (1981) The role of adenosine in the regulation of cerebral blood flow. *J. Cerebr. Blood Flow Metab.* **1**, 239–247.

Winsky, L. and Harvey, J. A. (1987) Effects of N^6-(L-phenylisopropyl)adenosine, caffeine, theophylline and rolipram on acquisition of conditioned responses in the rabbit. *J. Pharmacol. Exp. Ther.* **241**, 223–229.

Wojcik, W. J. and Neff, N. H. (1983) Differential location of adenosine A_1 and A_2 receptors in rat striatum. *Neurosci. Lett.* **41**, 55–60.

Yarbrough, G. G. and McGuffin-Clineschmidt, J. C. (1981) In vivo behavioral as-

sessment of central nervous system purinergic receptors. *Eur. J. Pharmacol.*
76, 137–144.

Yeung, S. H. and Green, R. D. (1984) [³H]5'-*N*-ethylcarboxamide adenosine binds
to both Ra and Ri adenosine receptors in rat striatum. *Naunyn-Schmiedebergs
Arch. Pharmacol.* **325,** 218–225.

Zielke, H. R. and Zielke, C. L. (1986) Lack of a sustained effect on catecholamines
or indoles in mouse brain after long term subcutaneous administration of caffeine or theophylline. *Life Sci.* **39,** 565–572.

CHAPTER 12

Adenosine
and Host Defense

Modulation Through Metabolism
and Receptor-Mediated Mechanisms

Bruce N. Cronstein
and Rochelle Hirschhorn

1. Introduction

Recent studies indicate that much as neurons communicate with each other by elaborating humors that activate specific cell surface receptors, the cells of the immune and host defense systems communicate through the medium of receptor–ligand interactions. Indeed, many of the receptors that are present on nervous tissue are present on neutrophils, lymphocytes, macrophages, and basophils (Goetzl, 1985).

Interest in the immunomodulatory effects of adenosine was sparked after the discovery that hereditary deficiency of the enzyme, adenosine deaminase (ADA), was associated with Severe Combined Immune Deficiency Disease (SCID) (Gilbert et al., 1972). It was quickly noted that adenosine and other substrates for this enzyme accumulated in the plasma and urine of affected patients (reviewed by Kredich and Hershfield, 1983; cf Hirschhorn et al., 1980b). The ob-

Adenosine and Adenosine Receptors Editor: Michael Williams ©1990 The Humana Press Inc.

servation that ADA deficiency was associated with immune defects was made shorly after Sattin and Rall (1970) demonstrated that adenosine interacted with specific receptors on neural cells. The majority of the evidence indicates that the immune malfunctions associated with ADA deficiency are not primarily related to the effects of adenosine acting through its receptors on lymphoid cells. However, investigation of the pathophysiologic defects in ADA deficiency has shed new light on the immunomodulatory role of adenosine.

Many laboratories have examined the effects of the purine nucleoside, adenosine, on the functions of both nonlymphoid cells and the lymphoid cells involved in immune or inflammatory responses. Nearly all of the studies to date indicate that adenosine modulates the function of monocytes, lymphocytes, neutrophils, mast cells, and basophils through interaction with adenosine A_2-type receptors present on the cell surface. The demonstration of similar receptors on all of these different cell types should not, in retrospect, have been surprising, because the cells involved in the immune and inflammatory response arise from a common stem cell source in the bone marrow. Moreover, the presence of adenosine receptors on monocytes, lymphocytes, neutrophils, basophils, and mast cells strongly suggests an important role for adenosine in modulating immune and inflammatory responses.

2. Purines and Immunodeficiency

Attention was first drawn to the relationship between purine metabolism and immune function with the demonstration by Giblett and colleagues in 1972 that two children suffering from SCID lacked ADA activity in their red blood cells (ADA–SCID). Since this initial observation, a deficiency of ADA activity has been found to be general and has been demonstrated in the peripheral blood lymphocytes and other tissues (Kredich and Hershfield, 1983).

SCID is a disease characterized by a marked susceptibility to infection. If left untreated, it usually results in death by the age of 1 or 2 years. Children suffering from this disease do not possess T cells, the lymphocytes that direct the cellular immune response, nor do they make antibodies. SCID is an inherited abnormality that is heterogeneous as to etiology; thus there are two distinct modes of inheritance.

Children may inherit the disorder either as an autosomal recessive trait or as an X-linked trait. Approximately half of the children suffering from the autosomal recessive form of this disease are deficient in ADA activity (approximately 20% of all cases of SCID; Hirschhorn and Hirschhorn, 1983; Hirschhorn et al., 1979).

In general, children suffering from ADA–SCID do not differ clinically from children suffering from SCID that is not associated with deficiency of ADA. Although the genetic defect can be demonstrated *in utero*, the clinical manifestations of SCID may not become apparent until several weeks or months after birth. The most overt clinical manifestations of SCID include overwhelming fungal, viral, or bacterial infections and failure to thrive. Lymphopenia can be found by routine clinical testing and there is an absence of nonmaternal antibodies (maternal antibodies may be present for as long as 6 mo after birth). Other associated findings include, in about 50% of patients, a characteristic bony lesion that is detected as flaring of the costochondral junctions on chest X-ray. Neurologic abnormalities occur in 10% of the patients (Hirschhorn et al., 1980a). Both renal and adrenal abnormalities have been described at autopsy, but the renal and adrenal pathology could be caused by the infections suffered by these children (Ratech et al., 1985).

The underlying immunodefiency in ADA–SCID is clinically apparent early in life, but may be progressive as T cell function deteriorates more rapidly than B cell function (Kredich and Hershfield, 1983). The pattern of progressive loss of immunologic function is consistent with the accumulation of metabolites that are more toxic to T lymphocytes than B lymphocytes. Indeed, both adenosine and deoxyadenosine are present in greatly increased concentrations in the plasma of children suffering from ADA–SCID, and deoxyadenosine is massively increased in the urine of these children as well (Mills et al., 1976; Simmonds et al., 1978; Kuttesch et al., 1978; Schmalstieg et al., 1978; Hirschhorn et al., 1982). Deoxy-ATP accumulates in the red blood cells (500-fold increase) and lymphocytes of children with ADA–SCID (Cohen et al., 1978; Donofrio et al., 1978; Coleman et al., 1978; Hirschhorn et al., 1980b). Hirschhorn et al., (1982) have also demonstrated modified adenine nucleosides that are either unique to or present in greatly increased amount in the urine of children with ADA–SCID. Many of the nucleoside substrates of ADA inhibit lymphocyte proliferation in vitro (Kredich and Hershfield, 1983; Carson and Seegmiller, 1976; Hirschhorn and Sela, 1977; Schwartz et al., 1978); the most potent inhibitor of T lymphocyte proliferaion is deoxyadenosine.

Hypotheses that explain the pathogenesis of the lymphospecific toxicity in ADA–SCID include the following: cellular depletion of phosphoribosylpyrophosphate (PRPP); pyrimidine starvation; inhibition by deoxy-ATP of ribonucleotide reductase (an enzyme required for DNA synthesis); induction of chromosome breakage by deoxyadenosine; and inactivation by deoxyadenosine of the enzyme s-adenosyl homocysteine hydrolase, resulting in accumulation of s-adenosyl homocysteine and interference with essential methylation reactions. The validity of these hypotheses has been tested by studying the concentrations of metabolites in cells from ADA–SCID patients, examining the effects of metabolites on cell cultures in vitro, and by in vivo studies of animals rendered pharmacologically ADA-deficient. The bulk of the evidence from such studies supports the pathophysiologic significance of accumulation of deoxy-ATP and deoxyadenosine with subsequent inhibition of ribonucleotide reductase and/or s-adenosyl homocysteine hydrolase. It is most likely that several different pathophysiologic mechanisms account for the lymphocyte defect in children suffering from ADA– SCID (Kredich and Herschfield, 1983).

Bone marrow transplantation from a histocompatible donor can provide a total, permanent cure for this disorder. This procedure not only replaces missing lymphocytes, but also leads to markedly diminshed concentrations of potentially toxic metabolites (Parkman et al., 1975). Because suitable bone marrow donors are not always available, some patients have been treated by transfusions of normal, irradiated red blood cells, which contain ADA activity, in a form of "enzyme replacement" therapy (Polmar et al., 1976). In a recent exciting development, several children have been successfully treated with infusions of polyethylene glycol (PEG)-treated ADA. The PEG treatment stabilizes the enzyme activity and diminishes its immunogenicity (Herschfield et al., 1987). Recently, "gene therapy" has been proposed as a potential treatment for these children, since only bone marrow cells, which are relatively accessible, would require genetic alteration.

Adenosine receptors probably do not play a major role in most of the manifestations of ADA deficiency. Only the neurologic manifestations found in a minority of children suffering from ADA deficiency have been ascribed to interaction of adenine nucleosides with adenosine receptors in the CNS (Hirschhorn et al., 1980a). However, the results of experiments designed to elucidate the mechanisms by which ADA deficiency leads to immunodeficiency first suggested the presence of adenosine receptors on lymphocytes.

3. Adenosine Receptors on Lymphocytes

As noted, interest in the effects of adenosine on lymphocyte function arose from the studies of ADA-deficient lymphocytes. Unfortunately, studies designed to determine the presence and function of adenosine receptors on lymphocytes have been complicated by the increasing number of lymphocyte subsets. Just as a myriad of different immunologic functions are ascribed to lymphocytes, so an increasing number of different subsets of lymphocytes have been described. Lymphocyte subsets are recognized morphologically (e.g., large granular lymphocytes), by functional differences (e.g., "helper" and "suppressor" cells) and by the presence of various protein markers on the plasma membrane of lymphocytes. Moreover, adenosine may affect lymphocytes from one animal species and not another; thus, the results of studies using lymphocytes of one animal species may not be comparable to results obtained studying cells from a different species (van De Griend et al., 1983). Lymphocytes mature and differentiate, and at least one study indicates that expression of adenosine receptors on lymphocytes may also depend on the state of differentiation of the cells (van De Griend et al., 1983). Moreover, adenosine is involved in multiple metabolic reactions in lymphocytes and other tissues; thus an excess of adenosine might affect the metabolism of the cell, leading to a change in its function. Despite these difficulties, a review of the literature reveals that, in general, adenosine interacts with A_2 receptors on the surface of some lymphocytes to suppress or dampen the immune response.

The first studies to demonstrate that adenosine interacts with a specific receptor on lymphocytes were performed in the mid-1970s. Results of these studies suggested that adenosine interacts with an external or R type of adenosine receptor. Wolberg and colleagues (1975) first demonstrated that adenosine stimulated an increase in cyclic AMP content of lymphocytes. In a subsequent study, these workers found that, in general, the ability of various adenosine analogs to inhibit lymphocyte cytotoxicity was correlated with the ability of the analogs to increase cellular content of cyclic AMP (Wolberg et al., 1978). Moreover, the analogs that stimulated cyclic AMP accumulation were those previously shown to be active at adenosine R receptors. The concentration of adenosine and its analogs that inhibited lymphocyte function or stimulated cyclic AMP accumulation were in the micromolar range (e.g., the ID_{50} of adenosine for lymphocyte function in the study of Wolberg et al., (1978) was 6.8 ± 2.5 μM). In 1978, two different laboratories reported that adenosine stimulated cyclic AMP accumulation in human peripheral blood lymphocytes through interaction with specific adenosine receptors on these cells (Marone et al., 1978; Schwartz et al., 1978), findings consistent with the presence of adenosine receptors on lymphocytes.

Subsequently, based on similar criteria, several laboratories have demonstrated that adenosine and its analogs interact with adenosine receptors on lymphoid cells derived from various species, including humans, rats, pigs, and guinea pigs (Nordeen and Young, 1978; Birch and Polmar, 1981, 1986; Moroz and Stevens, 1980; Bonnafous et al., 1979a,b, 1981; Fredholm, 1978, 1980; Nishida et al., 1984; Sandberg, 1983; van De Griend et al., 1983). The methylxanthines, theophylline and isobutylmethylxanthine, blocked the effects of adenosine, e.g., increasing intracellular cyclic AMP concentrations or altering lymphocyte function (Moroz and Stevens, 1980; Kammer and Rudolph 1984; Marone et al., 1978; Birch and Polmar, 1981; Fredholm et al., 1978). Additionally, in several of these studies, inhibition of adenosine uptake by dipyridamole either enhanced or did not diminish the effects of adenosine on either lymphocyte function or cyclic AMP accumulation (Marone et al., 1978; Fredholm et al., 1978). The order of agonist potency (5'N-ethylcarboxamidoadenosine [NECA]> adenosine> R-N^6-phenylisopropyladenosine [R-PIA]) found for stimulation of adenyl cyclase in lymphocytes by Bonnafous and colleagues (1981), is typical of that reported for A_2 adenosine receptors (Daly, 1982). In some studies, unfractionated lymphocytes were evaluated; in others, specific lymphocyte subsets were studied; hence, making direct comparisons is difficult.

In addition to altering cyclic AMP concentrations in lymphocytes, adenosine also modulates lymphocytes function. Lymphocytes, taken as a single population, perform a variety of different immunologic functions including synthesizing antibodies, helping or suppressing antibody synthesis, and lysing cells not of host origin or cells infected by viruses. The apparent multifunctional capacity of this single cell type reflects the actual heterogeneity of lymphocytes. Lymphocytes were first divided into T and B cells by virtue of the ability of T cells to form rosettes with sheep red blood cells and the ability of B cells to synthesize and secrete immunoglobulins (Roitt, 1986). A major advance in the study of the biology of lymphocytes has been the development of monoclonal antibodies capable of separating T lymphocytes into a variety of subpopulations by the presence or absence of various antigenic determinants on their surface. The different types of T cells defined by monoclonal antibodies also differ with respect to their function.

The in vitro assays of lymphocyte function fall into two major categories: mitogenic response and functional assays (e.g., lysis of target cells or antibody synthesis). In culture, lymphocytes undergo mitosis in response to antigens (e.g., tetanus toxoid), lectins, and other chemical stimuli (e.g., pokeweed mitogen, concanavalin A, and phytohemagglutinin) or irradiated lymphocytes (either foreign or autologous). In general, the motogenic response of lymphocytes to various stimuli reflects the overall competence of lymphoid cells to mount an immune response. Lymphocyte functions include synthesis of antibodies, which is measured directly. Antibody synthesis reflects a complex interplay between B lymphocytes, which synthesize antibodies, and T lymphocytes, which modulate synthesis. There are many other functional assays of lymphocyte function, such as lysis of target cells and ability to help or suppress antibody synthesis or another function.

While investigating the immune defect associated with adenosine deaminase deficiency, several laboratories demonstrated that adenosine inhibits the mitogenic response of human peripheral blood lymphocytes to phytohemagglutinin and concanavalin A (Carson and

Seegmiller, 1976; Hirschhorn and Sela, 1977; Schwartz et al., 1978). Subsequent studies have confirmed these observations (Bessler et al., 1981; Sandberg, 1984; Samet, 1986). More interesting, Nishida and colleagues (1984) have recently demonstrated that adenosine inhibits the mitogenic response of human peripheral blood lymphocytes to foreign lymphocytes (mixed lymphocyte reaction). It is not clear from these studies whether adenosine inhibits the mitogenic response of lymphocytes by engaging its receptor or by altered metabolism.

Adenosine also dramatically affects discrete lymphocyte functions. Adenosine inhibits lysis of target cells by murine lymphocytes (Wolberg et al., 1975,1978) and natural killer cell activity (lysis of a "standard" target cell) in human peripheral blood lymphocytes (Nishida et al., 1984). Several laboratories have reported either that adenosine-sensitive T lymphocytes suppress antibody synthesis in vitro or that adenosine diminishes the ability of T cells to "help" in vitro antibody synthesis (Moroz and Stevens, 1980; Birch and Polmar, 1981,1986; Kammer et al., 1983).

Functionally distinct subsets of lymphocytes can now be recognized by the presence of distinct cell surface antigens including T4, a marker for "helper/inducer" cells, and T8, a marker for "cytotoxic/ suppressor" cells. Some of these antigens, when linked by antibodies or other multivalent ligands, will undergo a process of "capping," in which the antigens migrate to one pole of the cell and are then internalized. In 1981, Birch and Polmar observed that adenosine stimulated an increase in the proportion of T lymphocytes bearing receptors for IgG, another marker of suppressor cells. In a subsequent study, it was reported (Birch et al., 1982) that adenosine increases the proportion of a subset of T cells bearing T8 antigen and receptors for IgG while it decreases the proportion of cells in this same subset expressing the T4 antigen. The effects of adenosine on T cell function are correlated with the change in surface antigens, i.e., diminished "helper" cell function. Kammer and Rudolph (1984) have observed that adenosine increases the rate at which lymphocytes cap T3, T4, and T8 antigens. Moreover, the effect of adenosine on capping is most likely mediated by stimulation of cyclic AMP accumulation, since Ro-20-1724, a nonmethylxanthine phosphodiesterase inhibitor, enhances the effects of adenosine. Schultz et al. (1988) have recently published the results of ligand binding studies which indicate that human T cells possess adenosine A_2 receptors.

Many of the studies cited above are consistent with the hypothesis that the functional effects of adenosine result from receptor–ligand interactions at the cell surface. However, recent studies indicate that only some of the effects of adenosine on lymphocyte function result from interaction of adenosine with its receptor on lymphocytes. Birch and Polmar (1986) reported that only diminished expression of T4 antigen by adenosine was mediated by interaction of adenosine with its receptor. Increased expression of T8 and receptors for IgG requires uptake of adenosine by the lymphocytes. It was further found that the effects of the purine on T helper function were mediated both by adenosine uptake and by interaction with adenosine receptors on the lymphocytes. Samet (1986) has recently suggested that adenosine does not inhibit the mitogenic response of murine splenocytes by interacting with adenosine receptors, since 8-phenyltheophylline does not reverse the effects of adenosine or 2-chloroadenosine.

Thus, adenosine receptors of the A_2 type are present on lymphocytes from different species of animals. Occupancy of adenosine receptors leads to alterations in the phenotypic expression of various lymphocyte surface markers, inhibits mitogenesis, and modulates lymphocyte function. The presence of adenosine receptors on the surface of lymphocytes suggests that, whereas some of the effects of adenosine are mediated by as yet unidentified intracellular events, at least some of the immunologic effects of adenosine are receptor-mediated.

4. Adenosine Receptors on Monocytes

Monocytes are multicomponent cells that are critically involved in directing the immune response and participating in host defense. These cells ingest invading bacteria, fungi, or foreign cells, and they secrete various proteins that are required for immune and inflammatory responses (e.g., complement components, Tumor Necrosis Factor (TNF) and Interleukin 1 [IL-1]) or contribute to inflammation (e.g., procoagulant activity). In addition, these cells, upon stimulation, are capable of generating various oxygen metabolites, including superoxide anion and H_2O_2, that are toxic to both invading pathogens and normal cells. The interaction of monocytes with lymphocytes is required for generation of appropriate immune responses. Results of recent studies indicate that monocytes also possess adenosine A_2 receptors. Activation of these receptors by appropriate ligands inhibits monocyte function.

The presence of adenosine A_2 receptors on monocytes was first demonstrated in 1984 (Lappin and Whaley, 1984). Adenosine and NECA (but not R-PIA), in a dose-dependent manner, inhibited the release of C2 (a complement component) by cultured human monocytes. In this study, theophylline partially reversed adenosine-mediated inhibition of monocyte function, but dipyridamole did not alter the effects of adenosine on monocyte function. Inhibition of C2 release by adenosine and its analogs correlated with stimulation of cyclic AMP accumulation in monocytes. Results of later studies indicate that adenosine and its analogs inhibit lysosomal enzyme release and superoxide anion generation by stimulated monocytes, confirm the presence of adenosine A_2 receptors on monocytes, and indicate that adenosine modulates superoxide anion generation by activating adenosine receptors (Riches et al., 1985; Elliott et al., 1986; Leonard et al., 1987).

As with lymphocytes, there may be multiple mechanisms by which adenosine modulates monocyte function. There is some evidence to suggest that adenosine mediates its effects on monocyte function by inhibiting transmethylation reactions. Pike and coworkers (1978) reported that high concentrations of adenosine inhibit human monocyte chemotaxis. In a subsequent study, these same investigators reported that adenosine inhibited superoxide anion generation by human monocytes (Pike and Snyderman, 1982). In these experiments, the effects of adenosine were studied in the presence of high concentrations of homocysteine and EHNA (an inhibitor of adenosine metabolism), and it was concluded that the mechanism by which adenosine inhibits chemotaxis and superoxide anion generation is inhibition of transmethylation reactions. A subsequent study by Sung and Silverstein (1985) did not confirm any correlation between inhibition of transmethylation reactions and inhibition of phagocytosis by murine macrophages. The possible role of adenosine receptors in mediating the effects of adenosine on monocytes was not evaluated.

Once monocytes migrate into tissues, they further differentiate into tissue macrophages with different characteristics that depend on the tissue in which they reside (e.g., Kupffer cells in the liver or pulmonary alveolar macrophages). Pulmonary alveolar macrophages are relatively accessible for sampling and study. Hasday and Sitrin (1985) have reported that adenosine analogs inhibit secretion of procoagulant activity and stimulate secretion of plasminogen activator by human alveolar macrophages in a manner consistent with the presence of adenosine A_2 receptors. Moreover, the role of endogenously secreted adenosine is further suggested by the similarity of the effects of dipyridamole to those of adenosine in monocytes.

It is apparent from the foregoing that adenosine has multiple effects on monocytes and that there are multiple mechanisms by which adenosine can modulate monocyte function. It is equally clear that the purine modulates monocyte function by activating A_2 receptors on their surface.

5. Adenosine Receptors on Neutrophils

Neutrophils are the most abundant circulating white blood cells and are usually the first cells to respond to an infectious or inflammatory stimulus. Like monocytes, neutrophils can phagocytose appropriately opsonized bacteria (coated with either specific antibodies or complement components) or other particulate matter. Upon stimulation, these cells may also secrete potentially toxic oxygen metabolites (e.g., O_2^-, H_2O_2, HOCl, OH•) and lysosomal enzymes into the supernatant medium. In order to reach extravascular foci of infection or inflammation, these cells, when stimulated, adhere to vascular endothelium and migrate towards the source of the chemoattractants. Recent studies indicate that neutrophils also possess A_2 receptors that, when activated, inhibit neutrophil function.

In 1980 it was reported (Marone et al., 1980) that neutrophils do not possess adenosine receptors, because adenosine did not inhibit release of β-glucuronidase (a constituent of azurophil granules) from stimulated neutrophils. Subsequently, adenosine was shown to inhibit generation of superoxide anion by stimulated neutrophil (Cronstein et al., 1983). Adenosine was unable to inhibit the release of lysozyme (an enzyme contained in the specific granules of neutrophils) and only moderately inhibited release of β-glucuronidase (Cronstein et al., 1983). Moreover, only extracellular adenosine was able to inhibit superoxide anion generation, since dipyridamole was found to enhance the effect of adenosine on neutrophil function. Subsequent studies confirmed the lack of inhibition by adenosine of specific granule release using a more specific and sensitive marker, Vitamin B_{12} binding protein (Cronstein et al., 1988a). Grinstein and Furuya (1986) have also reported that adenosine does not inhibit degranulation. Schrier and Imre (1986) and Schmeichel and Thomas (1987) have, however, questioned the functional specificity of adenosine inhibition with their recent report that, as opposed to earlier findings (Marone et al., 1980; Cronstein et al., 1983, 1987b; Grinstein and Furuya, 1986), adenosine analogs strikingly inhibit β-glucuronidase release. These differing results could arise from the use of different buffers in the studies, since some buffers either contain adenosine or may induce greater adenosine release than others (Cronstein et al., 1986; Cronstein, unpublished). Alternatively, the reported differences may result from the different techniques used to isolate the neutrophils, since it has been demonstrated that different isolation techniques alter the behavior of neutrophils (Haslett et al., 1985).

Subsequent studies have indicated that adenosine inhibits neutrophil function by activating A_2 receptors on neutrophils (Cronstein et al., 1985; Roberts et al., 1985a; Schrier and Imre, 1986). In all studies reported to date, NECA is the most potent inhibitor of superoxide anion generation by activated neutrophils. In addition, the effects of adenosine or NECA on superoxide anion generation by activated neutrophils can be blocked by xanthine adenosine antagonists.

Interestingly, adenosine does not inhibit neutrophil function in response to all stimuli. This possibly provides a clue as to the step in neutrophil activation that is inhibited by adenosine. Adenosine and its analogs do not inhibit generation of oxygen metabolites stimulated by the soluble agent phorbol myristate acetate, an agent that directly activates protein kinase C and thus bypasses stimulus–response coupling mechanisms at the neutrophil membrane (Cronstein et al., 1983). Additionally, the purine does not inhibit generation of oxygen metabolites stimulated by ingestion of either latex particles or immune complexes (Roberts et al., 1985a; Cronstein et al., 1987a). There is no explanation for these findings at the present time.

In addition to inhibiting superoxide anion generation, adenosine receptor agonists inhibit other neutrophil functions. Upon stimulation, neutrophils adhere to endothelial cells both in vivo and in vitro. The stable adenosine analogs, NECA and 2-chloroadenosine (2-CADO), prevent adherence of stimulated neutrophils to endothelial cells in a manner consistent with activation of A_2 receptors (Cronstein et al., 1986, 1987b). When complement is activated on the surface of neutrophils, the cells internalize those portions of the membrane that have been so attacked. 2-CADO can prevent the internalization of complement-coated membranes by neutrophils (Roberts et al., 1985b).

Surprisingly, activation of adenosine receptors on neutrophils can promote chemotaxis. Adenosine receptor activation can promote directed neutrophil migration induced by the activated complement components contained in zymosan treated serum (Rose et al., 1988). Adenosine has no apparent effect on random migration, however. These findings suggest that the purine, acting through its receptors, might influence migration of receptors for chemoattractants either within the plasma membrane or into the cell.

Pharmacologic studies of ligand receptor binding have been reported for neutrophil adenosine receptors (Cronstein et al., 1985). Binding of [^{3}H] NECA is similar to binding of [^{3}H] 2-CADO to whole neutrophils with approximately 10,000 receptor sites/cell and a K_d value of approximately 0.2 μM (Cronstein et al., 1985). Inhibition of binding by adenosine analogs correlates well with the agonist potency for superoxide anion generation; thus NECA > Adenosine> R-PIA is the order of activity of adenosine analogs for both inhibition of superoxide anion generation and for inhibition of [^{3}H] NECA binding. The ID_{50} of NECA for superoxide anion generation occurs at approximately 20% receptor occupancy. Recently, binding of [^{3}H] NECA to membrane preparations derived from HL-60 cells, a cell line derived from a promyelocytic leukemia, has been studied. [^{3}H] NECA binds to these membranes with similar characteristics as found for neutrophils (K_d = 1.5–4 μM, B_{max} = 40,000–70,000 receptor sites/cell; B.N. Cronstein, 1988b).

The mechanism by which activation of adenosine receptors alters neutrophil function has also been explored. Grinstein and Furuya (1986) have reported that receptor activation does not affect the activation of the Na^+/H^+ antiport system. Adenosine and its analogs stimulate a dose-related increase in cyclic AMP in the presence of Ro-20-1724, a non-methylxanthine phosphodiesterase inhibitor (Iannone et al., 1985; Cronstein et al., 1988a). However, the cyclic AMP increases that follow treatment of neutrophils with adenosine do not correlate temporally with the inhibition of superoxide anion generation (Cronstein et al., 1988a). Stimulation of neutrophils with the surrogate chemoattractant FMLP also leads to cyclic AMP accumulation in neutrophils, and NECA enhances the rise in cyclic AMP induced by FMLP (Cronstein et al., 1988a). The increase in cyclic AMP that follows FMLP stimulation is thought to be an autoregulatory loop within the neutrophil, since elevated cyclic AMP concentrations in the neutrophil are associated with inhibition of neutrophil function (Smolen et al., 1980). Other agents that stimulate cyclic AMP accumulation in the neutrophil, i.e., prostaglandins, inhibit all neutrophil functions, unlike adenosine, which is more selective in its effects. Studies using an inhibitor of protein kinase A (H-7) further demonstrate that cyclic AMP is not the second messenger for adenosine receptors in the neutrophil (Cronstein, unpublished data). Pasini et al. (1985) have reported that adenosine may act as a Ca^{2+} channel blocker in neutrophils. However, when measured directly, adenosine receptor activation does not appear to affect stimulated Ca^{2+}-fluxes in neutrophils, as indicated by the intracellular fluorescent Ca^{2+} probe, quin-2 (Cronstein et al., 1988a). Thus, like A$_2$ receptors in other mammalian tissues, those on neutrophils stimulate cyclic AMP accumulation. However, a lack of complete correlation between such increases and inhibition of function suggests that cyclic AMP may not play a pivotal role in neutrophil function. The mechanism(s) by which adenosine modulates either superoxide anion generation or chemotaxis is unclear and remains an area of active research.

6. Adenosine Receptors
on Mast Cells and Basophils

Basophils are peripheral blood granulocytes that migrate into the tissues, primarily the lung and gut, where they differentiate into mast cells. Mast cells, like tissue macrophages, take on different characteristics that depend on the tissue in which they reside. Both basophils and mast cells participate in acute hypersensitivity and allergic reactions (IgE-mediated reactions). Indeed, many of the manifestations of hypersensitivity are results of the release of mediators, such as histamine, from stimulated mast cells. Adenosine and its analogs modulate mast cell and basophil function through interaction with A_2 receptors on the cell surface, intracellular P receptors and inhibition of methylation reactions.

In 1978, Marquardt and colleagues reported that adenosine and 2-CADO enhance the release of histamine by rat mast cells stimulated by anti-IgE, concanavalin A, 48/80, and A23187, and that theophylline antagonized the effects of adenosine on mast cells. Many subsequent reports have confirmed the observations that adenosine and its analogs enhance the release of mediators by stimulated mast cells (Fredholm and Sydbom, 1980; Hughes et al., 1983; Holgate et al., 1980; Sydbom and Fredholm, 1982; Burt and Stanworth, 1983; Marquardt et al., 1984a; Marquardt and Wasserman, 1985) and that theophylline reverses the effects of adenosine on mast cells (Sydbom and Fredholm, 1982). Adenosine was also found to stimulate cyclic AMP accumulation in mast cells (Holgate et al., 1980; Marquardt et al., 1984a; Marquardt and Wasserman, 1985), although Sydbom and Fredholm (1982) could not demonstrate any effect of adenosine on cyclic AMP concentrations in mast cells. The order of potency for stimulation of cyclic AMP accumulation by adenosine analogs (NECA > adenosine > R-PIA) was characteristic of adenosine A_2-type receptors (Marquardt and Wasserman, 1985). Studies of [^{3}H] adenosine binding to mast cell membranes reveal a relatively high-affinity receptor with K_d of 24–28 nM (Marquardt et al., 1984a; Marquardt and Wasserman, 1985). The order of activity for inhibition of adenosine binding by adenosine analogs was also characteristic of the classical A_2 receptor. Moreover, Marquardt et al., (1984a) have demonstrated that mast cells cultured in the presence of aminophylline increase the number of adenosine receptors present on their membranes and become hyperresponsive to adenosine (Marquardt et al., 1986).

Although mast cells are derived from basophils, adenosine inhibits basophil function, in contrast to its enhancement of mast cell function. Adenosine inhibits release of histamine from stimulated human basophils (Marone et al., 1979, 1985; Morita et al., 1981; Church et al., 1983). Both the order of agonist potency for inhibition of basophil function and stimulation of cyclic AMP accumulation by adenosine receptors indicate that adenosine interacts with A_2-type receptors (Marone et al., 1979, 1985; Church et al., 1983). However, some confusion has arisen, in that Church and colleagues (1983) have reported that adenosine, acting through A_2 receptors, inhibits histamine release by stimulated basophils if the basophils are incubated with adenosine before stimulation, but enhances histamine release when adenosine is added to basophils after stimulation. For a further discussion of this issue, *see* Griffiths and Holgate (1990).

Thus, there are adenosine A_2 receptors on mast cells and basophils, activation of which may have opposing effects on basophil and mast cell function. It should be noted, however, that the basophils studied were of human origin, and the mast cells were obtained from rodents. The discrepancy between the reported effects of adenosine on basophils and mast cells may result, therefore, from either species differences or the process of differentiation that occurs after basophils enter the tissues.

Adenosine may modulate mast cell and basophil function by other mechanisms in addition to interaction with adenosine A_2 receptors. The purine may interact with P receptors to inhibit mast cell and basophil function (Burt and Stanworth, 1983; Church et al., 1983; Hughes and Church, 1986), and as with monocytes, high concentrations of adenosine may also affect basophil function by inhibiting transmethylation reactions (Morita et al., 1981).

7. Adenosine Receptors in Autoimmune Disorders

In general, activation of adenosine receptors on leukocytes diminishes immune and inflammatory responses. Thus, one might predict that diminished adenosine receptor function might underlie or contribute to diseases in which there is an excess of inflammation. Indeed, Mandler and coworkers (1982) reported that T lymphocytes from patients suffering from Systemic Lupus Erythematosus (SLE) do not respond appropriately to adenosine. Subsequent reports from this same laboratory have confirmed and expanded this observation (Kammer et al., 1983).

SLE is a systemic disease that usually occurs in young women of child-bearing age and is characterized by abnormal self-directed immune responses. Patients suffering from SLE may have inflammation of their pleural or peritoneal spaces, joints, blood vessels, kidneys, and skin. Affected individuals make antibodies to a wide variety of cellular and nuclear antigens. Many patients with SLE also suffer from seizure disorders, psychosis, or other nervous system manifestation.

Mandler and colleagues (1982) reported that adenosine did not diminish the ability of T cells from SLE patients to "help" antibody synthesis as it did in normal controls. Nonhydrolyzable cyclic AMP analogs modulated T cell function normally in cells from SLE patients. Thus, it was concluded that there was a defect in the ligand-receptor or receptor–adenylate cyclase interaction in T lymphocytes from such patients. Kammer et al., (1983) reported that there was improvement of defective adenosine responses in SLE patients when the disease was in remission. Kammer (1983) also found that adenosine, acting through its receptor on T lymphocytes, does not enhance capping of surface antigens in patients with SLE. However, this abnormality does not improve when the disease is in remission (Schultz et al., 1988). Although adenosine receptors do not appear to function normally in patients with SLE, it is not clear whether the receptor dysfunction is secondary to the disease or an intrinsic manifestation of SLE. Moreover, the subset of T cells in which the abnormality of adenosine receptors was described is not a homogeneous group of cells (e.g., these cells manifest different cell surface antigens).

It is tempting to speculate that abnormalities of adenosine receptors, either intrinsic or acquired, underlie some of the other manifestations of SLE. Some of the abnormalities of CNS function found in SLE patients might be explained by adenosine receptor dysfunction as well as by immune abnormalities. Adenosine receptor dysfunction in the peripheral arteries might also contribute to the cold- or stress-induced vasospasm of the digits (Raynaud's phenomenon) so common in patients suffering from SLE. However, in a recent study (Schultz et al., 1988), no difference between SLE patients and normals was found when ligand binding to A_2 receptors on T cells was studied.

8. Adenosine Is a Modulator of Inflammation—A Hypothesis

Many different cell types release adenosine into their surrounding medium in vitro. Indeed, neutrophils (Cronstein et al., 1983; Worku and Newby, 1983; Newby et al., 1983; Mann et al., 1986), lymphocytes (Fredholm et al., 1978; Sandberg, 1983), and mast cells and basophils (Fredholm and Sydbom, 1980; Marquardt et al., 1984b; Sydbom and Fredholm, 1982) all release adenosine into the medium. More important, adenosine release by some cells may be a cellular response to injury or stimulation. Newby and coworkers (1983) have reported that cultured cardiac myocytes also release adenosine and that the release of adenosine is proportional to the "energy charge" of the cells. Thus, myocytes that are injured or hypoxic would be expected to release more adenosine into the medium. Moreover, some noxious stimuli (H_2O_2) induce adenosine release (Ager and Gordon, 1984). Dying cells, such as those found at sites of tissue damage, lyse and release their intracellular stores of adenine nucleotides, which are rapidly metabolized by phosphatases and nucleotidases to adenosine (Pearson and Gordon, 1979). The high concentrations of adenosine present at or near sites of inflammation could prevent inflammatory cells from damaging the surrounding, unaffected tissues.

Since control of infection is the primary role of the host defense system, it is not surprising that there are pathways for elimination of adenosine in areas where leukocytes should function maximally. At sites of chronic infection and inflammation, such as pleural and peritoneal effusions caused by tuberculosis, ADA activity is greater than in noninfected fluids (Piras et al., 1978; Ocana et al., 1983; Pettersson et al., 1984; Martinez-Vazquez et al., 1986). Cronstein and coworkers (1987a) have recently shown that ADA may act as an opsonin and adhere to phagocytic particles (serum-treated zymosan particles). In addition, ADA activity is associated with phagocytic vacuoles in elicited murine peritoneal macrophages (Tritsch et al., 1985). Thus, the presence of ADA within the phagolysosome, by virtue of its enzymatic activity, would decrease adenosine concentrations where the cell contacts pathogenic organisms, without diminishing the extracellular adenosine concentration. Lower adenosine concentrations in the phagolysosome would permit secretion of greater quantities of potentially bactericidal agents, e.g., H_2O_2, at the site where they will be most beneficial. Higher adenosine concentrations in the extracellular milieu could prevent release of toxic mediators into the extracellular milieu, thus protecting surrounding undamaged tissues.

Conclusion

Adenosine and its analogs modulate the function of all types of leukocytes. Although there are many mechanisms by which adenosine can affect cellular function, it is clear that adenosine mediates at least some of its actions on leukocytes by interactions with A_2-type adenosine receptors. Adenosine A_1 receptors have not been demonstrated on leukocytes to date, but they may be present on white blood cells. The demonstration that adenosine is released from damaged cells or tissues suggests a novel role for adenosine as an endogenous antiinflammatory agent. The diminished immune function associated with adenosine excess (ADA deficiency) and the dysfunction of adenosine receptors on lymphocytes from patients with SLE further support the hypothesis that adenosine is an important regulator of both immune and inflammatory responses.

References

Ager, A. and Gordon, J. L. (1984) Differential effects of hydrogen peroxide on indices of endothelial cell function. *J. Exp. Med.* **159**, 592–603.

Bessler, H., Djaldetti, M., Kupfer, B., and Moroz, M. (1980) Two T lymphocyte subpopulations isolated from human peripheral blood following in vitro treatment with adenosine. *Biomedicine* **32**, 66–71.

Birch, R. E. and Polmar, S. H. (1981) Induction of Fc-gamma receptors on a subpopulation of human T lymphocytes by adenosine and impromidine, an H_2-histamine agonist. *Cell. Immunol.* **57**, 455–467.

Birch, R. E. and Polmar, S. H. (1986) Adenosine induced immunosuppression: The role of the adenosine receptor–adenylate cyclase interaction in the alteration of T-lymphocyte surface phenotype and immunoregulatory function. *Int. J. Immunopharmacol.* **8**, 329–337.

Birch, R. E., Rosenthal, A. K., and Polmar, S. H. (1982) Pharmacological modification of immunoregulatory T lymphocytes. II. Modulation of T lymphocytes. II. Modulation of T lymphocyte cell surface characteristics. *Clin. Exp. Immunol.* **48**, 231–238.

Bonnafous, J.-C., Dornand, J., and Mani, J.-C. (1979a) Adenosine-induced cyclic AMP increase in pig lymphocytes is not related to adenylate cyclase stimulation. *Biochim. Biophys. Acta* **587**, 180–191.

Bonnafous, J.-C., Dornand, J., and Mani, J.-C. (1979b) Hormone-like action of adenosine in mouse thymocytes and splenocytes. Evidence for the existence of membrane adenosine receptors coupled to adenylate cyclase. *FEBS Lett.* **107**, 95–99.

Bonnafous, J.-C., Dornand, J., Favero, J., and Mani, J.-C. (1981) Lymphocyte membrane adenosine receptors coupled to adenylate cyclase properties and occurrence in various lymphocyte subclasses. *J. Recept. Res.* **2**, 347–366.

Burt, D. S. and Stanworth, D. R. (1983) The effect of ribose and purine modified adenosine analogues on the secretion of histamine from rat mast cells induced by ionophore A23187. *Biochem. Pharmacol.* **32**, 2729–2732.

Carson, D. A. and Seegmiller, J. E. (1976) Effect of adenosine deaminase inhibition upon human lymphocyte blastogenesis. *J. Clin. Invest.* **57**, 274–282.

Church, M. K., Holgate, S. T., and Hughes, P. J. (1983) Adenosine inhibits and potentiates IgE-dependent histamine release from human basophils by an A_2-receptor mediated mechanism. *Br. J. Pharmacol.* **80**, 719–726.

Cohen, A. R., Hirschhorn, R., Horowitz, S. D., Rubinstein, A., Polmar, S. H., Hong, R., and Martin, D. W., Jr. (1978) Deoxyadenosine triphosphate as a potentially toxic metabolite in adenosine deaminase deficiency. *Proc. Natl. Acad. Sci. USA* **75**, 472–476.

Coleman, M. S., Donofrio, J., Hutton, J. J., Hahn, L., Daoud, A., Lampkin, B., and Dyminski, J. (1978) Identification and quantitation of adenine deoxynucleotides in erythrocytes of a patient with adenosine deaminase deficiency and severe combined immunodeficiency. *J. Biol. Chem.* **253**, 1619–1626.

Cronstein, B. N., Kramer, S. B., Weissmann, G., and Hirschhorn, R. (1983) Adenosine: A physiological modulator of superoxide anion generation by human neutrophils. *J. Exp. Med.* **158**, 1160–1177.

Cronstein, B. N., Rosenstein, E. D., Kramer, S. B., Weissman, G., and Hirschhorn, R. (1985) Adenosine: A physiologic modulator of superoxide anion generation by human neutrophils. Adenosine acts via an A_2 receptor on human neutrophils. *J. Immunol.* **135**, 1366–1371.

Cronstein, B. N., Levin, R. I., Belanoff, Weissmann, G., and Hirschhorn, R. (1986) Adenosine: An endogenous inhibitor of neutrophil-mediated injury to endothelial cells. *J. Clin. Invest.* **78**, 760–770.

Cronstein, B. N., Kubersky, S. M., Weissmann, G., and Hirschhorn, R. (1987a) Engagement of adenosine receptors inhibits hydrogen peroxide (H_2O_2) release by activated human neutrophils. *Clin. Immunol. Immunopath.* **42**, 76–85.

Cronstein, B. N., Rose, F. R., Levin, R. I., Recht, P. A., Hirschhorn, R., and Weissmann, G. (1987b) Engagement of adenosine A_2 receptors enhances neutrophil chemotaxis while inhibiting their sticking to endothelial cells. *Clin. Res.* **35**, 597a.

Cronstein, B. N., Kramer, S. B., Rosenstein, E. D., Korchak, H. M., Weissmann, G., Hirschhorn, R., and Buck, S. M. (1988a). Engagement of adenosine receptor raises cAMP alone and in synergy with engagement of the FMLP receptor and inhibits membrane depolarization. *Biochem. J.* **252**, 709–715.

Cronstein, B. N. (1988b) A human cell line of myeloid origin (HL60) possesses adenosine A_2 receptors. *Clin. Res.* **36**, 563a.

Daly, J. W. (1982) Adenosine receptors: Targets for future drugs. *J. Med. Chem.* **25**, 197–207.

Donofrio, J., Coleman, M. S., Hutton, J. J., Daoud, A., Lampkin, B., and Dyminski, J. (1978) Overproduction of adenine deoxynucleosides and deoxynucleotides in adenosine deaminase deficiency with severe combined immunodeficiency disease. *J. Clin. Invest.* **62**, 884–887.

Elliott, K. R. F., Stevenson, H. C., Miller, P. J., and Leonard, E. J. (1986) Synergistic action of adenosine and Fmet-leu-phe in raising cyclic AMP content of purified human monocytes. *Biochem. Biophys. Res. Commun.* **138**, 1376–1382.

Fredholm, B. B. and Sydbom, A. (1980) Are the anti-allergic actions of theophylline due to antagonism at the adenosine receptor? *Agents Actions* **10**, 145–147.

Fredholm, B. B., Sandberg, G., and Ernstrom, U. (1978) Cyclic AMP in freshly prepared thymocyte suspensions. Evidence for stimulation by endogenous adenosine. *Biochem. Pharmacol.* **27**, 2675–2682.

Giblett, E. R., Anderson, J. E., Cohen, F., Pollasa, B., and Meuwissen, H. J. (1972) Adenosine deaminase deficiency in two patients with severely impaired cellular immunity. *Lancet* **2**: 1067.

Goetzl, E. J. (ed.) (1985) Proceedings of a conference on neuomodulation of immunity and hypersensitivity. *J. Immunol.* **135** (Suppl.), 739s–863s.

Griffiths, T. L. and Holgate, S. T. (1990) The role of adenosine in respiratory physiology, in *Adenosine and Adenosine Receptors* (Williams, M., ed.) Humana Press, Clifton, New Jersey, this volume.

Grinstein, S. and Furuya, W. (1986) Cytoplasmic pH regulation in activated human neutrophils: Effects of adenosine and pertussis toxin on Na^+/H^+ exchange and metabolic acidification. *Biochim. Biophys. Acta* **889**, 301–309.

Hasday, J. D. and Sitrin, R. G. (1985) Modulation of alveolar macrophage effector functions by adenosine analogues. *Clin. Res.* **33**, 465a.

Haslett, C., Guthrie, L. A., Kopaniak, M. M., Johnston, R. B., Jr., and Henson, P. M. (1985) Modulation of multiple neutrophil functions by preparative methods or trace concentrations of bacterial lipopolysaccharide. *Am. J. Pathol.* **119**, 101–110.

Hershfield, M. S., Buckley, R. H., Greenberg, M. L., Melton, A. L., Schiff, R., Hatem, C., Kurtzberg, J., Markert, M. L., Kobayashi, R. H., Kobayashi, A. L., and Abuchoski, A. (1987) Treatment of adenosine deaminase deficiency with polyethylene glycol-modified adenosine deaminase. *New Engl. J. Med.* **316**, 589–596.

Hirschhorn, R. and Hirschhorn, K. (1983) Immunodeficiency disorders in *Principles and Practice of Medical Genetics* (Emery, A. E. and Rimoin, D. L., eds.) Churchill Livingstone, New York. pp. 1091–1108.

Hirschhorn, R. and Sela, E. (1977) Adenosine deaminase and immunodeficiency: An in vitro model. *Cell. Immunol.* **32**, 350–360.

Hirschhorn, R., Papageorgiou, P. S., Kesarwala, H. H., and Taft, L. T. (1980a) Amelioration of neurologic abnormalities after "enzyme replacement" in adenosine deaminase deficiency. *New Engl. J. Med.* **303**, 377–380.

Hirschhorn, R., Roegner, V., Rubinstein, A., and Papageorgiou, P. S. (1980b) Plasma deoxyadenosine, adenosine and erythrocyte deoxyATP are elevated at birth in an adenosine deaminase-deficient child. *J. Clin. Invest.* **65**, 768–771.

Hirschhorn, R., Vawter, G. F., Kirkpatrick, J. A., Jr., and Rosen, F. S. (1979) Adenosine deaminase deficiency: Frequency and comparative pathology in autosomally recessive severe combined immunodeficiency. *Clin. Immunol. Immunopathol.* **14**, 107–120.

Hirschhorn, R., Ratech, H., Rubinstein, A., Papageorgiou, P. S., Kesarwala, H., Gelfand, I., and Roegner-Maniscalco, V. (1982) Increased excretion of modified adenine nucleosides by children with adenosine deaminase deficiency. *Pediatr. Res.* **16**, 362–369.

Holgate, S. T., Lewis, R. A., and Austen, K. F. (1980) Role of adenylate cyclase in immunologic release of mediators from rat mast cells: Agonist and antagonist effects of purine- and ribose-modified adenosine analogs. *Proc. Natl. Acad. Sci. USA* **77**, 6800–6804.

Hughes, P. J. and Church, M. K. (1986) Inhibition of immunological and nonimmunological histamine release from human basophils by adenosine analogues that act at P-sites. *Biochem. Pharmacol.* **35**, 1809–1816.

Hughes, P. J., Holgate, S. T., Roath, S., and Church, M. K. (1983) The relationship between cyclic AMP changes and histamine release from basophil-rich human leucocytes. *Biochem. Pharmacol.* 2557–2563.

Iannone, M. A., Reynolds-Vaughn, R., Wolberg, C., and Zimmerman, T.P. (1985) Human neutrophils possess adenosine A_2 receptors. *Fed. Proc.* **44,** 580.

Kammer, G. M. (1983) Impaired T cell capping and receptor regeneration in active Systemic Lupus Erythematosus. Evidence for a disorder intrinsic to the T lymphocyte. *J. Clin. Invest.* **72,** 1686–1697.

Kammer, G. M. and Rudolph, S. A. (1984) Regulation of human T lymphocyte surface antigen mobility by purinergic receptors. *J. Immunol.* **133,** 3298–3302.

Kammer, G. M., Birch, R. E., and Polmar, S. H. (1983) Impaired immunoregulation in Systemic Lupus Erythematosus: Defective adenosine-induced suppressor T lymphocyte generation. *J. Immunol.* **130,** 1706–1712.

Kredich, N. M. and Hershfield, M. S. (1983) Immunodeficiency diseases caused by adenosine deaminase deficiency and purine nucleoside phosphorylase deficiency, in *The Metabolic Basis of Inherited Disease* (Stanbury, J., Wyngaarden, J. B., Frederickson, D. S., Goldstein, J. L., and Brown, M. S., eds.) McGraw Hill, New York. pp. 1157–1183.

Kuttesch, J. F., Schmalstieg, F. C., and Nelson, J. A. (1978) Analysis of adenosine and other adenine compounds in patients with immunodeficiency diseases. *J. Liquid Chromatog.* **1,** 97–109.

Lappin, D. and Whaley, K. (1984) Adenosine A_2 receptors on human monocytes modulate C_2 production. *Clin. Exp. Immunol.* **57,** 454–460.

Leonard, E. J., Shenai, A., and Skeel, A. (1987) Dynamics of chemotactic peptide-induced superoxide generation by human monocytes. *Inflammation* **11,** 229–240.

Mandler, R., Birch, R. E., Polmar, S. H., Kammer, G. M., and Rudolph, S. A. (1982) Abnormal adenosine-induced immunosuppression and cAMP metabolism in T lymphocytes of patients with Systemic Lupus Erythematosus. *Proc. Natl. Acad. Sci. USA* **79,** 7542–7546.

Mann, J. S., Renwick, A. G., and Holgate, S. T. (1986) Release of adenosine and its metabolites from activated human leucocytes. *Clin. Sci.* **70,** 461–468.

Marone, G., Findlay, S. R., and Lichtenstein, L. M. (1979) Adenosine receptor on human basophils: modulation of histamine release. *J. Immunol.* **123,** 1473–1477.

Marone, G., Plaut, M., and Lichtenstein, L. M. (1978) Characterization of a specific adenosine receptor on lymphocytes. *J. Immunol.* **121,** 2153–2159.

Marone, G., Thomas, L. L., and Lichtenstein, L. M. (1980) The role of agonists that activate adenylate cyclase in the control of cAMP metabolism and enzyme release by human polymorphonuclear leukocytes. *J. Immunol.* **125,** 2277–2283.

Marone, G., Vigorita, S., Antonelli, C., Torella, G., Genovese, A., and Condorelli, M. (1985) Evidence for an adenosine A_2/R_a receptor on human basophils. *Life Sci.* **36,** 339–345.

Marquardt, D. L. and Wasserman, S. I. (1985) [^{3}H]Adenosine binding to rat mast cells—pharmacologic and functional characterization. *Agents Action* **16,** 453–461.

Marquardt, D. L., Gruber, H. E., and Wasserman, S. I. (1986) Aminophylline exposure alters mouse bone marrow-derived mast cell adenosine responsiveness. *J. Allergy Clin. Immunol.* **78,** 462–469.

Marquardt, D. L., Parker, C. W., and Sullivan, T. J. (1978) Potentiation of mast cell mediator release by adenosine. *J. Immunol.* **120,** 871–878.

Marquardt, D. L., Walker, L. L., and Wasserman, S. I. (1984a) Adenosine receptors on mouse bone-marrow-derived mast cells: Functional significance and regulation by aminophylline. *J. Immunol.* **133,** 932–937.

Marquardt, D. L., Gruber, H. E., and Wasserman, S. I. (1984b) Adenosine release from stimulated mast cells. *Proc. Natl. Acad. Sci. USA* **81,** 6192–6196.

Martinez-Vazquez, J. M., Ocana, I., Ribera, E., Segura, R. M., and Pascual, C. (1986) Adenosine deaminase activity in the diagnosis of tuberculous peritonitis. *Gut* **27,** 1049–1053.

Mills, G. C., Schmalstieg, F. C., Trimmer, K. B., Goldman, A. S., and Goldblum, R. M. (1976) Purine metabolism in adenosine deaminase deficiency. *Proc. Natl. Acad. Sci. USA* **73,** 2867–2871.

Morita, Y., Chiang, P. K., and Siraganian, R. P. (1981) Effect of inhibitors of transmethylation on histamine release from human basophils. *Biochem. Pharmacol.* **30,** 785–791.

Moroz, C. and Stevens, R. H. (1980) Suppression of immunoglobulin production in normal human B lymphocytes by two T-cell subsets distinguished following in vitro treatment with adenosine. *Clin. Immunol. Immunopathol.* **15,** 44–51.

Newby, A. C., Holmquist, C. A., Illingworth, J., and Pearson, J. D. (1983) The control of adenosine concentration in polymorphonuclear leucocytes, cultured heart cells and isolated perfused heart from the cat. *Biochem. J.* **214,** 317–323.

Nishida, Y., Kamatani, N., Morito, T., and Miyamoto, T. (1984) Differential inhibition of lymphocyte function by 2-chloroadenosine. *Int. J. Immunopharmacol.* **6,** 335–338.

Nordeen, S. K. and Young, D. A. (1978) Refractoriness of the cyclic AMP response to adenosine and prostaglandin E_1 in thymic lymphocytes. Dependence on protein synthesis and energy-providing substrates. *J. Biol. Chem.* **253,** 1234–1239.

Ocana, I., Martinez-Vazquez, J. M., Segura, R. M., Fernandez-De-Sevilla, T., and Capdevila, J. A. (1983) Adenosine deaminase in pleural fluids. Test for diagnosis of tuberculous pleural effusion. *Chest* **84,** 51–53.

Parkman, R., Gelfand, E. W., Rosen, F. S., Sanderson, A., and Hirschhorn, R. (1975) Severe combined immunodeficiency and adenosine deaminase deficiency. *New Engl. J. Med.* **292,** 714–719.

Pasini, F. L., Cappechi, P. L., Orrico, A., Ceccatelli, L., and Di Perri, T. (1985) Adenosine inhibits polymorphonuclear leukocyte in vitro activation: A possible role as an endogenous calcium entry blocker. *J. Immunopharmacol.* **7,** 203–215.

Pearson, J. D. and Gordon, J. L. (1979) Vascular endothelial and smooth muscle cells in culture selectively release adenine nucleotides. *Nature* **281,** 384–386.

Pettersson, T., Klockars, M., and Weber, T. (1984) Pleural fluid adenosine deaminase in Rheumatoid Arthritis and Systemic Lupus Erythematosus. *Chest* **86**, 273.

Pike, M. C. and Snyderman, R. (1982) Transmethylation reactions regulate affinity and functional activity of chemotactic facto receptors on macrophages. *Cell* **28**, 107–114.

Pike, M. C., Kredich, N. M., and Snyderman, R. (1978) Requirement of S-adenosyl-L-methionine-mediated methylation for human monocyte chemotaxis. *Proc. Natl. Acad. Sci. USA* **75**, 3928–3932.

Piras, M. A., Gakis, C., Budroni, M., and Andreoni, G. (1978) Adenosine deaminase activity in pleural effusions: An aid to differential diagnosis. *Br. Med. J.* **2**, 1751,1752.

Polmar, S. H., Stern, R. C., Schwartz, A. L., Wetzler, E. M., Chase, P. A., and Hirschhorn, R. (1976) Enzyme replacement therapy for adenosine deaminase deficiency and severe combined immunodeficiency. *New Eng. J. Med.* **295**, 1337–1343.

Ratech, H., Greco, M. A., Gallo, G., Rimoin, D. L., Kamino, H., and Hirschhorn, R. (1985) Pathologic findings in adenosine-deaminase-deficient severe combined immunodeficiency. I. Kidney, adrenal and chrondro-osseous tissue alterations. *Am. J. Path.* **120**, 157–169.

Riches, D. W. H., Watkins, J. L., Henson, P. M., and Stanworth, D. R. (1985) Regulation of macrophage lysosomal secretion by adenosine, adenosine phosphate esters, and related structural analogues of adenosine. *J. Leuk. Biol.* **37**, 545–557.

Roberts, P. A., Newby, A. C., Hallett, M. B., and Campbell, A. K. (1985a) Inhibition by adenosine of reactive oxygen metabolite production by human polymorphonuclear leucocytes. *Biochem. J.* **227**, 669–674.

Roberts, P. A., Morgan, B. P., and Campbell, A. K. (1985b) 2-Chloroadenosine inhibits complement-induced reactive oxygen metabolite production and recovery of human polymorphonuclear leucocytes attacked by complement. *Biochem. Biophys. Res. Commun.* **126**, 692–697.

Roitt, I. (1986) *Essential Immunology,* 5th Ed. (Blackwell, Oxford, UK).

Rose, F. R., Hirschhorn, R., Weissmann, G., and Cronstein, B. N. (1988) Adenosine promotes neutrophil chemotaxis. *J. Exp. Med.* **167**, 1186–1194.

Samet, M. K. (1986) Evidence against functional adenosine receptors on murine lymphocytes. *Int. J. Immunopharmacol.* **8**, 179–188.

Sandberg, G. (1983) Regulation of thymocyte proliferation by endogenous adenosine and adenosine deaminase. *Int. J. Immunopharmacol.* **5**, 259–265.

Sattin, A. and Rall, T. W. (1970) The effect of adenosine and adenine nucleotides on the cyclic adenosine 3',5'-phosphate content of guinea pig cerebral cortex slices. *Mol. Pharmacol.* **6**, 12–23.

Schmalstieg, F. C., Mills, G. C., Nelson, J. A., and Goldblum, R. M. (1978) Limited effect of erythrocyte and plasma infusions in adenosine deaminase deficiency. *J. Pediatr.* **93**, 597-603.

Schmeichel, C. J. and Thomas, L. L. (1987) Methylxanthine bronchodilators potentiate multiple human neutrophil functions. *J. Immunol.* **138**, 1896–1903.

Schrier, D. J. and Imre, K. M. (1986) The effects of adenosine agonists on human
neutrophil function. *J. Immunol.* **137**, 3284–3289.

Schultz, L. A., Kammer, G .M., and Rudolph, S. A. (1988) Characterization of the
human T lymphocyte adenosine receptor: Comparison of normal and systemic lupus erythematosus cells. *FASEB J.* **2**, 244–250.

Schwartz, A. L., Stern, R. C., and Polmar, S. H. (1978) Demonstration of an adenosine receptor on human lymphocytes in vitro and its possible role in the adenosine deaminase-deficient form of severe combined immunodeficiency. *Clin.
Immunol. Immunopathol.* **9**, 499–505.

Simmonds, H. A., Sahota, A., Potter, C. F., and Cameron, J. S. (1978) Purine metabolism and immunodeficiency: Urinary purine excretion as a diagnostic
screening test in adenosine deaminase and purine nucleoside phosphorylase
deficiency. *Clin. Sci. Mol. Med.* **54**, 579–584.

Smolen, J. E., Korchak, H. M., and Weissmann, G. (1980) Increased levels of cylic
adenosine–3',5'-monophosphate in human polymorphonuclear leukocytes
after surface stimulation. *J. Clin. Invest.* **65**, 1072–1085.

Sung, S.-S. J. and Silverstein, S. C. (1985) Inhibition of macrophage phagocytosis
by methylation inhibitors. Lack of correlation of protein carboxymethylation
and phospholipid methylation with phagocytosis. *J. Biol. Chem.* **260**, 546–
554.

Sydbom, A. and Fredholm, B. B. (1982) On the mechanism by which theophylline
inhibits histamine release from rat mast cells. *Acta Physiol. Scand.* **114**, 243–
251.

Tritsch, G. L., Paolini, N. S., and Bielat, K. (1985) Adenosine deaminase activity
associated with phagocytic vacuoles. Cytochemical demonstration by electron microscopy. *Histochemistry* **82**, 281–285.

van De Griend, R. J., Astaldi, A., Wijermans, P., van Doorn, R., and Roos, D. (1983)
Low beta-adrenergic receptor concentration on human thymocytes. *Clin.
Exp. Immunol.* **53**, 53–60.

Wolberg, G., Zimmerman, T. P., Duncan, G. S., Singer, K. H., and Elion, G. B.
(1978) Inhibition of lymphocyte-mediated cytolysis by adenosine analogs.
Biochemical studies concerning mechanism of action. *Biochem. Pharmacol.*
27, 1487–1495.

Wolberg, G., Zimmerman, T. P., Hiemstra, K. Winston, M., and Chu, L.-C. (1975)
Adenosine inhibition of lymphocyte-mediated cytolysis: Possible role of cyclic adenosine monophosphate. *Science* **87**, 957–959.

Worku, Y. and Newby, A. C. (1983) The mechanism of adenosine production in rat
polymorphonuclear leucocytes. *Biochem. J.* **214**, 325–330.

CHAPTER 13

Adenosine Receptors

Future Vistas

Michael Williams

The field of adenosine-receptor-related research has grown in scope and complexity in the past decade, with three major international meetings on the subject (Baer and Drummond, 1979; Berne et al., 1983; Gerlach and Becker, 1987) and several others of no less importance (Stone, 1985; Stefanovich et al., 1985; Pelleg et al., 1987; Paton, 1988; Ribeiro, 1989; Jacobson et al., 1990). Receptor binding studies have provided preliminary evidence for new receptor subtypes in the brain (Bruns et al., 1987); new ligands have permitted a more in-depth evaluation of the distribution and function of the main A_1 and A_2 subclasses in both central and peripheral tissues (Williams and Jacobson, 1989). Molecular sizing of A_1 (Stiles et al., 1985) and A_2 (Barrington et al.,1989) receptors has provided the initiation point for the isolation and cloning of the two receptor subtypes. As with efforts in other receptor areas (Dixon et al., 1988; Kobilka et al., 1988), this will allow for a better understanding of the potential existence of new receptor subtypes that may be both species- and tissue-specific. Knowledge of the primary sequence of either receptor will then facilitate structure activity studies using point mutation approaches, and the ability to express either receptor will provide sufficient amounts for elucidating its three-dimensional structure. Elec-

Adenosine and Adenosine Receptors Editor: Michael Williams ©1990 The Humana Press Inc.

trophysiological and biochemical studies of expressed recep- tors in the frog oocyte system will also provide a means to elucidate the relationship of adenosine receptors to membrane G proteins, ion channels, and second-messenger systems (Cooper and Caldwell, 1989).

Considerable efforts in medicinal chemistry in the past decade from Warner Lambert (Bridges et al., 1988; Trivedi et al., 1989), CIBA-Geigy (Francis et al., 1988; Hutchison et al., 1989), Burroughs Wellcome (Daluge and Leighton, 1986; Krenitsky et al., 1987), and Nelson Research (Olsson and Thomson, 1986), as well as the National Institutes of Health (Daly et al., 1986; Jacobson, 1988), have resulted in a large number of potent and selective adenosine receptor agonists and antagonists. The agonists, without exception, have been structurally related to adenosine. However, using radioligand binding, several classes of nonxanthine antagonist heterocycles have been reported (Daly et al., 1988; Williams, 1989a).

The therapeutic potential of adenosine-related compounds has yet to be realized, primarily because the compounds tested to date have shown little superiority over other therapies currently available. Metrifudil (N^6-benzyl adenosine; Schaumann and Kutscha, 1972), although hypotensive, produced unacceptable CNS side effects. Similarly, the 140-fold selective A_2 receptor agonist, CGS 21680C, was no different in its effects on blood pressure and heart rate in the conscious SHR than the fivefold less active, fivefold A_2 selective agonist, CV 1808 (Hutchison et al., 1989). Both compounds, however, were free of the bradycardia usually observed following adenosine-agonist administration. Antipsychotic, analgesic, antiinflammatory, and sedative activity has been claimed for many new adenosine receptor agonists (Williams et al., 1989; Deckert and Gleiter, 1989), but human data on these compounds are not available. Given the marked species differences in adenosine-receptor-related phenomena (Williams, 1985; Ferkany et al., 1986; Stone et al., 1988) it will be important to extend the animal studies supporting these indications to humans at the earliest opportunity. In addition, given the complexity of effects elicited by the nucleoside, the reductionistic nature of many of the animal models used may not allow a realistic evaluation of the total spectrum of activity to be derived from a receptor-selective agonist. On the positive side, based on studies in human neutrophils

(Cronstein and Hirschhorn,1989), it is feasible that a hypotensive adenosine agonist may have additional benefits as an antiatherosclerotic agent (Mullane and Williams, 1989). Unfortunately, studies to prove such efficacy may take up to five years following drug approval. In the antiinflammatory area, the current unmet medical need may make an adenosine-related entity a beneficial therapeutic agent. On the negative side, however, the CNS side effects seen with adenosine-agonist administration (Schaumann and Kutscha, 1972; Sylven et al.,1986; Robertson et al., 1988), as well as the potential for effects on the immune system (Polmar et al., 1988; Cronstein and Hirshhorn,1989), would be deleterious for an antihypertensive agent even if there were additional benefit in its antiatherosclerotic activity. Conversely, a novel antipsychotic that reduced blood pressure via direct actions on the heart or indirectly via renal actions (Churchill and Bidani, 1989) would be a poor therapeutic agent.

Similar concerns have been expressed with adenosine antagonists (Williams, 1989a). Although both animal and human data would support a role for certain xanthines as cognitive enhancers (Jarvis and Williams, 1988;1989), the use of such agents, which are also cardiotonics, may be proscribed in the elderly target population, because of possible compromising effects on cardiovascular or renal function.

However, as previously discussed (Williams et al.,1989; Williams, 1989b), an apparent lack of selectivity has not affected the therapeutic potential of either aspirin or the HMG CoA reductase inhibitor mevinolin, or the research potential of any number of peptide receptor ligands. An open mind and appropriate clinical test paradigms are factors necessary to properly evaluate adenosine receptor ligands as therapeutic entities; much of the concern is based on isolated human studies using inappropriate doses and methods of administration or on animal data that has yet to be extrapolated to man or that has been generated with nonselective compounds. Furthermore, disease targets for novel agents based on adenosine receptor modulation should include diseases in which the uniqueness of paracrine effector agents may offer benefit. Continued emphasis on adenosine agonists as antihypertensives in the face of competing therapies, such as the angiotensin-converting enzyme inhibitors, has little logic.

A newer approach to adenosine therapy involves the use of agents that potentiate the actions of endogenous adenosine. More potent uptake blockers that are related to dipyridamole or mioflazine (Wauquier et al., 1987) represent one such approach, and the putative site/event-specific adenosine potentiator AICA riboside represents an alternate approach (Engler, 1987). Bruns and Fergus (1989) have also described some 2-amino-3-benzoylthiophenes, including PD 81,723, that are active as allosteric enhancers of A_1 receptor binding and function.

Whatever the approach, compounds are now available to test the adenosine hypothesis of neuromodulation. The crucial and necessary next step in this area will be to evaluate such compounds in disease models and thus move the adenosine-related research from an interesting and intensive area of basic research to one in which hypothesis testing has occurred. Whether such studies prove to be positive or negative, at the very least, the continued speculation as to the therapeutic potential of such agents will be replaced with tangible data.

References

Baer, H. P. and Drummond, G. I. (1979) *Physiological and Regulatory Functions of Adenosine and Adenine Nucleotides* (Raven, New York).

Barrington, W. W., Jacobson, K. A., Hutchison, A. J., Williams, M., and Stiles, G. L. (1989) Identification of the A_2 adenosine receptor binding subunit by photoaffinity crosslinking. *Proc. Natl. Acad. Sci. USA* **86,** 6572–6576.

Berne, R. M., Rall, T. W., and Rubio, R. (1983) *Regulatory Function of Adenosine* (Nijhoff, Boston).

Bridges, A. J., Bruns, R. F., Ortwine, D. G., Priebe, S. R., Szotek, D. L., and Trivedi, B. K. (1988) N^6-[2-(3,5-dimethoxyphenyl)-2-(2-methylphenyl)-ethyl adenosine and its uronamide derivatives. Novel adenosine agonists with both high affinity and high selectivity for the adenosine A_2 receptor. *J. Med. Chem.* **31,** 1282–1285.

Bruns, R. F. and Fergus, J. H. (1989) Allosteric enhancers of adenosine A_1 receptor binding and function, in *Adenosine Receptors in the Nervous System* (Riberio, J. A., ed.), Taylor and Francis, London, pp. 53–60.

Bruns, R. F., Lu, G. H., and Pugsley, T. A. (1987) Adenosine receptor subtypes, in *Topics and Perspectives in Adenosine Research* (Gerlach, E. and Becker, B. F., eds.), Springer-Verlag, Berlin, pp. 59–73.

Churchill, P. C. and Bidani, A. (1990) Adenosine and renal function, in *Adenosine and Adenosine Receptors* (Williams, M., ed.), Humana, Clifton, New Jersey, this volume.

Cooper, D. M. F. and Caldwell, K. K. (1990) Signal transduction mechanisms for adenosine, in *Adenosine and Adenosine Receptors* (Williams, M., ed.), Humana, Clifton, New Jersey, this volume.

Cronstein, B. N. and Hirschhorn, R. (1990) Adenosine and host defense, in *Adenosine and Adenosine Receptors* (Williams, M., ed.), Humana, Clifton, New Jersey, this volume.

Daluge, S. M. and Leighton, H. J. (1986) New 9-phenylxanthine compounds which antagonize the effects of adenosine in tissue preparations. European Patent Application 0 203 721.

Daly, J. W., Padget, W. L., and Shamim, M. T. (1986) Analogues of caffeine and theophylline: effects of structural alterations on affinity at adenosine receptors. *J. Med. Chem.* **29,** 1305–1308.

Daly, J. W., Horng, O., Padgett, W. L., Shamim, W. T., Jacobson, K. A., and Ukena, D. (1988) Non-xanthine heterocycles: activity as antagonists of A_1 and A_2 receptors. *Biochem. Pharmacol.* **37,** 655–664.

Deckert, J. and Gleiter, C. H. (1989) Adenosinergic psychopharmaceuticals? *Trends Pharmacol. Sci.* **10,** 99–100.

Dixon, R. A. F., Strader, C. D., and Sigal, I. S. (1988) Structure and function of G-protein coupled receptors. *Ann. Rep. Med. Chem.* **23,** 221–233.

Engler, R. (1987) Consequences of activation and adenosine-mediated inhibition of granulocytes during myocardial ischemia. *Fed. Proc.* **46,** 2407–2412.

Ferkany, J. W., Valentine, H. L., Stone, G. A., and Williams, M. (1986) Adenosine A1 receptors in mammalian brain: species differences in their interactions with agonists and antagonists. *Drug Dev. Res.* **9,** 85–93.

Francis, J. E., Cash, W. D., Psychoyos, S., Ghai, G., Wenk, P., Friedmann, R. C., Atkins, C., Warren, V., Furness, P., Hyun, J. L., Stone, G. A., Desai, M., and Williams, M. (1988) Structure activity profile of a series of novel triazoloquinazoline adenosine antagonists. *J. Med. Chem.* **31,** 1014–1020.

Gerlach, E. and Becker, B. F. (1987) *Topics and Perspectives in Adenosine Research* (Springer-Verlag, Berlin).

Hutchison, A. J., Webb, R. L., Oei, H. H., Ghai, G. R., Zimmerman, M. B., and Williams, M. (1989) CGS 21680C, an A_2 selective adenosine agonist with selective hypotensive activity. *J. Pharmacol. Exp. Ther.* **251,** 47–55.

Jacobson, K. A. (1988) Chemical approaches to the definition of adenosine receptors, in *Adenosine Receptors* (Cooper, D. M. F. and Londos, C., eds.), Liss, New York, pp. 1–26.

Jacobson, K. A., Daly, J. W., and Manganiello, V. V. (1990) *Purine Nucleosides and Nucleotides in Cell Signalling: Targets for New Drugs* (Springer-Verlag, New York).

Jarvis, M. F. and Williams, M. (1988) Adenosine antagonists as potential therapeutic agents. *Pharmacol. Biochem. Behav.* **29,** 433–441.

Jarvis, M. F. and Williams, M. (1990) Adenosine in central nervous system function, in *Adenosine and Adenosine Receptors* (Williams, M., ed.), Humana, Clifton, New Jersey, this volume.

Kobilka, B. K., Kobilka, T. S., Daniel, K., Regan, J. W., Caron, M. G., and Lefkowitz, R. J. (1988) Chimeric α_2-β_2-adrenergic receptors: delineation of domains involved in effector coupling and ligand binding specificity. *Science* **240**, 1310–1318.

Krenitsky, T. A., Rideout, J. L., and Koszalaka, G. W. (1987) New 7-anilinO–3(β-ribofuranosyl)-3H-imidazo(4,5-B) pyridine derivatives with analgesic, anti-inflammatory, antipyretic, hypotensive, vasoldilatory, antiprotozoal and antiviral activity. European Patent Application 0 260 852.

Mullane, K. M. and Williams, M. (1990) Adenosine and cardiovascular function, in *Adenosine and Adenosine Receptors* (Williams, M., ed.), Humana, Clifton, New Jersey, this volume.

Olsson, R. A. and Thomson, R. (1986) New N^6 monosubstituted adenosine derivatives useful as cardiovascular agents. WO 8504882.

Paton, D. M. (1988) *Adenosine and Adenine Nucleotides. Physiology and Pharmacology* (Taylor and Francis, London).

Pelleg, A., Michelson, E. L., and Dreifus, L. S. (1987) *Cardiac Electrophysiology and Pharmacology of Adenosine and ATP: Basic and Clinical Aspects* (Liss, New York).

Polmar, S. H., Fernandez-Mejia, C., and Bitch, R. E. (1988) Adenosine receptors: immunologic aspects, in *Adenosine Receptors* (Cooper, D. M. F. and Londos, C., eds.), Liss, New York , pp. 97–112.

Ribeiro, J. A. (1989) *Adenosine Receptors in The Nervous System* (Taylor and Francis, London).

Robertson, D., Biaggioni, I., and Tseng, C. J. (1988) Adenosine and cardiovascular control, in *Adenosine and Adenine Nucleotides. Physiology and Pharmacology* (Paton, D. M., ed.), Taylor and Francis, London, pp. 241–250.

Schaumann, E. and Kutscha, W. (1972) Klinsen pharmakologische untersuchungen mit einem neuen peroral wirksamen adenosinderivat. *Arzneimittel forsch.* **22**, 783–790.

Stefanovich, V., Rudolphi, K., and Schubert, P. (1985) *Adenosine: Receptor and Modulation of Cell Function* (IRL Press, Oxford).

Stiles, G. L., Daly, D. T., and Olsson, R. A. (1985) The A_1 adenosine receptor: identification of the binding subunit by cross-linking. *J. Biol. Chem.* **260**, 10806–10811.

Stone, G. A., Jarvis, M. F., Sills, M. A., Weeks, B., Snowhill, E. W., and Williams, M. (1988) Species differences in high-affinity adenosine A_2 binding sites in striatal membranes from mammalian brain. *Drug Dev. Res.* **15**, 31–46.

Stone, T. W. (1985) *Purines: Pharmacology and Physiological Roles* (VCH, Deerfield Beach, FL).

Sylven, C., Beermann, B., Jonzon, B., and Brandt, R. (1986) Angina-pectoris-like pain provoked by intravenous adenosine in healthy volunteers. *Br. Med. J.* **293**, 227–230.

Trivedi, B. K., Bridges, A. J., and Bruns, R. F. (1989) Structure activity relationships of adenosine A_1 and A_2 receptors, in *The Adenosine Receptors* (Williams, M., ed.), Humana, Clifton, NJ, in press.

Wauquier, A., Van Belle, H., Van Den Broeck, W. A. E., and Janssen, P. A. J. (1987) Sleep improvement in dogs after oral administration of mioflazine, a nucleoside transport inhibitor. *Psychopharmacology* **91**, 434–439.

Williams, M. (1985) Tissue and species differences in adenosine receptors and their possible relevance to drug development, in *Adenosine: Receptor and Modulation of Cell Function* (Stefanovich, V., Rudolphi, K., and Schubert, P., eds.), IRL Press, Oxford, pp. 73–86.

Williams, M. (1989) Adenosine antagonists. *Med. Res. Rev.* **9,** 219–243.

Williams, M. (1990) Adenosine receptors: historical perspective, in *Adenosine and Adenosine Receptors* (Williams, M., ed.), Humana, Clifton, New Jersey, this volume.

Williams, M. and Jacobson, K. A. (1990) Radioligand binding assays for adenosine receptors, in *Adenosine and Adenosine Receptors* (Williams, M., ed.), Humana, Clifton, New Jersey, this volume.

Williams, M., Hutchison, A. J., and Gschwend, H. (1989) Adenosine receptor ligands as therapeutic agents. *Curr. Patents,* **1,** 560–576.

Index